Structure-Property Relationships in Surface-Modified Ceramics

NATO ASI Series

Advanced Science Institutes Series

A Series presenting the results of activities sponsored by the NATO Science Committee, which aims at the dissemination of advanced scientific and technological knowledge, with a view to strengthening links between scientific communities.

The Series is published by an international board of publishers in conjunction with the NATO Scientific Affairs Division

A Life Sciences	Plenum Publishing Corporation
B Physics	London and New York
C Mathematical	Kluwer Academic Publishers
and Physical Sciences	Dordrecht, Boston and London
D Behavioural and Social Sciences	
E Applied Sciences	
F Computer and Systems Sciences	Springer-Verlag
G Ecological Sciences	Berlin, Heidelberg, New York, London,
H Cell Biology	Paris and Tokyo

Series E: Applied Sciences - Vol. 170

Structure-Property Relationships in Surface-Modified Ceramics

edited by

Carl J. McHargue
Oak Ridge National Laboratory,
Oak Ridge, Tennessee, U.S.A.

Ram Kossowsky
Pennsylvania State University,
State College, Pennsylvania, U.S.A.

and

Wolfgang O. Hofer
Kernforschungsanlage,
Jülich, F.R.G.

Kluwer Academic Publishers

Dordrecht / Boston / London

Published in cooperation with NATO Scientific Affairs Division

Proceedings of the NATO Advanced Study Institute on
Structure-Property Relationships in Surface-Modified Ceramics
Il Ciocco, Castelvecchio Pascoli (Tuscany), Italy
August 28 – September 9, 1988

Library of Congress Cataloging in Publication Data

```
Structure-property relationships in surface-modified ceramics / edited
  by Carl J. McHargue, Ram Kossowsky, Wolfgang O. Hofer.
      p.   cm. -- (NATO ASI series. Series E, Applied sciences ; vol.
170)
    Papers of a NATO Advanced Study Institute held at Il Ciocco
International Tourist and Conference Center, Castelvecchio Pascoli,
Tuscany, Italy, Aug. 28 to Sept. 9, 1988.
    "Published in cooperation with NATO Scientific Affairs Division."
    Includes indexes.
    ISBN 978-94-010-6931-1 (U.S.)
    1. Ceramic materials--Effect of radiation on--Congresses.  2. Ion
bombardment--Industrial applications--Congresses.   I. McHargue, C.
J.  II. Kossowsky, Ram.  III. Hofer, Wolfgang O., 1941-   .
IV. North Atlantic Treaty Organization.  Scientific Affairs
Division.  V. Series: NATO ASI series. Series E, Applied sciences ;
no. 170.
TA455.C43S78  1989
620.1'404228--dc20                                          89-34247
                                                               CIP
```

ISBN 978-94-010-6931-1   ISBN-13: 978-94-009-0983-0
DOI: 10.1007/978-94-009-0983-0

Published by Kluwer Academic Publishers,
P.O. Box 17, 3300 AA Dordrecht, The Netherlands.

Kluwer Academic Publishers incorporates the publishing programmes of
D. Reidel, Martinus Nijhoff, Dr W. Junk and MTP Press.

Sold and distributed in the U.S.A. and Canada
by Kluwer Academic Publishers,
101 Philip Drive, Norwell, MA 02061, U.S.A.

In all other countries, sold and distributed
by Kluwer Academic Publishers Group,
P.O. Box 322, 3300 AH Dordrecht, The Netherlands.

Printed on acid free paper

All Rights Reserved
© 1989 by Kluwer Academic Publishers.
Softcover reprint of the hardcover 1st edition 1989

No part of the material protected by this copyright notice may be reproduced or
utilized in any form or by any means, electronic or mechanical, including photo-
copying, recording or by any information storage and retrieval system, without written
permission from the copyright owner.

CONTENTS

PREFACE   ix

PART IV. ION-BEAM ASSISTED DEPOSITION

PREFACE

The use of ion beams for the modification of the structure and properties of the near-surface region of ceramics began in earnest in the early 1980s. Since the mechanical properties of such materials are dominated by surface flaws and the surface stress state, the use of surface modification techniques would appear to be an obvious application. As is often the case in research and development, most of the initial studies can be characterized as cataloging the response of various ceramic materials to a range of ion beam treatments. The systematic study of material and ion beam parameters is well underway and we are now designing experiments to provide specific information about the processing parameter - structure-property relationships.

This NATO-Advanced Study Institute was convened in order to assess our current state of knowledge in this field, to identify opportunities and needs for further research, and to identify the potential of such processes for technological application.

It became apparent that this class of inorganic compounds, loosely termed ceramics, presents many challenges to the understanding of ion-solid interactions, the relationships among ion-beam parameters, materials parameters, and the resulting structures, as well as relationships between structure and properties. In many instances, this understanding will represent a major extension of that learned from the study of metals and semiconductors.

The Institute was divided into four main areas. The first group of lectures dealt with the fundamentals of ion-solid interactions and the defect structure of compounds. Chemical and thermodynamic factors appear to be significantly more important than for simple metallic systems. The second group of lectures concerned the experimentally observed microstructural features and phase structures that develop during ion-beam processing of ceramics. The properties of ion-implanted and ion-beam-mixed ceramics were treated in Part III, and included discussions on mechanical, tribological, chemical, electrical, and optical properties. The final section was concerned with thin films and coatings prepared by ion-beam-assisted processes.

The Institute was held at Il Ciocco International Tourist and Conference Center, Castelvecchio Pascoli, (Tuscany) Italy during the period of August 28 to September 9, 1988. The setting and facilities were ideal for such a learning experience. The hotel possesses the proper blend of isolation that enhances group identification and interaction with the amenities of a first-class hotel and easy access to the cultural and historical attractions of Tuscany. The organizers are particularly grateful to Mr. Bruno Giannasi who served as our contact with the hotel staff and was instrumental in making possible our productive and successful meeting. The conference secretary was Ms. Lou M. Pyatt who also deserves much of the credit for our success and for the timely preparation of these proceedings.

x

The major support for the Institute was provided by the Scientific Affairs Division of NATO. Significant additional support was given by
        Metals and Ceramics Division, Oak Ridge National Laboratory
        Kernforschungsanlage, Jülich, FRG
        National Science Foundation, USA
        Natural Sciences and Engineering Council, Canada
        European Research Office-U.S. Army
        Construction Engineering Research Labortory-U.S. Army
        Nuclear Materials Committee of TMS and ASM
        Spire Corporation
        Nano Instruments, Inc.
        JNICE - Portugal
        TUBITAK - Turkey
        Ministry of Industry, Energy, and Technology - Greece

The home institutions of the directors provided much direct and indirect support for the planning and organization of the Institute. Their assistance, cooperation, and leniency are gratefully acknowledged.

Carl J. McHargue
Metals and Ceramics Division
Oak Ridge National Laboratory
P.O. Box 2008
Oak Ridge, Tennessee 37831-6118 USA

Ram Kossowsky
Applied Research Laboratory
Pennsylvania State University
Box 30
State College, Pennsylvania 16804 USA

Wolfgang O. Hofer
IPP, Kernforschungsanlage Jülich GmbH
Postfach 1913
D-5170 Jülich, Federal Republic of Germany

PROPERTIES OF CERAMICS AND THEIR STUDY BY COMPUTER SIMULATION METHODS

C. R. A. Catlow

Department of Chemistry, University of Keele,
Keele, Staffs. ST5 5BG, United Kingdom

1. INTRODUCTION

The defect physics and chemistry of non-metallic solids show a large
degree of diversity and complexity. Defects can be created by several
mechanisms and a wide range of defect species are possible particularly in
the structurally more complex materials. In this chapter we will first
review the basic nature of defects in insulating solids. We will then
describe the techniques that are available for simulating defects and
defect-dependent phenomena in these materials. The field of defect simulations in insulators has enjoyed considerable success over the last ten
years. Static and dynamical methods have been used, and the techniques have
been applied with success to both bulk and surface defects. This success
suggests that the methodology should be extended to the complex problems
posed by ion beam implanted ceramics. Complex defect reactions occur in
insulators, following radiation change as illustrated by our discussion
later in the article, of irradiation damage in NaCl. Even greater complexity can be expected for oxide ceramic materials.

2. DEFECTS AND DEFORMATION IN INSULATORS
2.1. Types of defects

It is useful to classify defects in ceramics into four principle categories.

(A) Point defects in which the defect is located at a single site or small
group of sites in the crystal. These comprise vacancies, i.e. atoms missing
from the regular lattice sites; interstitials, i.e. atoms present in sites
that are unoccupied in the perfect lattice; substitutionals, that is foreign
atoms present at perfect lattice sites; and clusters, i.e. aggregates
comprising several of the above. A simple example of the latter is given in
Fig. 1, which shows a complex of a divalent dopant ion and a cation vacancy
in NaCl.

(B) Extended defects, which have an indefinite extent in one or two dimensions. The commonest example is provided by dislocations which occur in all
crystalline materials. These defects whose nature is amply discussed in
several texts are "conservative defects," i.e. their formation does not
change the over-all chemical composition of the solid. In contrast, "non-
conservative" extended defects known as shear planes form in certain non-
stoichiometric oxides, e.g. TiO_{2-x} and WO_{3-x}. A diagramatic illustration of
such defects for the case of WO_3 is shown in Fig. 2, which shows a section
through the ReO_3 structure a distorted version of which is adopted by WO_3.
On reduction to form WO_{3-x} we might envisage that oxygen would be lost with
the formation of vacancies. If we imagine the latter being aligned to form
a vacancy disk as shown in Fig. 2(a), then by a subsequent process of shear
in which the bottom half of the crystal is displaced as shown, the vacancies

1

C.J. McHargue et al. (eds.), Structure-Property Relationships in Surface-Modified Ceramics, 1-25.
© *1989 by Kluwer Academic Publishers.*

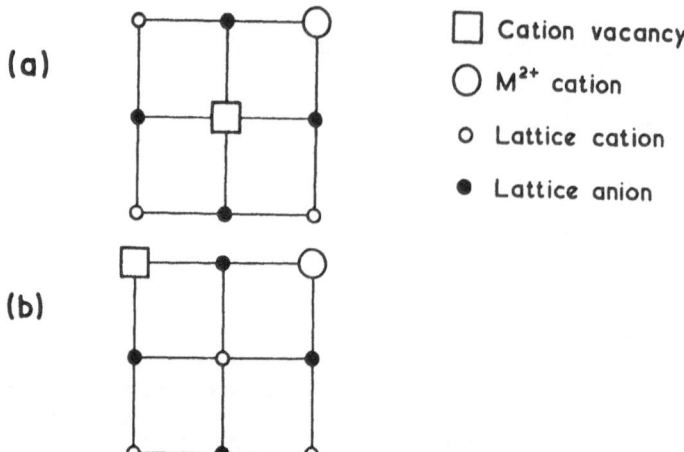

(a) □ Cation vacancy

○ M²⁺ cation

o Lattice cation

● Lattice anion

(b)

FIGURE 1. Dopant-defect paris in divalent cation doped rock salt structured halides: (a) n.n. pair (b) n.n.n. pair.

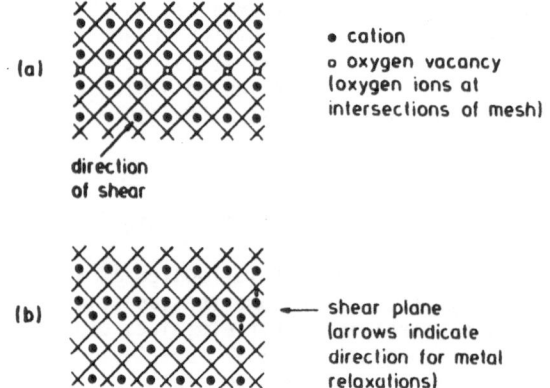

(a)

• cation
o oxygen vacancy (oxygen ions at intersections of mesh)

direction of shear

(b)

— shear plane (arrows indicate direction for metal relaxations)

FIGURE 2. Schematic illustration of shear-plane formation in ReO_3 structured oxides.

maybe eliminated with the creation of a planar fault on the cation sublattice. The formation of such defects clearly changes the oxygen to metal ratio.

(C) Surfaces and surface defects. A broad definition would include surfaces as a "defect" in that they are a discontinuity of the bulk crystal. More relevant, however, is the fact that well defined surface defects are now discussed in the ceramic literature; in particular, surface vacancy and impurity states are thought to be of importance.

(D) Nuclei of precipitates. A new phase precipitated into a host crystal cannot, of course, strictly be included in the classification of defect species. But there is a "fuzzy" division between nuclei of new phases (which may be defect aggregates) and precipitates, and for this reason we have included such species in our classification.

2.2. Defect creation mechanisms

The following types of mechanism are important in ceramic materials.

(A) Thermal Creation of Instrinsic Disorder, by which we mean the creation by thermally induced excitation of the disorder which is invariably present even in hypothetically pure materials at temperatures above absolute zero. Different types of intrinsic disorder reaction are possible and the defect concentrations are highly temperature dependent.

Themal generation of defects is best understood using the concept of defect reactions: the first, known as the Frenkel disorder reaction, involves the generation of interstitials by the displacement of ions from normal lattice sites to interstitial positions in the crystal structure. The Frenkel disorder reaction can therefore be written as:

$$I_L \rightarrow V_L + I_I ,$$

where I_L indicates an ion occupying a normal site; I_I is the interstitial species and V_L is the vacancy at the regular lattice position that is created by displacement of the ion I. The advantage of this way of describing the thermal generation of defects is that it can readily be shown that defect reactions are governed by the same chemical thermodynamic considerations as more conventional chemical reactions. Thus assuming that Frenkel disorder dominates, the equilibrium activities of vacancies and interstitials (written as a_V and a_I) are given by:

$$a_V a_I = k_F = \exp (-g_F/kT) , \qquad (1)$$

where k_F is the Frenkel disorder constant, and g_F is the free energy of formation of the Frenkel pair. The activity of perfect lattice ions may be taken as unity. Activities may of course be related to concentrations (x_V and x_I) via activity coefficients, i.e.

$$a_V = f_V x_V ; \quad a_I = f_I x_I . \qquad (2)$$

At very high dilutions it may be acceptable to take the activity coefficients as unity. Due, however, to the fact that defects are charged species and have therefore long-range Coulomb interactions, this is in general not an acceptable approximation; and the calculation of defect activity coefficients has been an important area of the theory of defects in solids.

The second type of defect reaction, known as Schottky disorder involves generation of vacancies by displacement of lattice ions to the surface of the crystal. For ionic materials Schottky disorder requires the formation of vacancies in stoichiometry ratios. Thus in 1:1 crystals, e.g. NaCl, equal concentrations of defects must be created; while in 2:1 crystals, e.g. CaF_2, cation and anion vacancies are created in the ratio 1:2. The Schottky disorder reaction for 1:1 ionic crystals may be written as:

$$I_L^+ + I_L^- \rightarrow V_L^+ + V_L^- , \qquad (3)$$

where the superscripts, + and -, indicate defects created at cation and anion sites respectively. Again, we may apply standard chemical thermodynamics to this equilibrium, giving for the vacancy activities, a_V^+ and a_V^-:

$$a_V + a_{L^-} = K_S = \exp (-g_S/kT) , \qquad (4)$$

where KS is the Schottky disorder equilibrium constant and g_S is the free energy of formation of the Schottky pair.

Since g_F and g_S are always finite (although possibly large), both Frenkel and Schottky reactions are always operative in a crystal. In practice, however, one type of disorder appears to dominate in a given crystal (although an unlikely coincidence in some material in the values of g_F and g_S cannot be ruled out). Crystals may be classified therefore according to whether the intrinsic disorder is of the Frenkel or Schottky type. Such a classification for the commoner materials is presented in Table 1.

TABLE 1
Dominant Intrinsic Disorder of Ionic Crystals

Compounds	Crystal structure	Dominant Intrinsic Disorder Reaction
NaCl (+ all isostructural alkali halides)	Rock-salt	Schottky
AgCl, AgBr	Rock-salt	Cation Frenkel
MgO (+ alkaline earth oxides)	Rock-salt	Schottky
MnO (+ other divalent transition metal oxides)	Rock-salt	Schottky
CaF_2 (+ other isostructural halides)	Fluorite	Anion Frenkel
UO_2 (+ other isostructural oxides)	Fluorite	Anion Frenkel
TiO_2 (+ isostructural oxides)	Rutile	(Schottky)
ZnO	Zinc-blende	(Cation Frenkel)
Al_2O_3 (+ isostructural oxides)	Corundum	(Schottky)

Note: reaction types given in brackets indicate that some uncertainty may still be associated with the nature of the predominant disorder.

(B) Chemical induction of defects. All real materials contain impurities, either accidentally present or deliberately introduced as dopants. If the dopant is an "aliovalent" species, i.e. with a different valence from the host lattice ions, then whether it is a substitutional (which is most commonly the case) or an interstitial, it must be "charge compensated" to retain electroneutrality. A simple example is given by the case of aluminium in magnesium oxide. Al^{3+} is invariably present even in high purity MgO; it substitutes for Mg^{2+} which disturbs the charge balance of the crystal. The Al substitutional, which may be denoted as $Al^{\cdot}$ (using the notation of Kroger and Vink) has an effective charge (i.e. charge relative to that of the same site in the perfect latice) of +1. It must be compensated by defects with an effective negative charge; and since the intrinsic disorder of the material is of the Schottky type, vacancies are created with one doubly charged cation vacancy (V_{Mg}) for every two Al^{3+} impurities.

A second example is given by Nb^{5+} doping of UO_2. The niobium substitutional $Nb_u^{\cdot}$ again must be compensated by defects with an effective negative charge. As UO_2 is a Frenkel disordered material, oxygen interstitials (O_I'') are created. The case of UO_2 also illustrates well how variable cation valence, which gives rise to non-stoichiometry in oxide has very similar consequences, as regards defect structures, to aliovalent doping. Thus oxidation of U^{*+} to U^{5+}, creating a quasi defect U_U leads to oxygen anion interstitial compensation. Indeed, non-stoichiometry is a very common cause of defect creation in oxides.

(C) Mechanical Creation of Defects. It is well known that mechanical deformation of materials may lead to defect creation, and indeed dislocations are commonly induced by mechanical treatment. Such processes may be important in ceramic materials as well as in metals.

(D) Radiation Damage. Radiation may induce defects via a variety of mechanisms. Implantation of course directly introduces impurities into materials. However, radiation creates lattice defects by "knock-on" or displacement processes which occur during the radiation induced cascade; in addition defects may be induced by photolysis where the primary act of the radiation is the creation of electron-hole pairs, which via secondary processes lead to defect generation. Examples for the case of NaCl will be presented in Section 3.

2.3. Defect aggregation

In all types of solid, defect interaction and aggregation processes become important at higher defect concentrations. This is, however, specially the case in ionic materials owing to the strong Coulomb interactions between defects. Defect clustering may be included within the mass-action formation discussed above. For example, the formation of the cluster shown in Fig. 1 may be described in terms of the defect reaction:

$$D_L = V_L \rightarrow C \ ,$$

where D_L indicates a dopant ion occupying a regular lattice site and C the cluster. The consequent mass action equation is given by:

$$\frac{a_C}{a_d a_v} = K_C = \exp \ (-g_C/kT) \ ,$$

where a_C and a_d are the activities of the cluster and dopant ion respectively and g_C is the free energy of cluster formation.

Highly complex modes of defect aggregation occur in many heavily doped and non-stoichiometric ionic materials. An example is shown in Fig. 3 for the case of $Fe_{1-x}O$, a grossly non stoichiometric compound in which cation vacancies are created to compensate the oxidation of lattice cations from Fe^{2+} to Fe^{3+}. It is now well established (see for example Chapter 16 of reference 2) that in these heavily defective oxides vacancies are stabilized by a mode of aggregation that involves the creation of additional cation interstitials as shown in Fig. 3.

2.4. Consequences and effects of defects

We have noted that defects radically alter the properties of solids. The properties influenced include mechanical, electrical, spectroscopic, and diffusional behavior. Many accounts of these phenomena are given elsewhere (1-3). Thus, we will concentrate on one aspect, viz the influence of defect structure on atomic transport in solids.

Atomic transport is manifested by the bulk properties of conductivity and diffusion and is nearly always effected by defects. Interstitial atoms may jump through the lattice, and lattice atoms may jump into vacant sites (hence effecting vacancy transport). As discussed below, these processes may be described using absolute rate theory. Given knowledge of the frequencies of these activated jumps, we may write the diffusion coefficient D, as:

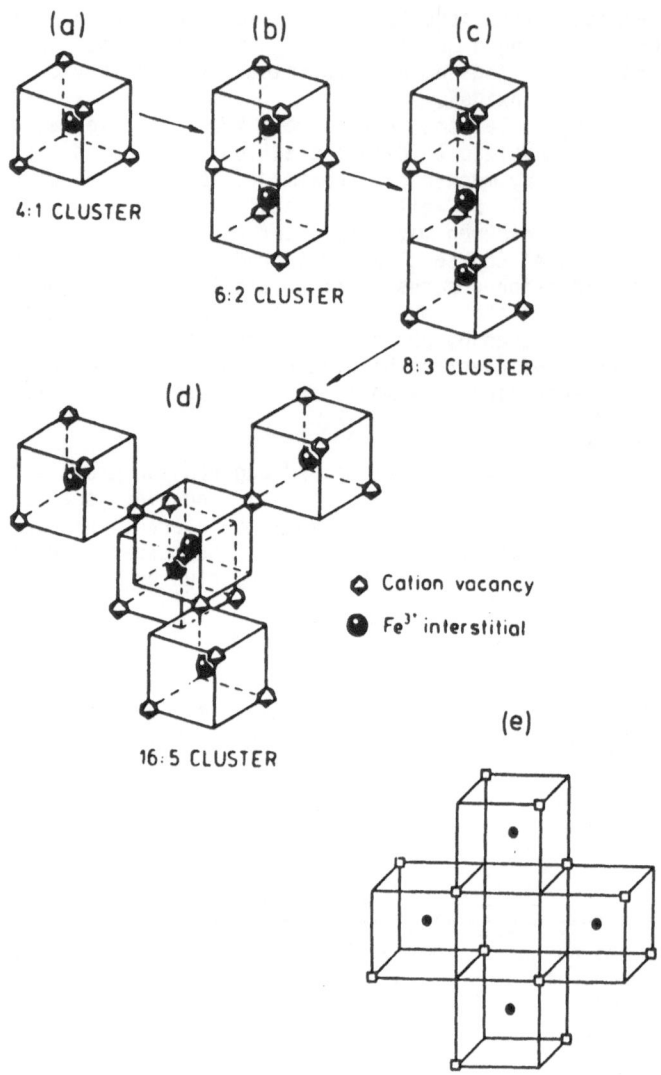

FIGURE 3. Vacancy-interstitial cluster models in $Fe_{1-x}O$.

$$D = 1/6 \ x\upsilon r^2 , \qquad (5)$$

where x is the defect concentration and r is the distance of the defect jump. If, as argued below, υ shows Arrhenius behavior, then since x is either fixed by the level of impurities (or of mechanical and irradiation damage), D itself will show Arrhenius behavior. Similar behavior will be shown by the electrical conductivity if it is effected by the same atomic transport mechanisms, which now transport charge rather than mass. In these circumstances the conductivity (σ) and diffusion coefficient (D) are related by the Nernst-Einstein relationship:

$$\frac{\sigma}{D} = \frac{ne^2}{fkT} , \tag{6}$$

where n is the number density of mobile ions, e is the electron charge, and f is a numerical factor (the correlation coefficient) which is close to unity and which depends on the migration mechanism.

The prediction of Arrhenius behavior is borne out experimentally. Figure 4 shows an Arrhenius plot for conductivity in KCl. Two linear regions are observed: that at higher temperature corresponds to normal intrinsic generation of defects; whereas at low temperatures the defect population is dominated by impurities. In ceramic materials, the low temperature, extrinsic region commonly dominates except at the highest temperatures, close to the melting point.

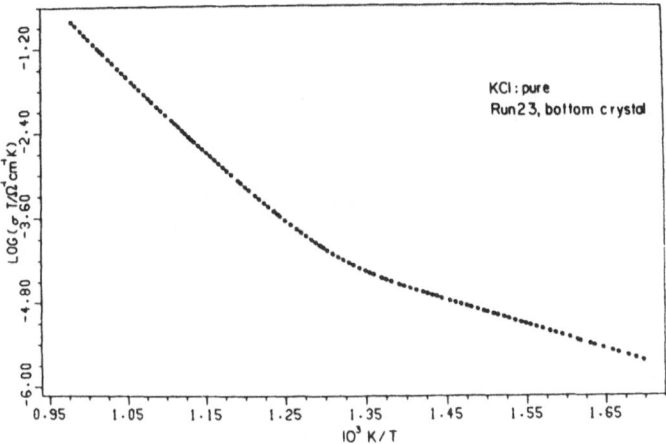

FIGURE 4. Arrhenius plot of conductivity for KCl (see Chapter by Jacobs in reference 2).

2.5. Defect structure of ceramics: summary

It should be clear from the above discussions that the defect properties of ceramics show complex features that are not encountered in parallel studies of metals. The variety of atom types leads to a corresponding variety in defect species. Low levels of impurities can lead to an enormous enhancement in defect concentrations and on defect dependent properties, e.g. diffusion. Defect aggregation is observed at all but the lowest defect concentrations. Because of this complexity, guidance from reliable theoretical techniques has played and continues to play a vital role in the development of this field. The remainder of this chapter will review the present state of theoretical studies of defects in ceramics.

3. DEFECT CALCULATIONS: TECHNIQUES

The methodology of defect calculations in ceramics has been extensively reviewed in recent years by the present author and others (1,4-6). For this reason the account here will concentrate on key features of the techniques; the reader is referred to references (1-6) for details. We shall concentrate on simulation techniques, i.e. methods which are based on models of the interatomic potentials describing the forces between the atoms or ions in the solids; these are then used in a variety of procedures based on classical rather than quantum physics. We will, however, refer to the emerging field of quantum mechanical studies of defects in ceramics. What quantities can be modelled using modern theoretical methods? Defect energies are now routinely calculated using Mott-Littleton techniques; such calculations involve a detailed simulation of the lattice relaxation around the defect. Defect entropies may also now be calculated by methods developed during the last few years by Jacobs, Gillan, Harding, Leslie, Mackrodt and Stoneham (7-10). The majority of calculations that have been performed yield defect parameters appropriate to zero temperature and to constant volume. But increasingly, the quasi-harmonic approximation is invoked to perform calculations for higher temperatures by using the lattice parameters appropriate to that temperature. Moreover, given estimates of the temperature variation of defect energies, it is possible to obtain, using standard thermodynamic relationships, values of the corresponding constant pressure quantities, further discussion of which is given below and in references (11) and (12).

The calculations referred to above are all based on lattice statics techniques, i.e. no explicit account of thermal motion is included. The methods are very flexible, and can be applied to defect formation and aggregation, and to the activated jump mechanisms which are generally responsible for defect transport. Moreover, the techniques can be extended to the study of surface and grain boundaries. The omission of explicit dynamical effects is, however, an approximation that becomes increasingly severe at high temperatures and for solids which show high atomic mobilities. For this reason there is increasing use of the molecular dynamics method which includes kinetic energy explicitly. In addition the Monte-Carlo method which is essentially a form of computational statistical mechanics is finding a role in the study of complex disordered materials.

3.1. Defect energy calculations: The Mott-Littleton methodology

The methods used here are conceptionally simple and were developed by Mott and Littleton (12) over fifty years ago. The defect is embedded in a surrounding region of crystal (containing typically 100-300 atoms of ions). Within this region, atomic coordinates are adjusted until all atoms are at zero force; this excercise obviously requires specification of the appropriate interatomic potentials. For more distant regions of the crystal, the polarization is calculated using expressions based on a continuum approximation. The interaction between the polarization field and the defect may be summed to infinity using analytical expressions. In practice, the calculations employ an interface region between the explicit and continuum regions, in which the interactions are calculated directly. Early work of Norgett (13) showed that it is possible to develop efficient computer codes based on the Mott-Littleton method. An important feature of this development was the use of matrix minimization methods in treating the inner explicit region. Norgett's original HADES code was generalized by Catlow et al. (14), and the latest "state-of-the-art" defect simulation code CASCADE written by Leslie (15), can handle defects in crystals of all symmetries and exploits the efficient vector processing facilities of, for example, the CRAY supercomputers. There is now abundant evidence that, given a sufficiently large inner region, and given high quality interatomic potentials, these methods are capable of yielding accurate defect energies in quantitative agreement with experiment. Examples will be given in Section 4 below, which will show that the methods are now indeed a predictive tool in defect physics.

3.2. Entropy calculations

Defect entropies have two components: first configurational terms arising from orientational and site degeneracy. For these we may use expressions of the type:

$$S = K \log_e(N) , \qquad (7)$$

where N is the number of orientations and the fractional site occupancy of the defect. Far more difficult to evalute is the vibrational entropy which arises from the perturbation by the defect of the vibrations of the surrounding lattice atoms. However, developments due to Gillan, Jacobs and Harding now allow such calculations to be performed. The vibrational entropy S_V is written as:

$$S_V = -K \ln \frac{\prod_{i=1}^{3N} \omega_i'}{\prod_{i=1}^{3N} \omega_i} + 3k(N' - N) [1 - \ln (\hbar/kT)]), \qquad (8)$$

where the primes indicate the perturbed vibrational frequencies. These are evaluated in a region of crystal surrounding the defect and compared to the unperturbed values. Difficulties may be encountered in the convergence of such calculations with expansion of the explicitly treated region. Details are given by Gillan and Jacobs (7). A general computer program, SHEOL, has been written by Harding (8).

Given knowledge of defect energies and entropies we may, of course, evaluate free energies. In this context we should note the important relationship for defect quantities, that is, $g_p = f_V$, where g_p is the Gibbs energy at constant pressure and f_V the Helmholtz constant volume quantity that is normally calculated. In addition we note the standard relationships:

$$h_p = u_V - V\beta T[\frac{du_V (V(T))}{dv} - T \frac{ds_V(V(T))}{dv}] \qquad (9a)$$

$$s_p = s_V - V\beta[\frac{du_V(V(T))}{dv} - T \frac{ds_V(V(T))}{dv}], \qquad (9b)$$

in which v is the unit cell volume and which allow us to convert constant volume to constant pressure quantities given knowledge of the thermal expansivity β and the derivatives du_V/dv and ds_V/dv. The latter may be calculated by varying the lattice parameter of the crystal as discussed in references (12) and (16).

Extensive studies of defect energies and entropies and their variation with temperature have been made within the confines of the quasi-harmonic approximation, by Harding and coworkers [see e.g. ref. (12)]. One important conclusion to follow from Harding's work is that generally h_p as measured experimentally at elevated temperatures is reasonably well approximated by u_V at 0 K. The same approximation does not, however, apply to the entropy term.

3.3. Defect migration energies

Thirty years ago, Vineyard (17) showed that absolute rate theory could be applied to defect transport processes and subsequent defect modelling studies by Gillan and Harding (18) and by de Lorenzi (19) have supported (and amplified) this approach. Thus, we can write the frequency of defect jumps using the expression:

$$\upsilon = \upsilon_0 \exp\left(-g_{ACT}/KT\right), \tag{10}$$

where g_{ACT} is the free energy of activation, i.e. the difference between the free energy of the saddle point for the migration process and that of the ground state of the defect. Given that the saddle point can be identified, g_{ACT} may therefore be evaluated using energy and entropy calculations of the type described above. For high symmetry crystals, identification of the saddle point is commonly straight-forward. For lower symmetry crystals, greater difficulties may be met, and a detailed search of the potential energy surface may be necessary. Examples of calculated activation energies will be given in Section 4.

3.4. Super-cell calculations

The methods described so far have concerned an isolated defect or defect cluster embedded in an infinite, perfect crystal. An alternative approach is to take a defect supercell, that is, an infinite periodically repeating array of defects; the repeat unit will normally be a supercell of the basic unit cell of the structure. Energy minimization techniques are then applied to the supercell. The lattice energy of the defect supercell may be compared with that of the corresponding perfect structure to yield the defect energy; similarly, by evaluating the phonon dispersion curves, the entropy of the defect supercell may be calculated and compared to that of the perfect lattice, thereby yielding the defect entropy.

For sufficiently large supercells, the resulting defect parameters should converge on the values obtained using the methods described in Section 3.1 and 3.2. And indeed, such convergence has been shown in recent studies of Leslie and coworkers (20).

3.5. Molecular dynamics calculations

The molecular dynamics (MD) method has emerged in recent years as one of the most powerful general computational methods in condensed matter science. The method is, again, conceptionally simple: an ensemble of particles containing several hundred (and in modern work, often several thousand) particles is defined. The ensemble is normally repeated infinitely by the application of periodic boundary conditions. Interatomic potentials must be specified, between the particles which simulate the component atoms of the system. The simulation is started by assigning all particle positions and velocities; and the simulation then proceeds by the solution of the classical equations of motion of the system in an iterative manner. A time step, Δt, is specified, after the lapse of which new coordinates and velocities are calculated. For infinitesimal Δt, the expression would be:

$$x_i(t + \Delta t) = x_i(t) + v_i(t)\,\Delta t \tag{11a}$$

$$v_i(t = \Delta t) = v_i(t) + \frac{F_i}{m_i}(t)\,\Delta t, \tag{11b}$$

where x_v, v_i, and m_i are the position, velocity and mass of the ith particle and $F_i(t)$ is the net force acting on the particle at time t. The above expressions are inaccurate for any finite Δt (which is typically

10^{-14} - 10^{-15} s) and in practice, more sophisticated updating algorithms including terms in higher powers of Δt are used. In the early stages of an MD simulation, which normally will involve several thousand time steps, the system equilibrates, i.e. equipartition of energy is achieved and the system acquires an equilibrium distribution of velocities. The second stage of the simulation is the production run; this proceeds again for several thousand time steps with the positions and velocities each time step being stored on tape or disk for subsequent analysis. This stored output from the simulation contains a record of the time evolution of the system over a period which is normally 10 - 1000 pico-seconds. We have discussed elsewhere (5) the scope and limitations of the MD method in the study of atomic transport in insulating solids. The method is intrinsically more powerful than the static methods discussed earlier in this section. Structural properties are derived from calculated radial distribution functions. Diffusion coefficients may be calculated directly from the variation of the root mean square displacement $< r^2_\alpha >$ of a particle of type α with time, for which we use the relationship:

$$< r^2_\alpha > = 6D_\alpha t + B_\alpha , \qquad (12)$$

where B_α is the temperature factor relating to the vibrational amptitude of atom α at its lattice site. In addition, a range of correlation factors, e.g. the velocity auto-correlation function and van Hove self-correlation function may be calculated; the latter is of particular interest and use as it enables the results of neutron scattering experiments to be simulated.

Despite this power and versatility, the method is limited in its application to transport in solids. The limitations arise from three main sources:

(i) Limited time scale. Even with the largest modern supercomputer, simulation periods much longer than 1000 psec in "real time" are rare due simply to the enormous computational demands of the longer simulations. In simulating transport processes it is therefore necessary that several migration processes occur within this time scale. This effectively confines the method to the study of transport in solids with high atomic mobilities. Techniques are available for the study of "rare events" but they require prior knowledge of the type of event (e.g. the migration mechanism).

(ii) If, as is normally the case, periodic boundary conditions are used, the system has no "surface", and, consequently Schottky disorder (which involves the displacement of lattice atoms to the surface of the crystal) cannot be generated naturally by the simulation. In Schottky disordered crystals, a fixed level of vacancies must be introduced at the start of the simulation. There is, of course, no such restriction for Frenkel disordered materials.

(iii) It is difficult to include the effects of electronic polarizability in an MD calculation as this increases by up to an order of magnitude the already high computational cost of the calculations. However, as discussed below, polarization energies make an important contribution to the total defect formation energy in ionic solids.

Despite these limitations, MD techniques are being increasingly used in solid state studies, and with growth in computer power, their role in the modelling of high temperature ceramics is certain to grow.

3.6. Monte Carlo molecular dynamics techniques

Molecular dynamics resembles MD methods in that a periodically repeated ensemble of particles is used and several thousand configurations of the system are generated. But unlike MD there is no time relation between successive configurations which are generated by making random moves of the component particles, with a variety of procedures being used to decide whether such a move should be accepted. In studies of atomic transport this type of sampling may be used to model systems with large numbers of different migration mechanisms, as will be discussed in Section 4.

3.7. Surface and grain boundary properties

As ceramic materials are in general polycrystalline, interfaces can exert a decisive influence on their properties. The types of simulation method described above have therefore been extended to the study of surfaces and grain boundaries and to the properties of defects and impurities at these interfaces. Notable contributions to this field have been made by Tasker and Duffy (21-23) and by Mackrodt and Colbourn (24). Calculations on perfect surfaces take a slab of crystal that is infinite in two dimensions. The slab is embedded on one side with the perfect crystal structure which extends to infinity; the other side simulates the surface. Energy minimization procedures are applied to the slab to yield the relaxed structure of the surface and the layers of atoms immediately below. Calculations show that there may be extensive relaxation (most commonly, "rumpling" which involves differential displacements of atoms perpendicular to the surface) in the surface layer, but that these displacements rapidly decay to zero on penetration into the bulk.

Grain boundary calculations also embed the interfaces with an explicitly relaxed slab which is, in turn, embedded on either side in an infinite perfect lattice. Calculations are reported on twist, twin, and tilt boundaries (23,25). Setting up the initial configuration of these calculations may be difficult even with high symmetry crystal structures.

Of greatest relevance and importance in the context of ceramic behavior are the calculations on defects and impurities at interfaces. These are performed by methods related to the Mott-Littleton method described in Section 3.1. After obtaining the relaxed surface or grain boundary structure, the defect is embedded in this structure and a surrounding region of atoms is relaxed to zero force, with the polarization of more distant regions being calculated using continuum theory. Notable applications of these methods have concerned the study of impurity segregations to surfaces and grain boundaries. The reader should refer to the papers of Tasker, Duffy, Colbourn and Mackrodt (22-27) for details.

3.8. Interatomic potentials

All the methods described above rely, as we have noted, on specified interatomic potentials. For ceramic materials, the potentials used to date have been based on the Born model of the solid; that is, the following assumptions have been made:

(i) The solid is an assembly of ions which are most often assumed to have integral charges (e.g. 2^+ for Mg, 2^- for O). The resulting Coulomb simulations must be modelled accurately. Straightforward real space summations converge slowly, and all modern methods use the Ewald procedure (28) which partitions the simulation into reciprocal and real space components both of which are rapidly convergent.

(ii) Short-range forces due to orbital overlap repulsion and dispersion and covalent terms are described by analytical functions. Most commonly these are of the 2-body, central force type, e.g. the well known Buckingham (or "6-exp") potential:

$$V(r) = A^- e^{r/\rho} - Cr^{-6} , \qquad (13)$$

although other functional forms, e.g. the Lennard-Jones and Morse, may be used. The restriction to two-body models is a major one, especially as we push the applications of simulation methods towards more covalent materials. Many-body terms may be included in a variety of ways, most simply by the use of bond bending energy functions, which add terms to the total energy of the crystal, of the type:

$$E\ (\theta) = \frac{1}{2}\ K\ (\theta - \theta_0)^2\ , \tag{14}$$

where θ is the bond angle of a specific trio of atoms (e.g. the O-Si-O bond angle) and θ_0 is the equilibrium value (e.g. the tetrahedral angle). These simple many-body terms have proved to be particularly effective in the modelling of silicate structures.

(iii) The effects of ionic polarization must be included in accurate modelling of the structures of defects in ceramics. Defects generally have non-zero charges and hence the surrounding lattice is polarized. The electronic component of the polarization energy is generally substantial. Earlier attempts to model this effect using point dipole models have now given way to shell model treatments, which, albeit crudely, model the physical basis of polarization as the displacement of valence shell electron density by the applied field; this is described in terms of the displacement of a massless shell from a core (in which the atomic mass is concentrated), the core and shell being connected by an harmonic spring.

Potential models must be parameterized; that is, values must be assigned to the variables in the analytical representation of the short-range potentials and to the variables in the shell model description of polarizability. Two strategies are employed here. In the first we use least-squares fitting of the parameters to the observed crystal properties (e.g. elastic and dielectric constants and phonon dispersion curves). This of course requires that the necessary crystal data are available. Moreover, the procedure only guarantees information on the interatomic potential at spacings close to those in the perfect lattice. For this reason theoretical methods have been developed for calculating short-range potentials. Considerable success has been enjoyed by electron gas methods (29-31), while direct calculations of interatomic potential surfaces by Hartree-Fock methods are now increasingly being used (32,33).

In concluding this section we note that potentials of the type we have described have, in addition to their success in modelling defect properties, had a considerable role in describing the structure and properties of perfect crystal structures. Crystals with highly complex structures, e.g. the "1:2:3" $YBa_2Cu_3O_7$ superconductor have been successfully modelled by energy minimization methods. Such techniques are therefore useful in the study of the complex crystal structures of many ceramic materials.

3.9. Quantum mechanical methods

We conclude our account of techniques in the theoretical defect physics of insulators by discussing briefly the applications of quantum mechanical (QM) methods. These methods include an explicit representation of the electronic structure of the defect by solving the Schrodinger equation at some level of approximation for the defect and usually a small number of surrounding atoms. The resulting quantum mechanically treated cluster may simply be "terminated" by adding surface hydrogen atoms to saturate unsatisfied valences. Alternative or additional procedures involve embedding the QM, cluster in an array of point charges which simulate the Madelung potential of the surrounding lattice. In calculations on metals and semiconductors supercell methods have been used in which the Schrodinger equation is solved for a periodic array of defects. The levels at which the Schrodinger equation is solved range from the approximate semi-empirical procedures, through ab-initio, Hartree-Fock to the configurational interaction (CI) methods which include electron correlation effects. We have discussed these and other technical aspects elsewhere (1).

When explicit information on the electron structure of defects is needed, there is, of course, no alternative to quantum mechanical calculations. However, in many applications in ceramics, information on energies, entropies, and structures of defects involving only closed-shell species, is all that is required. For this reason we expect that the simulation methods, which are far computationally cheaper, will continue to play a more important role. We should also note that in a recent and detailed study of Grimes and coworkers (34-36) on defects in MgO, it was shown that QM cluster methods essentially reproduced (and underwrote) the simulation techniques - a result which must give added confidence in the use of the latter.

4. APPLICATION OF MODELLING TECHNIQUES

Once more we note that detailed reviews of the application of defect simulations are available elsewhere; and we will confine our attention in this chapter to a limited number of recent applications of particular relevance to ceramic materials. The first concerns the predictive role of the calculation in quantitative studies of defect processes; the second concerns the applications of simulations in elucidating qualitative features of defect processes to illustrate which we will highlight the work undertaken by Diller et al. (37) on radiation damage in NaCl. Thirdly, we shall illustrate the success of MD simulations in yielding mechanistic features of defect transport. Finally, we illustrate, by example, the potential role of MC simulation in studying highly defective materials.

Restrictions of space prevents us from reviewing the valuable role of calculations on surface impurities and defects and we refer the reader to references (22-27).

4.1. Quantitative studies of defect parameters

The achievements here are best illustrated by the selection of results presented in Table 2. For simple ionic materials such as NaCl and CaF_2, calculations yield defect formation and migration energies that agree within the expected error range with the results of the best experimental studies. For ceramic materials, the paucity of accurate experimental data makes such comparisons more difficult. Here, however, theory may be truly predictive. MgO provides a good example. The calculated Schottky energy reported in Table 2 is high compared with early experimental studies. More recent work [see e.g. Wuensch (38) for a discussion] suggests a much more minor role for intrinsically generated defects in this material, and indicates Schottky energies more closely in accord with the experimental values.

The simulation studies of $BaTiO_3$ are also instructive. Calculations (39) indicated that in doping with high valence ions, for example, La (which substitutes for Ba) or Nb (which substitutes for Ti) charge compensation would be effected by the creation of vacancies at the Ti sites. This prediction, which at first seemed to be counter-intuitive has now received strong support from experimental studies.

In achieving quantitative agreement between experimental and calculated defect formation energies the inclusion of a detailed treatment of lattice relaxation is essential. In ceramic materials, lattice relaxation energies are large. In MgO, for example, values of 10 - 15 eV are typical for vacancy defects. In materials such as UO_2 where ionic charges are higher, even larger energies will be calculated.

To what extent is the quantitative success achieved in defect energy calcuations also obtained in calculations of entropies? Both calculate and experimental results are far less widely available for defect entropies.

TABLE 2(a). Defect energy calculations for simple ionic materials (details
and references given in reference 43d; experimental data in brackets)

Crystal	Process	Calculated Energy (eV)
NaCl	Schottky pair formation	2.4-2.7 (2.3-2.7)
	Cation vacancy migration	0.66 (0.7-0.8)
CaF$_2$	Anion Frenkel pair formation	2.6-2.7 (2.6-2.7)
	Anion vacancy activation	0.35 (0.5-0.6)
	Anion interstitial activation	0.91 (0.9-1.0)
MgO	Schottky pair formation	7.5-7.7 (5-7)
	Cation vacancy activation	1.8-2.2 (2.0-2.3)
NiO	Schottky pair formation	6-7
	Cation vacancy activation	1.86 (1.5)

(a) Schottky (eV per defect)		(b) Frenkel (eV per defect)	
BaTiO$_3$	2.29	Oxygen	4.49
TiO$_2$	2.90	Barium	5.94
BaO	2.58	Titanium	7.56

Nevertheless, where comparisons can be made the results are encouraging as
discussed by Harding in references (8) and (12).

4.2. Qualitative application of simulations

In illustrating this class of applications we take two examples from a
wide range of studies reported over the last ten years. The first relates
to ordering of extended defects; the second to the sequence of defect reactions that occurs in irradiated alkali halides.

4.2.1. Ordering of extended defects. The ordering of shear planes, which
as noted in Section 1, form in ceramic non-stoichiometric oxides has proved
to be one of the most fascinating topics in the defect chemistry of inorganic materials. Ordering with large repeat distance up to ~100 Å was shown
by high resolution electron microscopy studies. The origin of the effect
was unclear. Defect supercell calculations of Cormack et al. (40) on
ReO$_3$-structured oxides, revealed a variation of the shear plane energy with
interplanar spacing as shown in Fig. 5. The interaction energy curve is
attractive hence giving rise to defect ordering. Moreover, analysis of the
results of the calculations showed that the attractive forces arise from a
constructive interference of the elastic strain fields of planar defects.

16

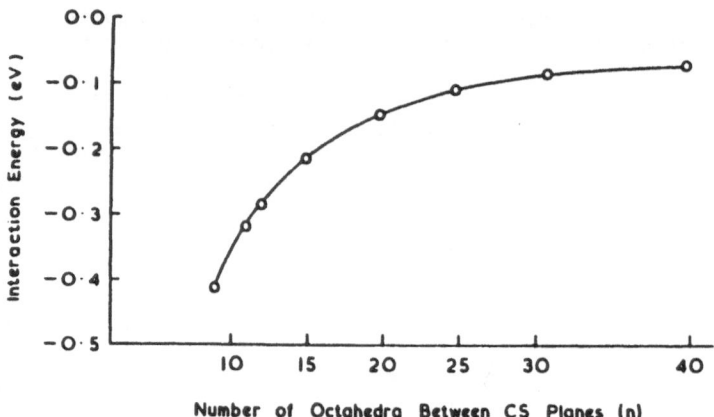

FIGURE 5. Variation of shear plane energy with spacing between shear-planes in ReO₃ structured oxides.

4.2.2. Radiolysis of the alkali halides. It has long been known that defects are created when alkali halide crystals are exposed to ionizing radiation. In the 1970's there developed an increasing understanding of the fundamental mechanisms of defect creation in those materials under irradiation. The role of calculations has been to illuminate the nature of the subsequent reactions which follow the initial defect creation process. It is well established that defects are products of the non-radiative decay of excitons which are bound electron-hole pairs created by the ionizing radiation. In the decay process which is discussed in detail by Stoneham (41), a halogen atom is ejected from its lattice site and, via a favorable collision sequence in the <110> direction, it comes to rest at a site that is well separated from the corresponding vacancy. The latter, as it still contains an electron, is an "F centre", whereas the interstitial atom binds strongly to a halogen anion to form an X_2^- halogen molecular species which could be expected to occupy either a <110> or <111> structure as shown in Fig. 6. Indeed, calculations show these structures to have very similar energies, and the fact that the <110> is generally the favored configuration is attributed to its stabilization by a small degree of covalent bonding with neighboring halide ions - a subtle feature that is not readily included in the simulations.

The major function of the simulation studies in this field has, however, been to assist our understanding of the types of defect reaction undergone by the H centre following its creation. To understand these we should first appreciate that the H centre is a mobile species; only a small displacement is needed for one of the component halogen atoms to rebind to a neighboring halide ion. H centres are therefore mobile at very low temperatures permitting the following processes to occur.

(i) Trapping by impurities. Calculations (37) showed that, for example, the H centre would be trapped by a Na^+ impurity in KCl. The resulting structure of the defect is shown in Fig. 7; the binding energy is calculated as ca. 0.8 eV. As discussed in refence (37), the configuration predicted by calculations is in good agreement with that inferred from e.p.r. studies.

(ii) Dimerization in which initially H centres form pairs which are bound loosely by the interaction of their elastic strain fields. For the most

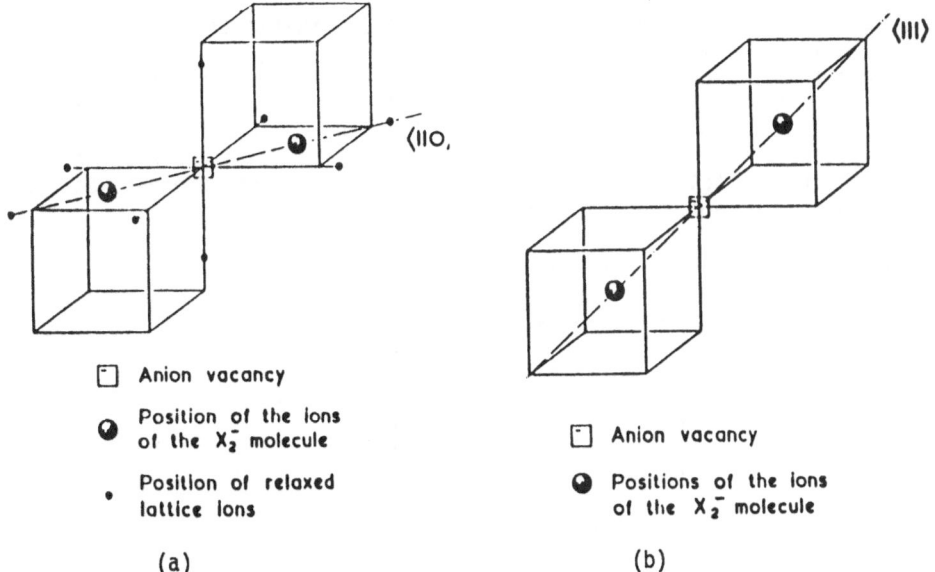

Anion vacancy

Position of the ions
of the X_2^- molecule

Position of relaxed
lattice ions

(a)

Anion vacancy

Positions of the ions
of the X_2^- molecule

(b)

FIGURE 6. (a) The <110> H centre. (In this and the following figures, corners of cubes, unless marked, indicate occupied lattice sites.) (b) The <111> H centre.

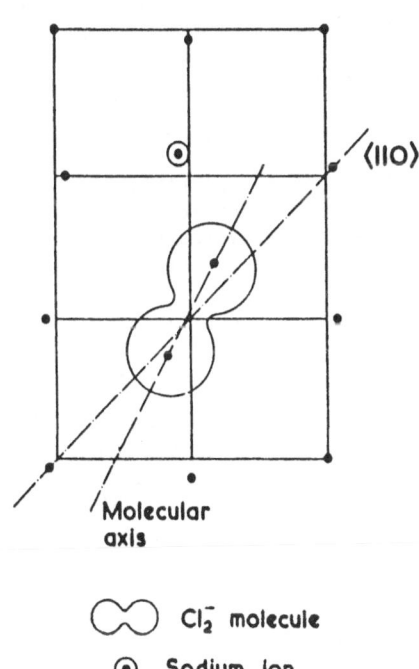

Cl_2^- molecule

Sodium ion

FIGURE 7. H centre adjacent to an Na^+ impurity in KCl. (The calculated positions of the ions are shown.)

strongly bound of such configurations in KCl, a binding energy of 0.15 eV
was calculated.

Simple chemical considerations would, however, suggest that this species
would be unstable and would be prone to collapse to yield a halogen molecular species. Again, calculations showed that such a process was, in
general, energetically favorable with the resulting X_2 molecule occupying a
pair of interstitial sites (see Fig. 8); and in KCl, this process was shown
to be favored by about 0.2 eV with respect to the di- H Centre.

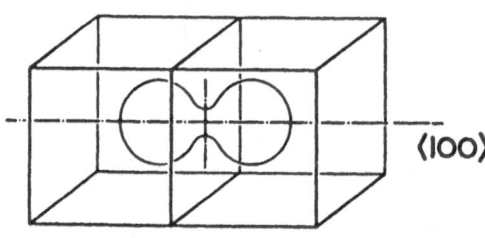

$\bigcirc\!\!\bigcirc$ X_2 molecule

FIGURE 8. Interstitial X_2 molecule.

(iv) Displacement by halogen molecules of lattice atoms to dislocation
loops. This reaction may be understood when we realize that halogen
molecules at interstitial sites (as shown in Fig. 8) will be highly
strained; and that this strain can be relieved by transferring the molecule
to a cation-anion vacancy complex as illustrated in Fig. 9. If the
di-vacancy/molecule complex were to be formed with displacement of lattice
ions to interstitial sites, the energy would be unfavorable. However,
displacement to a dislocation loop may occur in the interior of a crystal
and for sufficiently large loops is energetically equivalent to the accommodation of displaced ions at perfect lattice sites. Calculations show that
the latter process was energetically favored by 2.2 and 1.3 eV in NaCl and
KCl, respectively. We note that dislocation loops were observed in irradiated KCl by Hobbs et al. (42)

Figure 10 presents a schematic representation of the several processes
which a combination of experiment and calculations have shown to be important in irradiated alkali halides. We note that the F centres aggregate to
form metal colloids and that, at higher temperatures these colloids are
thought to anneal by reaction with the molecular halogen. The role of
calculations in this fascinating system has been to elucidate the energetic
feasibility of the various reactions that have been proposed.

4.3. Molecular dynamics simulation of ion transport mechanisms

The power of the MD techniques is perhaps most evident when it is used to
reveal the detailed mechanisms of ion migration in solids. In previous
reviews (5,43), we have described their use in studying atomic transport
mechanisms in several superionics including Li_3N and β''-Al_2O_3. Here we will
describe a recent application to an intriguing F^- ion conductor,
$RbBiF_4$ where MD studies of Cox (44,45) demonstrated the occurrence of highly
complex correlated migration mechanisms.

$RbBiF_4$ has, at higher temperatures, the fluorite structure (which can,
moreover, be quenched in at lower temperatures). The Rb and Bi ions are
distributed over the regular cation sites in the structure. There is no

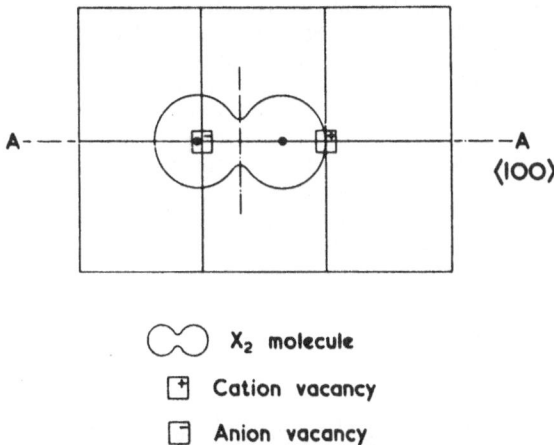

FIGURE 9. A molecule-vacancy pair complex. (The calculated equilibrium position of the molecule in NaCl is shown.)

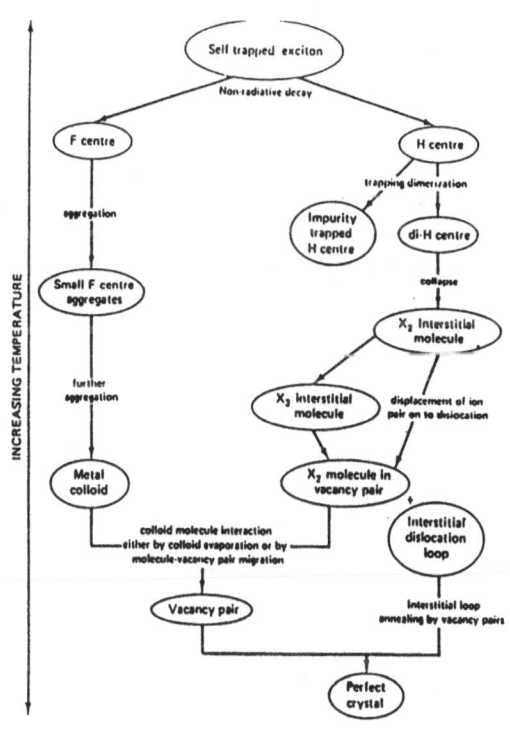

FIGURE 10. Radiation damage mechanisms and reactions in irradiated NaCl.

20

long-range ordering of this cation, although diffuse neutron scattering
studies (44) suggested that there may be appreciable short-range order. For
simplicity, the MD simulations assume a random cation distribution in a
large simulation box containing about 800 ions.

The MD simulations revealed fast F⁻ ion migration in accordance with the
observed superionic behavior of the material. Figure 11 shows plots of the
r.m.s. displacement of the ions vs time. The results were obtained for a
temperature of 450 K. The uppermost curve where diffusion is clearly
occurring refers to a Bi/Rb ratio of 1; the middle curve to one of 1.5; and
the lower curve to the ordered structure formed when the ratio reaches 3:1,
where diffusion has ceased.

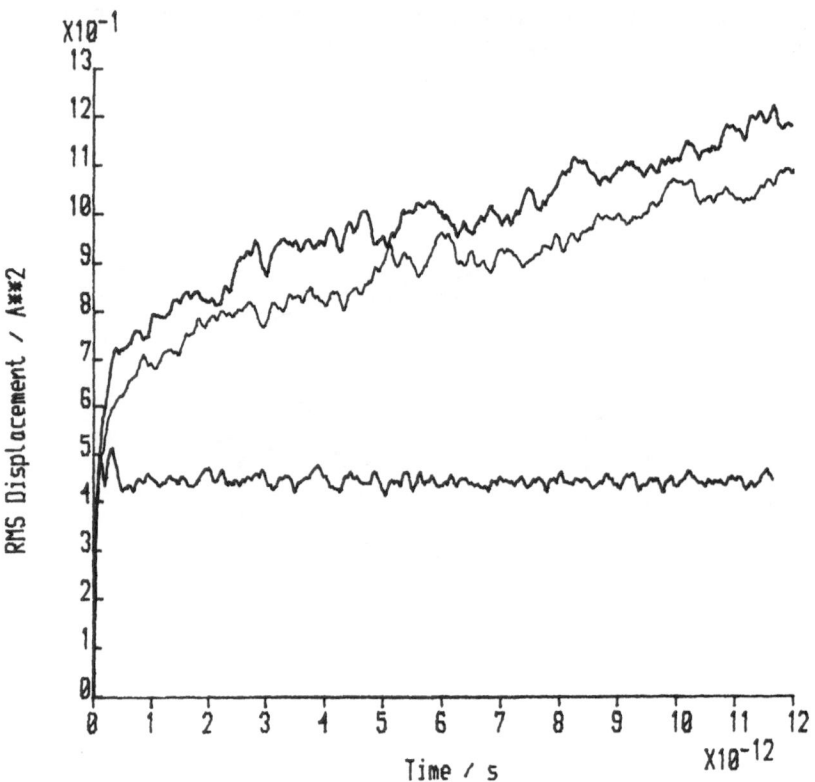

FIGURE 11. Root mean square displacements of F⁻ ions for various Rb/Bi
ratios in mixed Rb/Bi fluorides.

Cox made several studies of individual ion trajectories. He found that F⁻ ion migration occurs via a correlated or interstitially mechanism in which F⁻ interstitials displace neighboring lattice ions into interstitial sites. This is manifested by ion trajectories of the type shown in Fig. 12. However, far more complex correlated machanisms were revealed by the MD simulations. Several F⁻ ions move together in a concerted manner as illustrated in Fig. 13 which shows the trajectories of 3 F⁻ ions over a period of 10 ps. We note that correlated ionic motions have been demonstrated in simulation studies of other superionics including Li_3N and AgI.

4.4. Monte-Carlo simulation of ion transport in disordered materials

Extensive use has been made of the MC technique in simulating atomic transport in alloys (46,47). Here we illustrate the role of the technique by a recent application to oxygen transport in an oxygen-conducting ceramic material, Y/CeO_2, which has received considerable attention owing to its potential use as an electrolyte material in fuel cells.

CeO_2 again has the fluorite structure. It dissolves high concentrations of Y_2O_3 (and rare earth oxides). The trivalent cations substitute for the host lattice ions. The lower valence means that the Y'Ce defects are negatively charged and are compensated by oxygen vacancies. The latter are mobile giving rise to rapid oxygen ion transport in the material.

We might expect that with increasing dopant (and hence vacancy) concentrations, there would be a corresponding increase in conductivity. This is indeed observed at lower concentrations, but the simple prediction fails dramatically at higher dopant concentrations as illustrated in Fig. 14, where we note a marked decrease (by several orders of magnitude) of the conductivity after a maximum of about 6 mole % Y (ref. 48).

The aim of our MC study (49) of the material was to test whether we could reproduce and explain this remarkable effect. The MC simulation proceeded as follows:

(i) A simulation box was defined and periodic boundary conditions were applied, as with MD techniques. The box contained several thousand ions and dopants were introduced at random on the cation sites.

(ii) Vacancies were introduced by abstracting atoms from a certain number of sites, again at random.

(iii) Next a vacancy was selected at random as was a corresponding jump direction for the defect. The environment of the defect was examined and the jump frequency, W_j, obtained. Previous static lattice calculations had estimated activation energies for all possible dopant distributions in the first neighbor cation shell. Activation energies could therefore be calculated for each environment using the Arrhenius expression. To ensure maximum efficiency of the simulation, all jump frequencies were scaled so that the maximum value was unity.

(iv) A random number, R, in the range 0 to 1 was generated, and the value of the number compared with W_j. If $R < W_j$ the jump was deemed to be successful; if $R > W_j$ it was considered unsuccessful. This procedure effectively weights the probability of a jump according to its frequency.

(v) The procedures in steps 3 and 4 above were repeated several thousand times. In the initial stages of the simulation the system is equilibrated as in MD simulations. After equilibration the successive configurations generated by the simulation were used to follow the process of defect migration.

The results of the MC calculations are shown diagramatically in Fig. 15. Qualitatively the comparison between theory and experiment is satisfactory. The calculations reproduce the observed maximum in the conductivity at about the right dopant concentration. The quantitative agreement is less

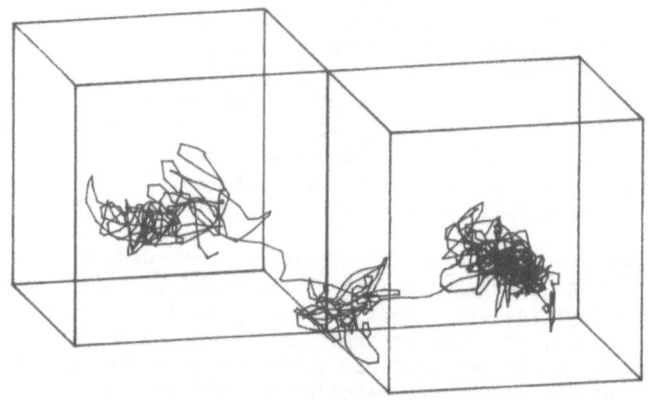

FIGURE 12. Trajectory of migrating ion in RbBiF₄. Lattice ions are at corners of cube. Interstitial sites are in centre.

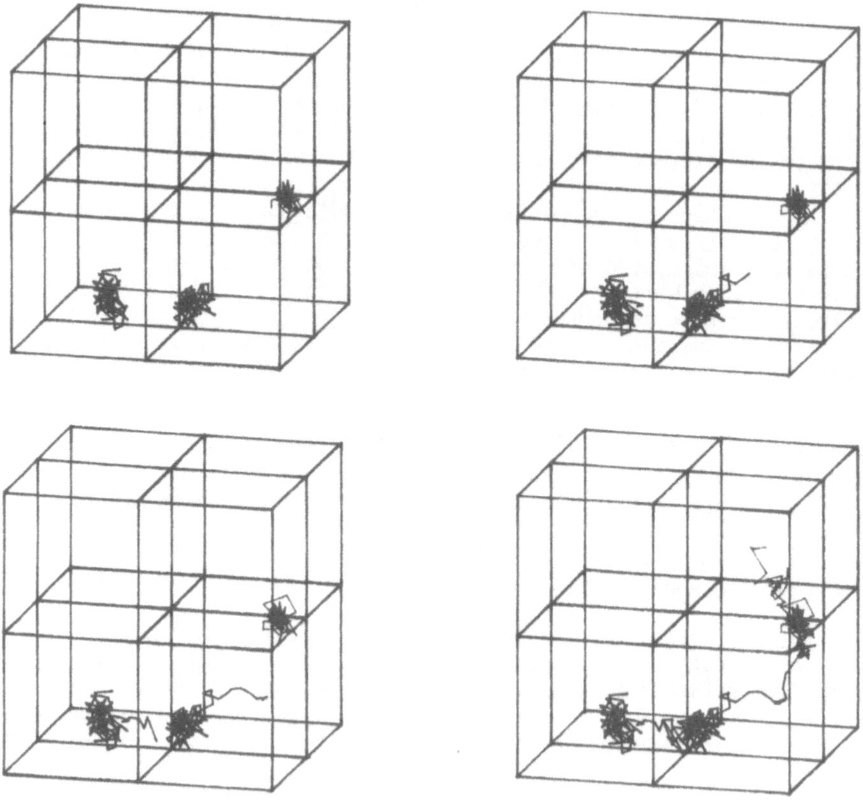

FIGURE 13. Trajectories of three ions in RbBiF₄.

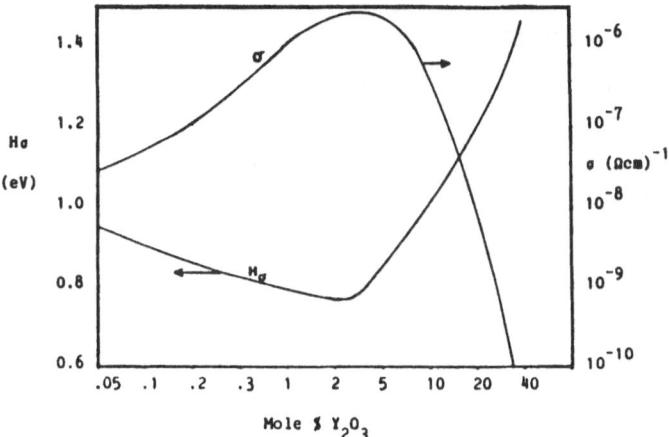

FIGURE 14. Plot of conductivity and Arrhenius enthalpy versus dopant concentration in CeO_2-Y_2O_3 (see ref. 48).

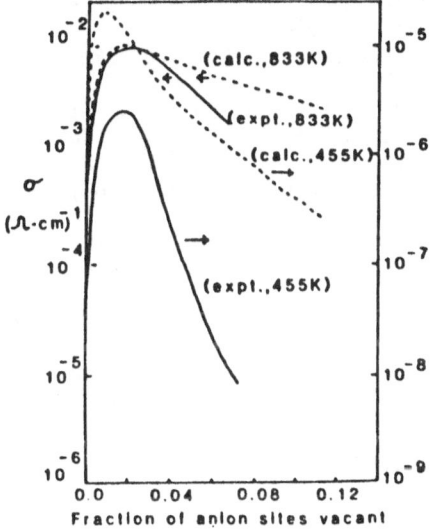

FIGURE 15. Calculated and experimental conductivities.

satisfactory, but is almost certainly as good as can be achieved with the
present simple model. Possibly of greater interest has been the insight
yielded by the simulations into the physical basis of the decrease in conductivity at high dopant concentrations. This is caused by the increasing
proportion of unprofitable vacancy jumps, i.e. jumps which do not effect
bulk transport of the oxygen, but which lead to, for example, vacancies
jumping around dopant ions or back and forth between a pair of oxygen sites.

   Further, more detailed calculations are needed on this intriguing system;
but the present study indicates the value of the MC approach in studying
transport in highly defective materials.

5. SUMMARY AND CONCLUSION

This chapter has discussed both the diversity of defect properties of ceramics, and the way in which these can be investigated by current computer modelling techniques. It should also be clear that the latter techniques could have a substantial role in the study of radiation damage in these materials. Study of the defect reactions which follow the initial damage process is one obvious area. But other applications such as a direct MD simulation of the displacement cascade caused by an implanted ion can be envisaged. Computer modelling of ion implanted ceramics is clearly an exciting topic for the future.

ACKNOWLEDGEMENTS

Much of the work described in this article has been supported by grants from the Science and Engineering Research Council and from the Theoretical Physics Division of the Atomic Energy Research Establishment, Harwell, U.K.

REFERENCES

1. Agullo Lopez F., CRA Catlow, and PD Townsend, Point Defects in Materials. London: Academic Press, 1988.
2. Beniere F and CRA Catlow (eds): Mass Transport in Solids. New York. Plenum Press, 1983.
3. Murch GE and AS Nowick (eds): Diffusion in Crystalline Solids. New York: Academic Press, 1984.
4. Catlow CRA and WC Mackrodt (eds): Computer Simulation of Solids. Berlin: Springer, 1982.
5. Catlow CRA: Ann. Rev. Mater. Sci. 16 (1986) 517.
6. Mackrodt WC: Transport in Non-Stoichiometric Compounds. G Petot Ervas, Hj Matzke, and C Monty (eds). Amsterdam: North Holland, 1984.
7. Gillan MJ and PWM Jacobs: Phys. Rev. B28 (1983) 759.
8. Harding JH: Phys. Rev. B32 (1985) 6861.
9. Harding JH and AM Stoneham: Phil. Mag. B34 (1983) 705.
10. Allan NL, WC Mackrodt, and M Leslie: Advances in Ceramics 23 (1988) 257.
11. Catlow CRA, J Corish, PWM Jacobs, and AB Lidiard: J. Phys. C. 14 (1981) L121.
12. Harding JH: Proc. Intern. Conf. Defects in Insulating Crystals, Parma, Italy, Cryst. Latt. Def. and Amorph. Mat., in press.
13. Norgett MJ: UKAEA Report AERE-R7650.
14. Catlow CRA, R James, WC Mackrodt, and RF Stewart: Phys. Rev. B25 (1982) 1006.
15. Leslie M: SERC, Daresbury Laboratory Report DL-SCI-TM31T, 1982.
16. Catlow CRA, J Corish, J Harding, and PWM Jacobs: Phil. Mag. A55 (1987) 481.
17. Vineyard G: J. Phys. Chem. Solids 3 (1957) 57.
18. Gillan MJ, JH Harding, and R-J Tarento: UKAEA Report AERE-M3494, 1985.
19. de Lorenzi G: Computer Simulation of Fluids, Polymers and Solids. CRA Catlow, SC Parker, and MJ Allen (eds). Dordrecht, Kluwer Academic Publishers, 1989.
20. Leslie M, JH Harding, and MJ Gillan: J. Phys. C. - in press.
21. Tasker PW: Phil. Mag. A39 (1979) 119.
22. Tasker PW: Surf. Sci. 87 (1979) 315.
23. Duffy DM and PW Tasker: Phil. Mag. A48 (1983) 155.
24. Colbourn EA, WC Mackrodt, and PW Tasker: Physica B 131 (1985) 41.
25. Duffy DM and PW Tasker: Phil. Mag. 50 (1984) 143.
26. Duffy DM and PW Tasker: Phil. Mag. 50 (1984) 155.
27. Mackrodt WC: Advances in Ceramics 23 (1988) 293.
28. Tosi MP: Solid State Physics 16 (1964) 1.

29. Wedepohl PT: Proc. Phys. Soc. 92 (1967) 79.
30. Gordon RG and YS Kim: J. Chem. Phys. 56 (1972) 3122.
31. Mackrodt WC and RF Stewart: J. Phys. C. 12 (1979) 431.
32. Mackrodt WC, RF Stewart, JC Cambell, and IM Hillier: J. Phys. Paris. 41 (1980) C7:64.
33. Saul P, CRA Catlow, and J Kendrick: Phil. Mag. 51 (1985) 107.
34. Grimes RW, CRA Catlow, and AM Stoneham: UKAEA Report AERE-M3687, 1988.
35. Grimes RW, CRA Catlow, AN Cormack, and AM Stoneham: Advances in Ceramics 23 (1988) 273.
36. Grimes RW: UKAEA Report - in press.
37. Catlow CRA, KM Diller, and LW Hobbs: Phil. Mag. 42 (1980) 123.
38. Wuensch BJ: Mass Transport in Solids. F Beniere and CRA Catlow (eds). New York: Plenum Press, 1983.
39. Lewis GV and CRA Catlow: J. Phys. Chem. Solids 47 (1971) 89.
40. Cormack AN, R Jones, PW Tasker, and CRA Catlow: J. Solid State Chem. 44 (1982) 174.
41. Stoneham AM: J. Phys. C. 7 (1974) 2476.
42. Hobbs LW, AE Hughes, and D Pooley: Proc. Roy. Soc. A332 (1973) 167.
43. Catlow CRA: Defects in Solids. AV Chadwick and M Terenzi (eds). New York: Plenum Press, 1986.
44. Cox PA: Ph.D. Thesis, University of Keele, 1988.
45. Cox PA and CRA Catlow: to be published.
46. Murch GE: Diffusion in Crystalline Solids. GE Murch and AS Nowick (eds). New York: Academic Press, 1984.
47. Murch GE: Phil. Mag. A46 (1982) 575.
48. Wang DY, DS Park, J Griffiths, and AS Nowick: Solid State Ionics 2 (1981) 95.
49. Murch GE, CRA Catlow, and AD Murray: Solid State Ionics 18/19 (1986) 186.

DISORDER, RANDOMNESS, AND AMORPHOUS PHASES

P. MAZZOLDI AND A. MIOTELLO*

Dipartimento di Fisica dell'Universita, Via Marzolo 8, 35131 Padova, Italy
*Dipartimento di Fisica dell'Universita, 38050 Povo (Trento), Italy

1. INTRODUCTION
   Ordered solids are experimentally well-characterized, and unified theoretical models for structure as well as for transport and relaxation processes
are available. However, a unified picture for disordered solids is not yet
available due to the lacking of general unified theoretical concepts, like
Bloch's theorem for regular lattices. In this paper, we will describe the
amorphous state, focusing attention on possible general criteria which
define the amorphous state. These criteria are both microscopic with nature
as, for example, the pair correlation function, and macroscopic in nature
like the thermodynamical aspects of the liquid-amorphous phase transition.
The main features that emerge from experimental investigations of structural
relations in glasses will be presented, and emphasis will be placed upon
low-frequency excitations whose nature is quite obscure. Transport properties of glasses (both atomic and electronic) will also be analyzed. As to
atomic transport, we will address the theoretical models which try to
explain ionic motion in disordered solids through cooperative atomic
rearrangeent. Theoretical models to describe transport that are derived
from general thermodynamical principles are presented. These models also
include correlation effects. As to the electronic transport, we will simply
discuss the effect of the disorder in inducing localization on otherwise
extended electronic states. Some concepts of the percolation theory will be
illustrated.

2. DISORDER
   An ordered system may be considered as a system where any observable quantity is periodic, according to the formal mathematical property:

$$F(r) = F(r + a_i) ,\qquad\qquad (1)$$

where r is any position vector and a_i are the Bravais lattice vectors. The
transition from a regular to a disordered system destroys such relationship;
that is, some physical properties are not invariant under translation.
However, some observable quantities, for example the spin "averaged"
electron density in magnetic structures, maintain relation (1) if the
"local" spin operator value changes from site to site.
   Obviously, it is necessary to define better the concept of disorder and we
consider two possible situations: topological disorder and continuous
disorder. Topological disorder is a physical situation where the relation
$F(r + R_i) \approx F(r)$ holds for values of $|r| < r_c$ [see Fig. 1(b)]. If this condition is not satisfied, continuum disorder occurs. Such concepts are
clarified in Fig. 1. All direct observational evidence about the atomic
arrangement in condensed matter is contained in the pair distribution

27

C.J. McHargue et al. (eds.), Structure-Property Relationships in Surface-Modified Ceramics, 27–45.
© 1989 by Kluwer Academic Publishers.

FIGURE 1. (a) Lattice order, (b) topological disorder, and (c) continuum disorder. [After Ziman (1)].

function $g(r_1, r_2)$. The $g(r_1, r_2)dV_1 dV_2$ is the probability of finding an atom centered on the point r_1 in the volume dV_1, and an atom at r_2 in the volume dV_2. If we assume that our sample is statistically homogeneous, the average density of atoms per unit volume must be constant, and the $g(r_1, r_2)$ function depends only on the relative vector separation r_{12} of points 1 and 2: $g(r_1, r_2) = g(r_{12})$. Obviously for a perfect crystal r_{12} corresponds to a Bravais lattice vector,

$$g(r_{12}) = n^{-1} \delta(r_{12} - a_i) ,$$

where n is the average density of atoms and δ the delta function.

Such concepts are illustrated in Fig. 2 for an ideal microcrystalline system. The introduction of the thermal atomic vibration modifies the g function as shown in Fig. 3.

The main features of the $g(r)$ function shown in Fig. 3 are the following: it is zero for a distance equal to the hard-core distance of an atom and then rises to a peak at a distance R_0, corresponding to the radius of the first coordination shell of atoms.

The area under this peak

$$z = \int g(r) \, 4\pi r^2 dr$$

is the coordination number of the structure, that is, the number of the nearest neighbors. Similarly the next peaks come from the second, third... coordination shell of next-neighbors. The peak broadening is due to the disorder (thermal, structural, ...) but the coordination number is not precisely defined and the peaks broaden and merge. When $g(r) \rightarrow 1$, such quantities become lost in the continuum background. In this context, it is convenient to introduce the total correlation function $h(r)$ which defines

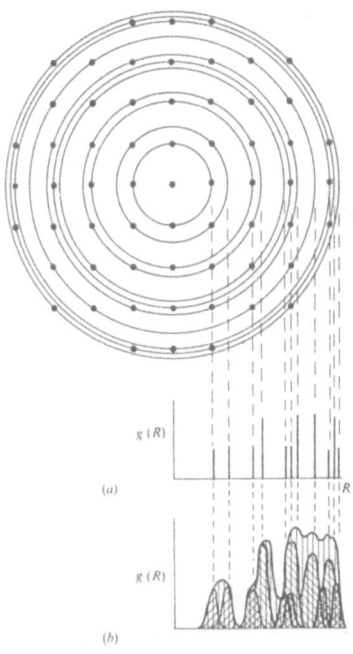

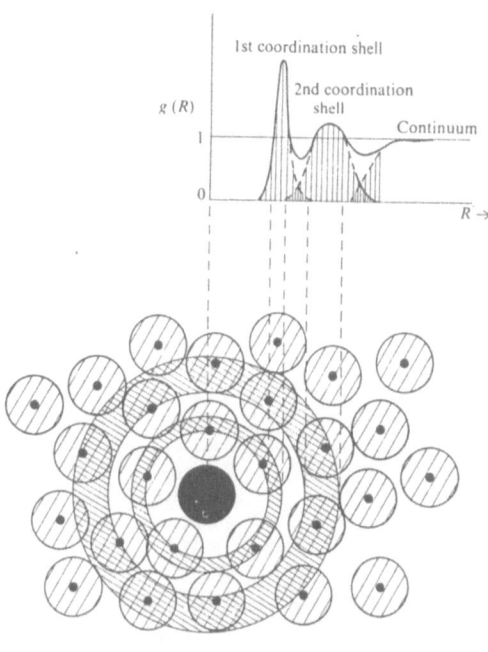

FIGURE 2. Pair distribution function for (a) ideal and (b) thermally broadened microcrystalline assembly. [After Ziman (1)].

FIGURE 3. "Features" of the pair distribution function for an amorphous solid. [After Ziman (1)].

the extent of local variation from statistical uniformity about any given atom of material: h(r) = g(r) - 1. The range of order, r_c, is empirically defined as a distance such that h(r) = o for $|r| > r_c$.

The radial distribution function, obtained by neutron scattering, of $Pd_{80}Si_{20}$ glass is shown in Fig. 4. The first peak splits into two sub-peaks A and B (at $r_A \simeq 2.4$ A and $r_B \simeq 2.8$ A) and the second peak presents three small humps. The sub-peak A is due to the Si-Pd pair correlation subjected to the strong chemical bond, while the sub-peak B is almost entirely due to the Pd-Pd pair correlation. This is confirmed by a comparison between the $Pd_{80}Si_{20}$ glass and the Pd_3Si crystalline compound. It is interesting to observe that the first peak splitting is not evident in the pair distribution function of liquid $Pd_{80}Si_{20}$ (see Fig. 5), although a slight hump appears in the position corresponding to the Si-Pd pair correlation of $Pd_{80}Si_{20}$ glass. The chemical short-range structure of $Pd_{80}Si_{20}$ alloys is essentially identical in the crystal, glass, and liquid states.

A variety of structural models, both conceptual and physical, have been proposed to describe the structure of amorphous solids. The general features of the radial distribution function arise quite naturally from the models. The structure of amorphous metals, ionic solids and molecular organic solids (and also some liquids) which are characterized by non-directional forces, can be described in terms of "dense random packing" (DRP) of hard spheres (3). More recently, for the structure of amorphous transition metal - metalloid alloys, Gaskell (4) proposed a model based on

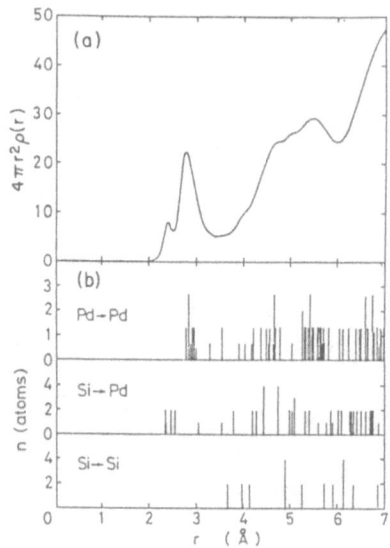

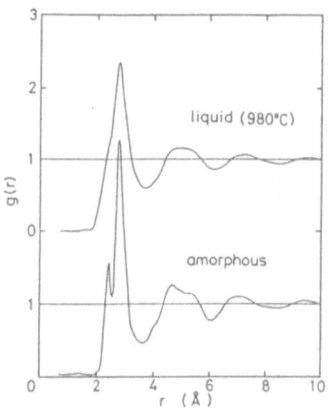

FIGURE 4. (a) Radial distribution
function of $Pd_{80}Si_{20}$ alloy glass
and (b) schematic crystal structure
of Pd_3Si orthorhombic-type compound.
[After Suzuki (2)].

FIGURE 5. Pair distribution function
of $Pd_{80}Si_{20}$ alloy glass at room
temperature and liquid at 980°C.
[After Suzuki (2)].

random packing of trigonal prisms. Non-metallic materials (excluding ionic
solids) which possess covalent, directional bonds are often described in
terms of a continuous random network (CRN). However, the g(r) arising from
these models, exhibits characteristics that are not inconsistent with hot
solid or perturbed microcrystal disorder. Discrimination on the basis of
the observable features of g(r) is very uncertain. The distinction between
amorphous solids and paracrystalline systems may be obtained by considering
some physical properties that present a more critical dependence on the statistical characteristics than does the pair correlation function.

3. THERMODYNAMICS OF GLASS TRANSITION

First of all, we want to recall the meaning of the order of a transition.
Following the Ehrenfest scheme, the order of a transition is the order of
the lowest derivative of the Gibbs free energy which shows a discontinuity
at the transition point.

When a liquid is cooled, one of two events is observed: crystallization
or glass formation (although some intermediate metastable states could be
present in some particular solidification processes).

The two extreme possibilities (amorphous or crystalline) can be evidenced
by analyzing the specific volume, V, (or other thermodynamical variables
such as entropy or enthalpy) as a function of temperature. A schematic
illustration is shown in Fig. 6. If the melt crystallizes on cooling, the
transition is characterized by a discontinuity, ΔV, at the melting point,
T_m. Such a marked change does not occur if the melt "supercools". Glass
formation is evidenced by a gradual change in slope of V versus T. Since
the transition to the glass is continuous, the glass transition temperature,

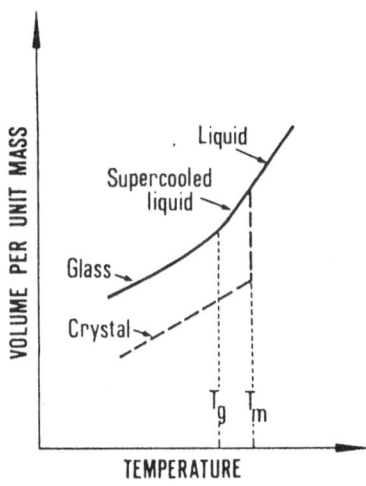

FIGURE 6. Schematic illustration of the relation between glassy, liquid, and solid states: variation of specific volume with temperature. T_m is the melting and T_g the glass-transition temperature. [After Mazzoldi and Arnold (5)].

T_g, cannot be defined for any particular glass and the term "transformation range" is used more than "transformation temperature". A "fictive" temperature, T_f is introduced, which is defined as that specific temperature obtained by the intersection of the extrapolated liquid and glass curves. At some temperature T below T_g, the glass structure reaches a metastable equilibrium configuration, which has a lower free energy than that of other liquid-like configurations. The lowest possible free energy is that of the crystalline material. The transition to the crystalline configuration from the liquid-like glass structure, at a temperature lower than T_f, occurs at an infinitely slow rate. T_g depends on the cooling rate, q, of the super-cooled liquid; an expression for it can be derived using the free-volume theory and this is

$$q = q_0 \exp\left[-1/c\ (1/T_g - 1/T_m)\right] ,$$

where c is a constant ($\simeq 3 \times 10^{-5}$) and q_0 a term which depends on the glass composition (for chalcogenide glasses, e.g., Se, $q_0 = 10^{23}$ Ks^{-1}; As_2S_3 $q_0 = 10^4$ Ks^{-1}). At T_g (Fig. 6) a discontinuity is present in derivative variables such as thermal expansion coefficient [$\alpha_T = (\delta \ln V/\delta T)_p$, bulk compressibility [$K_T = -(\delta \ln V/\delta P)_T$] and heat capacity at constant pressure [$C_p = T\ (\delta S/\delta T)_p$]. From this point of view, one could expect that the glass transition is of the second order. Consequently, the first-order extensive thermodynamic variables are continuous at the transition. However, let us prove that the liquid amorphous phase transition is not second order. Indeed the entropies of liquid (1) and glass (2) must be equal at the transition: $S_1 = S_2$ and also $dS_1 = dS_2$ for changes in temperature or pressure. In particular:

$$\left(\frac{\delta S_1}{\delta T}\right)_p dT + \left(\frac{\delta S_1}{\delta P}\right)_\tau dP = \left(\frac{\delta S_2}{\delta T}\right)_p dT + \left(\frac{\delta S_2}{\delta P}\right)_\tau dP .$$

Using some Maxwell's thermodynamics relations, we obtain for the shifts in transition temperature with pressure

$$\frac{dT_g}{dP} = TV \frac{\Delta\alpha_\tau}{\Delta C_p} \tag{3}$$

and

$$\frac{dT_g}{dP} = \frac{\Delta K_\tau}{\Delta\alpha_\tau} . \tag{4}$$

The experimental data concerning $\Delta\alpha_\tau$, ΔC_p, and ΔK_τ at the glass transition are in agreement (within the experimental errors) with the relation (3), but the values of $(\Delta K_\tau/\Delta\alpha_\tau)$ are generally appreciably higher than those of (dT_g/dP) (Eq. 4). This fact proves that the glass transition is not a simple second-order phase transition.

To conclude, we want to emphasize that Prigogine and Defay (6) showed that the ratio

$$R = \frac{\Delta K_\tau \Delta C_p}{TV(\Delta\alpha_\tau)^2}$$

is equal to unity if a simple ordering parameter determines the position of equilibrium in a relaxing system, but if more than one ordering parameter is responsible (as occurs in most glasses) then $R > 1$.

After these general considerations on the structure and thermodynamic features of a glass, we want now to consider the concept of a defect in an amorphous state.

4. DEFECTS IN AMORPHOUS SOLIDS

The concept of a defect in the structure of an amorphous solid might appear at first sight to be strange. We can introduce the concept of a defect in an amorphous state, if any given defect is defined with reference to some non-defective state. For non-crystalline solid, we have to consider an "ideal" amorphous structure as, for example, CRN for a covalent system, or a DRP of hard spheres for a metal.

We will now discuss some important defects which may be present in amorphous materials (3).

4.1. Dangling bond

This point defect is due to a broken or unsatisfied bond in a covalent solid. Such type of defect has no meaning in a solid formed from non-directional bonds, such as a metal, ionic salt or a rare gas. A single dangling bond normally contains one electron, is electrically neutral and is amphoteric. The electron occupancy can change in some particular conditions, and thus change the charge state of the center.

An isolated dangling bond cannot be present in a crystalline solid where crystallographic constraints preclude it. A simple molecular orbital picture may give the energy levels for electron states associated with an isolated dangling bond. The formation of a charged dangling bond is determined by electron-phonon interaction.

4.2. Vacancy

An atomic vacancy is the most simple form of structural defect in amorphous materials. Such a defect in covalent solids also contains a number of dangling bonds. Vacancies may occur either singly as monovacancies or as divacancies, clusters, etc. The vacancies may possess a variety of electron occupancies and hence charge states.

An example of a vacancy center in an amorphous system is the so-called "E'-center" produced in vitreous SiO_2 by x-ray, γ-ray or energetic particle irradiation. The E'-center may be described as an oxygen vacancy opposite an electron in a dangling Si sp^3 orbital (see Fig. 7). This atom displacement may occur through radiolytic processes. The energy required to create an electron-hole pair (about 9 eV in $α-SiO_2$), if converted to atomic motion through non-radioactive recombination, would be sufficient to overcome the Si-O bond strength of about 4.5 eV if concentrated at a single network site. The radiolytically displaced oxygen forms a new bond with a near-neighbor oxygen. This defect can also be formed by hole trapping at the Si-Si bond formed after the displacement of the bridging oxygen to a distant site due to collisional processes either by ions or other energetic particles. Hole trapping occurs at the Si-Si bond in the same manner as in the first example [Fig. 7(a)]. The formation of E'-like centers at low temperature where atomic, H, can migrate in fused silica is shown in Fig. 7(b). The growth of this defect requires the existence of a 3-coordinated Si precursor defect. The trapping of two electrons in an oxygen vacancy, i.e. a Si-V-Si configuration determines the formation of a non-paramagnetic defect called a B_2-center with an optical absorption band at 5.0 eV.

4.3. Density defects

Density defects are determined by an aggregate of a non-fixed number of vacancies (voids < 100 Å) or fluctuations in density (or free volume) which may be regarded as a vacancy distribution throughout the material.

The implication of density fluctuations on the diffusion mechanism in amorphous materials will be discussed in a following section.

Models which introduce a generalized form of dislocations in amorphous phase have been presented by many authors (3). However, the free-volume theory accounts for the morphology introduced by the dislocation concept. An additional type of line defect, associated with rotations, has been proposed and termed "disclination".

5. RELAXATION IN AMORPHOUS MATERIALS

Relaxation is defined as the approach to equilibrium of a system driven out of equilibrium by a perturbation (7). For an arbitrary time-dependence of the force X(t) (electric field, shear stress, and pressure) or constraint Y(t) (polarization or surface charge, shear strain, and volume) the response of the system is given by

$$Y(t) = \int_0^t \frac{dX(t')}{dt'} \, \Phi(t - t')dt' \tag{5}$$

and

$$X(t) = \int_0^t \frac{dY(t')}{dt'} \, \psi(t - t')dt' . \tag{6}$$

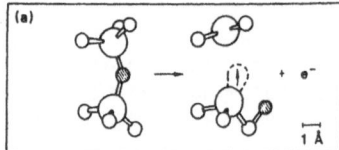

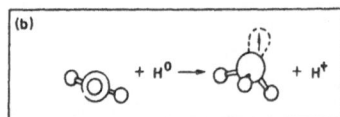

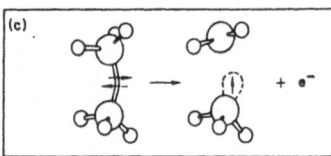

FIGURE 7. Model for the E' center in α-SiO₂: (a) Center generated by radiolytic processes at a normal lattice site; (b) center formed by interaction of 3-coordinated Si with migrant atomic H; (c) center formed by collisional-energy displacement of O to form vacancy. [After ref. (5)].

Equations (5) and (6) are mathematical expressions of Boltzmann's super-position principle that is applied to systems with linear responses or whose responses can be linearized. Linearity implies that response functions are unaffected by the level of force or constraint.

Before using Eqs. (5) and (6) for a particular problem (polarization due to an electric field), we want to comment on the function $\Phi(t)$, i.e. the relaxation function.

It is considered to be well-established that the relaxation function $\Phi(t)$ [as well as the retardation function $\psi(t)$] can be represented by the fractional exponential function $\Phi(t) = \exp - (t/\tau)^b$, where τ is the relaxation time and b is a constant, essentially temperature-independent. Such a function is called the Kohlrausch-Williams-Watts function. This function, even if it is not sufficiently supported by "ab initio" models, has been used successfully to describe dielectric, mechanical, structural and other kinds of relaxation. In many systems, τ shows an Arrhenius behavior given by the expression $\tau = \tau_0 \exp (U/k_BT)$ where U is an activation energy.

We want now to describe, by using Eq. (5), the electric field-induced polarization in amorphous materials.

Let us establish an analytical expression for $\epsilon''(\omega)$, the imaginary part of the dielectric constant. The dependence of the absorption current on time, following Eq. (5) is represented by the integral (with $b = 1$ in the relaxation function)

$$J(t) = g \int_{-\infty}^{t} \frac{dE}{dt'} e^{-\left(\frac{t-t'}{\tau}\right)} dt' \quad . \tag{7}$$

In the simplest case, when the field is sinusoidal, i.e.

$$E = E_m e^{i\omega t} ,$$

integration of Eq. (7) gives the expression

$$J(t) = g\left(\frac{\omega^2\tau^2}{1+\omega^2\tau^2}\right)e^{i\omega t}\, E_m + ig\left(\frac{\omega t}{1+\omega^2\tau^2}\right)e^{i\omega t}\, E_m \; . \tag{8}$$

The first term in the r.h.s. of Eq. (8) is in phase with the applied voltage and is known as the active component of the absorption current. The second term in r.h.s. of Eq. (8) is the current which leads the voltage by a phase angle $\pi/2$ and is the reactive (capacitative) component of the current. Now taking into account, the well-known relationship between capacitance and current and voltage amplitudes, it is easy to derive from the two terms in the r.h.s. of Eq. (8) the two relations:

$$\epsilon_r = 4\pi g\,\frac{\tau}{1+\omega^2\tau^2} \tag{9}$$

$$\epsilon_a = 4\pi g\,\frac{\omega t^2}{1+\omega^2\tau^2} \; . \tag{10}$$

Taking into account the phase shift of $\pi/2$ between ϵ_r and ϵ_a [similar to the phase shift between the first and second term in the r.h.s. of Eq. (8)] and using the symbols $\epsilon_r = \epsilon'$, $\epsilon_a\; \epsilon''$, we can write the permittivity, due to a slowly established polarization, as a complex quantity:

$$\epsilon^* = \epsilon' - i\epsilon'' \; . \tag{11}$$

The frequency dependences of ϵ' and ϵ'' are given by the functions $1/(1 + \omega^2\tau^2)$ and $\omega\tau/(1 + \omega^2\tau^2)$. These functions are shown graphically in Fig. 8.

In the following, we present a simple application of Eq. (10) to reproduce $\epsilon''(\omega)$ in a 30 Na_2O-70 SiO_2 glass at low frequency ($\simeq 10^{-3}$ Hz). Dielectric losses in this region of very low frequencies are attributed to polarization and relaxation of non-bridging oxygen-alkali metal complexes. It is generally assumed that in a glass (and this fact is a genuine feature of a disordered solid) many relaxation times, τ, operate in the presence of a disturbing force at a frequency, ω. Thus, Eq. (10) is usually written:

$$\epsilon''(\omega) = \frac{n_0 e^2\delta^2}{2\epsilon_0 k_B T} \int_0^\infty \frac{\omega\tau}{1+\omega^2\tau^2}\, f(\tau)d\ell n\tau \; , \tag{12}$$

where n_0 is the number density of non-bridging oxygen-alkali metal ion complexes, δ is the distance between the two positions of the alkali-ion involved in the polarization process, and $f(\tau)$ is a weighting function which gives the relative strength of a process with relaxation time, τ. In Fig. 9 we show along with the experimental points the theoretical curve obtained by Eq. (12) putting $\delta = 2\cdot3$ Å (the average distance between two non-bridging oxygens) and writing $f(\tau)/\tau$ as the sum of three equiprobable contributions (8).

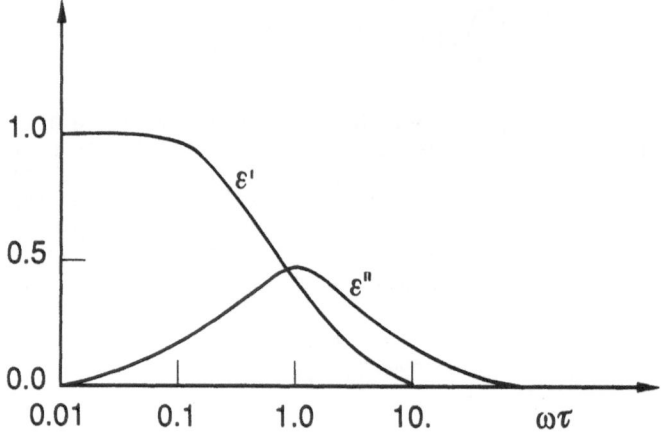

FIGURE 8. Schematic representation of ϵ' and ϵ'' as function of $(\omega\tau)$.

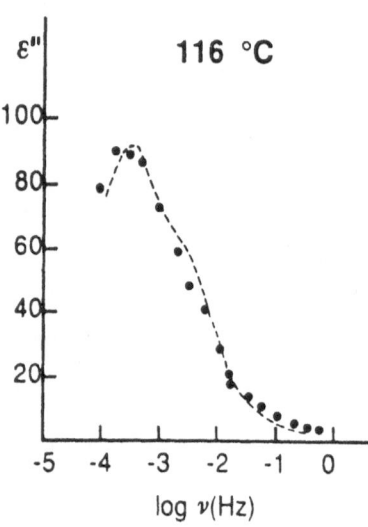

FIGURE 9. Imaginary part ϵ'' of the dielectric constant as a function of frequency ν at $T = 116°C$ in 30 Na_2O-70 SiO_2 glass. Full points: experimental results; dashed line: Eq. (12) [see ref. (8)].

6. EFFECT OF DISORDER IN THE TRANSPORT PROCESSES

6.1. Atomic diffusion

Many important properties of disordered solids are activation-type lattice processes, which, unlike electronic processes, have begun to be studied only in the last few years. From the available experimental results concerning these processes some general aspects emerge:

(1) The diffusion coefficients in disordered solids are considerably higher than in crystalline structures.

(2) There are indications that the activation energy does not follow the Arrhenius form.

(3) The diffusion properties are usually very sensitive to structural relaxation processes in disordered solids.

Disordered solids, such as amorphous semiconductors, amorphous magnetic materials, metallic and non-metallic aloys, have quite different properties.

However, if we consider the behavior of atomic particles, possible common features may be observed. These are due to the absence of periodicity in the spacial configuration of atoms and to the presence of a random component in the dependence of the potential energy, E(r), of the system on distance. Hence, in contrast to crystals, where the activation energy, E_c, for the diffusion jump is constant, there is a distribution of barriers in disordered solids. It is therefore reasonable to refer to the probability that the atom has an energy in the range of E to E + dE. As a consequence, the atomic diffusion coefficient in a disordered solid may be written as:

$$D = D_0 \exp \left(-\langle E \rangle / k_B T \right) , \qquad (13)$$

where D_0 is the pre-exponential factor which is independent of $\langle E \rangle$, T is the absolute temperature, and $\langle E \rangle$ is the approximately averaged activation energy.

As an example, let us analyze Eq. (13), looking at the experimental results for the Ag^{110} diffusion coefficient in amorphous Pd-19 at. % Si.

In Fig. 10 we show the Arrhenius plot of the Ag^{110} experimental diffusion coefficient (9). D_0 is $2 \cdot 10^{-6}$ cm²/sec and $\langle E \rangle$ = 1.3 eV.

Due to the supercooled nature of the amorphous alloy specimens, the parameters for diffusion may be expected to follow the pattern:

$$\langle E_l \rangle < \langle E_a \rangle < \langle E_c \rangle \text{ and } D_0^a < D_0^l < \langle D_0^c \rangle ,$$

where the a,c,l refer to the amorphous, crystalline, and liquid phases, respectively.

An inspection of literature concerning diffusivity data both in liquid and solid crystals leads to the conclusion that the values of D_0^a and $\langle E_a \rangle$ of Fig. 10 define the regime of diffusion in the amorphous metallic alloys as different from the liquid and crystalline solid phases.

It is instructive to examine the measured diffusion parameters in the light of the state of the atomic packing in the metallic amorphous phases. A dense random packing (DRP) model is now generally acceptable for amorphous metallic alloys. The structure of an amorphous alloy is visualized as random packing of the metal atoms analogous to the hard-sphere packing of Bernal (4) with the metalloid species filling in the larger voids.

The space filling may be described by a high density of tetrahedra and a lower concentration of larger voids.

According to the above model, the diffusion of the Ag tracer in the Pd-19 at. % Si amorphous alloy may occur in more than one manner:

(1) Interstitially if the larger voids are interconnected to some degree.

(2) Through a cooperative process involving an unspecified number of atoms utilizing the larger fraction of tetrahedral voids available in the DRP.

The interstitial process is unlikely since the void concentration is expected to be small, as evidenced by only a small density difference between crystalline and amorphous state of the alloy; as a consequence, voids cannot be easily interconnected to allow diffusion.

In the cooperative process, an atomic volume, equivalent to a vacancy in the crystalline structure, may be distributed among a cluster of atoms while retaining some sort of short-range order.

Thermal vibrations, below the glass transition temperature redistribute the free space in the DRP model, thereby, occasionally permitting atomic diffusion to occur whenever an atom finds itself adjacent to an "embryonic vacancy" which of course would be transient; the probability of its occurrence is rather small and the attempted atomic interchanges may be largely unsuccessful. In other words, diffusion would be highly correlated,

38

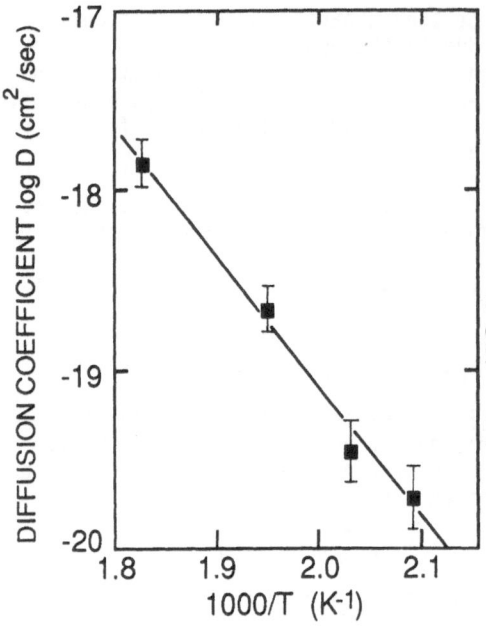

FIGURE 10. Arrhenius plot of Ag[110] diffusion coefficients in Pd-19 at. % Si amorphous specimens (9).

as in the free-volume model by Turnbull and Cohen (10), with an effective frequency much smaller than the normal Debye atomic frequency, thus, lowering D_0^α by orders of magnitude. The precise value of D_0^α would depend on the number of atoms involved in the cluster, the thermal vibrations, and the availability of the free space; energetically, the diffusion process would be easier in view of the larger atomic packing in the amorphous phase and a relaxed state of the saddle point through which atomic jumps are made.

6.2. Electron conductivity

The presence of topological (or other) disorder in an "ideal" random structure affects the electronic states in a variety of ways, as "tailing" of states into the gap at the band edges due to fluctuations in short-range order or chemical disorder. Disorder in the potentials experienced by the electrons determines spatially electron confinement to the vicinity of a single atomic site ("localized" states). Before discussing the localization, we will give some criteria to establish if the electron is localized or not. Three main criteria for localization have been established (3).

(i) Absence of electron diffusion at T = 0. If an electron is placed in an atomic site at the time t = 0, the state is considered localized if as t → ∝ the electron presents a finite probability proportional to $e^{-2\alpha r}$ (α is the localization probability) of remaining at distance, r, from the initial point, i.e. the electron has not diffused away. On the contrary the state is delocalized or extended if there is a finite chance of diffusion at T = 0.

(ii) The energy levels of the localized states are insensitive to boundary conditions.

(iii) A state is localized if the participation number P tends to zero, whereas, P has a finite value for an extended state. The participation number P is connected to the number of sites over which a localized wavefunction has significant amplitude, $P \propto [\Sigma_i \mid \psi_i \mid^2]^2 / [\Sigma_i \mid^4]$.

We will discuss in detail the localization problem in one dimension, considering that, in this case, the effect of the disorder appears clearer. Some considerations of two- and three-dimension systems will also be reported.

Let us consider a metal wire, with a cross-sectional area, A, and length, L, which contains disorder in comparison with the bulk material and with a finite conductivity at T = 0.

An electron wave-packet which diffuses along the length of the wire, does not depend on the boundary for the time, Δt, it takes to travel the wire length. On the basis of the uncertainty principle, the energy levels can be shifted only by an amount $\Delta E = h/\Delta t = hD/L^2$, where D is the electron diffusion constant.

Through the Einstein relation, we obtain the conductivity σ:

$$\sigma = e^2 \, D\left(\frac{dn}{dE}\right), \tag{14}$$

where dn/dE is the density of states per unit volume and so

$$\Delta E = \frac{\hbar}{d^2 R} \quad \frac{dE}{dN} , \tag{15}$$

where dE/dN is the average spacing between energy levels and R the wire resistance. If the energy shift, ΔE, due to a change in boundary conditions, is much less than the energy spacing, criterion (ii) for localization of states is satisfied. As a consequence, the resistance should increase much faster with increasing length than the expected linear dependence. Such a condition is satisfied if $R_C \gg h/e^2 \simeq 5$ kΩ. The localization length of the system is related to this resistance value. Recent work (12), considering instead of the diffusion time, the time between inelastic collisions, raised this critical resistance value to 36·5 kΩ.

Anderson et al. (11) proposed in 1980 that in one dimension the resistance should scale according to

$$R(L) = \left(\frac{\pi\hbar}{e^2}\right) \{\exp\,(\alpha L) - 1\} , \tag{16}$$

where α^{-1} is the localization length and L the length of wire at T = 0, or the distance between inelastic collisions at finite temperature.

Following the Anderson approach, the extra resistance associated with the localization should be inversely proportional to the cross sectional area A, of wire. Figure 11 shows the resistance as function of A^{-1}, at different temperatures, for Au-Pd alloy wires of several hundreds Angstrom diameter and a few micrometers in length.

In the case of two or three dimensions a variety of techniques, both numerical and analytical, have been developed. We report only the Anderson (13) work where the following simple tight-binding Hamiltonian is considered:

$$H = \Sigma_i \epsilon_i \mid i \rangle\langle i \mid + V \Sigma_{ij} \mid i \rangle\langle j \mid . \tag{17}$$

The basis states $\mid i \rangle$ are located on the sites of a <u>periodic</u> lattice and are coupled by constant nearest-neighbor interactions V.

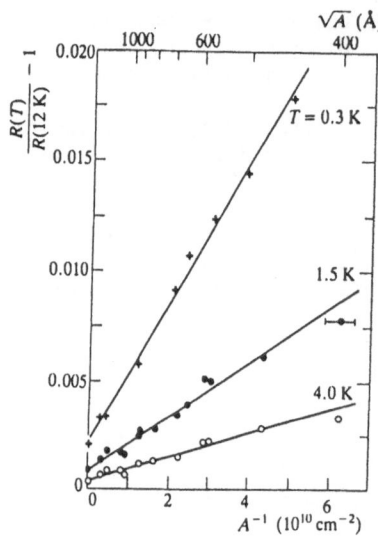

FIGURE 11. Plots of thin wires of $Au_{60}Pd_{40}$ resistance as a function of $1/A$, where A is the cross-sectional area, at several temperatures. [After Elliott (3)].

The disorder is introduced in the form of random site energies, ϵ_i, termed "diagonal disorder". The ϵ_i are taken to be uniformly distributed over a certain energy width W: $P(\epsilon) = 1/W$, $|\epsilon| < W/2$ or $P(\epsilon) = 0$, $|\epsilon| > W/2$, as indicated in Fig. 12.

The localization condition depends essentially on the competition between the two terms in the Hamiltonian. The first, in isolation, produces a localized state, whereas the second produced an extended state. The value of the ratio W/V decides which of these two effects is dominant.

Figure 13 shows wavefunction amplitudes for the Anderson model on a square lattice.

7. THERMODYNAMICAL APPROACH TO IONIC TRANSPORT IN GLASSES

In dielectric solids containing mobile ionic species (e.g. silicate glasses containing alkali metal ions) electrical currents are primarily due to ionic transport.

Many carriers are generally involved in the transport processes in glasses. These include ions, electrons, charged defects, phonons and photons. According to the Curie principle, transport processes of the same rank are coupled, i.e. the flux of each carrier is affected by the potential gradient of all the independent fluxes.

Let us consider a system having two independent charged species, i,j, and a dependent species, R, in the presence of a uniform temperature, T, distribution and an electric field. The fluxes of i and j species have the following general expression:

$$\begin{pmatrix} J_i \\ J_j \end{pmatrix} = - \begin{pmatrix} L_{ii} & L_{ij} \\ L_{ji} & L_{jj} \end{pmatrix} \begin{pmatrix} \Delta(\tilde{\mu}_i - \tilde{\mu}_r)/T \\ \Delta(\tilde{\mu}_j - \tilde{\mu}_r/T \end{pmatrix} \tag{18}$$

where $\tilde{\mu}$ is the electrochemical potenital defined as

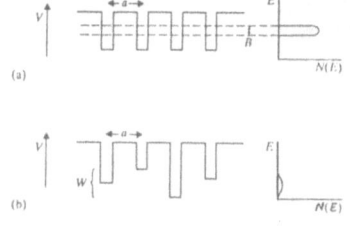

FIGURE 12. Representation of the potential wells (a) for a crystalline lattice, (b) in the Anderson model, and the density of states expected for a tight-binding model. [After Elliott (3)].

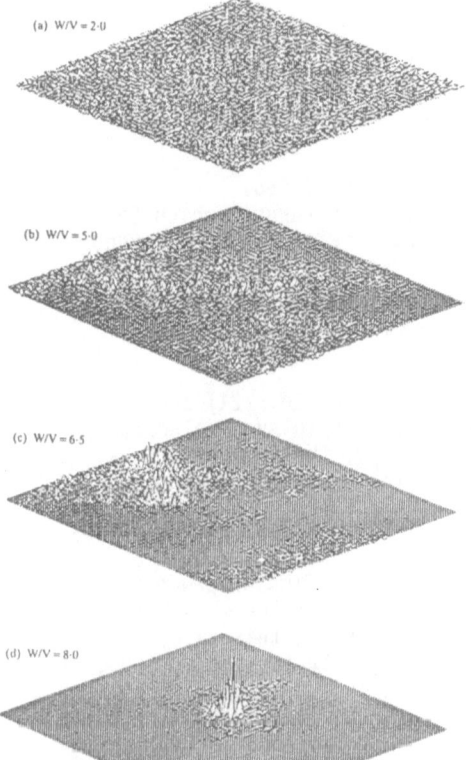

FIGURE 13. Wavefunction amplitudes for the Anderson model on a square lattice, showing (a) extended and (b) localized states. [After Elliott (3)].

$$\tilde{\mu} = \mu + q\Phi \tag{19}$$

μ is the chemical potential, q the charge of the carrier, and Φ the electric potential in which the charges migrate.

The matrix of L's coefficients (the phenomenological parameters which connect fluxes and forces) is required to be symmetric according to Onsager's reciprocal relations, since we utilize independent fluxes and potentials.

By use of the Gibbs-Duhem relation, the quantity $\tilde{\mu}_R$ may be eliminated in Eq. (18). Moreover since the independent species are i and j (n_i and n_j being their atomic concentrations), the chemical potential may be written as $\mu_i = \mu_i (n_i, n_j)$ and

$$\frac{d\mu_i}{dx} = \left(\frac{\delta\mu_i}{\delta n_i}\right)\left(\frac{dn_i}{dx}\right) + \left(\frac{\delta\mu_i}{\delta n_j}\right)\left(\frac{dn_j}{dx}\right). \tag{20}$$

A similar expression holds for $d\mu_j/dx$.

Putting relation 20 in Eq. (18), the continuity equations for n_i and n_j become (after an appropriate arrangement of the terms):

$$\frac{\delta n_i}{\delta t} = D_{ii}\left(\frac{\delta^2 n_i}{\delta x^2}\right) + D_{ij}\left(\frac{\delta^2 n_j}{\delta x^2}\right) - u_i\left(\frac{\delta}{\delta x}\right)(n_iE)$$

$$\tag{21}$$

$$\frac{n_j}{\delta t} = D_{jj}\left(\frac{\delta^2 n_j}{\delta x^2}\right) + D_{ji}\left(\frac{\delta^2 n_i}{\delta x^2}\right) - u_j\left(\frac{\delta}{\delta x}\right)(n_jE)$$

where D are diffusion coefficients, and u = mobilities.

Following a calculation scheme similar to that through which Eq. (21) has been obtained, an expression for ionic conductivity may be derived, namely:

$$\sigma = \frac{L_{ii}}{T}q_i^2 + \frac{L_{jj}}{T}q_j^2 + q_iq_j\ 2\frac{L_{ij}}{T} = \sigma_i + \sigma_j + 2\sigma_{ij}\ . \tag{22}$$

In the following we show an application both of Eq. (21) and of Eq. (22). During electron irradiation of soda-lime glasses in Auger-electron spectroscopy (AES) experiments at low electron energy, the Auger Na signal decreases while the Ca signal increases. The Na depletion in the first few surface layers is accompanied by an accumulation at a depth roughly corresponding to the electron range. Such an effect has been satisfactorily described by integration of the continuity equation for the ordinary and electric field (due to charged particle irradiation) assisted diffusion processes. Such an approach cannot explain the "anomalous" increase of the Ca AES signal during the electron irradiation. Indeed Ca^{2+} migrates, in this case, against the electric field direction. Nevertheless continuity equations, (Eqs. 21) including correlation effects, applied to Na and Ca species (indicated by i,j in subscripts), reproduce satisfactorily the experimental results.

In Fig. 14(a) we show the experimental and calculated LMM Auger signals of K and Ca during the AES of K-Ca glass. In Fig. 14(b) we report the experimental and calculated LMM Auger signals for the Ca element during AES of Na-Ca glass. The full lines are the calculated Auger signals which include the mean free path of the Auger electrons.

The experimental LMM Auger signal of Na has not been reported because of very fast signal decay while in the calculation the constant parameters of Eq. (21) were chosen in such a way as to determine the experimental disappearance of the Na signal in a time interval less than 1 min. The calculated D_{ii} and D_{jj} diffusion constants are some orders of magnitude greater than those expected for ordinary processes in glasses of similar composition, indicating an enhanced diffusion process during irradiation due to the breaking of the electronic bonds.

A value $D_{ij} = 0$ is obtained, indicating that the Ca migration does not influence the alkali diffusion. The D_{ji} value, which determines the influence of Na on Ca diffusion, is of the same order of magnitude as D_{ii} and D_{jj}.

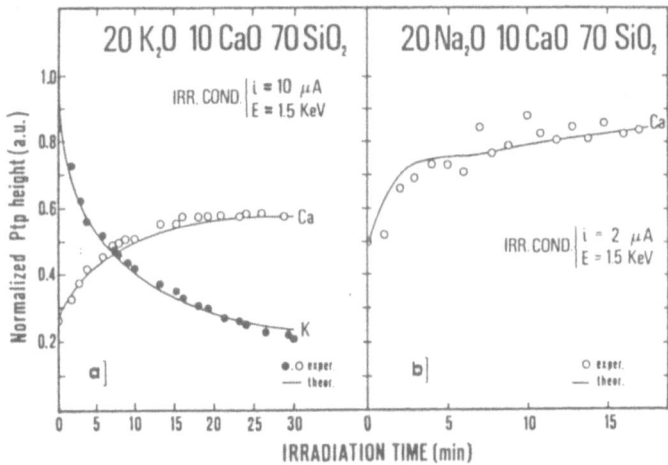

FIGURE 14. Experimental and calculated alkali and calcium Auger peak heights as function of electron bombardment time (14).

Equation (22) has been applied to the analysis of the mixed-alkali effect in the electrical conductivity of mixed-alkali silicate glasses. The conductivity is observed to fall rapidly upon small additions of foreign alkali to each single-alkali glass without important structural changes. An exponential scaling behavior of σ_i and σ_j (the conductivity of single-alkali species) as a function of their concentration has been introduced, as observed in single alkali silicate glasses (15). σ_{ij}, which is determined only by two-temperature independent-free parameters, describes the i-j interaction. In Fig. 15 we show the experimental and calculated values of ionic conductivity for a $Na_2O-K_2O-SiO_2$ glass as a function of temperature (the molar contents of alkali oxides is 24%). The dashed lines represent the theoretical curves calculated from Eq. (22) and utilizing a best-fit procedure. Figure 16 shows the Arrhenius plot of σ_{ij}.

This theoretical approach has also been applied to a large number of mixed alkali glasses with a good agreement between experimental and calculated values of the conductivity (17).

8. ASPECTS OF PERCOLATION THEORY

Percolation arguments play a varied and significant part in the discussion of many properties of disordered systems. Here we will only emphasize that percolation arguments may prove the Anderson's localization criteria. Let us stress, first of all, some concepts pertinent to percolation theory by quoting the site percolation problem of classical probability theory: "atoms" are distributed at random on the sites of a regular lattice in such a way that any given site has a probability, p, of occupation; what is the probability, P(p), that a given atom belongs to an infinite cluster? In Fig. 17(a) we illustrate the site problem (on square lattice) with no percolation path crossing this block (p = 0.5) and in Fig. 17(b) we illustrate the site problem showing a percolation path (p = 0.6). The counterpart of the site percolation problem is the bond percolation problem defined as follows: imagine each site of the lattice to be occupied and lines drawn (random) between neighboring lattice sites. Then each line can be an open bond with probability, p, or a closed bond with probability (1-p).

44

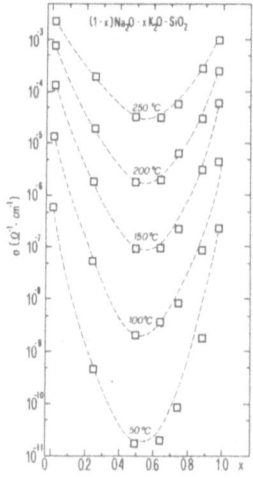

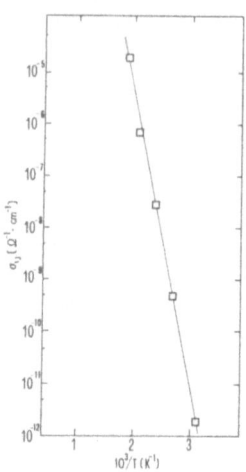

FIGURE 15. Electrical conductivity of (1-x)Na₂O x K₂O-SiO₂d glass at different temperatures. The squares are the experimental data and the lines the theoretical results (16).

FIGURE 16. Arrhenius plot of σ_{ij} (16).

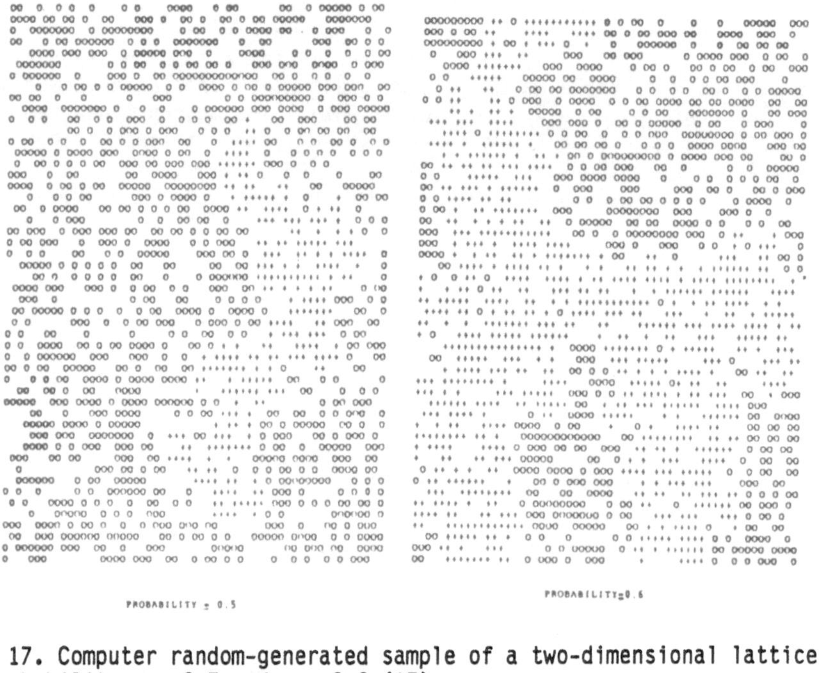

PROBABILITY = 0.5

PROBABILITY=0.6

FIGURE 17. Computer random-generated sample of a two-dimensional lattice with probability p = 0.5 and p = 0.6 (15).

The companion problem of bond percolation applies to a lattice where a
fraction, p, of the "bonds" are favorable (i.e. open to traversal, not
blocked, etc.). The function, P(p), then refers to the probability that a
given favorable bond is part of an infinite cluster linked by such a bond.
For all the physical models in which physicists are generally interested, it
can be shown that the above definition of the percolation probability function P(p) is equivalent to definition derived from other concepts such as
the existence of connected favorable paths extending from one boundary to
another of an infinitely thick specimen. The fundamental theorem of percolation theory proves that the percolation probability P(p) is of measure
zero for $p < p_c$ where the critical concentraiton p_c is characteristic of the
lattice type.

Given these preliminary concepts pertinent to the general percolation
theory, let us briefly illustrate how with percolation arguments the
Anderson's localization criteria for electronic transport may be derived.
Indeed we may say that an electron of energy, ϵ, cannot easily pass through
a site whose "atomic" level ϵ_e differs from ϵ by more than say V. Out of
the whole range W of the disorder potential only a fraction $p \approx 2V/W$ of the
lattice sites may be deemed favorable. From the point of view of the
electron, therefore, we have a percolation problem; the electron will be
localized if, of course:

$$p < p_c \approx 1/z ,$$

(where z is the atomic coordination number) and so $V/W \approx 1/2z$, i.e. a localization condition which is not very far from that quoted above. This is
only a simple example of how percolation concepts are powerful (and elegant
too) in a physics context.

REFERENCES

1. Ziman JM: Models of Disorder. Cambridge: Cambridge University Press, 1979.
2. Suzuki K: Methods of Experimental Physics. DL Price and K. Sköld (eds). San Diego: Academic Press, Inc., Vol. 23, Part B, Ch. 12, 1987.
3. Elliott SR: Physics of Amorphous Materials. London: Longman, 1983.
4. Gaskell PH: J. Non-Cryst. Solids 32 (1979) 207.
5. Mazzoldi P and Arnold GW: Ion Beam Modification of Insulators. Amsterdam: Elsevier, Ch. 5, 1987.
6. Prigogine I and Defay R: Chemical Thermodynamics. London: Longman Greens, 1954.
7. Rekhson S: J. Non-Cryst. Solids 95/96 (1987) 131.
8. Miotello A, Mazzoldi P and Toigo F: submitted for publication, 1988.
9. Gupta D, Tu KN, and Asai KW: Phys. Rev. Lett. 35 (1975) 796.
10. Turnbull D and Cohen MH: J. Chem. Phys. 51 (1970) 3038.
11. Anderson PW, Thouless DJ, Abrahams E, and Fisher DS: Phys. Rev. B 22 (1980) 3519.
12. Thouless DJ: Sol. State Comm. 34 (1980) 683.
13. Anderson PW: Phys. Rev. 109 (1958) 1492.
14. Miotello A and Mazzoldi P: Phys. Rev. Lett. 54 (1985) 1675.
15. Mazzoldi P and Miotello A: J. Non-Cryst. Soloids 95/96 (1987) 897.
16. Miotello A and Mazzoldi P: J. Non-Cryst. Solids 94 (1987) 181.
17. De Marchi G, Mazzoldi P, and Miotello A: J. Non-Cryst. Solids, in press.

THE DISPLACEMENT CASCADE IN SOLIDS

DON M. PARKIN

Center for Materials Science*
Los Alamos National Laboratory
Los Alamos, NM 87545

1. INTRODUCTION

The displacement cascade in a solid occurs when an atom in the material is displaced from its site by an incident damaging particle and moves through the material producing additional displacements. This process is of central importance in the study of radiation effects in solids. The use of ion beams to modify the near-surface regions of materials is an important area of research and technology where knowledge of the displacement cascade is required for a broad range of ion irradiation parameters and technologically important and complex materials. In general we use a displacement cascade model to calculate an exposure parameter such as displacements per atom (dpa) as a method of comparing different irradiations and as a radiation-induced defect source term in modeling the effects of the irradiation.

The displacement cascade in monatomic materials has received the most extensive study. The majority of these investigations have employed some variant of the binary collision approximation. The early and influential work of Kinchin and Pease (1,2) used this approximation to calculate the number of displacements produced in a collision cascade of hard spheres. This study formed the basis for much subsequent work, and the most widely accepted method currently used to estimate numbers of displacements is a modified form of their original result (3). Computer-based methods have allowed the inclusion of realistic crystal lattice structure, temperature, and other properties of the material under study. These calculation methods have advanced our fundamental understanding of the displacement cascade. They have also permitted calculations that include geometries and conditions important in ion beam modification of materials.

The more general case of displacement cascades in polyatomic materials, those consisting of two or more elements, has received much less study. These materials, which are more representative of the class of materials used in application, introduce many new parameters into the problem including elemental composition, material stoichiometry, complex crystal structure, bonding type (for example, metal, ionic, covalent), and additional types of defects. The majority of authors who have investigated this problem have restricted their studies to limited values of the important parameters (4–13). As a result, little information about general behavior has emerged.

The most comprehensive investigations are the computer-based approaches of Parkin and Coulter (14–18) and computer calculations using the TRIM Monte Carlo methods (19). Parkin and Coulter derived a set of integrodifferential equations in the binary collision approximation using the methods of Lindhard et al. (20) that provide a fairly complete

* This work was performed under the auspices of the U.S. Department of Energy.

47

C.J. McHargue et al. (eds.), Structure-Property Relationships in Surface-Modified Ceramics, 47–60.
© 1989 by Kluwer Academic Publishers.

characterization of the nature of the displacement cascade. The TRIM code has been widely distributed by its authors and in its most recent versions can be used to calculate properties of the displacement cascade similar to those obtained by Parkin and Coulter. In addition, important spatial information about the cascade is part of the TRIM calculations. To a certain extent, the two approaches complement each other. The integrodifferential method gives results for all energies in one calculation. This provides, for example, the number of displacements as a function of energy for a selected set of material input parameters. On the other hand, a TRIM calcualtion yields results for only one energy per run but gives spatial information. The important aspects of elemental composition, material stoichiometry, threshold effects, bonding type, and irradiating particle and energy can be investigated with these two methods. However, both approaches use a random amorphous material model so that crystal structure effects are included only approximately in the results.

In this paper we will review the integrodifferential equation approach of Parkin and Coulter with a focus on identifying the principal characteristics of the displacement cascade. Selected results from TRIM calculations will also be included. Following this discussion, the methods presented will be applied to investigating the effects of irradiation on the $Y(RE)Ba_2Cu_3O_7$ superconductors [the term Y(RE) stands for yttrium or rare earth].

2. THE DISPLACEMENT CASCADE

2.1. Integral equations for damage energy

The monatomic damage energy of Lindhard et al. (20–22), usually as approximated by Robinson (23), plays a central role in the internationally accepted procedure for calculating the number of displacements in monatomic materials (3). It has at times been applied to polyatomic materials by using an average charge number and mass number for the material in evaluation of the function. Although this is certain to be a reasonable procedure when the ranges of target A- and Z-values are small, it is less clear how accurate it is in general. We start our discussion by developing the polyatomic damage energy (14).

Suppose that a given target material contains N_i atoms per unit volume with charge and mass numbers Z_i and A_i, respectively, where $i = 1, \ldots n$. Let $d\sigma_{ij}(E,T)/dT$ be the differential scattering cross section for an atom of type i and kinetic energy E to elastically transfer energy T to an atom of type j that is initially at rest, and let $S_i(E)$ be the electronic stopping power for atoms of type i in the target. Now let $\nu_i(E)$ be the damage energy that will be deposited in the target by an atom of type i with initial energy E. Following the reasoning of Lindhard et al., one notes that as the atom of type i moves a distance dx through the target, it will lose energy by elastic collision with other atoms; but by its definition, the damage energy must remain unchanged in this process. One concludes that $\nu_i(E)$ must satisfy the equation

$$s_i(E)\nu_i'(E) = \sum_{j=1}^{n} f_i \int_0^{M_{ij}E} dT \, \frac{d\sigma_{ij}(E,T)}{dT}$$
$$\times \left[\nu_j(T) + \nu_i(E - T) - \nu_i(E) \right] \quad , \tag{1}$$

where the prime denotes the differential with respect to E and where $s_i(E) \equiv S_i(E)/N$ and $f_i = N_i/N$, with N the total number of atoms per unit volume, and

$$M_{ij} = \frac{4A_i A_j}{(A_i + A_j)^2} \quad .$$

Using the universal scattering cross section and electronic stopping power formula of LSS (20–21), Eq. (1) can be numerically integrated on the computer to provide damage energies. For a material with n elements there are n values of the damage energy, each one yielding the damage energy deposited in the material by an energetic ion of the nth type. Damage energies for foreign projectiles in a material can also be calculated to a very good approximation. If the material $Mg_xO_yHe_z$ is input with $x = 1$, $y = 1$, and $z = 10^{-9}$, for example, the damage energy for helium in MgO can be obtained.

Coulter and Parkin (CP) fit the numerically determined damage energy functions to Robinson-type functions (23) of the form

$$\nu(E) = \frac{E}{1 + C_1 E^{0.15} + C_2 E^{0.75} + C_3 E} \cdot \qquad (2)$$

The use of the power 0.15 rather than Robinson's 1/6 was necessitated by the use of a different approximation to the LSS $f(\xi)$ function. This fact also explains why damage energies in monatomic materials calculated by the Robinson formula yield values of $\nu(E)$ about 6% higher than the CP results.

The effects of placing an atom in a binary target material are illustrated in Fig. 1, which compares the atom's damage efficiency $\nu(E)/E$ in a monatomic target of atoms of its own kind ("self-atoms") with that in the binary target material. In this case, the calculations appropriate for Y_2O_3 are given. The damage efficiency for the heavy atom yttrium in Y_2O_3 is everywhere less than for yttrium in monatomic yttrium. For the lighter oxygen, the damage efficiency in Y_2O_3 is lower than in oxygen at low energies, but higher at high energies. Overall, however, the "self-atom" damage efficiencies are similar to the corresponding binary material ones.

In a set of calculations on the material $U_xZr_{1-x}C$, CP found that the damage efficiencies for values of x from 0.02 to 0.98 depended only weakly on x. This observation shows that the damage efficiency (energy) for any one of the constituents is not strongly determined by stoichiometry.

In summary, the damage energy calculations produced two important findings. First, in rough general terms, combining different atom types in a polyatomic material reduces damage efficiencies relative to monatomic materials. Second, deviation from stoichiometric behavior in the damage energies casts serious doubt on procedures in which the material is represented by average Z and A. In fact, CP found that as a general rule, damage energies in polyatomic materials are better represented by monatomic "self-atom" values than for average Z- and A-values.

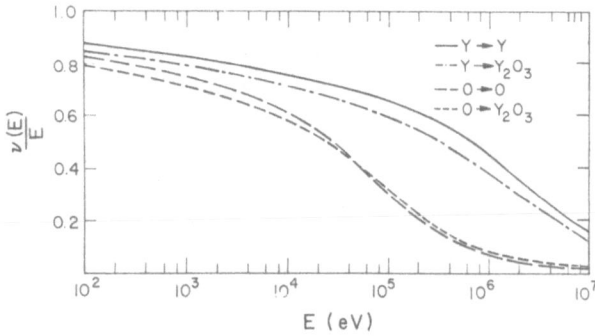

FIGURE 1. Damage efficiency, $\nu(E)/E$ versus energy for yttrium and oxygen in Y_2O_3 and yttrium in yttrium and oxygen in oxygen.

2.2. Equations for the net displacement functions

The damage energy described in the previous section provides information on the partition of energy between the limiting value of the kinetic energy and the electronic energy loss. It does not, however, give direct information about the number of displacements. Using the same methods of LSS, we can derive integrodifferential equations that directly yield the number of displacements. The net displacement function $g_{ij}(E)$ as defined by Parkin and Coulter (PC) is the average number of type j atoms displaced and not recaptured in subsequent replacement collisions in a displacement cascade initiated by a primary knock-on atom (PKA) of type i and energy E (15). With the convention that $g_{ii}(E)$ counts the PKA, the equation for the net displacement function is

$$s_i(E)\frac{dg_{ij}(E)}{dE} = \sum_k f_k \int_0^{M_{ik}E} dT\, \frac{d\sigma_{ik}(E,T)}{dT}$$
$$\times \{\rho_k(T)g_{kj}(T - E_k^b) + [1 - \rho_k(T)\lambda_{jk}(E - T)]$$
$$\times g_{ij}(E - T) - g_{ij}(E)\} \quad . \tag{3}$$

The first term on the right-hand side $[g_{kj}(T - E_k^b)]$ gives the contribution from a type k atom that is displaced by a type i atom, and the second term gives the contribution of subsequent effects of the type i atom $[g_{ij}(E - T)]$, provided it is not captured [probability $1 - \rho_k(T)\lambda_{jk}(E - T)$] in the current collision. The term E_k^b is the binding energy lost as the type j atom is displaced. The third term on the right and the term on the left arise from motion of atom i with electronic energy loss but no atomic collision.

Equation (3) has been written using a general notation for the displacement probability ρ_k and the capture probability λ_{ij}; however, the simple sharp-threshold forms

$$\rho_k(T) = \begin{cases} 0, & T < E_k^d \\ 1, & T > E_k^d \end{cases} \tag{4}$$

and

$$\lambda_{ik}(E) = \begin{cases} 1, & E < E_{ik}^{\mathrm{cap}} \\ 0, & E > E_{ik}^{\mathrm{cap}} \end{cases} \tag{5}$$

are used in all the calculations. Here, E_k^d, termed the displacement threshold, is the average kinetic energy a type k atom must receive to be displaced from its site. The term E_{ik}^{cap}, called the capture energy, is the average residual energy of a type i atom that has displaced a type k atom, below which it will be trapped in the type k site. The value of E_{ik}^{cap} is set equal to E_i^d in the normal Kinchin-Pease model for monatomic materials; for polyatomic materials, however, the possibility that $E_{ik}^{\mathrm{cap}} \neq E_j^d$ for $i \neq j$ must be included.

2.2.1. Results for the net displacement functions.

The nature of the displacement cascade does not lend itself to a description by a universal formula depending in a simple fashion on model and material parameters, except for the special case of monatomic materials. A generalization of the modified Kinchin-Pease formula, however, is useful in discussing the properties of $g_{ij}(E)$. Let us define a set of displacement efficiencies k_{ij} by the relation

$$g_{ij} \equiv \delta_{ij} + \frac{k_{ij}f_j\nu_i}{E_j^d} \quad . \tag{6}$$

The Kronecker delta on the right-hand side of Eq. (6) gives explicity the threshold behavior of g_{ij} caused by the PKA itself. The component k_{ij} is thus the efficiency for displacement of atoms other than the initial recoil and vanishes at threshold. Equation (6) can be written as

$$k_{ij} = \frac{\bar{g}_{ij} E_j^d}{f_j \nu_i} \quad , \tag{7}$$

where $\bar{g}_{ij} \equiv g_{ij} - \delta_{ij}$. Since E_j^d and f_j are constant, the energy dependence of k_{ij} reflects any failure of the damage energy to describe the nature of polyatomic displacement cascades. If the energy dependence of g_{ij} is accurately described by ν_i, then k_{ij} is a proportionality constant relating the net number of secondary displacements to the damage energy multiplied by a term containing the displacement energy and the stoichiometry.

As given by Eq. (3), g_{ij} is strictly the net number of displaced atoms, or the number of interstitials, of type j. In monatomic materials $g = v$, where v is the number of vacancies. In a polyatomic material we cannot assume that $g_{ij} = v_{ij}$ because of the possibility of disordering replacements. A relationship between g_{ij} and v_{ij} will be shown later.

To indicate the various parameters used in the calculation, PC adopted the following notation: materials and their corresponding parameters will be given as

$$A_x B_y \ldots (E_{11}^{\mathrm{cap}}, E_{12}^{\mathrm{cap}}, \ldots, E_{21}^{\mathrm{cap}}, E_{22}^{\mathrm{cap}}, \ldots) \quad .$$

where for a diatomic material we have $A_x B_y (E_{11}^{\mathrm{cap}}, E_{12}^{\mathrm{cap}}, E_{21}^{\mathrm{cap}}, E_{22}^{\mathrm{cap}})$. Because PC used $E_i^d = E_{ij}^{\mathrm{cap}}$ and $E_k^b = 0$, this notation contains information on all parameters used in a specific calculation. The assumption that $E_i^d = E_{ij}^{\mathrm{cap}}$ is equivalent to assuming that the normal Kinchin-Pease model for monatomic materials holds for the corresponding sublattice.

The energy dependence of the k_{ij} can be characterized by data for MgO(62,00,00,62), a material for which $M_h/M_l \approx 1$, where h and l refer to the heavy and light atoms, respectively. The data in Fig. 2 show that k_{ij} is a strong function of E from threshold to several thousand electron volts, when it reaches some value dependent on the material parameters and becomes roughly constant for all higher energies. The TaO ($M_h/M_l \gg 1$) results in Fig. 3 show the same general behavior except for the significant difference that the strong energy dependence of k_{ij} persists to about 100 keV. When k_{ij} is roughly constant, $k_{ih} > k_{il}$ and $k_{ih}/k_{il} = 1.6$ for TaO. Note in comparision that for monatomic materials in the energy-independent region, $k = 0.4$ (16).

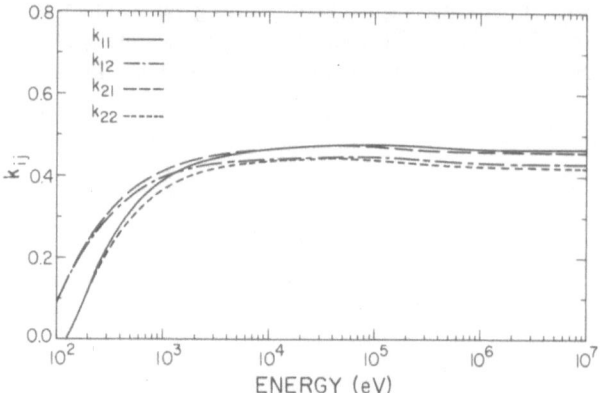

FIGURE 2. Values of the displacement efficiencies k_{ij} as functions of energy for the material MgO(62,00,00,62).

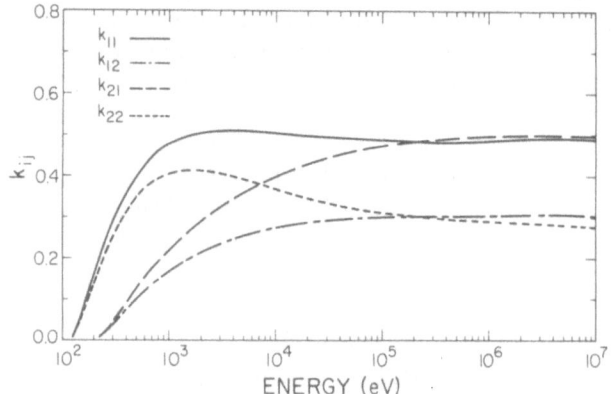

FIGURE 3. Values of the displacement efficiencies k_{ij} as functions of energy for the material TaO(60,60,60,60).

These results illustrate that the nature of the displacement cascade is a function of both material type and energy. The number of displacements is approximately proportional to the damage energy only beyond a certain energy that depends on the mass ratio: a few thousand electron volts for $M_h/M_l \approx 1$ and about 10^5 eV for $M_h/M_l \gg 1$. Further, the limiting values of k_{ij} are material dependent.

In the damage energy region, when k_{ij} is approximately constant, the cascade is independant of PKA type, that is, $k_{ij} \rightarrow k_j$. Thus, only in the damage energy region is the average cascade structure independent of the type of irradiating ion or PKA energy. We can write Eq. (6) as

$$g_{ij} \approx \frac{k_j f_j \nu_i}{E_j^d} \quad .$$

(8)

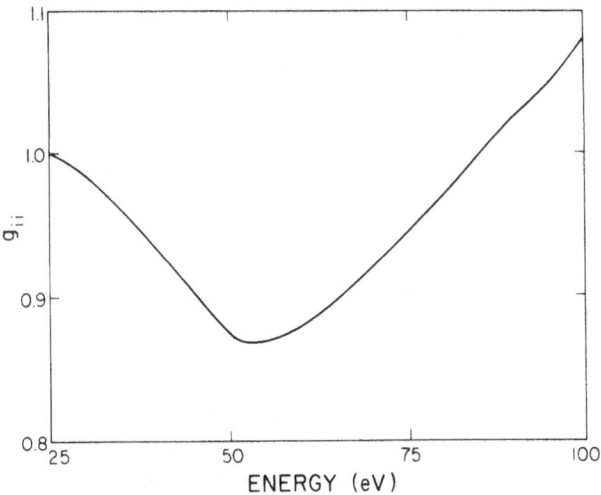

FIGURE 4. Values of the net displacement
function g_{ii} as a function of energy for the
material Cu(25,25,25,25).

The role of the threshold energy is explicitly shown in Eqs. (6) to (8). The capture energy
is also an important parameter in determining g_{ij} and in determining the value of k_{ij}. The
near-threshold behavior is controlled by both the threshold and capture energies, and this
behavior propagates to all energies. The values of k_{ii} are suppressed near threshold when
$i \neq j$ capture occurs, as can be seen in Fig. 4, where g_{ii} for Cu(25,25,25,25) is shown.
As the PKA energy increases from threshold, g_{ii} decreases, and at roughly $2E_j^d$ it reaches
a minimum value of 0.87. This minimum value reflects the number of displacements that
result in $i \neq j$ capture in the low-energy residual of the cascade.

For the damage energy region, we can now in an approximate way separate the role of
$i \neq j$ capture energy. PC found that to a good approximation

$$k_{ij} \approx g_{jj}(\min)k_{ij}(0) \quad . \tag{9}$$

where the parameters $k_{ij}(0)$ are the values of k_{ij} for the case $A_x B_y(E_1^d, 00, 00, E_2^d)$ and the
$g_{ij}(\min)$ are the minimum values of g_{ij} for $A_x B_y(E_{11}^{\mathrm{cap}}, E_{12}^{\mathrm{cap}}, E_{21}^{\mathrm{cap}}, E_{22}^{\mathrm{cap}})$.

This result reveals much about the displacement cascade. The term $k_{ij}(0)$ reflects the
energy transferred to each sublattice as all atoms slow down. Interaction between sublattices
is limited to collision events since there is no $i \neq j$ capture. Interaction between sublattices
is described by the $i \neq j$ capture that determines $g_{ij}(\min)$. Since bonding type is represented
in this model by the values of E_j^d and E_{ij}^{cap}, we now have a description of how it affects
g_{ij}. For example, in ionic crystals it is likely that cations and anions would not be captured
on opposing sites so that E_{ij}^{cap}, $i \neq j$, would be rather low or zero.

The mass ratio, independent of material parameters such as bonding type, has an important
effect on $i \neq j$ capture. PC showed that for $M_h/M_l \geq 4$, $i \neq j$ capture essentially does

not occur for kinematic reasons. The kinematically derived maximum mass ratio for which $i \neq j$ capture can occur is 5.8 for the case $E_{ij}^{\text{cap}} = E_j^d$. Based upon these results, we can say that for materials with mass ratio less than about 4, details of the interaction of the various atom types in the threshold energy range are critical to describing the displacement cascade. For materials with mass ratio greater than 4, less information is required.

The net displacement function includes only the number of atoms or interstitials that are not subsequently captured at some vacant site. If atoms are captured only at vacancies of the same type ($i = j$ capture), then the number of interstitials g_{ij} is equal to the number of vacancies v_{ij}. However, when $i \neq j$ capture occurs, the result is a type i vacancy with a type j interstitial, and $g_{ij} \neq v_{ij}$. In this case we can use Eq. (9) to estimate v_{ij}. Let $g_{ij}(0)$ be the net displacement function calculated for $E_{ij}^{\text{cap}} = 0$, $i \neq j$, and let k be the atom index different from j. Now $g_{ik}(0)$ is the number of type k interstitials, and $g_{ij}(0)$ can be considered the number of type j vacancies for this parameter choice. When E_{ij}^{cap} becomes nonzero, the number of type k interstitials will undergo a fractional reduction of $1 - g_{kk}(\text{min})$ according to the previous discussion; the total reduction is $g_{ik}(0)\,[1 - g_{kk}(\text{min})]$. Since the type k interstitials that disappear do so by capturing on type j sites and thus eliminating type j vacancies, one concludes that

$$v_{ij} = g_{ij}(0) - g_{ik}(0)[1 - g_{kk}(\text{min})] \quad . \qquad k \neq j \tag{10}$$

PC showed that Eq. (10) estimates v_{ij} to an accuracy of a few per cent in all the cases they studied.

One consequence of the difference between g_{ij} and v_{ij} is that even without the influence of external sinks, one will usually generate nonequilibrium numbers of interstitials and vacancies—deriving from disordering recombination processes—in displacement cascades. Using Eqs. (9) and (10), we can express the interstitial excess as

$$\begin{aligned} \bar{I}_{ij} = g_{ij} - v_{ij} &= g_{ij}(0)[g_{jj}(\text{min}) - 1] \\ &\quad + g_{ik}(0)[1 - g_{kk}(\text{min})] \quad . \qquad k \neq j \end{aligned} \tag{11}$$

For the equal-threshold case and small mass ratio, Eq. (11) shows that there will be an excess of heavy-atom vacancies and light-atom interstitials.

The cascade stoichiometry can be written as

$$s_i = \frac{\bar{g}_{ij}}{\bar{g}_{ik}} = \frac{k_{ij}}{k_{ik}} \cdot \frac{E_k^d}{E_j^d} \cdot \frac{f_j}{f_k} \quad . \qquad j \neq k \tag{12}$$

The term f_j/f_k is the stoichiometric ratio for the material. Only if the product of the terms in k_{ij} and E_j^d equals unity is the cascade stoichiometry the same as that of the material. Usually the condition is not obtained even when $E_j^d = E_k^d$. Thus the general condition is that the stoichiometry of the displacement cascade is not the same as that of the material. Equal stoichiometry is achieved only in special cases.

TRIM calculations using code versions that follow all recoils can be used to obtain g_{ij} for the case $E^{\text{cap}} = E^d$. The values of g_{ij} tend to be several tens of percentage points higher than those calculated with the PC method described here. Note, however, that TRIM uses the Biersack-Ziegler potential, which is more realistic than the Thomas-Fermi potential as used by LSS.

2.3. Y(RE)Ba$_2$Cu$_3$O$_7$ Superconductors—An Example

In the previous sections we outlined and reviewed a number of important characteristics of the displacement cascade and their relationships to material parameters. The new high-temperature superconductors have potential applications in radiation environments covering the spectrum from ionizing radiation through fast electrons, neutrons, low-energy light and heavy ions, to very high energy light ions. In addition, they are oxide ceramics for which ion beam modification in thin-film form promises to be an important processing tool. They make an excellent example for applying the results discussed in this work.

The analysis given here was based primarily on diatomic materials, whereas the superconductors contain four elements. If we view the three metals as a group, then the formula can be written as M$_6$O$_7$. The mass ratios for oxygen compared with copper, barium, yttrium, and the rare earth gadolinium are 4, 8.5, 5.6, and 9.8, respectively. Among the metals, the ratios vary from 1.1 to 2.5. Simplistically, then, we would expect that the k_{ij} would behave in a manner similar to that of TaO ($M_h/M_l = 11$) in Fig. 3, where M_h is equated to a tantalum atom. Collectively, the results with metal atoms would be similar to those with MgO.

The results for YBa$_2$Cu$_3$O$_7$ with $E^d = E^{cap} = 60$ eV are shown in Fig. 5 for yttrium PKAs and in Fig. 6 for oxygen PKAs. Data not shown for the other elements are similar. In the damage energy region, the supposition is reasonably correct. The displacement efficiency for the metal atoms are greater than for oxygen and the differences are similar to that in TaO. Also the relative values of the k_{ij} for the metal atoms are nearly the same as for MgO. There is one significant difference, however. The damage energy region is reached at a few thousand electron volts, similar to MgO, and not near 10^5 eV as for TaO. Thus we conclude that the cascade structure is roughly constant and independent of PKA type for energies above a few thousand electron volts. In this energy range, the cascade stoichiometry will be near that of the material for the metals but be different in oxygen. PKA or irradiating ion energies above a few thousand electron volts cover essentially all relevant radiation environments except low-energy electrons and gamma rays. TRIM calculations for 500-keV oxygen ions yield a displacement cascade stoichiometry of Y$_1$Ba$_2$Cu$_3$O$_6$ and PC calculations for energies relevant to fast neutrons yield Y$_1$Ba$_2$Cu$_3$O$_5$.

With this simple characterization of the displacement cascade, let us review published observations of radiation effects in the Y(RE)Ba$_2$Cu$_3$O$_7$ materials. The available data can be summarized for the present needs by the following list of statements.

1. No significant damage is observed for ionizing radiation or gamma rays (24,25).

2. Electron irradiation suggests that $E^d \approx 20$ eV for oxygen (26,27).

3. Oxygen displacements affect twin structures but little else (27,28).

4. Fast neutrons, thermal neutrons, and high-energy protons reduce the critical temperature (29–31).

5. Superconductivity is destroyed and an amorphous phase formed after 500-keV oxygen, 1-MeV electron, and 300-keV electron irradiation (32–34).

6. Grain boundary amorphization occurs faster than in grain amorphization with similar dose ratios for 500-keV oxygen and 1-MeV electron irradiation (32,33).

7. Grain boundary amorphization occurs faster for YBa$_2$Cu$_3$O$_7$ than for GdBa$_2$Cu$_3$O$_7$ (33).

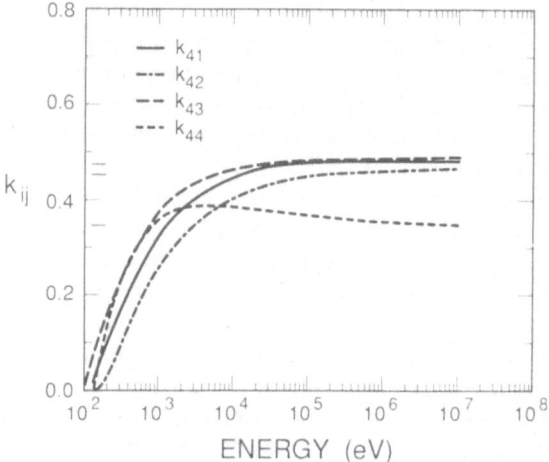

FIGURE 5. Values of the displacement efficiencies k_{ij} for an yttrium PKA in $YBa_2Cu_3O_7$ with $E_j^d = E_{ij}^{cap} = 60$ eV.

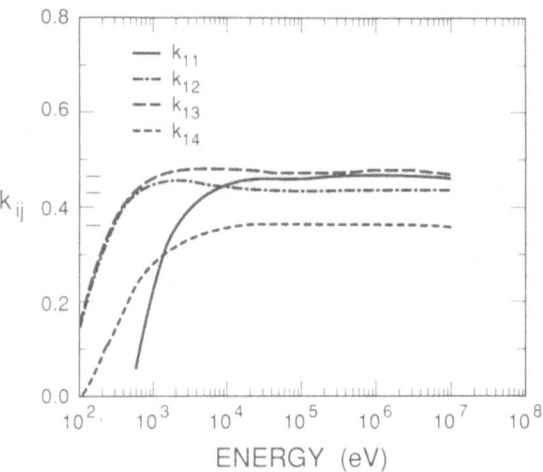

FIGURE 6. Values of the displacement efficiencies k_{ij} for an oxygen PKA in YBa_2Cu_3O with $E_j^d = E_{ij}^{cap} = 60$ eV.

From these results we can form a hypothesis about the damage mechanism and its relationship to the displacement cascade. The new superconductors respond to radiation like a metal, not an oxide. Thus the displacement cascade is the damage mechanism of importance. Oxygen displacements occur near 20 eV, but since oxygen displacements produce only little effect, the observations of reduced T_c, total loss of superconductivity, and amorphization are due primarily to damage on the metal atom sublattices. The metal atom part of the cascade must have similar characteristics over the entire recoil energy range that occurs for 300-keV electrons to 500-keV oxygen ions. This energy range includes both the threshold and damage energy regions as discussed above (also see Figs. 5 and 6).

This hypothesis suggests than an important displacement cascade parameter is the stoichiometry. We defined the cascade stoichiometry $s_i(E)$ in Eq. (12), but for general irradiation conditions, it is useful to define the average cascade stoichiometry for type j displacements as

$$S_j = \sum_i P_i \frac{\int d\sigma_i/dT \cdot s_i(T) \cdot dT}{\int d\sigma_i/dT \cdot dT} \quad , \tag{13}$$

where P_i is the reative probability of producing a PKA of type i, and $d\sigma_i/dT$ is the differential scattering cross section. For ion irradiations $S_j = s_i(E)$.

The electron irradiation results on $YBa_2Cu_3O_7$ and $GdBa_2Cu_3O_7$ both show amorphization at grain boundaries but with the gadolinium material becoming amorphous at a dose roughly three times that of yttrium (33). This result shows that the yttrium or rare earth atom is an important part of the cascade damage. To produce a cascade of metal atoms with 300-keV electrons requires a displacement threshold for the metal atoms of approximately 20 eV. For low-energy electron irradiations, a principal source of metal atom displacements is the oxygen cascade. Values of $s_i(T)$ for oxygen cascades in $YBa_1Cu_3O_7$ obtained from TRIM calculations are shown in Fig. 7. The reference atom is copper and its stoichiometric value has been set to its material value of 3.

Nastasi et al. (33), using Eq. 13, Fig. 7, and the McKenly-Feshbach displacement cross section, calculated the stoichiometry of the displacement cascade to be $Y_{0.4}Ba_{0.3}Cu_3O_7$ and $Gd_{0.1}Ba_{0.3}Cu_3O_7$ for 1-MeV electrons. The factor of 4 difference between yttrium and gadolinium displacements is in reasonable agreement with the observed factor of 3 in amorphization doses. They concluded that either the cascade stoichiometry or the absolute number of yttrium or gadolinium displacements was the important factor.

In this example, the characterisitics of the displacement cascade were used to obtain an initial picture of how displacement damage would occur in the $Y(RE)Ba_2Cu_3O_7$ materials under displacing irradiation. Specific experimental observations were then examined and a hypothesis proposed that included elements of both sets of information. The final conclusion drawn regarding the role of the yttrium or rare earth atom was supported by both the calculated cascade characteristics and the experimental data. The displacement cascade description obtained by the calculational methods briefly reviewed here provided necessary information used in forming the conclusion.

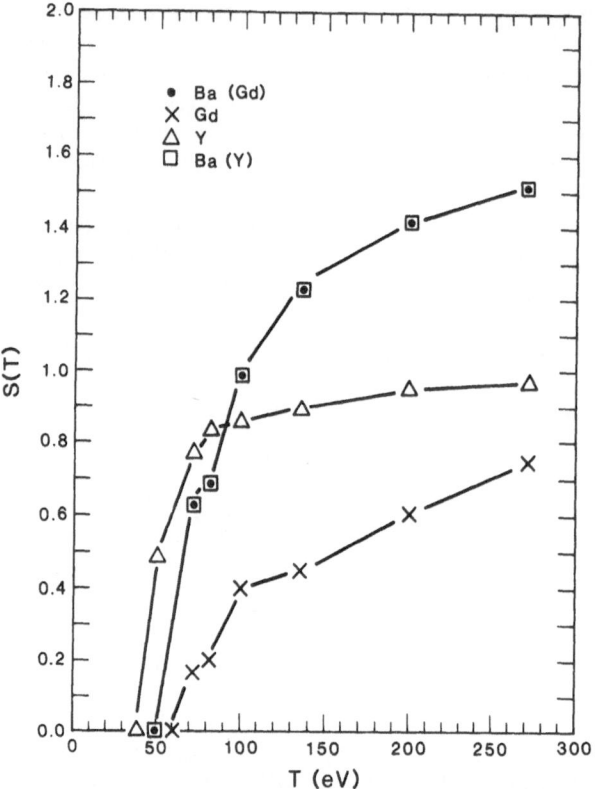

FIGURE 7. TRIM calculations of the ratio of displaced yttrium, barium, and gadolinium atoms to copper atoms, $S(T)$, produced by oxygen recoils of energy T, in the materials $GdBa_2Cu_3O_7$ (Gd) and $YBa_2Cu_3O_7$ (Y). Displacement values for yttrium, barium, and gadolinium have been normalized by setting the copper displacement stoichiometry at 3.

3. SUMMARY

The nature of the displacement cascade in solids is, in general, a function of the PKA type, PKA energy, and the material. For monatomic materials, the modified Kinchin-Pease formula is the most widely accepted method of calculating the number of displacements. Through various approximations it is often extended to the more general case of polyatomic materials. The discussion presented here shows that the principal characteristics of the displacement cascade in polyatomic materials can only be obtained from a model that explicitly includes the relevant material parameters. Both the integrodifferential equations of Parkin and Coulter and TRIM calculations can be used to obtain data for specific materials and irradiation conditions. The more complete information thus obtained increases our understanding and adds to our description of the interaction between a material and a radiation field. This information also more completely describes the defect production source term that is often used in modeling a materials response to irradiation.

REFERENCES

1. Kinchin, G. H., and Pease, R. S., *J. Nucl. Energy* **1**, 200 (1955).
2. Kinchin, G. H., and Pease R. S., *Rep. Prog. Phys.* **18** (1955).
3. Norgett, M. H., Robinson, M. T., and Torrens, I. M., *Nucl. Eng. Design* **33**, 50 (1975); **33**, 91 (1975).
4. Baroody, E. M., *Phys. Rev.* **112**, 1571 (1958).
5. Baroody, E. M., *Phys. Rev.* **116**, 1418 (1959).
6. Kostin, M. D., *J. Appl. Phys.* **36**, 850 (1966).
7. Kostin, M. D., *J. Appl. Phys.* **37**, 3801 (1966).
8. Felder, R. M., and Kostin, M. D., *J. Appl. Phys.* **37**, 791 (1966).
9. Felder, R. M., *J. Phys. Chem. Solids* **28**, 1383 (1967).
10. Winterbon, K. B., *Nucl. Sci. Eng.* **53**, 261 (1974).
11. Ishino, S., and Matsutani, Y., *J. Appl. Phys.* **48**, 1822 (1977).
12. Soullard, J., and Alamo A., *Radiat. Effects* **38**, 133 (1978).
13. Anderson, N., and Sigmund, P., *Kgl. Daske Videnskab. Selskab, Mat-fys. Medd.* **39**(3) (1974).
14. Coutler, C. A., and Parkin, D. M., *J. Nucl. Mater.* **88**, 249 (1980).
15. Parkin, D. M., and Coulter, C. A., *J. Nucl. Mater.* **101**, 261 (1981).
16. Coulter, C. A., and Parkin, D. M., *J. Nucl. Mater.* **95**, 193 (1980).
17. Parkin, D. M., and Coulter, C. A., *J. Nucl. Mater.* **103/104**, 1315 (1981).
18. Parkin, D. M., and Coulter, C. A., *J. Nucl. Mater.* **85/86**, 611 (1979).
19. Zeigler, J. F., Biersack, J. P., and Littmark, U., *The Stopping and Range of Ions in Solids*, Vol. 1 of *The Stopping and Range of Ions in Matter*, ed. Zeigler, J. F.: New York: Permagon: 1985.
20. Lindhard, J., Nielson, V., Schaff, M., and Thomson, P. V., *Kgl. Danske Videnskab. Selskab, Mat-fys. Medd.* **33**(10) (1963).
21. Lindhard, J., Schaff, M., and Schriott, H. E., *Kgl. Danske Videnskab. Selskab, Mat-fys. Medd.* **33**(14) (1963).
22. Lindhard, J., Nielson, V., and Schaff, M., *Kgl. Danske Videnskab. Selskab, Mat-fsy. Medd.* **36**(10) (1068).
23. Robinson, M. T., in Proc. BNES Nuclear Fusion Reactors Conf. at Culham Laboratory: 1969: p. 364.
24. Maisch, W. G., Summers, G. P., Campbell A. B., Dale, C. J., Ritter, J. C., Knudson, A. R., Elam, W. T., Herman, H., Kirkland, J. P. Neiser, R. A., and Osofsky, M. S., *IEEE Trans. Nucl. Sci.* **NS-34**, 1782 (1987).
25. Bohandy, J., Suter, J., Kim, B. F., Moorjani, K., and Adrian, F. J., *Appl. Phys. Lett.* **51**, 2161 (1987).
26. Kirk, M. A., Bahir, M. C., Liu, J. Z., Lam, D. J., and Weber, H. W., to be published.
27. Mitchell, T. E., Roy, T., Schwarz, R. B., Smith J. F., and Wohlleben, D., *J. Elec. Micros. Techn.* **8**, 317 (1988).
28. Clinard, F. W., private communication.
29. Cost, J. R., Willis, J. O., Thompson, J. D., and Peterson, D. E., *Phys. Rev.* **B37**, 1563 (1988).
30. Hastings, I. J., Palmer, B. J. F., Gin, A. Y. G., Scoberg, J. A., Thatcher, J. C., and Towner, J., to be published.
31. Willis, J. O., Cooke, D. W., Brown, R. D., Cost, J. R., Smith, J. F., Smith, J. L., Aikin, R. M., and Maez, M., to be published.

32. Clark, G. J., LeGoues, F. K., Marwick, A. D., Laibowitz, R. B., and Koch, R., *Appl. Phys. Lett.* **51**, 1462 (1987).
33. Nastasi, M., Parkin, D. M., Zocco, T. G., Koihe, J., and Okamoto, P. R., to be published.
34. Clark, G. J., Marwick, A. D., LeGoues, F., Laibowitz, R. B., Koch, R., and Madakson, P., to be published.

ION ENERGY DISSIPATION AND SPUTTERING DURING BOMBARDMENT OF MULTICOMPONENT MATERIALS

Jørgen Schou

Association-EURATOM, Risø National Laboratory, Physics Department,
DK-4000 Roskilde, Denmark

I. INTRODUCTION

Ion-slowing down, energy loss and sputtering for elemental as well as multicomponent materials have been treated comprehensively within the latest twenty years. In particular, the large number of existing computer codes have turned out to be reliable in many of the predictions. The aim with the present work has not been to discuss the areas thoroughly, but largely to follow basic ideas. Readers who might want details about these subjects are referred to the many existing reviews, which are mentioned in the text in connection with each section. The present work does not include any description of the historical development or of the experimental procedures. These topics are decribed in many of the reviews.

This review concentrates on the energy loss of charged particles, and on the neutral particle emission that evolves from the energy deposition in the material. In fact, the common themes for all sections are the energy loss mechanisms and the different channels through which the energy is released for particle emission. Effect from monocrystalline targets will not be treated. Charged particle or cluster emission will not be included systematically either.

The main features for the energy loss and the subsequent energy conversion are shown in Fig. 1. The energy loss quantities are treated in greater detail in Sect. II. We note that damage and particle emission are closely connected concepts. Displacements induced by the primary beam in the surface region lead indispensably to particle ejection as well as damage. The two features, electronic damage and electronic sputtering, are observed for non-conducting materials. An interesting aspect for these features is that they may behave very differently as a component in a chemical compound or a mixture from that in an elemental target. In addition, Fig. 1 shows phenomena such as electronic sputtering, photon emission, chemical changes, and electronic damage that occur with a considerable time delay of nanoseconds up to seconds compared with the other ones. These features are discussed in connection to sputtering of insulating materials in Sect. IV.

The references in this work are representative for the treatment, but in no way a comprehensive list of all important work in these areas.

II. STOPPING POWER FOR ELEMENTAL MATERIALS

The stopping power dE/dx of a medium for a particle beam is the average energy loss of the particles after passing a thin layer. It is conveniently expressed as

$$-\frac{dE}{dx} = N S(E) \qquad (\text{II}.1)$$

where N is the number density of atoms in the medium, and S the stopping cross section. The dependence of the stopping power on the kinetic energy E of the particle has been emphasized in the stopping cross section. The stopping power is expressed in units of energy × [length]$^{-1}$, and the cross section energy × area. Nevertheless, the stopping power is often tabulated in units as eV/10^{15} atoms/cm^2 or MeV/g/cm^2. These units are equivalent to the stopping cross section, since the

C.J. McHargue et al. (eds.), Structure-Property Relationships in Surface-Modified Ceramics, 61–102.
© 1989 by Kluwer Academic Publishers.

62

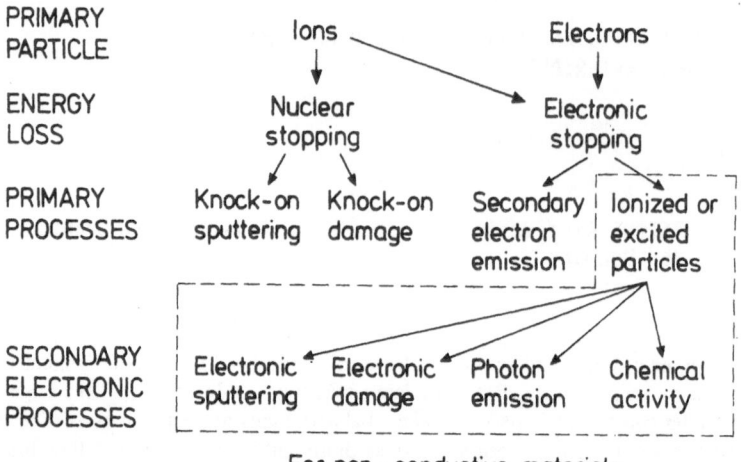

Fig. 1. A schematic survey of different processes associated with the energy loss of the primary particles, (Direct knock-on sputtering by relativistic electrons is not included).

dependence of the stopping power on the density enters only via N (apart from a density effect that is important only in the ultrarelativistic region). According to the treatment of Lindhard et al. (1963), the collisions are separated in those between the primary particle and the nuclei and those between the particle and the electronic system of the atoms. Consequently, the stopping crossing section is split up in two parts:

$$S(E) = S_e(E) + S_n(E)$$

(II.2)

where S_e is the electronic and S_n the nuclear stopping cross section. The phrase "nuclear stopping" is firmly established, but is somewhat misleading as most of the collisions take place between screened nuclei including electrons rather than bare nuclei.

Since the stopping power is one of the important quantities in all problems of particle slowing-down, it has been treated thoroughly by several authors. Bohr's work (Bohr (1948)) is one of the milestones in stopping theory, but within the latest twenty years other comprehensive reviews (Sigmund (1975)) or similar treatments have appeared, e.g. Bonderup (1981), Cruz (1986), Ziegler et al. (1985), Bichsel (1988) and Sigmund (1988a). Some of the recent tabulations of stopping powers are accompanied by a general introduction to stopping theory, e.g. ICRU (1984), Andersen and Ziegler (1977) and Janni (1982). The particular field, stopping of heavy relativistic particles has been covered by Ahlen (1980) in a comprehensive review.

A survey of the two stopping cross sections is shown in Fig. 2. The division of the energy range in four regimes is somewhat artificial, since some theories cover more than one regime. Nevertheless, the classification is convenient for the characteristics of the total stopping power. The nuclear stopping power is important only in the low-energy regime. In the intermediate, high-energy, and relativistic regime it is typically a factor of 2 M_p/m_e ($\approx$ 4000) lower than the electronic stopping power (Lindhard et al. (1963), Sigmund (1975)). M_p is the proton mass and m_e the electron mass.

A relevant parameter for the interaction between collision partners in the low-energy and intermediate regimes is "Lindhard's" parameter (Lindhard et al. (1968)):

$$\varepsilon = \frac{E_r}{Z_1 Z_2 e^2 / a_L}.$$

(II.3)

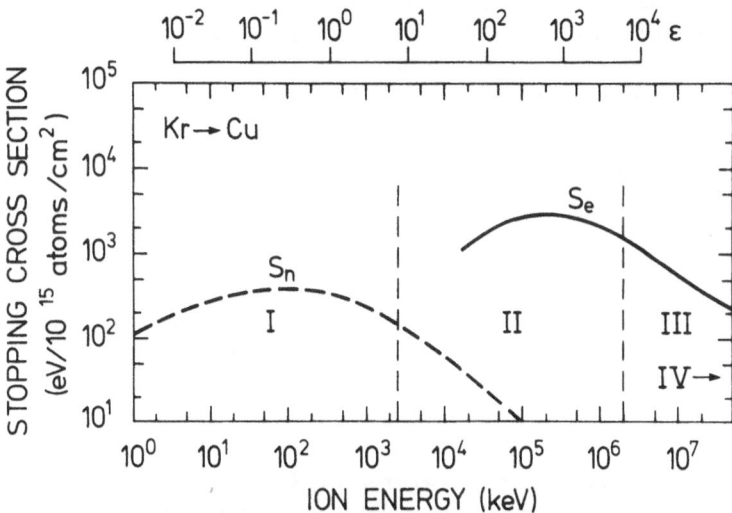

Fig. 2. The stopping cross sections as a function of ion energy in regime I, II and III. S_n, nuclear stopping cross section, Eq. (II-6), S_e, electronic stopping cross section from Ziegler (1980). The reduced energy ε, Eq. (II-5) is shown as well. (I) low-energy, (II) intermediate, (III) high-energy and (IV) ultrarelativistic regime.

Z_1, M_1 and Z_2, M_2 are the atomic numbers and masses of the primary ion and of the target material, respectively. a_L (see below) is the screening radius and $E_r = M_2E/(M_1 + M_2)$ the relative energy of the colliding particles. Lindhard et al. (1968) suggested that

$$a_L = 0.8853a_o(Z_1^{2/3} + Z_2^{2/3})^{-1/2} \,, \tag{II.4}$$

where a_o is the Bohr-radius (0.0529 nm) and 0.8853 a constant derived from the Thomas-Fermi potential between interacting atoms (Lindhard et al. (1968)).

The quantity ε turned out to be a convenient scaling unit for the energy. A expression frequently used is

$$\varepsilon = 32.53M_2E/[Z_1Z_2(M_1 + M_2)(Z_1^{2/3} + Z_2^{2/3})^{1/2}] \,, \tag{II.5}$$

where the ion energy E is in keV and the masses M_1 and M_2 are in amu. A screening radius similar to Eq. (II.4) was introduced by Firsov (1959). However, since the Firsov radius deviates less than about 10 per cent from that of Lindhard, Lindhard's screening radius will be utilized for the evaluation of ε everywhere in the present work. Lindhard et al. (1968) showed on the basis of the Thomas-Fermi model that the nuclear stopping power $NS_n(E)$ for all elemental beam ion-target combinations could be expressed as

$$NS_n(E) = \frac{\pi a_L^2 \gamma N}{(\varepsilon/E)} s_n(\varepsilon) \,, \tag{II.6}$$

where

$$\gamma = 4M_1M_2/(M_1 + M_2)^2 \tag{II.7}$$

is determined by the maximum energy transfer γE from a particle with energy E and mass M_1 to a particle at rest with mass M_2. The important property of the relation (II-6) is that the nuclear stopping power for all combinations depends on the primary energy E via the dimensionless reduced nuclear stopping s_n. The maximum of $s_n(\varepsilon)$ is around 0.35 for values of ε about 0.3, and the overall shape is shown by NS_n in Fig. 2. The absolute magnitude of NS_n is primarily determined by the constant ratio (ε/E) in the denominator. Heavy beam atoms incident on a heavy target lead to a combination with a small ratio, and in turn, a high nuclear stopping power.

However, experiments in the late seventies, e.g. Andersen et al. (1975), demonstrated that the screening determined by the Thomas-Fermi function for a neutral atom falls off too slowly for large interatomic distances. Wilson et al. (1977) improved the interatomic potential to a Molière-like screening function that showed a faster fall-off at large distances. In particular, the combination krypton incident on carbon was investigated as a representative case. Based on this Kr-C potential these authors presented a simple approximation for the reduced nuclear stopping power:

$$s_n(\varepsilon) = \left[\frac{1}{2}ln(1+\varepsilon)\right] / \left(\varepsilon + 0.10718\varepsilon^{0.37544}\right) . \tag{II.8}$$

This procedure has recently been refined by Ziegler et al. (1985). However, the straightforward Kr-C-expression (II-8) shows the expected asymptotic behaviour for Rutherford scattering at $\varepsilon >> 1$ and falls off faster than the corresponding Thomas-Fermi stopping power for $\varepsilon << 1$ as required. Wilson et al. (1977) utilized Firsov's screening radius which is slightly less than the one of Lindhard.

The electronic stopping power at low velocities in regimes I and II may be determined from the Lindhard-Scharff treatment (Lindhard and Scharff (1961)). Although more precise tabulations have emerged (Andersen and Ziegler (1977), Ziegler (1977), (1980) and (1985)), the Lindhard-Scharff value is still a convenient reference standard and the predictions of the treatment are usually correct to within a factor of two. The electronic stopping power is proportional to velocity and may be expressed as

$$s_e(\varepsilon) = k_L\varepsilon^{1/2} . \tag{II.9}$$

s_e is the reduced (dimensionless) electronic stopping power corresponding to Eq. (II-6) and

$$k_L = \frac{0.0793Z_1^{2/3}Z_2^{1/2}(M_1 + M_2)^{3/2}}{(Z_1^{2/3} + Z_2^{2/3})^{3/4} M_1^{3/2} M_2^{1/2}} \tag{II.10}$$

where M_1 and M_2 are in amu. A similar velocity dependence of the electronic stopping power was obtained by Firsov (1959), but experimental work within the last twentyfive years indicate that the exponent in Eq. (9) may be different from 1/2.

The electronic stopping power in the energy region above the stopping power maximum is treated by Bethe-Bloch theory (Ahlen (1980) and Sigmund (1988a)). A convenient expression is the velocity-dependent stopping cross section

$$S_e(E) = \frac{4\pi e^4 Z_1^2 Z_2}{m_e v^2} L(v) , \tag{II.11}$$

where Bethe's stopping number essentially is expressed as

$$L(v) = ln\left(\frac{2m_e v^2}{I}\right) - ln\left(1 - \frac{v^2}{c^2}\right) - \frac{v^2}{c^2} - \frac{C}{Z_2} - \frac{\delta}{2} . \tag{II.12}$$

I is the target mean-excitation potential, C/Z_2 the shell correction and δ an ultra-relativistic density correction (Sternheimer and Peierls (1971)) which otherwise will be neglected in the present context. The mean excitation potential is defined as

$$lnI = \sum f_n \ln E_n ,$$ (II.13)

where E_n are all possible energy transitions of the target atom and f_n the corresponding dipole-oscillator strengths. The calculations of I are usually based on a statistical model of the target atom. The simplest result is the well-known Bloch's rule (Sigmund (1975))

$$I = I_o Z_2 ,$$ (II.14)

where $I_o \approx 10$ eV.

The shell correction accounts for the deviations from the requirement in Bethe's derivation that the projectile velocity should be much larger than that of the bound electrons.

Bethe's calculation is based on a first-order quantal perturbation treatment. The result is an asymptotic expression that is valid for small values of Bohr's parameter (1948)

$$\kappa = \frac{2Z_1 e^2 v_B}{\hbar v}$$ (II.15)

A more comprehensive treatment by Bloch leads to the inclusion of an additive term

$$\Delta L = \psi(1) - Re\psi(1 + i\kappa/2)$$ (II.16)

in Eq. (II-14). In the quantum mechanical limit $\varkappa << 1$ Bloch's results goes over into the one of Bethe. For ions of high atomic number or of low velocities Bloch's treatment is more correct than that of Bethe (Andersen et al. (1977), and Sigmund (1988a)). The substantial change is that Z_1 enters into the stopping number, and thus, that the position of the maximum depends on the atomic number of the beam ion.

A problem that is closely related to the stopping power is the charge state of the penetrating ions. The projectile electrons with orbital electron speeds exceeding v will stick to the projectile while the slower ones, belonging to the outer shells, will be stripped. According to Bohr's estimate (1948) the average charge state of heavy ions is

$$Z_1^* \approx Z_1^{1/3} v/v_B$$ (II.17)

in the velocity range where $1 << Z_1^* < Z_1/2$, see e.g. Sigmund (1975). This value of charge state means that the average ion charge increases with the ion velocity. However, regardless of the initial charge state that may be far from the average state, an ion beam will approach the equilibrium charge distribution after the passage of a thin layer. A comprehensive treatment of the charge states of ion beams has been presented by Betz (1972) and Betz et al. (1988).

There have been a few attempts on treatments that comprise the electronic stopping power on both sides of the stopping power maximum (Brice (1975) and Cruz (1986)). The most extensive compilations have been carried out by Ziegler and coauthors (Andersen and Ziegler (1977), Ziegler (1977), Ziegler (1980) and Ziegler et al. (1985)). These stopping power tabulations which are the only comprehensive ones within the latest twenty years have had an overwhelming influence on the literature ranging from particle slowing-down to ion implantation studies.

Apart from the proton stopping powers these tabulations are based on an empirical charge state evaluation for Z_1^*. This charge state is assumed to be completely independent of Z_2. The electronic stopping is determined by the local density approximation so that

$$S_e(E) = \int_V \hat{I}(v,\rho)(Z_1^*(v))^2 \rho(\vec{r})d\vec{r} \quad , \quad \int_V \rho(\vec{r})d\vec{r} = Z_2$$ (II.18)

(Ziegler (1980)). $\hat{I}$ is the stopping interaction function of a projectile of a unit charge with a free electron gas of spatially varying density ρ(r) over the volume V of the target atom. The electronic density is normalized so that the stopping power obtained by Eq. (II-18) is approximated by the Bethe-stopping power (II, 11 and 12) with the sum of the mean excitation potential and the shell corrections as an adjustable parameter so that the best possible agreement with the experimental data is obtained. The stated overall accuracy of the tables is 2%, which is clearly unrealistic. There are no experimental points with which the reliability of the tabulations might have been tested for primary energies from 20 MeV/amu up to 1000 MeV/amu. Furthermore, the majority of the data points are obtained with ions of atomic numbers below $Z_1 = 20$. The two apparent deficiencies of the tabulations: i) the use of an empirical standard charge state evaluation, and ii) the application of the Bethe formula without the Bloch-correction (Eq. II-16) mean that the inaccuracy is expected to increase with both the atomic number of the ion and the primary energy. Nevertheless, these stopping power tabulations will be applied in the present work, since there is no comparable alternative.

In energy dissipation problems one may encounter not only the electronic and nuclear stopping powers, but also the integrated quantities $\eta(E)$ and $\nu(E)$. $\eta(E)$ is the total energy ultimately deposited in the electronic system by the primary ion and by possible recoils, and $\nu(E)$ is the corresponding energy which ends up in atomic motion (Lindhard et al. (1963)). Once the energy is transferred to the electronic system, it is unavailable for atomic motion in a time-scale comparable to the slowing-down of the primary ion and the recoils. The reason is the extremely small efficiency of energy transfer from an atom to an electron, (corresponding to Eq. (II-7)) and in turn back from an electron to an atom. Therefore,

$$\eta(E) + \nu(E) = E .$$

(II.19)

Winterbon (1975) has made extensive tabulations of $\nu(E)$ as well of other related quantities.

III. ENERGY LOSS IN MULTICOMPONENT MATERIALS

The stopping power in multicomponent materials is occasionally different from the expected value. Usually this value is determined by Bragg's rule which expresses the total stopping cross section in a weighted sum of the contributions (Sigmund (1975)):

$$S(E) = N_j S_j(E)$$

(III.1)

N_j is the atomic number density and S_j the stopping cross section of the j'th element. The nuclear stopping cross section of multicomponent materials behaves strictly according to Bragg's rule (at least in the usual approximation of binary collisions). The electronic stopping cross section is influenced by changes in the electronic structure as a result of chemical binding. Since only the outermost electron shell changes, the total change in the stopping cross section depends on the relative contribution of this shell to the full value.

The original statement by Bragg and Kleeman (1905) was that the decrease in range relative to air of MeV α-particles in a compound is determined by a suitable summation over the constituent atoms. The accuracy of Bragg's rule has been discussed by Sigmund (1975), Kreutz et al. (1980), and by Thwaites (1983), (1985) and (1987).

Changes in the electronic structure occur not only from the effect of molecular binding in mononuclear gases and of the chemical binding in compounds, but also as a result of the physical state of the material. Therefore, the gas-solid difference will be included partly in the following treatment. A schematic survey of the different types of collisions is shown in Fig. 3.

Of course, Bragg's rule has to break down for any condensed material or compound at a sufficiently high experimental accuracy. Deviations from the additivity is expected mostly for

i) comparatively low energies where the relative contribution from valence electrons to the stopping power becomes large, and

ii) for light elements where the valence electrons constitute a major fraction of the total number of electrons.

ATOMS   MOLECULE

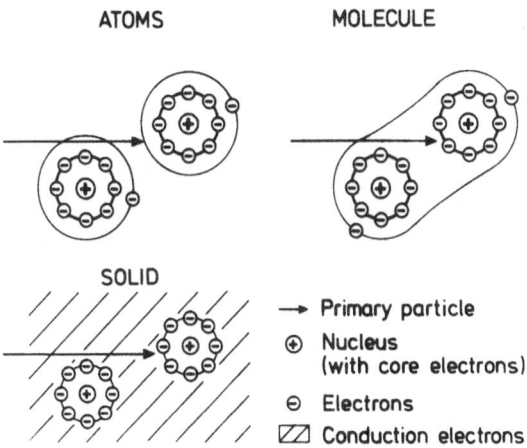

SOLID

→ Primary particle

⊕ Nucleus
  (with core electrons)

⊖ Electrons

▨ Conduction electrons

Fig. 3. A schematic survey of ion-atom collisions with different state of the outermost electrons.

For light ions at the stopping maximum the deviation rarely exceeds 10 per cent. For energies above the stopping power maximum the changes in the binding from chemical effects or condensation enter into the mean excitation potential I alone, Eq. (II-13). Therefore, the tabulations of I usually contain a value for the solid state, the gaseous state, and one or more for the atoms in a chemical compound (see e.g. the ICRU-table (1984)).

Occasionally, an analogous additivity rule is formulated for the mean excitation energy of a molecule (ICRU (1984) and Thwaites (1987)):

$$ln\,I = \sum N_j Z_j \ln I_j / (\sum N_j Z_j) \tag{III.2}$$

where I_j is the mean excitation energy of the j'th element. The application of this additivity requires the choice of appropriate values for the mean excitation potentials of the constituents. The simplest procedure, often used in the past, is to take the same I-values for the atomic constituents of a compound as for the corresponding elements. This introduces some error because of the neglect of molecular binding effects. An additional error may be incurred when elemental I-values for gases (e.g. oxygen or nitrogen) are applied to the constituents of solid compounds. The accuracy of the additivity rule may be improved by assigning values of I depending on the type of compound and on the chemical state.

An instructive procedure has been applied by Sabin and Oddershede (1987). The mean excitation potential is decomposed into orbital mean energies

$$ln\,I = f_{core} \ln I_{core} + (1 - f_{core}) \ln I_{val} \tag{III.3}$$

from the core electrons and the valence electrons. An analysis of carbon hydrogen molecules led the authors to the conclusion that the type of chemical bonds determined the mean excitation energy. They obtained the result that the C-C bond was assigned to a higher value I(C-C) than that of the double bond I(C=C) which in turn was larger than the carbonhydrogen bond value I(CH). In terms of energy loss it means that, for example, the stopping cross section of a carbon atom in a double-bond compound is larger than that in a single-bond one. These statements agree well with experimental results from Powers et al. (1972).

This procedure requires a comprehensive experimental data set, and generally it is difficult to characterize the chemical bonds in greater detail. However, Eq. (III-3) allows us to separate the contribution I_{val} from the valence electrons. Apart from the lightest materials this is usually a minor fraction of the total mean excitation potential. This division of the mean excitation potential into contributions from the individual shells (Sigmund (1982) and Sabin and Oddershede (1982)) is a substantial, simplifying feature in stopping theory.

In Table 1 the recommended values of the mean excitation potentials of common constituents are listed. Generally, the values are larger than or equal to the elemental values. Zieglers compilations (1980) result in values that are somewhat smaller than the ICRU-values for aluminium, silicon and copper. The ICRU-values are supported by the experimental results for 10- to 30-MeV protons incident on light compounds (Tschalär and Bischel (1968)). The values derived for the oxides were significantly larger than those evaluated from Bragg's rule on the basis of the elemental values.

TABLE 1.

	ICRU elemental value	ICRU value in solid oxides, carbides	Ziegler values (solids)
H	19.2 (gas)	19.2	17.1
C	780 (graphite)	81	79
N	82.0 (gas)	82	83.8
O	95.0 (gas)	106	105.5
Al	166	188	162
Si	173	195	159
Cl	174 (gas)	180	187
Cu	322	364	330

At velocities much lower than considered above, the total deviation from Bragg's rule is primarily caused by the finite velocities of the projectiles compared to the valence electrons. This effect is pronounced for energies about the stopping power maximum, i.e. for primary velocities comparable to velocities of the most loosely bound electrons.

The deviation from Bragg's rule have been demonstrated by Powers et al. (1984) in a striking measurement of the energy loss in gaseous amine compounds. In this way no gas/solid differences enter into the analysis. Their results shown in Fig. 4 show that the atomic stopping cross section of nitrogen molecules in the gas phase (triple bond) is larger than the stopping cross section of nitrogen atoms with single bonds in the organic compounds. Furthermore, the stopping cross section of the latter atoms peaks at a substantially higher energy than the stopping cross section of the triple-bonded nitrogen atoms. At higher primary energies the difference between the stopping cross sections is reduced. Similar features have been observed previously for the stopping cross sections of single- and double-bonded carbon and oxygen atoms (Powers et al. (1984)).

The systematic studies on solid oxides demonstrate clearly that the stopping cross section of oxygen atoms in oxides is significantly less than that for gaseous oxygen (Feng et al. (1974) and Ziegler et al. (1975)). These authors indicate a reduction of the stopping cross section from 6 to 20 per cent for 1-2 MeV helium ions. There is no absolute agreement about the magnitude of the deviation from Bragg's rule of the oxides in the literature (Thwaites (1983)), but apparently the majority of the contributions support a more or less pronounced decrease from the value obtained by Bragg's rule.

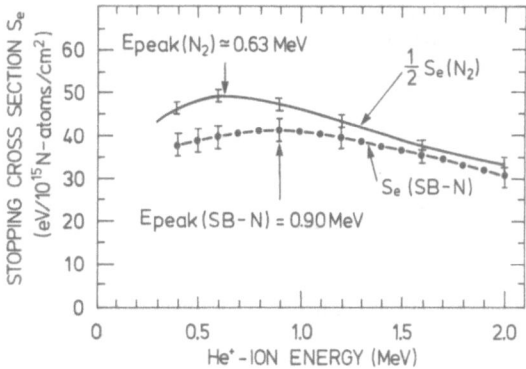

Fig. 4. The electronic stopping cross section for nitrogen atoms in a molecular gas and for single-bonds in organic gases. Data from Powers et al. (1984).

For metallic alloys no systematic deviations from Bragg's rule have been observed (Feng et al. (1973)). This is in agreement with the point that deviations only occur if the state of the valence electrons change from the element to the mixture.

The gas-solid difference is a result of the difference in electronic structure of the valence electrons (Shiles et al. (1980)). Obviously, only the latter term in Eq. (III-3) depends on the presence of conduction electrons or other solid state features (Sabin and Oddershede (1988)). This may be understood in the context of the local density approximation, in which the mean excitation energy I is expressed by

$$ ln\,I = \frac{1}{Z_2} \int_V \rho(\vec{r})\,ln[\gamma\hbar\omega_p(\vec{r})]d\vec{r}\,, \tag{III.4}$$

where

$$ \omega_p(\vec{r}) = (4\pi\rho(\vec{r})e^2/m_e)^{1/2} \tag{III.5}$$

is the local plasma frequency and γ a constant ($\approx \sqrt{2}$) (Bonderup (1967)). The density of the conduction electrons is larger than that of the corresponding electrons in atomic orbitals. This means that the plasma frequency is enhanced, and in turn that the contribution from the logarithmic factor in the integral increases.

The gas-solid difference has been observed by many groups (Thwaites (1985)). Generally, the stopping cross section for the atoms in the solids are less than that for the gases as expected from the considerations above. The difference is typically from 10 to 2 per cent for He-ions about or above the stopping power maximum. At lower energies, corresponding to projectile velocities less than the outermost electrons, the difference may increase considerably (Børgesen (1985) and Oddershede et al. (1983)).

Unfortunately, most of the experiments have been performed with light ions up to $Z_1 = 3$ with velocities larger than the Bohr-velocity v_B. A consistent picture that includes heavy ions as well as light low-energy ions does not exist presently. Ab initio calculations of the stopping cross section or the mean excitation potential are difficult, if not impossible. Only the simplest ones, e.g. water (Geertsen et al. (1986)) have been studied by such calculations or by the evaluation from optical data (Ashley (1982)). Calculations of the stopping cross section based on the application of a harmonic oscillator (Sigmund and Haagerup (1986)) may be a promising future tool for chemical compounds.

IV. SPUTTERING OF ELEMENTAL MATERIALS
IV-A. Sputtering as an erosion process

The **erosion** of solids by energetic particles may be caused by a variety of mechanisms that even may work simultaneously:

i)   beam-induced evaporation
ii) knock-on sputtering
iii) electronic sputtering, and
iv) desorption of thin layers.

Strictly speaking, desorption is the erosion of mono- or multilayers of a material deposited on a different substrate. This process has been included because of the close relationship between desorption via electronic processes and electronic sputtering. The ejection of material by knock-on or electronic sputtering as well as desorption processes is the result of **single particle impact**. In contrast, the material evaporates as a consequence of **beam heating** in beam-induced evaporation. These four types of erosion are shown schematically in Fig. 5. This division is incomplete, and does not contain chemical sputtering for example. This type of erosion occurs as the result of chemical activity between the implanted ions and the target particles (Roth (1983) and (1986a)) or between a reactive gas and the target particles (Oostra and de Vries (1987) and Coburn and Winters (1987)).

Erosion or sputtering have been covered by several recent reviews (Sigmund (1981) and (1987a), Szymonski (1982), Kelly (1984), Andersen and Bay (1981), Roth (1986b), Hofer (1986), Andersen (1987) and (1988), Zalm (1988) and Hofer (1989)).

Only polycrystalline targets will be discussed in the following. Sputtering of single crystalline targets has been reviewed by Robinson (1981) and Roosendaal (1981).

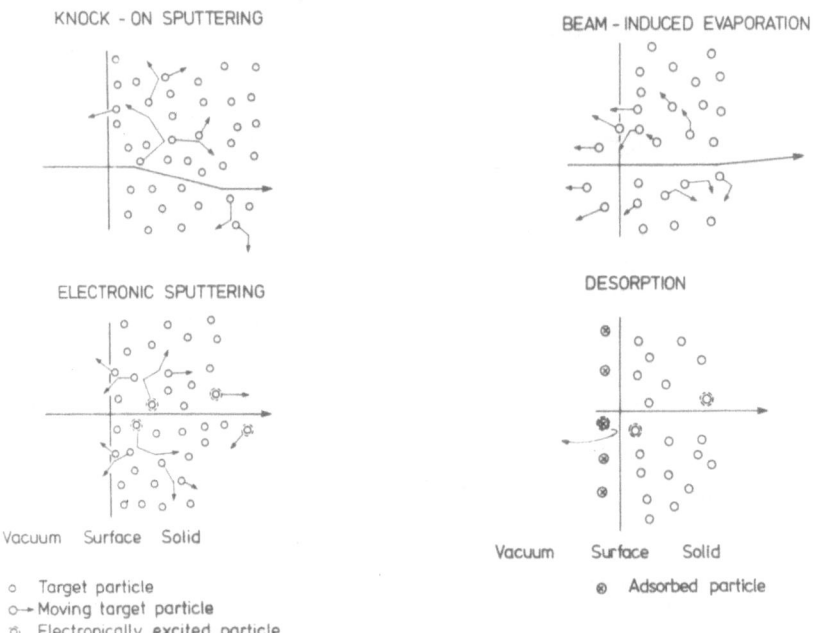

Fig. 5. Schematic survey of different types of erosion.

Beam-induced evaporation occurs typically for hot targets that are additionally heated by particle beams. A characteristic feature of sputtering is the relative insensivity to target temperature apart from those close to the melting point. Recent measurements by Besocke et al. (1982), Hofer et al. (1983) and Bohdansky et al. (1987) demonstrate that the yield increases very little with target temperature even up to T 0.8 T_m, where T_m is the melting temperature.

For frozen gases the effect of additional beam heating is fairly pronounced because of the low binding energy of the atoms or molecules in the solid. Even at relatively low temperatures the extra heating deposited by the beam leads to enhanced evaporation because of the strong, nonlinear dependence of the evaporation on the target temperature. Alternatively, the beam intensity is so high that beam heating alone causes evaporation. Both cases have been studied for the ices (Schou et al. (1984) and Schou (1987)). Figure 6 demonstrates that the yield increases sharply above a characteristic threshold temperature. This quantity is clearly correlated with the sublimation energy U_o, which is much larger for sulphur dioxide (0.37 eV) for example than for xenon (0.164 eV). All the materials show a clear low-temperature behaviour with a yield determined by processes other than evaporation.

Beam-induced evaporation has been treated by Sigmund and Szymonski (1984) and Urbassek and Sigmund (1984) on the basis of a general heat conduction treatment. The total yield Y_{tot} is a sum of the ordinary low- temperature yield Y (from knock-on or electronic sputtering) and a term that accounts for the yield increase by evaporation:

$$Y_{tot} = Y + \frac{1}{J}(\phi(T_a + \Delta T_{eff}) - \phi(T_a)) , \tag{IV.1}$$

where J is the current density, T_a the temperature of the ambient target and ΔT_{eff} the average temperature rise of the target as a result of beam heating. The evaporation rate (number of atoms evaporated per unit time and area)

$$\phi = A_\phi T^{-1/2} \exp(-U_o/k_B T) \tag{IV.2}$$

is then enhanced drastically by an increase ΔT_{eff} from the beam because the total temperature T_a + ΔT_{eff} enters as an argument in the exponential (A_φ is a constant and k_B Boltzmann's constant).

Another temperature effect appears if the chemical activity suddenly increases above a characteristic temperature. In particular, this phenomenon has been studied for hydrogen-implanted carbon targets (Roth (1986a)).

IV-B. Sputtering of elemental materials

Sputtering is the result of direct collisions between the primary and target atoms (knock-on sputtering) or electronic transitions (electronic sputtering). **Knock-on sputtering** has also been named collisional, ballistic or nuclear sputtering. As a matter of fact, electronic sputtering originates from collisions as well. The term "ballistic" may indicate a macroscopic process, and nuclear sputtering a mechanism that is caused by nuclei rather than atoms. Since both terms are misleading, we will apply the concept knock-on sputtering in the following. The total energy that is available for knock-on sputtering, is the elastically deposited energy $\nu(E)$ from Eq. (II-19). $\nu(E)$ is the integral of the spatial distribution $F_D(E,x, \cos\theta)$ of energy deposited in atomic motion by a primary of energy E and angle of incidence θ (Winterbon et al. (1970)). Similarly, $\eta(E)$ is the integral of the spatial distribution of energy D_e deposited in electronic excitations:

$$E = \int F_D(E, x, \cos\theta)dx + \int D_e(E, x, \cos\theta)dx$$

$$= \nu(E) + \eta(E) \tag{IV.3}$$

Electronic sputtering is caused by electronic transitions in which kinetic energy for atomic motion is liberated. This energy release requires the existence of repulsive potentials during the

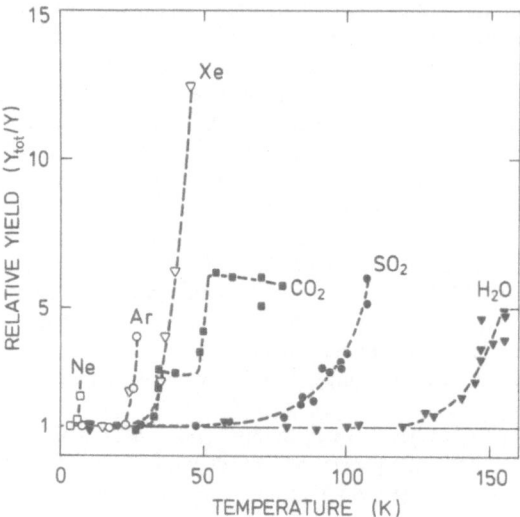

Fig. 6. Beam-induced evaporation caused by external heating for frozen gases and water ice. From Schou (1987).

electronic deexcitation, and has been observed only for non-metallic materials. The energy available for electronic sputtering is the quantity $\eta(E)$ from Eqs. (II-19) and (IV-3) (Szymonski (1982) and Schou (1987)).

Let us consider some of the important features of knock-on sputtering as a starting point. The energy required for the atomic motion of the target particles is transferred from the fast primary to the target atoms through direct collisions. The struck atoms may initiate collision cascades via secondary and higher order collisions. The direct momentum transfer to the atoms of the material does not depend on whether the material is a metal, a volatile insulator or a gas, since the velocity of the incident particle is much higher than the velocities of the atoms of the material. The state of the material plays a role alone in the last stage of the cascade, when the kinetic energy of the moving atoms becomes comparable to the binding energies of target atoms. Therefore, the standard theory for knock-on sputtering for metals may largely be extended to semiconductors and insulators.

The standard theory of sputtering is Sigmund's analytical treatment (1969) based on Boltzmann transport theory. The semi-infinite target is embedded in an infinite medium of the same material. The primary particles initiate their motion in the plane x = 0 that corresponds to the surface (Fig. 7). The target particles set in motion may be ejected, i.e. enter the negative half space, if their kinetic energy perpendicular to the surface is larger than the surface binding energy. In the theory a planar surface barrier at x = 0 is used with the sublimation energy as surface binding energy U_0. The solution for the Boltzmann-equation is obtained in the limit of high primary energy compared to instantaneous energy of the moving target atoms. This limit corresponds to an isotropic motion of the atoms in the solid. The treatment considers primarily low collision densities, i.e. the struck atoms in the cascade are at rest before the collision.

The backsputtering yield Y from a plane surface is given by (Sigmund (1969) and (1981))

$$Y = \Lambda F_D(E, 0, \cos\theta) \tag{IV.4}$$

where $F_D(E,0,\cos\theta)$ is the surface value of the spatial distribution of energy deposited into atomic motion. The constant

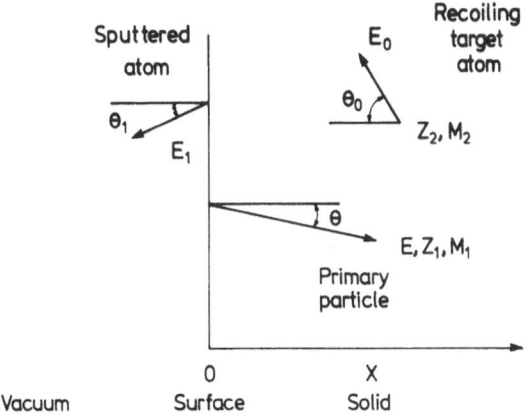

Fig. 7. The geometry of a sputtering event.

$$\Lambda = 3/(4\pi^2 N C_o U_o) = 0.042/(N U_o [\text{Å}^2])$$ (IV.5)

depends only on the properties of the target material, the sublimation energy U_o and the atomic density N. The cross section C_o ($\sim 1.81 \cdot 10^{-20}$ m^2) originates from a previously proposed application of a low-energy interatomic Born- Mayer potential and is common for all target materials in this approximation. The surface value $F_D(E,0,\cos\theta)$ depends on the mass M_1, the atomic number Z_1, the energy E and the direction θ of the incident ion, and on the target properties M_2, Z_2 and the density N.

From dimensional arguments alone one may express

$$F_D(E, 0, \cos\theta) = \alpha N S_n(E)$$ (IV.6)

where α is a dimensionless function that varies with the mass ratio M_2/M_1 and angle of incidence θ.

α is relatively insensitive to variations of the primary energy. At perpendicular incidence for beam atoms much heavier than the target atoms $\alpha \approx 0.17$. This low value compared with unity means that a major fraction of the deposited energy is transported away from the surface by recoiling target atoms. Empirical and theoretical curves have been presented by Andersen and Bay (1975) and Sigmund (1981). In most of these evaluations the electronic stopping power was small and, therefore, neglected. For the case of light ions the inelastic energy loss can be so large that α is reduced compared to the standard value given by these authors. Winterbon's tabulations (1975) includes the effect of the electronic stopping power on the value of α.

The combination of Eqs. (IV-4) and (6) gives the well-known formula for the sputtering yield (Sigmund (1969)):

$$Y = \Lambda \alpha N S_n(E)$$ (IV.7)

in which no adjustable parameters enter. The agreement between this result and experimental data for low collision densities turned out to be convincingly good apart from a few materials. Experimental results for krypton ions on copper as well as argon ions on nickel is shown in Fig. 8. One notes the agreement in absolut magnitude as well as energy dependence in both cases. The dependence on the beam atom is demonstrated in Fig. 9 for the yield from silicon. For the dependence on M_2 and Z_2, the agreement with Eq. (IV-7) is very good as well. Since it is not possible to vary the sublimation energy for a real target we have to rely on results on computer simulations for this parameter. Indeed, calculations by Hou and Robinson (1979) demonstrate that

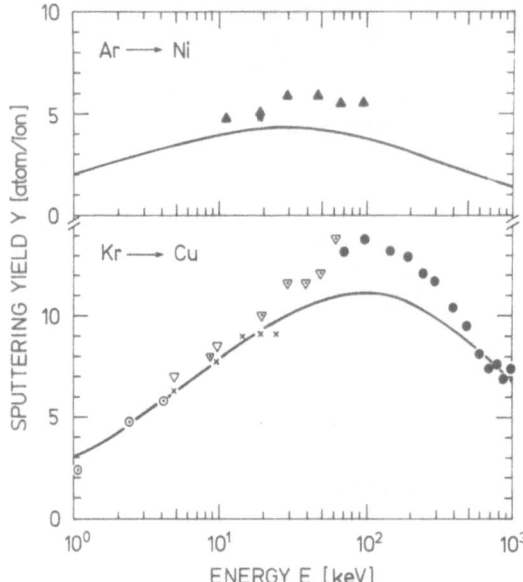

Fig. 8. Sputtering yield from Kr⁺ on Cu and Ar⁺ on Ni. Experimental data from Sigmund (1987a) and Andersen and Bay (1981). The solid line is the yield predicted by Eq. (IV-4). The nuclear stopping is from Eqs. (II-6) and (II-8).

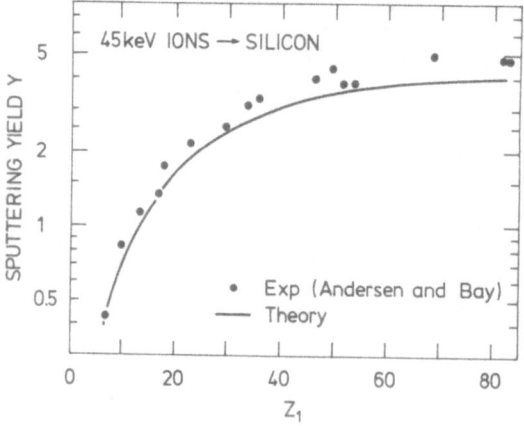

Fig. 9. Experimental data from Andersen and Bay (1975). The solid line is the theoretical yield, comp. Fig. IV.4.

the yield is roughly inversely proportional to the assumed surface binding energy U_o as predicted by the standard theory. Recently, Kelly (1987) has pointed out that the surface binding energy may depend on the character of the different sites for the atoms prior to the ejection. This means that the yield in Eqs. (IV-4,5,7) is expected to be determined by a distribution of binding energies rather than a single value of U_o. However, the agreement between the predictions of the linear collision-cascade theory and the experimentally obtained energy distributions does not lead to the need for any refinements for the surface binding model.

The derivation of the sputtering yield Eq. (IV-7) is based on an integration over the internal flux of particles that arrive at the surface $x = 0$ (Sigmund (1981)). The material constant

$$\Lambda = \frac{\Gamma_m}{2} \int_{U_o} \frac{dE_o}{E_o \, N S_n(E_o)} \left(1 - \frac{U_o}{E_o}\right) \tag{IV.8}$$

depends on the instantaneous nuclear stopping power $NS_n(E_0)$ for the slowly moving target atoms of kinetic energy E_0, and a constant Γ_m that is determined from the interatomic potential of the atoms. The planar surface binding leads to the factor $(1 - U_0/E_0)$ in the integrand.

Actually, Eqs. (IV-4,8) demonstrate that sputtering may be visualized as a three-stage process, in which the production of moving target atoms by the primary ion is the first stage (the factor $F_D(E,0,\cos\theta)$). The second is the migration of the atoms set in motion (the factor $(E_0 NS_n(E_0))^{-1}$) and the third stage is the passage of the surface barrier (the factor $(1 - U_0/E_0)$).

The analytical calculations were performed with a power cross section

$$d\sigma(E,T) = C_m E^{-m} T^{-1-m} dT , \qquad (IV.9)$$

for the energy transfer T up to $T_{max} = \gamma E$ to an atom at rest. This general approximation for the cross section was originally introduced by Lindhard et al. (1968). m varies slowly from m = 1 at high energies to $m \approx 0$ at very low energies. The origin of this dependence is the power approximation $R^{-1/m}$ to an interatomic potential, where R is the distance between colliding atoms. C_m is a constant that depends on m (Lindhard et al. (1968)). The corresponding stopping cross section for atoms of equal masses ($M_1 = M_2$) is

$$S_n(E) = (1 - m)^{-1} C_m E^{1-2m} . \qquad (IV.10)$$

The integration of the emitted energy spectrum as a function of the internal energy E_0 leads to

$$\Lambda = \frac{\Gamma_m}{8(1 - m)(1 - 2m)NS_n(U_o)} , \qquad (IV.11)$$

from which the usual expression for Λ in Eq. (IV-6) is obtained. The exponent m that characterizes the interaction was in the original treatment (Sigmund 1969) m = 0, which means $\Gamma_0 = 6/\pi^2$. Thus, one arrives at the result that the sputtering yield is determined essentially by the ratio of the nuclear stopping power of the primary to that of a recoil atom at very low energy. A similar relationship for the electronic stopping powers in secondary electron emission was indicated by Bethe (1941), (Schou (1988)).

The complete energy and angular spectrum is expressed by

$$d^2Y/dE_1 d\Omega_1 = k_m \frac{E_1}{(E_1 + U_o)^{3-2m}} |\cos\theta_1| , \qquad (IV.12)$$

where

$$k_m = F_D(E,0,\cos\theta)(1/4\pi)\Gamma_m(1 - m)/(NC_m) . \qquad (IV.13)$$

The solid ejection angle Ω, is determined by the polar angle θ_1 and an azimuthal angle, E_1 is the energy of the ejected particles measured after the passage of the surface barrier. For a fixed angle θ_1 of exit it means that the spectrum has a maximum at

$$E_1 = U_o/(2 - 2m) \qquad (IV.14)$$

and falls off as E_1^{-2+2m} at high exit energies.

Thompson (1968) and Farmery and Thompson (1968) determined an energy distribution of this shape for m = 0 twenty years ago. The characteristic asymptotic E_1^{-2}-behaviour has been observed in many cases for metals (Demkowski et al. (1987), Husinsky et al. (1987)) and for insulators (Chrisey et al. (1988), Schou (1987), and Haring (1984)). For metals it turns out that the maximum appears very close to one-half of the sublimation energy. For insulators the surface binding energy U_0 often corresponds to an energy somewhat lower than the sublimation energy. Results for vanadium and sulphur are shown in Fig. 10.

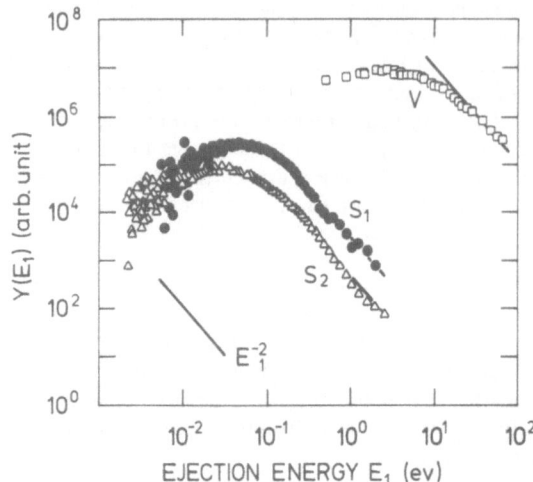

Fig. 10. The energy distribution of sputtered particles from sulphur (Chrisey et al. (1988)) and vanadium (Demkowski et al. (1987)). S_1, sulphur atoms, and S_2, sulphur molecules, ejected as a result of 44 keV Ar^+ bombardment. V atoms from 2 keV Ar^+ bombardment. Solid lines indicate an E_1^{-2}-dependence.

The value m = 0 is lower than that from the realistic Kr-C interatomic potential (m $\approx$ 1/6) and that for the Thomas-Fermi potential (m $\approx$ 1/4). Nevertheless, it leads to a convincing agreement with the experimental results. However, recent investigations indicate that m $\approx$ 0.1. (Vicanek et al. (1988)).

The angular distribution of the ejected particles is a cosine-function according to Eq. (IV-13). This distribution is rarely found experimentally. At low primary energies below 200-300 eV the angular distribution is less peaked along the direction of the surface normal than the cosine distribution.

Above ion energies of a few keV the anisotropy of the cascades becomes increasingly less important (Sigmund (1987a)). However, most of the existing angular distributions are much narrower than the predicted cosine-function (Rödelsberger and Scharmann (1976), Andersen et al. (1985) and Dumke et al. (1983)). The angular distribution was $\cos^n\theta$ where the value of the exponent usually ranged from 1.0 up to 2.0. For a liquid GaIn eutectic even the value n = 4 has been observed. Sigmund (1987a) suggested that the deviation from the cosine-function was caused partly by the scattering at the surface. Because of the missing scattering centers in the negative halfspace, the ejected atom experiences a mean deflection toward the surface normal. At the standard conditions the exponent is enhanced from n = 1 to n = 1.55 by this effect.

The linear collision-cascade theory has been a convincing treatment that predicts the important features of sputtering fairly well. No other theoretical treatment has reached a comparable level. The original theory (Sigmund (1969)) has been extended in several directions (Sigmund (1981) and (1987a)).

Alternative treatments exist (M. Urbassek (1984), Williams (1981)). The energy spectra calculated for a semi-infinite target by Urbassek (1984) turned out to be similar to the spectrum from linear collision-cascade theory (Eq. IV-12) except at the lowest primary energies.

IV-C. Outside the linear collision-cascade regime

The treatment considered in the previous sections ceases to be applicable if either the recoil cascades become too dense or if the energy transfer to the recoil atoms is so small that no cascades develop. The first case, the **spike** regime is characterized by a high collision density so that the moving particles have a large probability of striking other target particles that already have been set in motion. In the latter case, the **single knock-on** regime, the recoil atoms receive sufficiently high energy to get sputtered, but not to generate recoil cascades. These two regimes as well as the linear collision-cascade regime are shown schematically in Fig. 11.

The spike regime occurs typically at heavy ion bombardment or at bombardment by molecular ions (Thompson 1981). At heavy ion bombardment the yield is enhanced from that predicted by linear collision-cascade theory at the primary energies around the maximum of the nuclear stopping power (Bay et al. (1976)). The high nuclear stopping power means that the energy density in the cascade area becomes so high that an elastic spike develops. The molecular ions dissociate upon impact under an equal sharing of energy. The simultaneous overlap of the cascades from each of the atoms leads to dense cascades. The measured yields per **atom** from molecules are up to one order of magnitude larger than that from atomic ions (Andersen and Bay (1973), Olivia-Floria et al. (1979), Johar and Thompson (1979) and Hofer et al. (1983)).

The theoretical treatments for the elastic spike have largely been based on heat conduction theory (Vineyard (1976), Johnson and Evatt (1980), Sigmund and Claussen (1981), Claussen (1982), Szymonski and Poradzisz (1982)). An instructive evaluation was performed for cylindrical symmetry by Sigmund and Claussen (1981). This cylindrical spike model refers to an idealized system for which the energy of the primary ion is deposited at a constant rate along the ion track. The yield from the spike increases more rapidly than the square of the surface value $F_D(E,0,1)$ of the deposited energy, and the spectrum of the sputtered particles peaks at an energy lower than the value $U_0/2$ from linear collision-cascade theory (Sigmund and Claussen (1981)). The shift of the maximum of the energy spectrum was observed by Szymonski and de Vries (1977) and by Ahmad et al. (1980). A striking transition from a sputtering yield proportional to $F_D(E,0,1)$ to a yield approximately proportional to the cube of F_D at sufficiently high values of F_D was demonstrated by Thompson and collaborators (Thompson (1981) and Stevanovic et al. (1984)) for noble metal targets as well as for a frozen xenon target. In xenon the transition from the linear collision-cascade to the spike regime takes places for values of F_D close to 50 eV/10^{15} atoms/cm^2, whereas the threshold for the spike generation is from 100-350 eV/10^{15} atoms/cm^2 for these metals. The threshold depends on the sublimation energy U_0 which is about one order of magnitude less for solid xenon than for the metals. This trend is in qualitative agreement with the predictions from Sigmund and Claussen (1981).

An interesting feature in the experimental energy spectra is the pronounced E_1^{-2}-tail. These neutrals of relatively high energy escape from the cascade area during the development of a linear collision-cascade which is established within 10^{-13}s. The elastic spike develops from the linear collision-cascade to a time 10^{-12} s and lasts to about 10^{-11} s after the ion impact (Claussen (1982)). This is the reason why sputtered particles are ejected by both kinds of mechanism.

The single knock-on regime has been discussed by Andersen and Bay (1981). A typical case is light-ion sputtering of metals (Roth 1980). The strong decrease of the yield with decreasing energy was explained by Behrisch et al. (1979) on the basis of extensive computer calculations. For large mass ratios ($M_2/M_1 \gtrsim 5$) the threshold for sputtering is given by

$$E_{th} = U_o(1 - \gamma)\gamma ,$$
(IV.15)

where $\gamma = 4 M_1 M_2/(M_1 + M_2)^2$ is the maximum efficiency of energy transfer (Eq. II-7). This result leads to the interpretation that sputtering close to the threshold is caused by reflected projectiles on their way out through the surface. In fact, Bay et al. (1977) and Bohdansky et al. (1980) showed that the energy dependence of the sputtering yield was fairly well described by a universal yield formula, in which a dimensionless energy E/E_{th} alone enters. The yield Y turned out to be well approximated by

$$Y(E) = Q(M_1, M_2, U_o)(E/E_{th})^{1/4}(1 - E_{th}/E)^{7/2}.$$
(IV.16)

This expression leads to a yield equal to zero for the primary energy $E = E_{th}$. In addition, the energy dependence was found to work quite well up to primary energies around 20 E_{th} for the

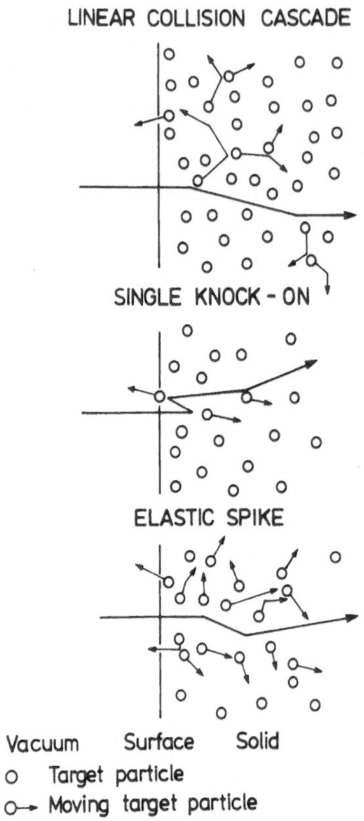

LINEAR COLLISION CASCADE

SINGLE KNOCK - ON

Fig. 11. Schematic survey of different types of erosion. The design of the figure is from Sigmund (1981).

ELASTIC SPIKE

Vacuum   Surface   Solid
o   Target particle
o→ Moving target particle

materials investigated. The precise values of the function Q as well as a refinement of the threshold concept is described by Andersen and Bay (1981) and Roth (1986b).

Littmark and Fedder (1982) succeeded in showing that the absolute yields from heavy target bombarded by light ions could be fairly well predicted, even though they assumed that the primary recoils alone are responsible for sputtering. Computer simulations by Biersack and Eckstein (1984) have shown that the contribution from secondary recoils increases with the primary energy, and that this latter contribution exceeds that from primary recoils at energies about 1 keV for primary helium ions.

IV-D. Electronic sputtering of elemental materials

Electronic sputtering, i.e. sputtering via electronic transitions, is the dominant erosion process for insulators at temperatures below the beam-induced evaporation regime during light ion bombardment at sufficiently high primary energies. About and above the maximum for the electronic stopping power the nuclear stopping power is so small that the knock-on sputtering is insignificant. On the other hand, the two kinds of sputtering processes are occasionally comparable at medium energies, and at low energies even primary helium ions may produce dominant knock-on sputtering.

The electronic sputtering yield is closely correlated to the electronic stopping power. In fact, this connection has encouraged the use of the term "electronic sputtering". During many of the early experiments about ten years ago the correlation to the electronic stopping power was immediately established, but the detailed electronic transitions have been satisfactorily explored for only a few insulating elements.

Electronic sputtering requires the existence of repulsive potentials during the electronic deexcitation. The repulsion converts electronically deposited energy to translational energy of the target particles. In metals such potentials are not active because of the presence of delocalized conduction electrons. Typically the repulsive potential in non-conductive materials is generated by pairs of atoms in metastable electronically excited states. For semiconductors electronic sputtering may occur by photon-bombardment from high-intensity lasers (Itoh (1987)). Of course, in this particular example there is no momentum transfer from the primary photons to the target atoms.

Electron bombardment represents a case similar to photon bombardment. Sputtering induced by primary electrons is a clear example of electronic sputtering. At non-relativistic energies the cross section for sputtering by direct electron-nucleus collisions is so low that this type of sputtering is insignificant even for the most volatile solids (Townsend (1983) and Schou et al. (1986)). However, at relativistic energies from about 0.5 MeV and above the electrons possess sufficient momentum to directly displace target atoms by a collision (Cherns et al. (1976) and Cherns (1979)). This type of sputtering is parallel to knock-on sputtering by ions.

Only few insulating elements have been studied comprehensively. Apart from sulphur (Chrisey et al. (1988) and Torrisi et al. (1986)) only the frozen gases of the least chemically active elements have been studied (Schou (1987) and Brown and Johnson (1986)). Among these ices the sputtering of solid rare gases and solid nitrogen have been explored fairly well, whereas no investigations of sputtering have been carried out for reactive materials such as solid chlorine or flourine.

The frozen gases have turned out to be of major importance for the understanding of electronic sputtering. The reason is that the yields are generally large and the electronic deexcitations are frequently well-known. Although the deexcitation channels in the solid phase differ somewhat from those in the gas phase, the knowledge of the electronic structure and deexcitation dynamics in the solid rare gases (Schwentner et al. (1985) and Zimmerer (1987)) and nitrogen (Ohler et al. (1977) is fairly comprehensive. This means that the processes in which kinetic energy is liberated, have been largely identified.

The experiments in which the condensed gases have been irradiated by electrons or MeV light ions, have clearly demonstrated that only a minor fraction of the electronically deposited energy is converted to kinetic energy available for sputtering. Most of the energy is emitted as luminescence from atomic deexcitations in solid nitrogen and from atomic and molecular deexcitations in the solid rare gases. In addition, the deexcitation takes place as multiphonon processes which apparently do not contribute to sputtering. However, the details of the relaxation via multiphonon production are not yet fully explained. For the solid rare gases one arrives at the result that only about 10 per cent of the deposited electronic energy become available as kinetic energy for the atoms.

The processes leading to electronic sputtering of solid argon have been particularly well investigated (Reimann et al. (1984a), Reimann et al. (1988), Schou (1987), Pedrys et al. (1988) and Ellegaard et al. (1988)). A schematic representation of the important transitions in solid argon is shown in Fig. 12. The incident beam particles create atomic holes Ar^+ along the track. The initial holes become self-trapped as a molecular holes Ar_2^+. When this ion captures an electron, the molecular ion dissociates into a highly excited atom Ar^* and a ground state atom. The excited atom becomes self-trapped as a molecular exciton Ar_2^* that finally deexcites to the repulsive ground state after nanoseconds or microseconds depending on the orientation of the spin. By this ground state repulsion about 1.2 eV is liberated. This energy is so much larger than the sublimation energy of 80 meV that the repulsing rare gas atoms initiate a low-energy cascade (Garrison and Johnson (1985), Schou(1987) and Ellegaard et al. (1988)). Reimann et al. (1984a) and (1988) suggest that the dissociative recombination of the molecular ion Ar_2^+ may release energy as well. In this process the ion captures an electron, but the neutral molecule dissociates into atoms immediately. The process supplies the repulsing atoms with 1-2 eV. How important this additional process is, is for the moment unclear. Apart from small variations the deexcitation mechanism and the production of sputtered atoms seem to be similar for all solid rare gases.

The strong luminescence from the decay of the molecular exciton Ar_2^* turned out to be important for identifying the energy releasing process. Reimann et al. (1984a) showed that the electronic excitations leading to sputtering had a similar diffusion length as those that produce luminescence from the exciton Ar$_2^*$. This observation makes it plausible that this transition (the M-band in Fig. 12) is responsible for the emission of the photons as well as the non- radiative ground state repulsion. This deexcitation scheme has been convincingly demonstrated by Pedrys et al. (1988) and O'Shaughnessy et al. (1988) who observed a peak that corresponds to this ground state repulsion, in the energy distribution of sputtered xenon and krypton atoms.

A survey of the energy dependence of the sputtering yield for solid argon and a few data points for polycrystalline sulphur are shown in Fig. 13. The difference in the magnitude of the yield is primarily determined by the surface binding energy. Chrisey et al. (1988) derived a value $U_o = 0.19$ eV that is much larger than the value 0.060 eV determined for solid argon by Pedrys et al. (1988). For values of the electronic stopping cross section below 70 eV/10^{15} Ar/cm^2 the yield is almost proportional to the stopping cross section. This corresponds to energies above the maximum of the stopping power peak. At energies considerably below the maximum the yield may be approximated by a linear function of the stopping cross section as well, but with a larger efficiency than above the maximum. This feature has not yet been explained satisfactorily. At the highest excitation densities for energies above the stopping power maximum the yield increases faster than linearly. Reimann et al.(1984a) and (1988) suggest that the excitations interact at high excitation densities. The early attempts to describe the yield by a quadratic dependence on the stopping power similar to that for water ice (Section VI-B) turned out to be misleading.

The enhancement of the yield has to be considered in connection with the observations of large yields for heavy ions incident on a volatile solid as for example solid argon and krypton in the keV regime (Boring et al. (1987)). The yields of the order 10^3 as well as the approximative quadratic dependence of the yield on the surface value $F_D(E,0,1)$ are indicative of an elastic collision spike. Therefore, the analogy leads us to the straight-forward concept, an **electronic** spike, which occurs for high electronic excitation densities. However, the details in the development of the spike for elemental materials as well as for the chemical compounds considered in Section VI is not understood for the moment.

Crisey et al. (1988) suggest that the yields of sulphur are quadratic rather than linear functions of the electronic stopping cross section. This relationship which has been derived on the basis of few points, indicates cooperative effects between the electronic excitations or overlap of the low-energy cascades.

The ejection of particles is not initiated by cascades produced by the incoming particle of energy E, but by low-energy cascades produced by the average energy release E_s per electron-hole pair. It means that the energy E_s for electronic sputtering corresponds to the primary energy E for knock-on sputtering.

The existing spectra from elemental frozen gases do not show a pronounced E_1^{-2}-behaviour as for example those shown in Fig. 10. The reason is that the energy release E_s does not exceed a few eV, since the complete collision-cascade behaviour appears in the limit of large ratios of E_s/E_1. Nevertheless, the existing spectra may be approximated quite well by a linear collision- cascade in the lower part of the spectrum (Pedrys et al. (1988)).

In some cases electronic and knock-on sputtering take place simultaneously, and it has been possible occasionally to determine the relative contributions of electronic and knock-on sputtering (Stevanovic et al. (1984), Chrisey et al. (1986) and (1988), and Schou et al. (1988)).

Ellegaard et al. (1986) derived an expression for the electronic sputtering yield in the absence of mobile excitations. The yield Y is determined by

$$Y = \Lambda \frac{1}{2} D_e(E, 0, \cos\theta)(E_s/W) \tag{IV.17}$$

where $D_e(E,0,\cos\theta)$ as usually is the surface value of the electronically deposited energy, E_s the non-radiative energy release from electronic deexcitations per electron-hole pair, and W the energy required to make an electron-hole pair (Schou (1987)). Λ is the material constant that enters the knock-on case as well (Eqs. (IV. 4-5) and (7-8)). The main assumption is that the liberated energy

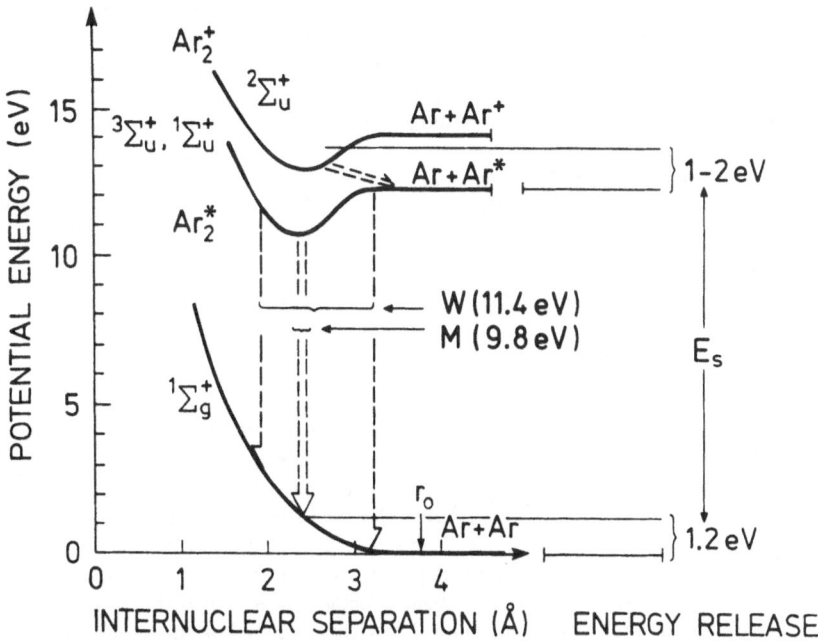

Fig. 12. Schematic representation of the important transitions in solid Ar. The energy levels are from the solid. The transitions that may lead to energy release are indicated by dashed arrows.

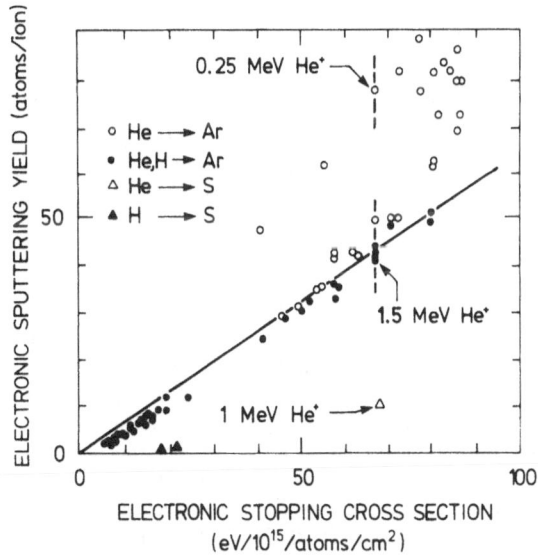

Fig. 13. The electronic sputtering yield for light ion bombardment of solid Ar and S. The solid line is merely drawn to guide the eye. The dashed line indicate the value for energies on both sides of the stopping power maximum. ○, Besenbacher et al. (1981), ●, Reimann et al. (1988), △, Torrisi et al. (1986), and ▲, Chrisey et al. (1988).

E_s is sufficient to initiate a low-energy cascade. Since these non-radiative transitions are completely isotropically distributed (as shown in Fig. 5), one arrives of the factor 1/2 in Eq. (IV-17). The formula is based on the assumption that the energy E_s is released in one event, for example by two repulsing atoms each with the energy 1/2 E_s, but extensions of Eq. (IV-17) are possible. A similar expression is indicated by Brown and Johnson (1986) and Garrison and Johnson (1985).

For the simple case of a light MeV-ion one has as a very good approximation that the surface value $D_e(E,0,1)$ is equal to the electronic stopping power $NS_e(E)$. For electrons the surface value is somewhat larger than $NS_e(E)$, typically a factor of 1.5-2 for the light materials up to solid argon (Schou (1988)).

In analogy to Eq. (IV-6) one may express D_e as a product of a dimensionless quantity β and the electronic stopping power (Schou (1980)).

$$D_e(E,0,1) = \beta \, NS_e(E) \tag{IV.18}$$

Similar to α β is a slowly varying function of the primary energy.

Equation (IV-17) and the corresponding expression for the knock-on sputtering yield enable us to estimate the efficiency of the processes, i.e. how well are the energy deposition F_D and D_e exploited for the production of sputtered particles. Electronic sputtering is clearly a factor of 1/2 (E_s/W) less efficient than knock-on sputtering. For solid nitrogen this quantity is about 0.04. Actually, the simple formula (IV-17) is inapplicable to materials with mobile excitations, for example, as solid rare gases. Nevertheless, it provides us with an estimate of the order of magnitude. The yield estimated in this way is about a factor of 2 less than the experimental data for energies above the stopping power maximum. The yield from the low-energy side of the maximum is considerably larger than the high energy data. Besenbacher et al. (1981) as well as Reimann et al. (1988) point out that the discrepancy between the high-energy and the low-energy data may be partly explained by considering the charge states of the projectiles.

Although Eq. (IV-17) predicts the electronic sputtering yield in a feasible manner, there are obvious limitations in its applicability. The formula applies only for low excitation densities, where the sputtering is produced by individual non-radiative transitions without overlap of the low-energy cascades. The lack of sufficient primary data for the electronic deexcitations, e.g. E_s, makes it often difficult to utilize Eq. (IV-17) even at the low excitation densities. This is the reason why no estimate of the yield for sulphur has been made. Furthermore, recent measurements on sulphur by Torrisi et al. (1988) may indicate that the sputtering is not induced at low excitation densities, but only at high excitation densities. This high excitation regime corresponds to the primary energies for which a quadratic dependence of the yield on the stopping power is observed.

A few remarks on **electronically induced desorption** may now be appropriate. Although desorption takes place for adsorbates of material deposited on a different substrate, there are obvious similarities between electronic sputtering of an elemental condensed gas and desorption of monolayers of the gas deposited on a metal. The important models for desorption, for example the MGR-model, were deloped for ion emission rather than for emission of neutrals (Gomer (1983) and Menzel (1986)). This Menzel-Gomer-Redhead model was modified by Antoniewicz (1980) in such a way that the emission of neutral desorbed particles could be explained as well. In this model the transfer of energy from electronic excitation to atomic motion occurs by neutralization of an ionized adsorbed atom. The initial excitation leads to formation of an ion, which moves inward (Fig. 5), makes a transition to a neutral weakly bonding curve and is then desorbed as a neutral atom. In electronic sputtering the parallel case is the dissociative recombination of the trapped molecular ion, eg. Ar_2^+. This process in an insulating multilayer requires a hot electron, which may be a photoelectron, a secondary or a free substrate electron. The cascade multiplication, however, is a clear indication of sputtering rather than desorption of repulsing species. The cascade processes are dominant in electronic sputtering of volatile materials as, for example solid argon, whereas the particle ejection for electronic sputtering of the less volatile xenon largely may be regarded as a desorption-like process (Pedrys et al. (1988) and O'Shaughnessy et al. (1988)).

V. KNOCK-ON SPUTTERING FROM MULTICOMPONENT MATERIALS
V-A. Introduction

Sputtering from multicomponent targets exhibits a number of features that do not occur for elemental targets. Of course, the yield of a component depends on the concentration of that particular component, but also the presence of the other components may influence the yield. A complicating feature is that the surface concentration needs not to be identical to that of the bulk, and that the primary beam immediately induces changes in the surface layers.

For high fluences the composition of the sputtered flux has to become identical to that of the bulk because of mass conservation. However, even though a stationary situation with regard to the sputtered flux or the signals of the surface concentration sooner or later may be reached, the sputtering of multicomponent targets is characterized by transients in the composition of the flux or in the surface region. This means that **preferential ejection** of some of the components takes place. A particularly interesting situation occurs, if the surface layer from which the sputtered particles originate, becomes enriched in one of the components compared to the bulk composition.

The main features of sputtering from multicomponent systems have recently been reviewed by Shimizu (1987), Andersen (1984) and Betz and Wehner (1983). These reviews contain a comprehensive data collection in addition to previous reports on sputtering of multicomponent systems by Coburn (1979) and Kelly (1978), (1979), (1980a) and (1980b).

Figure 14 shows an example of multicomponent sputtering from Andersen et al. (1984). A homogenous CuPt alloy was bombarded by keV argon ions. This measurement is particularly interesting, since the sputtering yield of the two components was determined in contrast to many similar experiments in which only the depletion of one or more components in the surface was studied. One notes the transient in the sputtering yields up to a thickness of 10^{18} atoms/cm^2, and that the partial yield of copper is much larger than that of platinum during the initial erosion. The ratio of the yields does not reach unity as expected because the collection angle might not have been fully representative of the average flux according to the authors. However, the stoichiometric ratio was indeed obtained in a number of related measurements (Andersen et al. (1984)). The layer which has to be sputtered away to reach the stationary case is not directly correlated to the primary energy or the particle range. Nevertheless, one notes that the transient for the lowest energies is somewhat smaller than that for the highest energy.

Preferential ejection includes **preferential sputtering** that occurs whenever the composition of the sputtered particles differs from that of the outermost layers of the target. The sputtered particles originate mainly from the outermost 5 Å ($\sim 2.5 \cdot 10^{15}$ atoms/cm^2). Andersen (1979) and (1984) pointed out that preferential sputtering may be identified relative to concentrations in these outermost layers rather than the bulk.

In the following let us apply index A, B, C... for the properties of the components. Preferential sputtering for a binary medium occurs whenever

$$Y_A/Y_B \neq \bar{c}_A/\bar{c}_B \qquad\qquad (V.1)$$

where the concentration values $\bar{c}_A$ and $\bar{c}_B$ are properly weighted averages over the depth of origin of the sputtered particles (Andersen (1984)). Y_A, Y_B are the yield of A-atoms and B-atoms, respectively. A major disadvantage by this criterion is that the distribution of the depth of origin is not accurately determined for any system. On the other hand, it allows us to simplify the discussion of preferential sputtering to the partial yields and the concentration of the relevant surface layers. An even more detailed scheme has been proposed by Sigmund et al. (1982).

A major difficulty in identifying preferential sputtering rather than preferential ejection originates from the experimental technique that has been used for surface characterization. The depth sensitivity of these techniques is shown in Fig. 15. Low energy ion scattering spectroscopy (ISS) traces the very surface, Auger electron spectroscopy (AES) samples the composition over a depth that may considerably exceed the depth of origin. Rutherford backscattering spectrometry (RBS) yields the composition up to a depth more than two orders of magnitude greater than that of the origin of the sputtered particles. Secondary ion emission spectrometry (SIMS) yields information

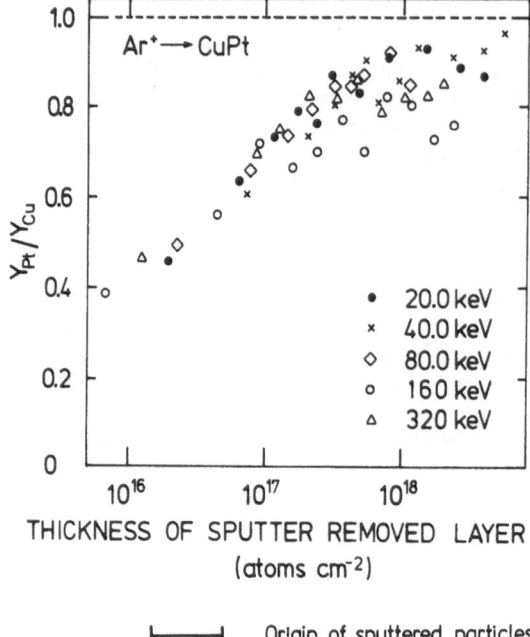

Fig. 14. The ratio of the partial sputtering yields Y_{Pt}/Y_{Cu} versus the thickness of the sputtered layer for Ar^+-ion bombardment of CuPt alloys.

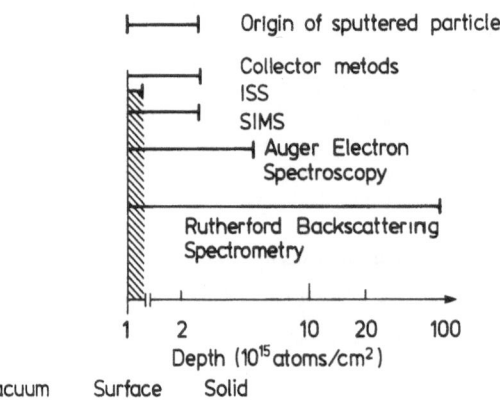

Fig. 15. The information depth for different techniques used to measure changes in the surface composition. From Betz and Wehner (1983).

on the composition corresponding to the depth of origin but the ionization probability for the ejected particles varies strongly from one material to another. In addition, the initial concentrations may be changed by the particle bombardment in a subsequent SIMS-profiling study. Therefore, it is not surprising that many of the surface studies led to deviations from each other in the composition. Several authors (Andersen (1984) and Shimizu (1987)) have pointed out that a strong concentration gradient may occur, so that the observed surface composition or the apparent concentration profiles depend on the method that has been utilized.

It is obvious that preferential ejection is not caused by preferential sputtering alone, if other chemical or beam-induced mechanisms change the surface concentration of one or more of the components relative to the bulk concentration. Experiments in which the concentration profiles of a compound sample differ from that of the bulk (e.g. Liau et al. (1978)), show only that an efficient transport process of matter out of or deeper into the solid is operative. The processes other than preferential sputtering will be discussed in the last part of this section.

Usually, a total yield Y

$$Y = \sum_j Y_j \tag{V.2}$$

is defined in the literature (Betz and Wehner (1983)). However, this sum of partial yields is feasible alone in the cases for which the partial yields are measured simultaneously.

V-B. Expectations for preferential sputtering

One might expect intuitively that the partial yield Y_A from a multicomponent target is expressed as

$$Y_A = c_A Y_{AA}$$

where Y_{AA} is the corresponding yield from an elemental A-target. We consider the simple case of a constant surface concentration $\bar{c}_A = c_A$. Of course, the equation is valid for the two extremes $c_A = 0$ or $c_A = 1$ but for all other values the internal flux deviates frequently from that of the pure element. The energy sharing of the constituents of a compound was systematically investigated by Andersen and Sigmund (1974). On the basis of this treatment the ratio of the partial yields from a binary compound becomes

$$Y_A/Y_B = f_{AB}(c_A/c_B) = (M_B/M_A)^{2m}(U_{BM}/U_{AM})^{1-2m}(c_A/c_B) \tag{V.3}$$

f_{AB} is an enhancement factor, M_A the atomic mass, and U_{AM} the value of the surface binding energy for an atom A in the mixture. (The index 0 will be omitted in this section). Equation (V-3) presented by Sigmund (1981) and discussed in Sigmund (1980) predicts:

i) the lightest component to sputter preferentially for equal surface binding, and
ii) the component with the smallest surface binding to sputter preferentially for equal masses.

However, since recent investigations indicate that $m \approx 0.1$ (Vicanek et al. (1988)) the non-stoichcometric effect is much more pronounced for different surface potentials than for differences in mass. Nevertheless, the mass factor is responsible for preferential sputtering of isotopes for which the surface binding is nearly equal (Sigmund (1987b)).

For an elastic spike Sigmund (1981) obtained a different expression for the ratio of the partial yields, although the basic trend is similar. However, the treatment did not include any dependence on time or lateral distance to the point of impact as for example Sigmund and Claussen (1981) and Sigmund and Szymonski (1984).

V-C. The dependence on the mass

The mass dependence of the partial yields was recently examined by Sigmund (1987b). Although the results are based on a treatment restricted to small mass differences, the conclusions about the dependence on the mass are indeed appropriate for the general case as well. The procedure was encouraged by the observation that isotope effects vary approximately linearly with the mass (Olson et al. (1979)).

Let us consider the mean mass

$$M = \sum c_i M_i \quad (i = A, B, ...) \tag{V.4}$$

so that

$$M_i = M + \Delta M_i . \tag{V.5}$$

The important feature is now that the efficiency of the energy transfer γ_{AB} (Eq. (II-7)) is unity up to first order in the mass differences.

The cross section for an A-atom to hit a B-atom is correspondingly:

$$d\sigma_{AB}(E,T) = \left(1 + m\frac{\Delta M_A - \Delta M_B}{M}\right) d\sigma \, , \, 0 < T \le E \tag{V.6}$$

where $d\sigma$ is the usual power cross section for a monatomic medium (IV-9). Equation (V-6) holds only for isotopes of the same element, but the extension to neighbor elements was included in the treatment (Sigmund (1987b)).

Let us now assume that the A-atom is **lighter** than the B-atom, so that

$$\Delta M_A < \Delta M_B \tag{V.7}$$

This means that

$$d\sigma_{AB} < d\sigma_{BA} \tag{V.8}$$

This inequality leads to two simple statements:

i) Once in motion the light atoms experience a smaller stopping power than the heavier ones do, so that they travel longer than the average, and

ii) an atom of low mass has a higher probability of being set in motion than heavier ones. It turns out that both effects contribute equally to preferential sputtering.

Before going on let us consider the slowing-down flux $G(E,E_o)$ in a monatomic medium. G is defined as the average number of atoms moving in an energy interval $[E_o, E_o + dE_o]$ generated by a constant flux ψ of projectiles per unit time of initial energy E. This flux is determined by

$$G(E,E_o) = \frac{\psi}{v_o \, NS_n(E_o)} \int_{E_o}^{E} F(E,E')dE' \tag{V.9}$$

where $F(E, E')\,dE'$ is the recoil density, i.e. the mean number of recoiling atoms generated with an energy in the interval $[E', E' + dE']$ by a primary of energy E (Sigmund (1981)). In the monatomic medium the recoil density was determined as a standard result as

$$F(E,E_o) \approx \nu(E)\Gamma_m E_o^{-2} \tag{V.10}$$

In the following the mass M_A need not to be smaller than M_B. For multicomponent targets the slowing down flux of B-atoms generated by a primary flux of A-atoms per unit time of energy E becomes

$$G_{AB}(E,E_o) = \frac{\psi}{v_{oB} \, NS_{n,B}(E_o)} \int_{E_o}^{E} F_{AB}(E,E')dE' \tag{V.11}$$

where $v_{OB} = (2\,E_o/M_B)^{\frac{1}{2}}$ is the velocity of the moving B-atoms and

$$NS_{n,B}(E_o) = N\left(\frac{M_B}{M}\right)^m S_n(E_o) \tag{V.12}$$

the nuclear stopping power for a B-atom moving in the isotopic mixture. S_n is the usual nuclear stopping cross section for the monatomic medium with atoms of mass M. The recoil density F_{AB} of B-atoms set in motion with an energy around E' by an A-atom of energy E is expressed as

$$F_{AB} = c_B \left(\frac{M}{M_B} \right)^m F(E, E_o) \qquad (V.13)$$

up to first order in ΔM for the isotopic mixture. $F(E, E_o)$ is the usual recoil density for a monatomic medium.

We note that the internal flux G_{AB} receives equal contributions $(M/M_B)^m$ from the slowing-down and from the production. The result derived by Andersen and Sigmund (1974) for targets of binary components

$$\frac{v_{OA} G_{jA}}{v_{OB} G_{jB}} = \frac{c_A}{c_B} \left(\frac{M_B}{M_A} \right)^{2m} , \qquad (V.14)$$

emerges from Eqs. (V-11), (V-12) and (V-13).

Before completing the discussion let us shortly regard the dependence on the primary particle. We note that the properties of the primary particle solely enters into the recoil density F via the energy $\nu(E)$ deposited into atomic motion. In the elastic collision region $(\nu(E) \simeq E)$ only the primary energy E enters. This means that the fluxes could have been initiated by any primary and not only by A-atoms, and that for example

$$G_{AB}(E, E_o)dE_o = G_{1B}(E, E_o)dE_o \qquad (V.15)$$

corresponding to the usual incoming atom of mass M_1 and atomic number Z_1. Equation (V-15) was a main result in Andersen and Sigmund (1974).

The prediction of a preferential flux of the lightest component in isotopic mixtures or near-neighbor mixtures has been difficult to trace in experiments because of the possible influence of the mass dependence on the ionization probability. Only low-fluence experiments or computer simulations are feasible in this context. Shapiro et al. (1988) indicate a much stronger dependence on the mass than predicted by Sigmund (1987b) on the basis of Monte-Carlo simulations for artificial copper, but the controversy is not yet resolved (Andersen (1988)).

V-D. The dependence on the surface binding energy

Let us consider the simple case of slightly different masses and atomic numbers. Only small deviations ΔZ_j from the mean atomic number Z are included in analogy to Eqs. (V-4) and (V-5).

The recoil density is now

$$F_{AB} = c_B \left(\frac{M}{M_B} \right)^m \left(\frac{Z_B}{Z} \right)^{b_m} F(E, E_o) \qquad (V.16)$$

where b_m is a constant that depends on m alone. The stopping power for a B- atom in motion is

$$N S_{n,B}(E_o) = \left(\frac{M_B}{M} \right)^m \left(\frac{Z_B}{Z} \right)^{b_m} N S(E_o) \qquad (V.17)$$

similar to Eq. (V-12). The partial yield Y_B is determined by an integral over the product of the slowing-down density $G_{AB}(E, E_o)$ and the velocity v_{OB} over all possible escape directions according to the treatment in Sec. IV. The apparently redundant inclusion of the dependence on the atomic number in Eqs. (V-16) and (V-17) serves to demonstrate that the recoil density $F(E, E_o)$ entering the yield evaluation is the standard density for a monatomic medium characterized by the mean quantities M and Z. We note that the dependence of the slowing down density G_{AB} on the atomic number cancels out, and that the surface value $F_D(E,0,\cos\theta)$ of the deposited energy emerges from the recoil density $F(E,E_o)$ (Sigmund (1981)). In the elastic-collision region it means that the density and the surface value are identical for, for example $G_{AB}(E,E_o)$ and $G_{BA}(E,E_o)$. The ratio of the two partial yields becomes then

$$\frac{Y_{BM}}{Y_{AM}} = \left(\frac{c_B}{c_A}\right)\left(\frac{M_A}{M_B}\right)^{2m}\frac{NC_m U_{AM}^{1-2m}}{NC_m U_{BM}^{1-2m}} \tag{V.18}$$

where the last fraction merely is the ratio of the material constant Λ for the two values of the surface binding energy. Therefore, the factor $(U_{AM}/U_{BM})^{1-2m}$ in the yield ratio is a simple consequence of the different surface potential.

A similar relationship emerges from most of the existing treatments. Kelly and Harrison (1985) present a quantity $(U_{AM}/U_{BM})(M_A/M_B)^{2m}$ $(= f_{BA})$ for which the exponent m is set equal to zero for the surface binding factor, but not for the mass factor.

The problem of determining the surface binding energy for the constituents in a multicomponent target is by no means trivial. In general, one may expect that the binding energies of the component depend on the composition (Andersen (1984)). Thermodynamic arguments encouraged Kelly (1978) and (1980b) to express the surface binding energies for a random binary alloy as a linear combination of the internal nearest-neighbor bond strengths. For a few alloys one may avoid such considerations by evaluating the binding energy U_{AM} from experimental energy distributions, provided that the sputtering takes place in the linear collision-cascade regime. Szymonski (1980a) found for a Cu-Zn alloy that the binding energy for the most volatile element zinc increased relative to the elemental value and that the value for copper decreased similarly. A similar observation was made by Oechsner and Bartella (1980) for Ni-W alloy and by Schorn et al. (1988) for a Cu-Li alloy. A systematic study of the binding energy for chromium atoms in different matrices has been performed by Husinsky et al. (1987) and by Wucher and Oeschner (1987) for oxidized niobium and tantal surfaces.

An isolated test of the dependence on the surface binding energy as predicted by Eq. (V-3) seems difficult. The reason is that a low surface binding energy is expected to lead to surface segregation of the pertinent component. In such a case preferential ejection may be caused by the enrichment of the segregated component at the outermost surface layers rather than by preferential sputtering.

V-E. Light ion sputtering of multicomponent materials

For light ion bombardment surface enrichments have been observed especially at low ion energies (Roth (1986b)). This is a consequence of the strong difference in the sputtering yield for different target masses as the ion energy approaches the threshold as described in Sec. (IV-C). These features have been demonstrated for example by Varga and Taglauer (1982) or Roth et al. (1980) in experiments as well as by computer simulations by Eckstein and Biersack (1985) and Eckstein and Möller (1985).

V-F. Information from the angular distribution

It was pointed out by Sigmund and coauthors (1982) that the concentration profile may influence the angular distribution. A surface layer enriched in one component is sputtered in the entire half-space, while atoms originating from deeper layers are preferentially emitted in a narrow cone around the surface normal. This effect has been exploited by Andersen et al. (1983) and (1984) and Ichimura et al. (1981) for studies of ion-induced segregation.

V-G. Beam-induced effects

The surface is immediately modified by even a short ion bombardment. Preferential sputtering is only one out of a variety of processes (Andersen (1984), Wiedersich (1985)). An altered surface layer can be formed during ion bombardment by a combination of the different processes. The surface concentrations $\bar{c}_A$ and $\bar{c}_B$ of the two components in a binary material change until a balance between the sputtering loss and the possible mechanisms of particle transport to the surface has been reached:

$$\bar{c}_A/\bar{c}_B = (1/f_{AB})(c_A/c_B) \qquad\qquad (V.19)$$

The typical effects are shown schematically in Fig. 16.

Recoil implantation is as sputtering a collisional effect. This effect alone means that the surface becomes depleted in the lightest component at high ion energies. The atoms becomes implanted to larger depths. The theoretical treatments have been carried out by Sigmund (1979) and (1988b).

Cascade mixing takes place as a result of the random motion of the higher- order recoils within the collision cascade. This collisional effect has been comprehensively discussed by Andersen (1979), Hofer and Liffmark (1979) and Sigmund and Gras-Marti (1980). Andersen (1984) argues that the mixing is not an efficient feeding mechanism for preferential sputtering of a component, because the displacement energy E_d is one order of magnitude larger then the surface binding energy. This effect is primarily important for the determination of the depth resolution.

Diffusion may be an important process in supplying the lightest atoms to the surface layers from the entire damaged region. Ordinary diffusion constants depend only on temperature and not on the irradiation conditions (Andersen (1984)). Ho (1978) concluded that the diffusion was substantially enhanced by the irradiation. This radiation enhanced diffusion (RED) implies that the radiation enhances the concentration of mobile vacancies considerably over the thermodynamic equilibrium value. Consequently, the diffusion coefficient is increased over a depth comparable to the range of the projectiles, and the enhancement is proportional to the flux. In most of the experiments the striking increase of the diffusion coefficients was studied (Eltoukhy and Greene (1980) and Rivaud et al. (1982)). At elevated temperatures competing transport mechanisms may occur Shimizu (1987) and Rehn et al. (1985).

Surface segregation (Gibbsian segregation) plays an important role in providing the surface with atoms in order to replace the preferentially sputtered component. The difference in surface binding energies constitutes the driving force of the particle transport, since the system gains energy if a strongly bound atom at the surface exchanges its position with a weakly bound one from the interior. Usually the loosest bound component will be enriched in the first one or two atomic layers (Andersen (1984)). The systematics have been discussed by for example Abraham and Brundle (1981).

The pronounced defect production during ion bombardment means that the defect migration processes are important. The same defects which are responsible for RED can also produce **radiation induced segregation** (RIS). In general, vacancy and/or interstitial defects in alloys preferentially migrate via particular alloying elements. Because of this preferential coupling of some alloying elements to the defect fluxes, certain elements will be swept into and other elements out of local regions which experience a net influx or outflow of defects. In this manner, RIS can generate concentration gradients during irradiation of homogeneous alloys even in the absence of preferential sputtering and Gibbsian segregation. The effects have been described in several reviews by Johnson and Lam (1978), Lam et al. (1978), Kirschner (1985), and Rehn and Lam (1987).

These effects which may operate simultaneously, participate in a complex interplay. This has encouraged theoretical studies (Lam and Wiedersich (1987)), in which these effects have been incorporated. However, the complexity means that the task of identifying different contributions to preferential ejection is non-trivial.

VI. ELECTRONIC SPUTTERING OF MULTICOMPONENT MATERIALS
VI-A. Introduction

Electronic sputtering is important for insulating multicomponent materials as well as for elemental insulators. However, the presence of different excited species in a solid target during irradiation leads to a high chemical activity that has not been observed for elemental targets. The formation of new radicals or even stable molecules that were not present in the original matrix, makes it difficult to identify the deexcitation channels. Let us classify the targets into four groups

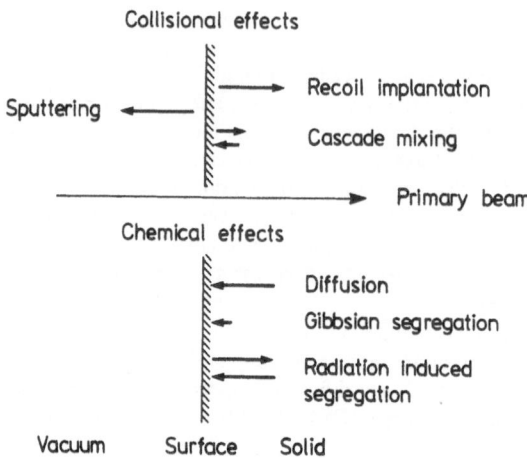

Fig. 16. A survey of the different processes during ion bombardment. The design is taken from Andersen (1984).

i)   Frozen gases (H_2O, SO_2, CO...)
ii)   Alkali halides,
iii) Room temperature oxides etc. (SiO_2, Al_2O_3...) and
iv)   Organic materials.

A common feature is the luminescence during or after electronic irradiation. The competition between photon emission and non-radiative transitions which supply kinetic energy for defect production or sputtering, is just as pronounced for these materials as for solid argon considered in Sect. IV. The possible mechanisms for electronic sputtering are so different that the groups will be discussed separately. The organic materials will not be included here. The erosion of these materials has been reviewed by Sundqvist and Macfarlane (1985), Johnson (1987), Sundqvist (1989) and Wien (1989).

VI-B. Sputtering of condensed heteronuclear gases

The group of ices from room temperature liquids or gases has been comprehensively studied because of the relevance for erosion of dust grains or planetary surfaces in space (Johnson et al. (1984) and Rössler (1986)). In particular, the measurements have been concentrated on water, sulphur dioxide, carbon monoxide and methane ices. The mechanisms for electronic sputtering or the production of electronic defects in the ices are not well- known, and large invidual differences are expected for the deexcitation channels in the materials. Nevertheless, sputtering of water ice exhibits a number of features that are representative for the condensed gases.

The sputtering yield of ice is shown in Fig. 17. This figure includes data obtained with protons as well as medium-light ions up to flourine ions. The curves show a plateau at the energy, where the nuclear stopping power is important. The erosion process in this regime is knock-on sputtering. The interesting feature is the peak of the yield for proton energies about 100 keV and for the heavier ions about 200 keV/amu. The behaviour is similar to that of the electronic stopping power, and we note that the maximum yield of electronic sputtering is one or two orders of magnitude larger than the yield from knock-on sputtering, dependent on the projectile.

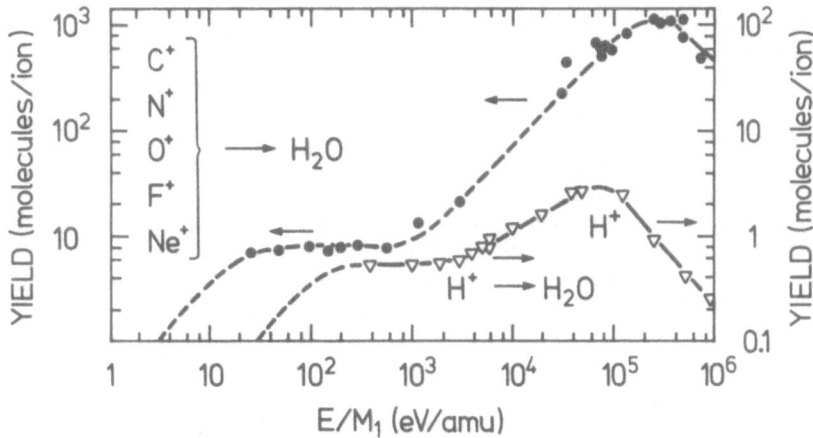

Fig. 17. A survey of the sputtering data for water ice bombarded by H^+, C^+, N^+, O^+ and Ne^+-ions. From Johnson (1987). The lines are drawn to guide the eye.

Some of the main features of electronic sputtering of water ice are summarized in the following:

1) The electronic yield is proportional to the square of the electronic stopping power for light ions. For medium light ions the yield increases even faster than the square of the electronic stopping power.

2) The mass spectrum of the sputtered particles shows the parent molecule as well as a number of other molecules or atoms. Radicals and molecules of a composition different from the ordinary fragmentation pattern have been observed.

3) The energy spectrum of the sputtered particles shows an E_1^{-2}-behaviour for the high-energy tail in many cases, but the maximum is at energies considerably below one-half of the sublimation energy.

4) The sputtering yield does not depend on the ice film thickness.

5) Luminescence from radiative transitions of impurities or of the host molecules in the ice has been observed.

The **stopping power dependence** for light ions was observed by Brown et al. (1980a). The data for MeV protons and helium ions indicate a clear quadratic dependence on the stopping power whereas data with MeV fluorine ions from Seiberling et al. (1982) and Cooper and Tombrello (1984) show a dependence which is proportional rather to the forth power of the stopping power than the square. The quadratic dependence means that cooperative effects between the individual electronic excitations are important (Brown and Johnson (1986), Schou (1987), and Johnson (1987)).

The **mass spectrum** shows a dominant peak for H_2O, but the yield of hydrogen molecules is comparable (Haring et al. (1984a)). The relative amounts of the species depend on the type of primary ion. Even particles which do not occur via a simple decomposition as for example molecular oxygen, were observed. Reimann et al. (1984b) studied transients in the mass spectra and the influence of the substrate temperature on the transients. The mass spectrum of the ice reflects the high chemical activity between the electronically excited atoms or molecules.

The **energy spectrum** is for all species peaked at an energy much less than the sublimation energy of about 0.5 eV. Reimann et al. (1984b) found that the emitted heavy water molecules resulting from MeV helium ion bombardment surprisingly well could be described by a linear collision-cascade behaviour (Eq. (V-12)) with a surface binding energy of $U_0 \approx 0.05$ eV. A similar observation of a maximum at very low energy for keV hydrogen ion bombardment was observed by Haring et al. (1984b). These authors observed as well a high-energy tail from emitted oxygen molecules, that may be represented by an E_1^{-2}-tail. All these observations indicate that a low-energy collision-cascade may be responsible for the particle ejection. The process that initiate this cascade is not known, but a dissociative transition of a radical or a molecule formed of species from two neighbor excitations is a possible candidate. The maximum for the energy distribution of ejected water molecules is shifted to an energy much lower than one-half of the sublimation energy because of energy loss to vibrational excitation.

The **yield dependence** on the **thickness** was measured by Brown et al. (1980b) with Rutherford backscattering spectrometry. They did not observe any variations in the yield from thicknesses of 400 Å up to 2000 Å. However, recent results based on a complicated procedure for infrared absorption by Rocard et al. (1986) and Benit et al. (1987) may indicate a thickness dependence of yield.

Luminescence induced by electron bombardment has been observed in the visible region by Trotman et al. (1986). The strongest peak has been identified, but the origin of all the features is not yet determined. It turned out that guest atoms or molecules within an impurity level even far below one per mille has been responsible for many transitions that previously were attributed to the host molecules. This impurity-induced radiation is well known from the solid rare gases and nitrogen (Schwentner et al. (1985), Zimmerer (1987) and Oehler (1977)).

Many of the condensed gases investigated show trends different from these features. In particular, the mass spectrum of sulphur hexaflouride, (Pedrys et al. (1984)) and methane (Pedrys et al. (1986) and Brown et al. (1987a)) shows oscillations in the relative abundance as a function of the atomic concentration of the lightest component. In solid methane transients in the mass spectra as well as carbon enrichment were observed (Brown et al. (1987b) and Foti et al. (1987)).

These considerations demonstrate that the theoretical treatment is difficult, and that one may not expect a common "universal" mechanism for the sputtering of condensed heteronuclear gases.

The individual differences in the electronic excitations lead to substantial differences in the behaviour of the electronic sputtering yield. Nevertheless, some attempts on comprehensive treatments have been performed, for example by Watson and Tombrello (1985). The current status of the theories has recently been reviewed by Johnson (1987).

VI-C. Sputtering of alkali halides

The electronic sputtering for alkali halides is a well-established feature, although the deexcitation mechanism is not known in greater detail. The sputtering process as well as the deexcitation mechanism have been described by Itoh (1976) and (1987), Townsend (1983), Townsend and Lama (1983), Szymonski and de Vries (1983), Williams et al. (1986), and by Avouris et al. (1987).

The yield induced by electron bombardment has turned out to be of a considerable magnitude compared with the corresponding yield from for example condensed gases (Sect. VI-B). Szymonski et al. (1985) measured a yield of 14 molec./electron for sodium chloride bombarded by 0.5 keV electrons with normal incidence for a target temperature of about 600 K. The alkali and halogen atoms are emitted independently, and the yield in units of molecules per electron is merely a convenience for comparisons. Yields of similar magnitude from other alkali halides have been reported by Townsend (1983).

An interesting feature for alkali halides is the strong competition between luminescence and sputtering. An example of 0.5 keV electrons incident on sodium chloride from Szymonski et al. (1988) is shown in Fig. 18. One notes that the luminescence and the sputtering show a simultaneous enhancement and reduction with decreasing temperature. The luminescence originates from recombination of an F-center (an electron trapped of a vacant halogen site) and an H-center (a molecular ion located in a single-halogen site). Migrating H-centers may reach the surface and induce particle ejection by relaxation at the surface. Results similar to those in Fig. 18 have been obtained previously by Townsend et al. (1976).

Electronic sputtering of alkali halides is closely related to defect production. In contrast to condensed gases or many other materials the behaviour of these electronic defects is relatively well

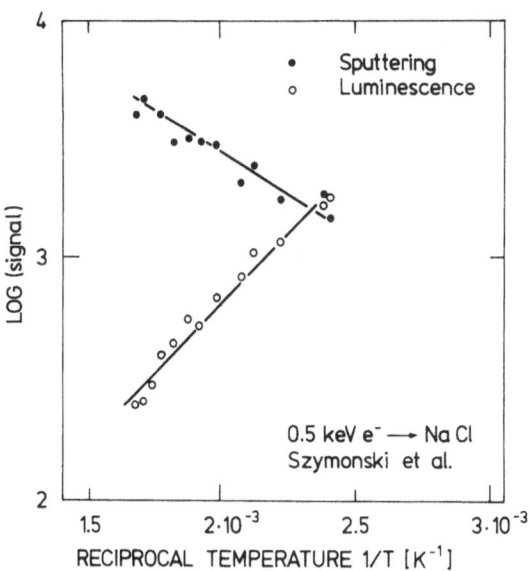

Fig. 18. Simultaneous data for sputtering and luminescence in arbitrary units from (100) NaCl bombarded by 0.5 keV electrons versus the reciprocal temperature. The crystal was bombarded at normal incidence and by a beam of current density from 22 to 200 µA/cm². The sputtering yield was determined at temperatures where Y(Na) = Y(Cl). The luminescence yield was determined from the broad band around 400 nm (Fig. 2 in Postawa et al. (1987)). From Szymonski et al. (1988).

known (Perez (1976), Itoh (1976), Townsend (1983) and Thevenard (1989)). The electronic excitations in the halogen sublattice are mobile, and the ejection of the halogen atom is driven by the electronic deexcitation at the surface.

The alkali halides show the remarkable feature that electronic sputtering takes place for the halide component alone. The surface becomes gradually enriched with the alkali atoms, and the erosion may cease eventually, unless the temperature (for example produced by a high beam current density) is so high that the alkali atoms evaporate just as fast as the halogen atoms get sputtered. Therefore, electronic sputtering takes place solely under certain circumstances, e.g. at a high target temperature, and it is meaningless to introduce a sputtering yield for alkali halides, unless the temperature and the current density is specified. At low temperatures the sputtering is determined largely by the evaporation rate of the metal.

As mentioned above the halogen and the metal atoms are emitted as atoms rather than as molecules. A weak signal of alkali halide molecules was observed only for crystals of CsI-structure. The emission of diatomic halogen molecules was reported by Szymonski et al. (1985) never to exceed a contribution of about one-tenth of the total emission.

The energy distribution of the emitted particles is a Maxwell-Boltzmann distribution corresponding to the target temperature for the metal atoms as well as for a considerable fraction of the halogen atoms. At low target temperatures (< 400 K) a clear non-thermal component occurs in the energy spectrum of the halogen atoms ranging from 0.1 eV to 0.5 eV (Overeijnder et al. (1978)). Szymonski (1980b), and Szymonski and de Vries (1983) suggested that the non-thermal part of the yield originates from focused replacement sequences along the <110> row of the crystal. Then the ejection of the halogen atom takes place when the long range movement of H-centres along <110> directions intersects the surface. The authors suggested that the thermal component of halogen atoms originated from thermally diffusing H-centres (Itoh (1976)) after termination of the replacement sequence. If the migrating centre reaches the surface, the halogen atom can evaporate and contribute to the thermal component of the sputtering.

Since the cohensive energy of the atoms in the lattice is about 1 eV and the liberated energy is not much larger than that, a collision cascade does not develop. Nevertheless, the erosion is usually named electronic sputtering because of the initial energy deposition into the electronic system of the alkali halides in terms of ionizations and excitations. Indeed, the total sputtering yield induced by electron bombardment was observed by Al Jammal et al. (1973) to be closely related to the behaviour of the spatial distribution $D_e(E,x,cos)$ of energy deposited into the electronic system (conf. Sect. IV D).

Sputtering of alkali halides becomes more complicated during ion bombardment than electron bombardment because of the simultaneous effect of electronic and knock-on processes (Biersack and Santner (1976), Husinsky et al. (1988) and Poradzisz et al. (1988)). The metal-halide balance at the surface, which governs the erosion, is now determined by a combination of knock-on sputtering and evaporation.

Many of these features have been observed for other materials, e.g. alkali earth halides, as well (Betz and Husinsky (1988)). Overeijnder et al. (1978) observed that the halogen emission can be stopped completely from silver bromide and lead iodide, after a metal layer has been formed on the surface of the salt.

VI-D. Sputtering of oxides and related compounds

The electronic sputtering of room temperature oxides and related materials has not been studied comprehensively as for example for alkali halides. During electron bombardment of many oxides oxygen release has been detected (Pantano and Madey (1981) and Townsend and Lama (1983)), but the yields have not yet been measured systematically. The mechanisms suggested are related to those from electron stimulated ion desorption (Madey et al. (1984) and Knotek (1984)).

Strong luminescence and absorption have been observed in the oxides (Tanimura et al. (1988), Crawford (1983)). Electron irradiation of silicon dioxide produces a blue luminescence which has been anti-correlated with the rate of damage formation over the temperature range 77 to 300 K by Jaque and Townsend (1981). This means probably that the defect production mechanism leads to sputtering as well, similar to the features observed for alkali halides.

Room temperature oxides and uranium hexafluoride have been investigated in a number of experiments on sputtering with heavy MeV-ions (Seiberling et al. (1980) and (1982), Griffith et al. (1980), Qiu et al. (1982) and (1983), and Meins et al. (1983)). The partial sputtering yields from sapphire and lithium niobate bombarded by MeV chlorine ions were measured by Qiu et al. (1982). The results (Fig. 19) demonstrate clearly that the process is electronic rather than knock-on sputtering. In addition to the interest in electronic sputtering from these materials the major aim was to explore the mechanism of track formation in insulating materials.

The models for creating permanent damage along the track of heavy ions were partly developed more than twenty years ago (Fleischer et al. (1965)). Nevertheless, the details are not yet explained (Walker (1982), Fischer and Spohr (1983), Balanzat et al. (1988)). The registration demands not only physical damage because of high excitation density, but requires a subsequent chemical etching as well.

The track from heavy MeV particles, e.g. fission fragments, are produced via ionizations (Morgan and Van Vliet (1970) and Matzke (1982)). The direct collisions with the atoms do not play any role at these energies. Fleischer et al. (1965) suggested that the damage was created by a Coulomb explosion in which the ionized atoms would be driven away from their original positions into the surrounding lattice. These authors correlated the threshold for track formation with the mechanical strength of the material. Other authors (Sigrist and Balzer (1977)) have emphazised the connection between thermal conductivity and the threshold of the electronic stopping power for track formation.

Haff (1976) suggested that the Coulomb explosion may lead to sputtering as well. The ejected particles need not to be charged, but may be neutrals initiated by the violent repulsions of the ions in the track. The possible sputtering as well as the formation of the track are shown in Fig. 20.

In order to test the connection between track formation and sputtering a number of experiments were performed by Seiberling et al. (1980), and Qiu et al. (1982) and (1983). It turned out that the high-energy sputtering yield from sapphire and lithium niobate bombarded by chlorine ions is fairly similar Fig. 19, even though the threshold in electronic stopping power for track formation differs by more than a factor of two. Similar experiments with crystalline and amorphous silicon dioxide with a large difference in thermal diffusivity did not show any significant differences in the electronic sputtering yield for MeV chlorine ions. These experiments demonstrate that the probable connection between sputtering and track formation cannot be expected to be a universal feature for these materials. The models for sputtering of insulators have recently been review by Wien (1989).

VII. CONCLUSION

The **energy loss** processes during ion slowing-down are closely connected to damage production and sputtering for both pure elements and multicomponent materials. The general behaviour of the stopping power of pure elements is known relatively well, although there exist ion-target combinations and energy regimes, for which the stopping power predictions are uncertain.

A complicating feature for multicomponent targets is the deviation of the electronic stopping power from the additivity of the stopping power of the constituents, the so-called **Bragg's rule**. The deviations are largest for target elements of low atomic number and for low-energy projectiles. Very little is known about the possible deviations for heavy ions.

The energy loss to the nuclei described by the nuclear stopping power leads to **knock-on** (ordinary) sputtering, in which target particles are ejected as a result of the collisions initiated by the primary. The behaviour of the sputtering yield for multicomponent targets is generally much less known than the yield from pure elements. For multicomponent materials a number of additional parameters, for example the primary ion fluence and the concentration of the components, influence the sputtering.

Sputtering via electronic transitions takes place for insulating materials, in which the kinetic energy for the particles in motion stems from electronic deexcitations to repulsive states. This **electronic** sputtering may be correlated to the electronic stopping power in analogy with the connection between knock-on sputtering and nuclear stopping power. Electronic sputtering has been explored very little compared to knock-on sputtering. It seems likely that no universal mechanism is responsible for electronic sputtering, and that the mechanism depends strongly on the particular type of the material and of the primary particles.

ACKNOWLEDGEMENTS

The author thanks H.H. Andersen, W.O. Hofer, P. Sigmund and B. Stenum for comments to the manuscript. The author has appreciated discussions with J. Oddershede, A. Perez, M. Szymonski, H. Sørensen, and K. Wien. The author thanks B. Stenum for assistance with the preparation of the drawings and M. Szymonski for the opportunity to use his unpublished result in the present work.

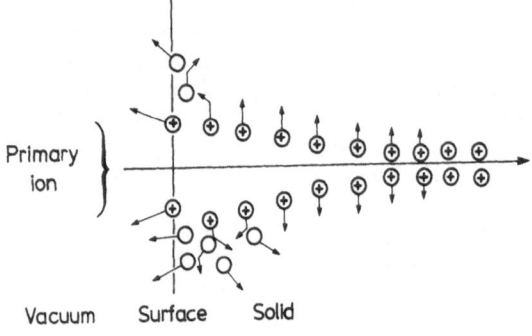

Fig. 19. A schematic survey of a track and sputtering production by a Coulomb explosion.

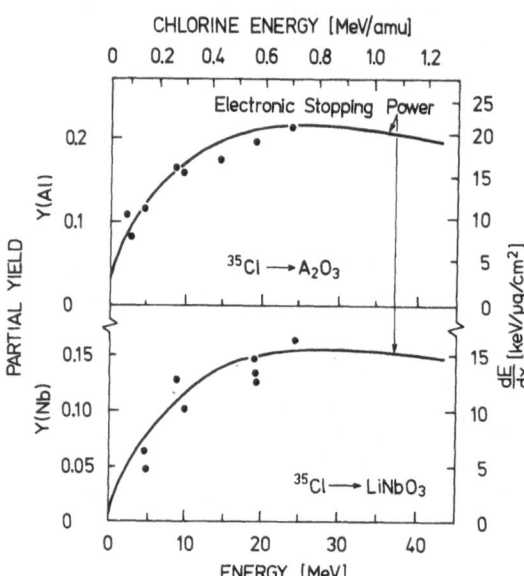

Fig. 20. Electronic sputtering yield from Cl⁺-ions on LiNbO₃ and sapphire. Partial yields, Y(Nb) and Y(Al) were determined. Solid line, the electronic stopping power evaluated from Ziegler (1980) on the basis of Bragg's rule. Data from Qiu et al. (1982).

REFERENCES

F.F. Abraham and C.R. Brundle (1981) J. Vac. Science. Techn. 18, 506.

S.P. Ahlen (1980), Rev. Mod. Phys. 52, 121.

S. Ahmad, B.W. Farmery and M.W. Thompson (1980) Nucl. Instr. Meth. 170, 327.

Y. Al Jammal, D. Pooley and P.D. Townsend (1973) J. Phys. C 6, 247.

H.H. Andersen and H. Bay (1973) Rad. Effects 19, 139.

H.H. Andersen and H.L. Bay (1975). J. Appl. Phys. 46, 1919.

H.H. Andersen, J. Bøttiger and H. Wolder Jørgensen (1975). Appl. Phys. Lett. 26, 678.

H.H. Andersen and J.F. Ziegler (1977) **Hydrogen stopping powers and ranges in all elements** (Pergamon, New York).

H.H. Andersen, J.F. Bak, H. Knudsen and B.R. Nielsen (1977), Phys. Rev. A 16, 1929.

H.H. Andersen (1979) J. Vac. Sci. Techn. 16, 770.

H.H. Andersen and H.L. Bay (1981) in: **Sputtering by particle bombardment I** ed. R. Behrisch (Springer, Berlin-Heidelberg) p. 145.

H.H. Andersen, B. Stenum, T. Sørensen and H.J. Whitlow (1983) Nucl. Instr. Meth. 209-210, 487.

H.H. Andersen (1984) in: **Ion implantation and beam processing**, eds. J.M. Williams and J.M. Poate (Academic, Australia) p. 127.

H.H. Andersen, B. Stenum, T. Sørensen and H.J. Whitlow (1984) Nucl. Instr. Meth. B 2, 601.

H.H. Andersen, B. Stenum, T. Sørensen and H.J. Whitlow (1985) Nucl. Instr. Meth. B 6, 459.

H.H. Andersen (1987) Nucl. Instr. Meth. B 18, 321.

H.H. Andersen (1988) Nucl. Instr. Meth. B 33, 466.

N. Andersen and P. Sigmund (1974) Mat. Fys. Medd. Dan. Vid. Selsk. 39, No.3.

P.R. Antoniewicz (1980) Phys. Rev. B 21, 3811.

J.C. Ashley (1982) Rad. Research 89, 25.

Ph. Avouris, F. Bozso and R.E. Walkup (1987) Nucl. Instr. Meth. B 27, 136.

E. Balanzat, J.C. Jousset and M. Toulemonde (1988) Nucl. Instr. Meth. B 32, 368.

H.L. Bay, H.H. Andersen, W.O. Hofer and O. Nielsen (1976) Nucl. Instr. Meth. 132, 301.

H.L. Bay, J. Roth and J. Bohdansky (1977) J. Appl. Phys. 48, 4722.

R. Behrisch, G. Maderlechner, B.M.U. Scherzer and M.T. Robinson (1979) Appl. Phys. 18, 391.

J. Benit, J.-P. Bibring, S. Della-Negra, Y le Beyec, M. Mendenhall, F. Rocard and K. Standing (1987) Nucl. Instr. Meth. B 19-20, 838.

F. Besenbacher, J. Bøttiger, O. Graversen, J.L. Hansen and H. Sørensen (1981) Nucl. Instr. Meth. 191, 221.

K. Besocke, S. Berger, W.O. Hofer and U. Littmark (1982) Rad. Effects 66, 35.

H.A. Bethe (1941) Phys. Rev. 59, 940.

G. Betz and G.K. Wehner (1983) in: **Sputtering by particle bombardment II**, ed. R. Behrisch (Springer, Berlin-Heidelberg) p. 11.

G. Betz and W. Husinsky (1988) Nucl. Instr. Meth. B 32, 331.

H. Betz (1972) Rev. Mod. Physics 44, 465.

H.-D. Betz, R. Höppler, R. Schramm and W. Oswald (1988) Nucl. Instr. Meth. B 33, 185.

H. Bichsel (1988) Rev. Mod. Physics 60, 663.

J.P. Biersack and E. Santner (1976) Nucl. Instr. Meth. 132, 229.

J.P. Biersack and W. Eckstein (1984) Appl. Phys. A 34, 73.

J. Bohdansky, J. Roth and H.L. Bay (1980) J. Appl. Phys. 51, 2861.

J. Bohdansky, H. Lindner, E. Hechtl, A.P. Martinelli and J. Roth (1987) Nucl. Instr. Meth. B 18, 509.

N. Bohr (1948) Mat. Fys. Medd. Dan. Vid. Selsk. 18, No. 8.

E. Bonderup (1967) Mat. Fys. Medd. Dan. Vid. Selsk. 35, No. 17.

E. Bonderup (1981) **Penetration of charged particles through matter**, Lecture notes, Institute of Physics, University of Århus, DK-Århus 8000 C.

P. Børgesen (1985) Nucl. Instr. Meth. B 12, 73.

J.W. Boring, D.J. O'Shaughnessy and J.A. Phipps (1987) Nucl. Instr. Meth. B 18, 613.

W.H. Bragg and R. Kleeman (1905), Phil. Mag. 10, 318.

D.K. Brice (1975): **Ion implantation range and energy distributions**, ed. by D.K. Brice (Plenum, New York) vol. 1.

W.L. Brown, W.M. Augustyniak, E. Brody, B. Cooper, L.J. Lanzerotti, A. Ramirez, R. Evatt and R.E. Johnson (1980a) Nucl. Instr. Meth. 170, 321.

W.L. Brown, W.M. Augustynik, L.J. Lanzerotti, R.E. Johnson and R. Evatt (1980b) Phys. Rev. Lett. 45, 1632.

W.L. Brown and R.E. Johnson (1986) Nucl. Instr. Meth. B 13, 295.

W.L. Brown, L.J. Lanzerotti, J.E. Bower and K.J. Marcantonio (1987a) Nucl. Instr. Meth. B 24-25, 512.

W.L. Brown, G. Foti, L.J. Lanzerotti, J.E. Bower and R.E. Johnson (1987b) Nucl. Instr. Meth. B 19-20, 899.

98

D. Cherns, F.J. Minter and R.S. Nelson (1976) Nucl. Instr. Meth. **132**, 369.

D. Cherns (1979) Surf. Science **90**, 339.

C. Claussen (1982) Thesis, University of Odense, DK-5230 Odense, (unpublished).

J.W. Coburn (1979) Thin solid films **64**, 371.

J.W. Coburn and H.F. Winters (1987) Nucl. Instr. Meth. B **27**, 243.

B.H. Cooper and T.A. Tombrello (1984) Rad. Effects **80**, 203.

J.H. Crawford, Jr. (1983) Semiconductors and Insulators **5**, 599.

D.B. Crisey, J.W. Boring, J.A. Phipps, R.E. Johnson and W.L. Brown (1986) Nucl. Instr. Meth. B **13**, 360.

D.B. Crisey, J.W. Boring, R.E. Johnson and J.A. Phipps (1988) Surf. Science **195**, 594.

S.A. Cruz (1986) Rad. Effects **88**, 159.

J. Dembowski, H. Oechsner, Y. Yamamura and M. Urbassek (1987) Nucl. Instr. Meth. B **18**, 464.

M.F. Dumke, T.A. Tombrello, R. Weller, R.M. Housley and E.H. Cirlin (1983) Surf. Science **124**, 407.

A.V. Eltoukhy and J.E. Greene (1980) J. Appl. Phys. **51**, 4444.

W. Eckstein and J.P. Biersack (1985) Appl. Phys. A **37**, 95.

W. Eckstein and W. Möller (1985) Nucl. Instr. Meth. B **7-8**, 727.

O. Ellegaard, J. Schou, H. Sørensen and P. Børgesen (1986) Surf. Science **167**, 474.

O. Ellegaard, R. Pedrys, J. Schou, H. Sørensen and P. Børgesen (1988) Appl. Phys. A **46**, 305.

J.S.-Y. Feng, W.K. Chu and M.-A. Nicolet (1973) Thin solid films **19**, 227.

J.S.-Y. Feng, W.K. Chu and M.-A. Nicolet (1974) Phys. Rev. B **10**, 3781.

O.B. Firsov (1959) Zh. Eksp. Teor. Fiz. **36** (1517) and Sov. Phys. JETP **7**, 1076.

B.E. Fischer and R. Spohr (1983) Rev. Mod. Phys. **55**, 907.

R.L. Fleischer, P.B. Price and R.M. Walker (1965) J. Appl. Phys. **36**, 3645.

B.W. Farmery and M.W. Thompson (1968) Phil. Mag. **18**, 415.

G. Foti, L. Calcagno, F.Z. Zhou and G. Strazzulla (1987) Nucl. Instr. Meth. **24-25**, 522.

B.J. Garrison and R.E. Johnson (1985) Surf. Science **148**, 388.

J. Geertsen, J. Oddershede and R. Sabin (1986) Phys. Rev. A **34**, 1104.

R. Gomer (1983), in: **Desorption induced by electronic transitions**, DIET I, eds. N.H. Tolk, M.M. Traum, J.C. Tully and T.E. Madey (Springer), p. 40.

J.E. Griffith, R.A. Weller, L.E. Seiberling and T.A. Tombrello (1980) Rad. Effects **51**, 223.

P.K. Haff (1976) Appl. Phys. Lett. **29**, 473.

R. Haring (1984) Thesis, FOM-Institute of Atomic And Molecular Physics, Amsterdam.

R.A. Haring, R. Pedrys, D.J. Oostra, A. Haring and A.E. de Vries (1984a) Nucl. Instr. Meth. B **5**, 476.

R.A. Haring, R. Pedrys, D.J. Oostra, A. Haring and A.E. de Vries (1984b) Nucl. Instr. Meth. B **5**, 483.

P.S. Ho (1978) Surf. Science **72**, 253.

W.O. Hofer and U. Littmark (1979) Phys. Lett. A **71**, 457.

W.O. Hofer, K. Besocke and B. Stritzker (1983) Appl. Phys. A **30**, 83.

W.O. Hofer (1986) in: **Erosion and growth of solids stimulated by atom and ion beams**, eds. G. Carter, G. Kiriakidis and J.L. Whitton (Nijhoff, Netherlands) p.1.

W.O. Hofer (1989) in: **Sputtering by particle bombardment III** , eds. R. Behrisch and K. Wittmaack (Springer) (in press).

M. Hou and M.T. Robinson (1979) Appl. Phys. **18**, 381.

W. Husinsky, P. Wurz, B. Strehl and G. Betz (1987) Nucl. Instr. Meth. B **18**, 452.

W. Husinsky, P. Wurz, K. Mader, E. Wolfrum, B. Strehl, G. Betz, R.F. Haglund Jr., A.V. Barnes and N.H. Tolk (1988) Nucl. Instr. Meth. B **33**, 824.

S. Ichimura, M. Shikata and R. Shimiza (1981) Surf. Science **108**, L 393.

ICRU (1984) Stopping powers for electrons and positrons, Report 37, ICRU-publications, 7910 Woodmont Avenue, Bethesda, MD-20814, USA.

N. Itoh (1976) Nucl. Instr. Meth. **132**, 201.

N. Itoh (1987) Nucl. Instr. Meth. B 27, 155.

J.F. Janni (1982) Atomic and Nucl. Data Tables, 27, 147.

F. Jaque and P.D. Townsend (1981) Nucl. Instr. Meth. 182-183, 781.

S.S. Johar and D.A. Thompson (1979) Surf. Science 90, 319.

R.A. Johnson and N.Q. Lam (1978) J. Nucl. Mat. 69, 424.

R.E. Johnson and R. Evatt (1980) Rad. Effects 52, 187.

R.E. Johnson, L.J. Lanzerotti and W.L. Brown (1984) Adv. Space Res. 4, 41.

R.E. Johnson (1987) Int. J. Mass. Spec. Ion. Proc. 78, 357.

R. Kelly (1978) Nucl. Instr. Meth. 149, 553.

R. Kelly (1979) Surf. Science 90, 280.

R. Kelly (1980a), in: Proceedings of the Symposium on Sputtering in Perchtoldsdorf, eds. P. Varga, G. Betz and F.P. Viehböck (Inst. Allgem. Physik, Technische Universität, Vienna) p. 390.

R. Kelly (1980b) Surf. Science 100, 85.

R. Kelly (1984) Rad. Effects 80, 273.

R. Kelly and D.E. Harrison (1985) Mat. Sci. Eng. 69, 449.

R. Kelly (1987) Nucl. Instr. Meth. B 18, 388.

J. Kirschner (1985) Nucl. Instr. Meth. B 7-8, 742.

M.L. Knotek (1984) Rep. Prog. Phys. 47, 1499.

R. Kreutz, W. Neuwirth and W. Pietsch (1980) Phys. Rev. A 22, 2606.

N.Q. Lam, P.R. Okamato and R.A. Johnson (1978) J. Nucl. Mat. 78, 408.

N.Q. Lam and H. Wiedersich (1987) Nucl. Instr. Meth. B 18, 471.

Z.L. Liau, J.W. Mayer, W.L. Brown and J.M. Poate (1978) J. Appl. Phys. 49, 5295.

J.J. Lindhard and M. Scharff (1961) Phys. Rev. 124, 128.

J. Lindhard, V. Nielsen, M. Scharff and P.V. Thomsen (1963) Mat. Fys. Medd. Dan. Vid. Selsk. 33, No. 10.

J. Lindhard, V. Nielsen and M. Scharff (1968) Mat. Fys. Medd. Dan. Vid. Selsk. 36, No. 10.

U. Littmark and S. Fedder (1982) Nucl. Instr. Meth. 194, 607.

T.E. Madey, D.E. Ramaker and R. Stockbauer (1984) Ann. Rev. Phys. Chem. 35, 215.

Hj. Matzke (1982) Rad. Effects 64, 3.

C.K. Meins, J.E. Griffith, Y. Qiu, M.H. Mendenhall, L.E. Seiberling, and T.A. Tombrello (1983) Rad. Effects 71, 13.

D. Menzel (1986) Nucl. Instr. Meth. B 13, 507.

D.V. Morgan and D. van Vliet (1970) Contemp. Phys. 11, 173.

J. Oddershede, J.R. Sabin, and P. Sigmund (1983) Phys. Rev. Lett. 51, 1332.

H. Oechsner and J. Bartella (1980) In: Proceedings of the VII Conf. Atomic Coll. in Solids, Moscow, 1979 (Moscow State University Publishing House, U.S.S.R.) Vol. 2, p. 327.

O. Oehler, D.A. Smith and K. Dressler (1977) J. Chem. Phys. 66, 2097.

A.R. Olivia-Florio, E.V. Alonso, R.A. Baragiola, J. Ferron and M.M. Jakas (1979) Rad. Effects Lett. 50, 3.

R.R. Olson, M.E. King, and G.K. Wehner (1979) J. Appl. Phys. 50, 3677.

D.J. Oostra and A.E. de Vries (1987) Nucl. Instr. Meth. B 18, 618.

D.J. O'Shaughnessy, J.W. Boring, S. Cui and R.E. Johnson (1988) Phys. Rev. Lett. 61, 1635.

H. Overeijnder, M. Szymonski, A. Haring and A.E. de Vries (1978) Rad. Effects 36, 63.

C.G. Pantano and T.E. Madey (1981) Applic. Surf. Science 7, 115.

R. Pedrys, R.A. Haring, A. Haring and A.E. de Vries (1984) Nucl. Instr. Meth. B 2, 573.

R. Pedrys, D.J. Oostra, R.A. Haring, L. Calcagno, A. Haring and A.E. de Vries (1986) Nucl. Instr. Meth. B 17, 15.

R. Pedrys, D.J. Oostra, A. Haring, A.E. de Vries and J. Schou (1988) Nucl. Instr. Meth. B 33, 840.

A. Perez, J. Davenas and C.H.S. Dupuy, Nucl. Instr. Meth. 132, 219.

A. Poradzisz, Z. Postawa, J. Rutkowski and M. Szymonski (1988) Nucl. Instr. Meth. B 33, 830.

100

Z. Postawa, J. Rutkowski, A. Poradzisz, P. Czuba and M. Szymonski (1987) Nucl. Instr. Meth. B 18, 574.

D. Powers, W.K. Chu, R.J. Robinson and A.S. Lodhi (1972) Phys. Rev. A6, 1425.

D. Powers, H.G. Olson and R. Gowda (1984) J. Appl. Phys. 55, 1274.

Y. Qiu, J.E. Griffith and T.A. Tombrello (1982) Rad. Effects 64, 111.

Y. Qiu, J.E. Griffith, W.J. Meng and T.A. Tombrello (1983) Rad. Effects 70, 231.

L.E. Rehn, R.S. Auerback and P.R. Okamato (1985) Mat. Sci. Eng. 64, 1.

L.E. Rehn and N.Q. Lam (1987) J. Mater. Eng. 9, 205.

C.T. Reimann, R.E. Johnson and W.L. Brown (1984a) Phys. Rev. Lett. 53, 600.

C.T. Reimann, J.W. Boring, R.E. Johnson, J.W. Garrett, K.R. Farmer, W.L. Brown, K.J. Marcantonio and W.M. Angustyniak (1984b) Surf. Science 147, 227.

C.T. Reimann, W.L. Brown and R.E. Johnson (1988) Phys. Rev. B 37, 1455.

L. Rivaud, A.H. Eltoukhy, and J.E. Greene (1981) Rad. Effects 61, 83.

M.T. Robinson (1981) in: **Sputtering by particle bombardment**, I, ed. R. Behrisch (Springer, Berlin-Heidelberg) p. 73.

F. Rocard, J. Benit, J.-P. Bibring, D. Ledu and R. Meunier (1986) Rad. Effects 99, 97.

K. Rödelsperger and A. Scharmann (1976) Nucl. Instr. Meth. 132, 355.

H.E. Roosendaal (1981) in: **Sputtering by particle bombardment** I, ed. R. Behrisch (Springer, Berlin-Heidelberg) p. 219.

K. Rössler (1986) Rad. Effect 99, 21.

J. Roth (1980), in: **Proceedings of the Symposium on Sputtering in Perchtoldsdorf**, eds. P. Varga, G. Betz and F.P. Viehböck, (Inst. Allgem. Physik, Technische Universität, Vienna), p. 773.

J. Roth, J. Bohdansky and A.P. Martinelli (1980) Rad. Effects 48, 213.

J. Roth (1983), in: **Sputtering by particle bombardment**, II, ed. R. Behrisch (Springer, Berlin Heidelberg) p. 91.

J. Roth (1986a) in: **Physics of plasma-wall interactions in controlled fusion**, eds. D.E. Post and R. Behrisch (Plenum, New York-London) p. 389.

J. Roth (1986b) in previous ref., p. 351.

J.R. Sabin and J. Oddershede (1982) Phys. Rev. A. 26, 3209.

J.R. Sabin and J. Oddershede (1987) Nucl. Instr. Meth. B 27, 280.

J.R. Sabin and J. Oddershede (1988) to be published.

R.P. Schorn, M.A. Zaki Ewiss and E. Hintz (1988) Appl. Phys. A 46, 291.

J. Schou (1980) Phys. Rev. B 22, 2141.

J. Schou, H. Sørensen and P. Børgesen (1984) Nucl. Instr. Meth. B 5, 44.

J. Schou, P. Børgesen, O. Ellegaard, H. Sørensen and C. Claussen (1986) Phys. Rev. B 34, 93.

J. Schou (1987) Nucl. Instr. Meth. B 27, 188.

J. Schou (1988) Scanning Microscopy 2, 607.

J. Schou, O. Ellegaard, H. Sørensen and R. Pedrys (1988) Nucl. Instr. Meth. B 33, 808.

N. Schwentner, E.-E. Koch and J. Jortner, (1985) **Electronic excitations in condensed rare gases** (Springer, Berlin-Heidelberg).

L.E. Seiberling, J.E. Griffith and T.A. Tombrello (1980) Rad. Effects 52, 201.

L.E. Seiberling, C.K. Meins, B.H. Cooper, J.E. Griffith, M.H. Mendenhall and T.A. Tombrello (1982) Nucl. Instr. Meth. 198, 17.

M.H. Shapiro, T.A. Tombrello and D.E. Harrison Jr. (1988) Nucl. Instr. Meth. B 30, 152.

E. Shiles, T. Sasaki, M. Inokuti and D.Y. Smith (1980) Phys. Rev. B 22, 1612.

R. Shimizu (1987) Nucl. Instr. Meth. B 18, 486.

P. Sigmund (1969) Phys. Rev. 184, 383 and 187, 768.

P. Sigmund (1975) In: **Radiation damage processes in materials** , ed. C.H.S. Dupuy (Noordhoff, Leyden) p. 3.

P. Sigmund (1979) J. Appl. Phys. 50, 7261.

P. Sigmund (1980) J. Vac. Sci. Techn. 17, 396.

P. Sigmund and A. Gras-marti (1980) Nucl. Instr. Meth. **168**, 389.

P. Sigmund (1981) in: **Sputtering by particle bombardment** I, ed. R. Behrisch (Springer, Berlin-Heidelberg) p.9.

P. Sigmund and C. Claussen (1981) J. Appl. Phys. **52**, 990.

P. Sigmund (1982) Phys. Rev. A **26**, 2497.

P. Sigmund, A. Oliva and G. Falcone (1982) Nucl. Instr. Meth. **194**, 541 and Nucl. Instr. Meth. B **9**, 354 (1985).

P. Sigmund and M. Szymonski (1984), Appl. Phys. A **33**, 141.

P. Sigmund and U. Haagerup (1986) Phys. Rev. A **34**, 892 and Phys. Rev. A **35**, 3965.

P. Sigmund (1987a) Nucl. Instr. Meth. B **27**, 1.

P. Sigmund (1987b) Nucl. Instr. Meth. B **18**, 375.

P. Sigmund (1988a) Unpublished manuscript, available at the author. (Lecture at NATO Advanced Study Institute, Viana do Castelo, Portugal, Aug.-Sept. 1987).

P. Sigmund (1988b) Nucl. Instr. Meth. B **34**, 15.

A. Sigrist and R. Balzer (1977) Helv. Physica Acta **50**, 49.

R.M. Sternheimer and R.F. Peierls (1971) Phys. Rev. B **3**, 3681.

D.V. Stevanovic, D.A. Thompson and J.A. Davies (1984) Nucl. Instr. Meth. B **1**, 315.

B. Sundqvist and R.D. Macfarlane (1985) Mass. Spectr. Rev. **4**, 421.

B. Sundqvist (1989) In: **Sputtering by particle bombardment**, III, eds. R. Behrisch and K. Wittmaack, (Springer) (in press).

M. Szymonski and A.E. de Vries (1977) Phys. Lett. **63** A, 359.

M. Szymonski (1980a) Appl. Phys. **23**, 89.

M. Szymonski (1980b) Rad. Effects **52**, 9.

M. Szymonski (1982) Nucl. Instr. Meth. **194**, 523.

M. Szymonski and A. Poradzisz (1982) Appl. Phys. A **28**, 175.

M. Szymonski and A.E. de Vries (1983) in: **Desorption induced by electronic transitions**, DIET I, eds. N.H. Tolk, M.M. Traum, J.C. Tully and T.E. Madey (Springer) p. 216.

M. Szymonski, J. Rutkowski, A. Poradzisz and Z. Postawa (1985) in: **Desorption induced by electronic transitions**, DIET II, eds. W. Brenig and D. Menzel (Springer) p. 160.

M. Szymonski, J. Rutkowski and Z. Postawa (1988). To be published.

K. Tanimura, C. Itoh and N. Itoh (1988) J. Phys. C **21**, 1869.

P. Thevenard (1989), these proceedings.

D.A. Thompson (1981) Rad. Effects. **56**, 105.

M.W. Thompson (1968) Phil. Mag. **18**, 377.

D.I. Thwaites (1983) Rad. Research **95**, 495.

D.I. Thwaites (1985) Nucl. Instr. Meth. B **12**, 84.

D.I. Thwaites (1987) Nucl. instr. Meth. B **27**, 293.

L. Torrisi, S. Coffa, G. Foti and G. Strazzulla (1986) Rad. Effects **100**, 61.

L. Torrisi, S. Coffa, G. Foti, R.E. Johnson, D.B. Chrisey and J.W. Boring (1988) Phys. Rev. B **38**, 1516.

P. Townsend, R. Browning, D.J. Garlant, J.C. Kelly, A. Mahjoobi, A.J. Michael and M. Saidoh (1976) Rad. Effect **30**, 55.

P.D. Townsend and F. Lama (1983) In: **Desorption induced by electronic transitions**, DIET I, eds. N.H. Tolk, M.M. Traum, J.C. Tully and T.E. Madey (Springer) p. 220.

P.D. Townsend (1983) in: **Sputtering by particle bombardment**, II, ed. R. Behrisch (Springer, Berelin Heidelberg), p. 147.

S.M. Trotman, T.I. Quickenden and D.F. Sangster (1986). J. Chem. Phys. **85**, 2555.

C. Tschalär and H. Bischel (1986) Phys. Rev. **175**, 476.

M. Urbassek (1984) Nucl. Instr. Meth. B **4**, 356.

M. Urbassek and P. Sigmund (1984) Appl. Phys. A **35**, 19.

P. Varga and E. Taglauer (1982) J. Nucl. Mat. **111-112**, 726.

M. Vicanek, J.J. Jimenez-Rodrigues and P. Sigmund (1988) to be published.

G.H. Vineyard (1976) Rad. Effects **29**, 245.

R.M. Walker (1982) Rad. Effects **65**, 131.

C.C. Watson and T.A. Tombrello (1985) Rad. Eff. **89**, 263.

H. Wiedersich (1985) Nucl. Instr. Meth. B **7-8**, 1.

K. Wien (1989) Rad. Effects Def. Solids (in press).

M.M.R. Williams (1981) Phil. Mag. A **43**, 1221.

R.T. Williams, K.S. Song, W.L. Faust, C.H. Leung (1986) Phys. Rev. B **33**, 7232.

W.D. Wilson, L.G. Haggmark, and J.P. Biersack (1977) Phys. Rev. B **15**, 2458.

K.B. Winterbon, P. Sigmund and J.B. Sanders (1970) Mat. Fys. Medd. Dan. Vid. Selsk. **37**, No. 14.

K.B. Winterbon (1975): "**Ion implanation range and energy distributions**", ed. by D.K. Brice (Plenum, New York), vol. 2.

A. Wucher and H. Oechsner (1987) Nucl. Instr. Meth. B **18**, 458.

P.C. Zalm (1988) Surf. Interf. Anal. **11**, 1.

J.F. Ziegler, W.K. Chu and J.S.-Y. Feng (1975) Appl. Phys. Lett. **27**, 387.

J.F. Ziegler (1977) **Helium stopping powers and ranges in all elemental matter** (Pergamon, New York).

J.F. Ziegler (1980) **Handbook of stopping cross section for energetic ions in all elements** (Pergamon, New York).

J.F. Ziegler, J.P. Biersack and U. Littmark (1985). **The stopping and range of ions in solids** (Pergamon, New York).

G. Zimmerer (1987), in: **Excited state spectroscopy in solids**, ed. M. Manfredi (North-Holland, Amsterdam) p. 37.

MECHANISMS OF ION BEAM MIXING

F.W. SARIS

FOM-INSTITUTE FOR ATOMIC AND MOLECULAR PHYSICS,
KRUISLAAN 407, 1098 SJ AMSTERDAM, THE NETHERLANDS

1. INTRODUCTION

Research on Ion Beam Mixing was triggered in 1973 when Van der Weg et al. (1) discovered Paladium Silicide formation during Ar ion irradiation of thin Pd films on Si substrates. This could not be explained by recoil implantation, for every Ar ion more than 100 Pd and Si atoms were mixed. Ion beam mixing was found to be not only an efficient way to introduce foreign atoms into subsurface layers, the atomic concentrations achieved are much higher than obtainable by direct ion implantation. Sputtering limits the maximum concentration of foreign atoms that can be implanted in surface layers to about 10-25%. In ion beam mixing, for every ion several hundred atoms of the substrate and the overlayer are intermixed. Therefore the required ion doses are not so high that the sputtering limit is reached.

This has stimulated much fundamental research concerning the mechanism of ion beam mixing. Originally most of the attention was focussed on ballistic effects in the primary and secondary recoil events triggered by the ion beam. However, comparison between systems of different atomic species, that could be expected to have the same ballistic effects, surprisingly showed completely different degrees of mixing. Thus it was realized that thermodynamic and chemical effects play an important role in the ion-induced collision cascades. Further, experiments at widely different target temperatures revealed the importance of radiation enhanced diffusion. So, at present three different mechanisms, operating during ion beam mixing, have been identified: 1) ballistic or recoil effects, 2) cascade mixing, 3) radiation enhanced diffusion.

Thanks to computer simulations we know that the slowing down of energetic ions in a solid and the evolution of disturbances are transient processes for which different time regimes can be identified. As the energetic ions penetrate a solid, they transfer part of the kinetic energy by colliding with target atoms in primary collisions. This results in what is called recoil mixing. The later-generation collisions produce many low energy recoils, which induce small displacements in random directions. This is the cascade mixing process, lasting about 10^{-12} s after the projectile struck the target. At the end of this prompt regime, the collision cascade will have generated a non-equilibrium number of defects. If the ambient temperature is sufficiently high, these defects can be mobile, causing more diffusion and mixing. This temperature dependent radiation-enhanced diffusion may lead to segregation, precipitation or solute trapping, depending on the specific properties of the materials involved and the defect generation applied. Together, these transient processes often result in metastable phase formation and novel materials properties. This has stimulated much research in ion beam mixing also outside the area of silicide formation. At present, our understanding of the fundamentals of ion mixing is based on work in the area of metal bilayer systems.

It is the purpose of this paper to introduce the different mechanisms of ion beam mixing. Specific examples are chosen from the literature to illustrate qualitatively, and where possible also quantitatively, the relative importance of recoil mixing, cascade mixing and radiation

C.J. McHargue et al. (eds.), Structure-Property Relationships in Surface-Modified Ceramics, 103–116.
© *1989 by Kluwer Academic Publishers.*

enhanced diffusion. It will be shown that our understanding of ion bombardment phenomena in solids has made significant progress in recent years. Yet, most of the work sofar was concerned with metal-metal systems or metal-semiconductors. Investigations of insulators and especially ceramics have started relatively recently. Yet, it is hoped that the fundamental aspects observed in metal-metal systems may be used as guidelines for predicting the effects of irradiation of ceramics. Although this may work for the degree of ion beam mixing that will occur in a given system, it should not be expected that the metastable phases formed can be predicted. Although equilibrium phase diagrams have some predictive power for metal-metal systems, it is still too early to formulate general rules for phase formation. In this paper first the three different ion mixing mechanisms are illustrated, then the metastable phase formation will be discussed. This is not a comprehensive review for such reviews have been published recently elsewhere (2,3,4).

2. BALLISTIC OR RECOIL MIXING

Careful studies with thin markers have given evidence for, and quantitative measure of ballistic mixing effects. Fig. 1 shows the Rutherford Backscattering profiles of W markers in a Cu matrix, before and after Xe ion bombardment (5). The Cu sample has a shallow W marker at 300 Å depth and a deep W marker at 1600 Å depth. The Xe ion energy is 460 keV and was chosen such that the projectile comes to rest beyond the shallow marker, but does not reach the deep W marker. Fig. 1 shows that this deep marker has shifted to shallower depth as a result of

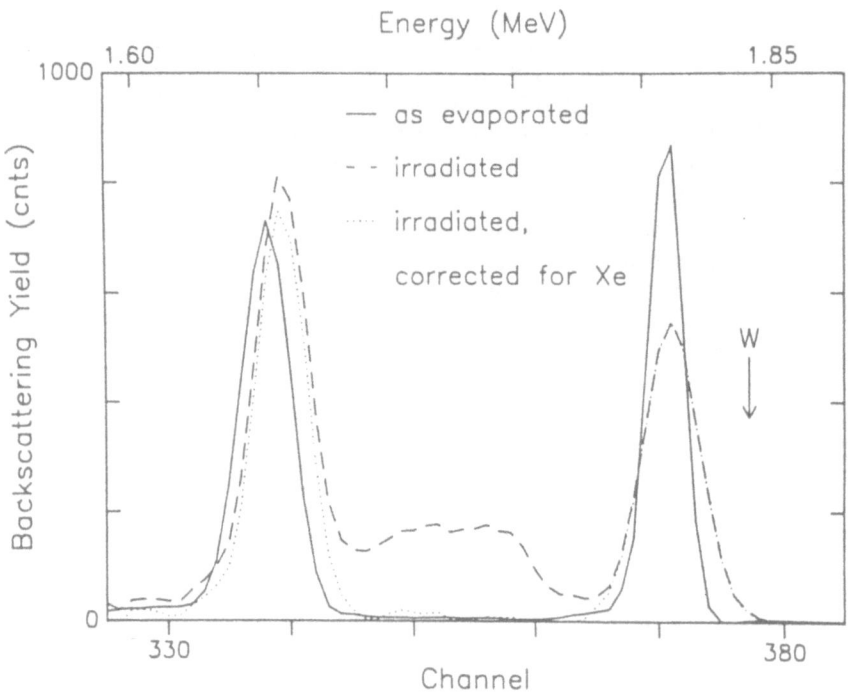

FIGURE 1. W signal of 2 MeV He RBS spectra of virgin and implanted parts of 4.3 Å thin W films in Cu. Dose = 5.5×10^{15} at./cm^2, 460 keV Xe. Irradiation and RBS analysis performed at room temperature.

Xe bombardment. This is due to the sputtering of Cu by the Xe irradiation. The shallow W marker, however, does not seem to shift in the RBS spectra. This means that the distance between the two W markers has shrunk. Since the deep marker in the Cu matrix is not reached by the Xe ions the distance between the two markers can only shrink because the shallow marker shifts inward as a result of ion irradiation. The amount of marker shift of the shallow marker is about equal to the thickness sputtered away by the ion dose. Since the shift is directed inward and the surface layer is thinned down, the shallow marker does not seem to shift position in the RBS spectrum of fig. 1.

Similar marker shifts have been observed for quite different marker/substrate combinations (6). Indeed, fig. 2 shows that for Au markers in Cu and the same Xe dose the distance between the two markers has shrunk by the same amount as for W markers (5). This is expected if ballistic effects dominate, for W and Au are rather similar in mass and very different from Cu. So a 460 keV Xe ion will transfer the same amount of kinetic energy to W or to Au atoms and on average much less to Cu atoms. Also Ta markers in Cu were found to behave identically (5) and again the mass differences with the projectile and substrate atoms are such that the kinetic energy transfer favours the marker atoms.

Initially, the explanation for marker shifts was sought in the so-called matrix relocation effect described quantitatively by Littmark (7). He pointed out that vacancies and interstitials produced by an incoming projectile are not distributed uniformly throughout the matrix. In the core of the collision cascade the concentration of vacancies will be enhanced, whereas

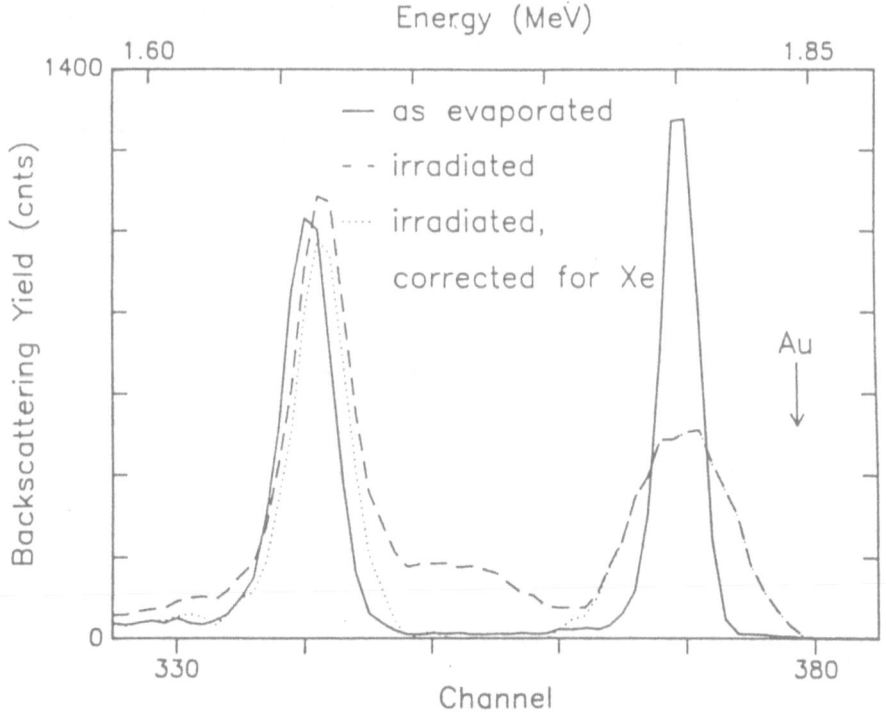

FIGURE 2. Au signal of 2 MeV He RBS spectra of virgin and implanted parts of 5.7 Å thin Au films in Cu. Dose = 5.4×10^{15} at./cm^2, 460 keV Xe. Irradiation and RBS analysis performed at room temperature.

interstitials will be distributed along the outer regions of the collision cascade. According to Littmark, after the collision cascade is over a matrix relocation should occur in order to keep its density uniform. Thus a marker layer which happens to be located in between the regions where vacancies and where interstitials dominate, will shift towards the region of vacancies due to the matrix relocation. This means that if the core of the collision cascade lies beyond the depth of the marker, it will shift to greater depth as a result of irradiation. On the other hand it is predicted that if the marker lies beyond the depth of the collision cascade, the marker should shift to shallower depth. This effect, however, was not corroborated by experiment (5). Markers were found to shift to greater depths always.

An alternative explanation came from computer simulations by Rouche (8) who found that the distribution of recoils leads to a shift in depth of heavy recoils relative to light recoil atoms. This comes about because the momentum distribution of primary recoils, produced by an ion penetrating a solid, is not isotropic and heavy recoils will be distributed more forward into a light matrix (note that light atoms may be reflected by heavy ones, but not vice versa). So a simple ballistic or recoil effect has been observed to occur during heavy ion bombardment of marker layers. The effect is small, typically the W, Au or Ta markers in Cu shift about 40 Å for 460 keV Xe ion dose of 5.5×10^{15} cm^{-2}. The shift is always directed inward irrespective of the Xe ion energy, provided it is high enough to affect the marker. The shift disappears if the thickness of the marker layers is increased beyond a few monolayers. This must be due to the fact that the recoil distance is only a few interatomic spacings, and if the marker is more than a few monolayers thick the W recoils mainly collide with other heavy W atoms instead of light Cu atoms and the anisotropic effects are absent. These observations concerning the relative distribution of displacements are consistent also with the work described by Parker in these proceedings.

Closer examination of the marker experiments of fig. 1 and 2 reveals an interesting difference. The broadening of the shallow Au marker as a result of Xe ion irradiation is much larger than the broadening of the W marker in Cu. This cannot be explained by a simple kinematic effect, for the Xe ion will transfer the same kinetic energy to Au and to W since their masses are nearly equal. The observed difference in the degree of ion mixing must be due to chemical effects operating during the collision cascade. This is the subject of the next section.

3. CASCADE MIXING

Fig. 3 shows a comparison of the extent of ion beam mixing in the systems Cu-W, Cu-Ta and Cu-Au. On the basis of purely ballistic arguments these systems are expected to show practically the same behaviour upon ion bombardment, since W, Ta and Au are positioned very close to each other in the periodic system. The heats of formation (at the 1-1 composition) of these systems are different: +24 kJ/mole for Cu-W, 0 kJ/mole for Cu-Ta and −10 kJ/mole in the case of Cu-Au (21). Contrary to the layer shifts, a clear correlation between film broadening and heat of mixing ΔH_m has been observed. The FWHM of the Gaussian Au profile in Cu after irradiation at room temperature amounts to 350 Å. For the Ta film this value is 250 Å, whereas for W only a broadening to a FWHM of 175 Å has been observed. In fig. 3 the mixing para‐meter $Dt/\phi F_D$ has been plotted as function of ΔH_m for room temperature experiments as well as 10 K irradiation with Xe ions and Ne ions. In general the mixing increases with the chemical driving force in the three systems.

Similar observations were made for totally different systems in many laboratories. Fig. 4 shows a rather elegant example of chemical effects in ion mixing of Hf, Ta, W, Ir and Pt over‐layers with Zr and Pd substrates (9). Whereas Pt shows the largest degree of mixing with Zr and Hf the least amount, the reverse is observed for Pd targets. Hf shows the largest amount of

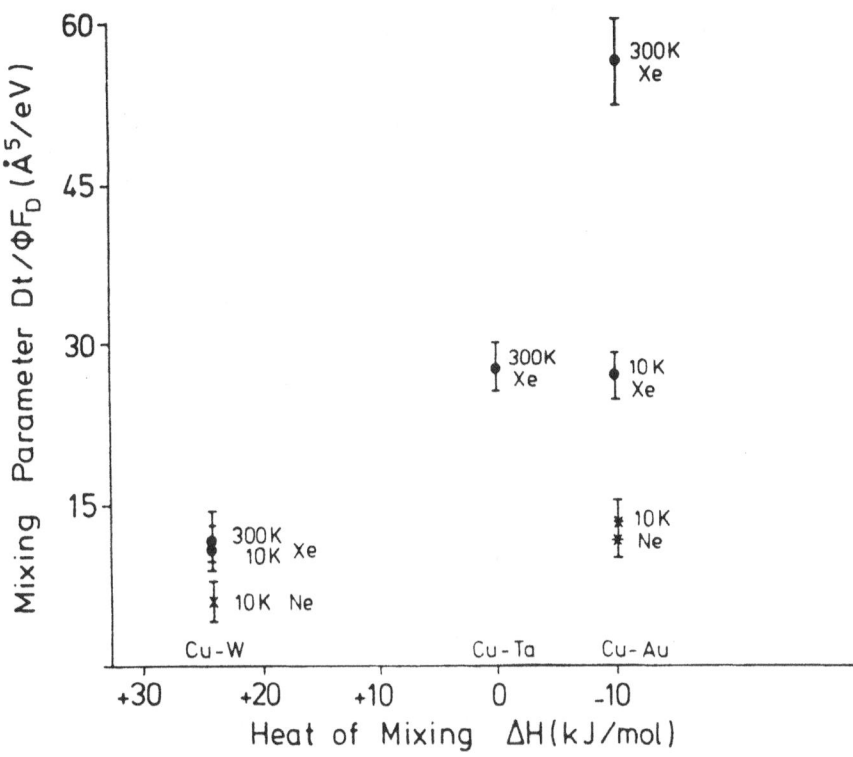

FIGURE 3. Thin film mixing parameters as a function of heat of mixing. RBS analysis performed at room temperature. Film thicknesses were between 4 and 6 Å, irradiation doses amounted to $4.4–5.5 \times 10^{15}$ at./cm^2 for Xe and 4×10^{16} at./cm^2 for Ne. Ion energies were 450 keV (RT) and 400 keV (10 K) for Xe, and 65 keV for Ne.

mixing with Pd and Pt the smallest amount. This is, indeed, in very nice agreement with what one would expect considering the heats of mixing in each of these metal-bilayer systems.

Also for the mixing of various metal overlayers with Au and with Pt substrates a clear correlation with the heat of mixing has been observed (10), see fig. 5. Moreover the Caltech group pointed out that not only the heats of mixing should be considered but also the cohesive energy of the materials involved. There exists a striking correlation between the mixing rate and the average cohesive energy of each bilayer (11), see fig. 6. The latter is the arithmetic average of the heat of sublimation of the pure elements. For the systems in fig. 6 the heat mixing is near zero. The heat of mixing effect and the cohesive energy effect emphasize the need for a thermodynamic approach to problems in ion mixing. Therefore, it is only natural to incorporate the idea of an energy spike into a phenomenological description of ion mixing (Johnsson et al., 12). Immediately after an energetic ion penetrates a solid the number of recoil atoms will increase sharply and thus the projectile energy is gradually shared by many target atoms in the collision cascade. In a small volume one expects equipartition of energy to take place and a relatively high energy spike is formed. When the average energy of all target atoms in the spike has decreased to approximately 1 eV chemical effects will come into play. Systems with large heat of mixing will have large driving forces inside the spike. If the heat of mixing is near zero

108

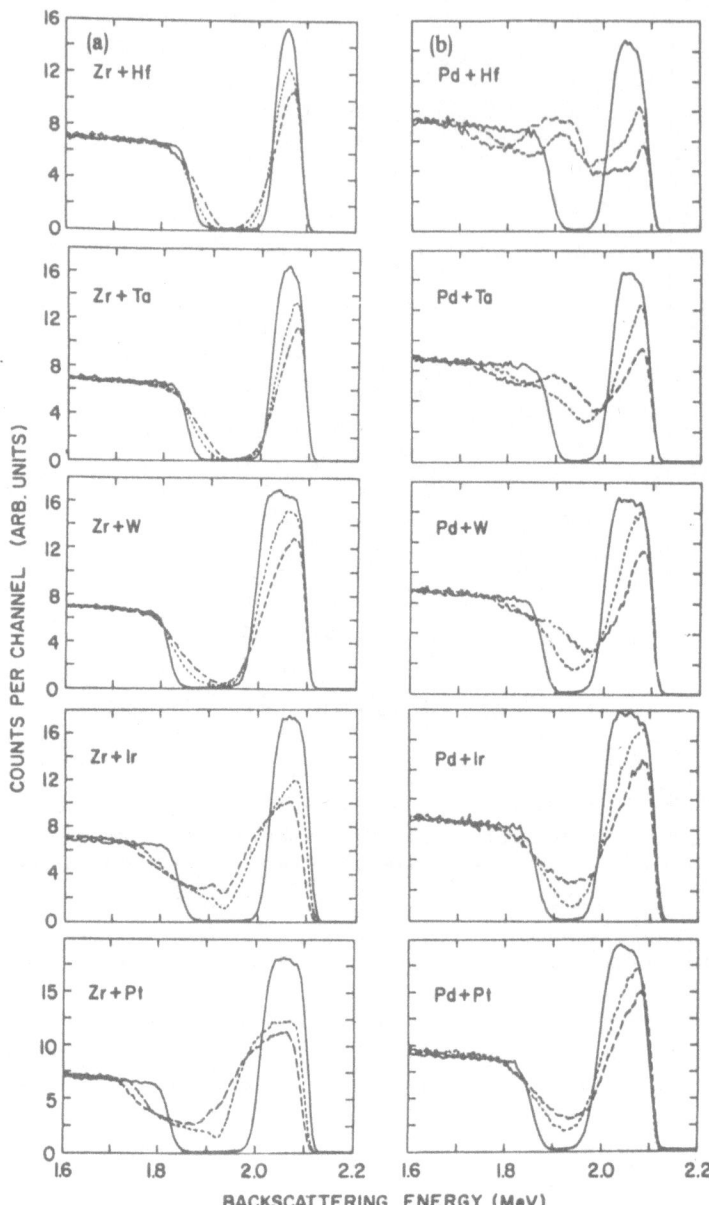

FIGURE 4. RBS spectra showing the extent of Kr ion beam mixing by doses of 1×10^{16} Kr/cm^2 and 2×10^{16} Kr/cm^2 with reference to unirradiated samples (a). Compare the interaction of Zr with a variety of metals when the chemical affinity with Zr increases down the page from Hf to Pt. (b) similarly compare the interactions with Pd where the affinity increases from Pt up to Hf.

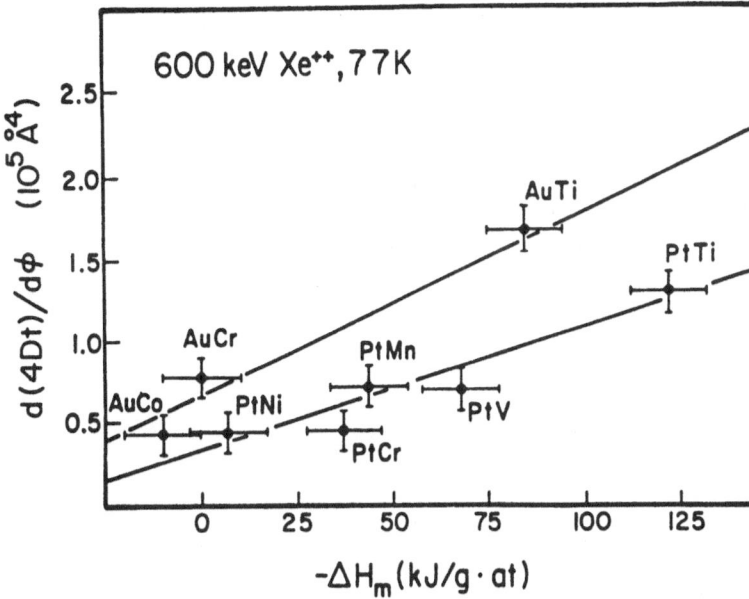

FIGURE 5. Correlation between the mixing rate d(4Dt)/dφ and Miedema's heat of mixing ΔH_m for bilayers irradiated with 600 keV Xe at 77 K.

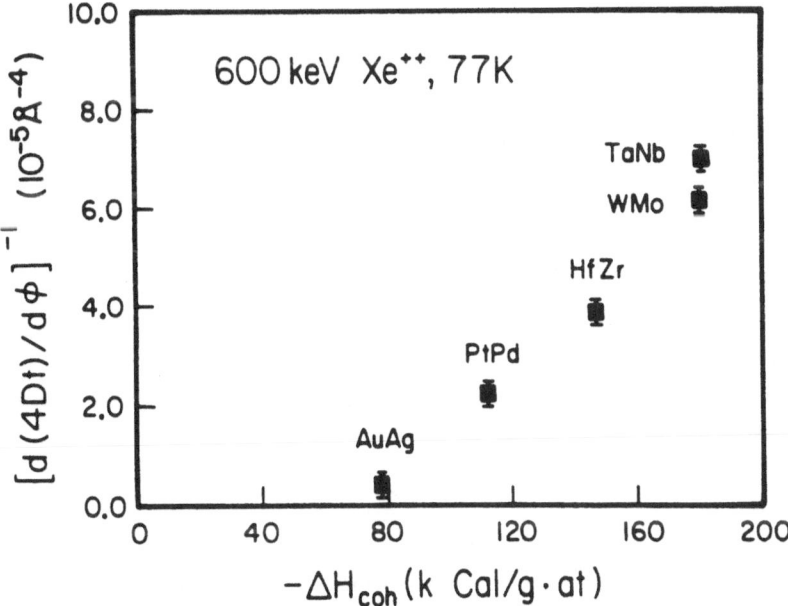

FIGURE 6. Correlation between the mixing rate d(4Dt)/dφ and the average cohesive energy ΔH_{coh} for systems with a near zero heat of mixing.

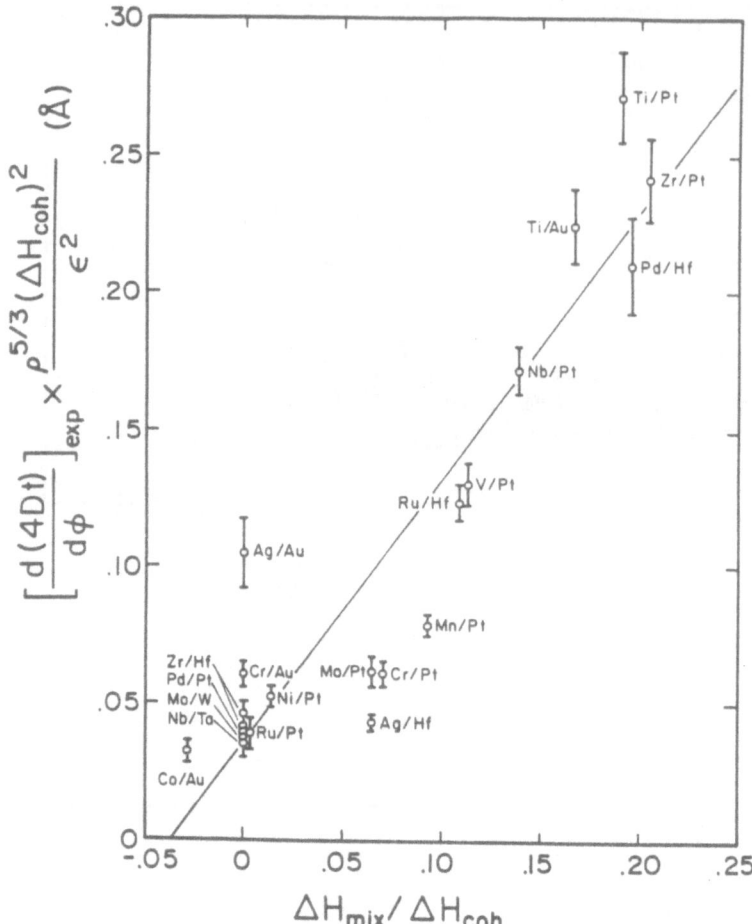

FIGURE 7. Summary of results of ion mixing of metallic bilayers.

or even positive there is no driving force except for ballistic effects. Large cohesive energies may counteract the tendency to mix, whereas small cohesive energies do not exert a mixing barrier. This semi-empirical model has been put into an equation based on Vineyard's classical work concerning thermal spikes (13). The Caltech group has come up with the expression:

$$\frac{d(4Dt)}{d\phi} = \frac{K_1 \varepsilon^2}{\rho^{5/3}(\Delta H_{coh})^2} \left(1 + K_2 \frac{\Delta H_m}{\Delta H_{coh}}\right)$$

Experimental support (14) for its validity is shown in fig. 7. It is shown that many different systems obey the above equation, in which ε is the energy/atom put into the spike and ρ its atomic density. There is a quadratic dependence on $(1/\Delta H_{coh})$ which may be rationalized by the fact that mixing is not only dependent on atomic displacement but also migration is necessary and both will be dependent on $(1/\Delta H_{coh})$. In addition, the equation shows the ratio of

$(\Delta H_m/\Delta H_{coh})$ because these two parameters counteractive: a negative heat of mixing in a system is a good driving force, but its action is hindered by the materials cohesive energy. In any case fig. 7 lends strong support for the phenomenological model for ion mixing. Of course, it does not prove the concept of spikes, let alone thermal spikes. Whereas it is reasonable to assume that chemical effects are important when the mean energy of all the atoms in the collision cascade has become of the order of 1 eV, we do not have any evidence for thermal equilibrium nor do we know in what phase (gas, liquid or solid) the material is inside the energy spike.

Eventually, after about 10^{-11} sec the region of the solid in which the collision cascade has occured must come into equilibrium with the ambient. At that stage the concentration of displaced atoms will have decreased significantly but the defect concentration will still be larger than in the surrounding material. This may lead to additional mass transport through radiation enhanced diffusion, the subject of the next section.

4. RADIATION ENHANCED DIFFUSION

It has taken the ion-beam community quite some time to notice the temperature dependence of ion mixing. Yet, this is a very strong effect and different target temperatures and beam

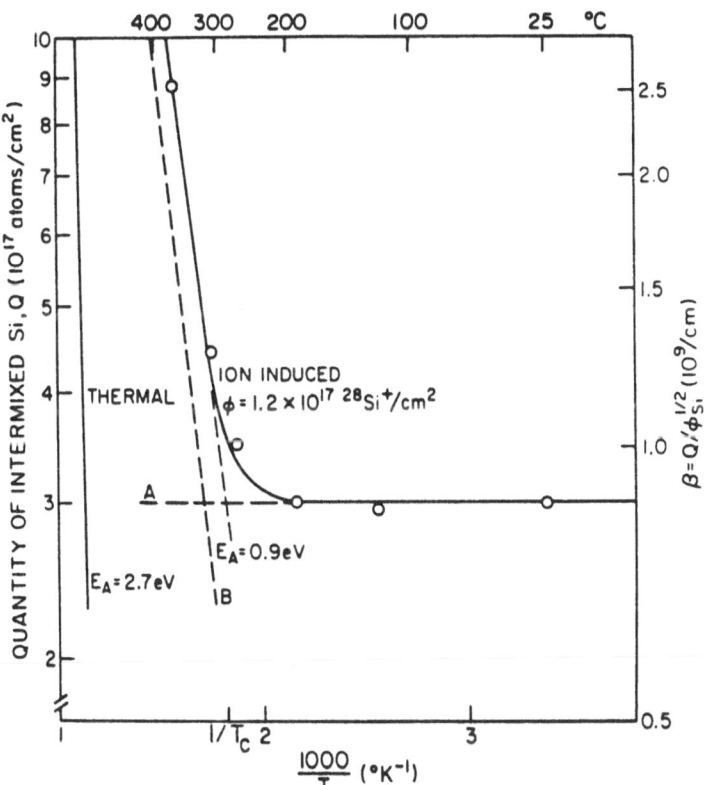

FIGURE 8. Logarithm of the quantity of intermixed silicon versus reciprocal temperature for a dose of 1.2×10^{17} Si+/cm^2.

heating have made it difficult to compare results from different laboratories and find any systematics. Fig. 8 gives an example of one of the early ion mixing studies as function of substrate temperature (15). Si+ ions have been used to mix Nb with Si in the temperature range of 25°C to 400°C. One can distinguish a temperature independent regime (A), from room temperature up to about 250°C, in which the quantity of intermixed Si is constant and equal to 3×10^{17} cm^{-2} for a beam dose of 1.2×10^{17} cm^{-2}. Above 250°C target temperature the quantity of intermixed Si increases dramatically with temperature to a value of 8.5×10^{17} cm^{-2} at 375°C. Note that this temperature dependent intermixing sets in at a lower temperature than what is necessary for thermal diffusion of Si in Nb. From the slope in fig. 8 one obtains an activation energy for the temperature dependent ion mixing which is equal to $E_A = 0.9$ eV, much smaller than the activation energy for purely thermal diffusion which has an activation energy $E_A = 2.7$ eV. This difference comes from the presence of the ion beam, which produces defects that need a high enough ambient temperature for migration but the defects do not have to be created thermally. So, the ion mixing which was discussed in the previous sections was

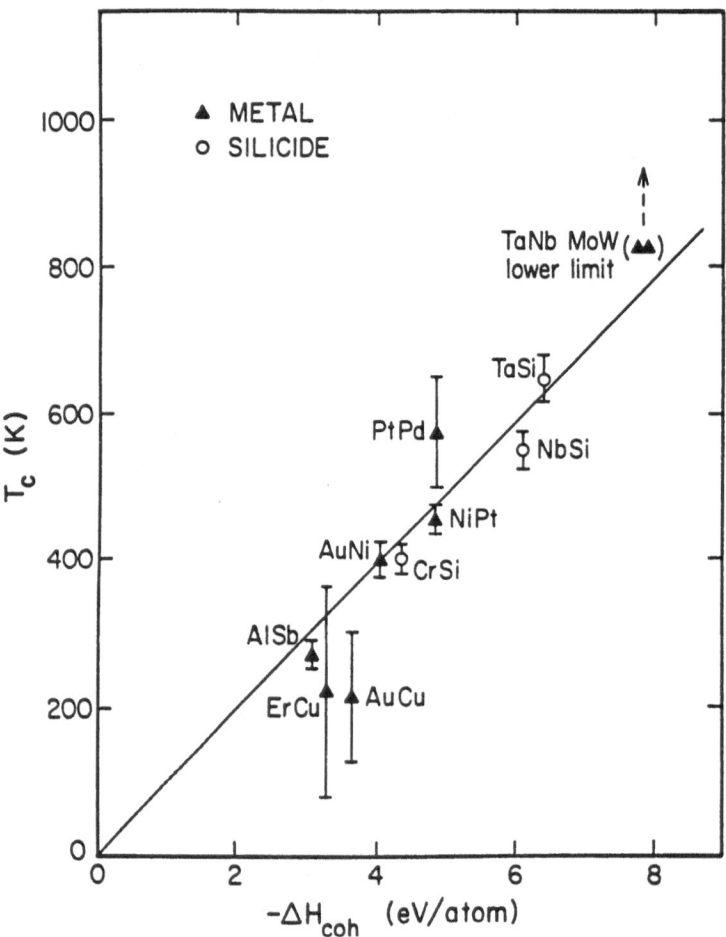

FIGURE 9. Correlation between the average cohesive energy ΔH_{coh} and the critical temperature T_c at which radiation-enhanced diffusion becomes dominant.

performed in the temperature independent regime A whereas at higher temperature, in regime B, ion mixing is always dominated by radiation enhanced diffusion.

Many systematic studies have been made of mass transport in the presence of radiation defects. In one system radiation enhanced precipitation may occur whereas in another radiation enhanced segregation dominates. Foreign atoms may diffuse into a substrate material along with the defect flux, but the inverse has been observed also. Not only the atom migration depends strongly on the kind and mobility of defects, also final location of foreign atoms in the ion irradiated matrix is affected by solute trapping. Unfortunately, although radiation enhanced diffusion dominates ion mixing in many systems especially at elevated temperatures, there are no general rules (4). The degree of mixing and the final state of a system depends in detail on the species involved and on the kind of radiation applied.

Although the outcome of radiation enhanced diffusion for individual systems may not be predicted, it is possible to know when radiation enhanced diffusion effects will dominate ion mixing. Clearly the cohesive energy of a material determines the temperature at which defects may be expected to become mobile. By measuring the quantity of intermixed atoms as function of substrate temperature one can determine the critical temperature, T_c, above which ion mixing is temperature dependent (in fig. 8 $T_c = 250°C$). Fig. 9 shows a nice correlation between measured critical temperatures, for metals as well as for silicide systems, as function of the cohesive energy (16). This is of importance especially if one is interested in metastable phase formation by ion mixing and radiation enhanced diffusion effects are to be avoided. For many systems, as fig. 9 shows, active target cooling during ion bombardment should be applied if long range diffusion and equilibration is to be avoided.

5. METASTABLE PHASE FORMATION

Ion bombardment allows the formation of metastable phases that cannot be formed in any other way. Ion implantation is also a convenient way of micro-alloying of a large variety of materials for metallurgical investiga-

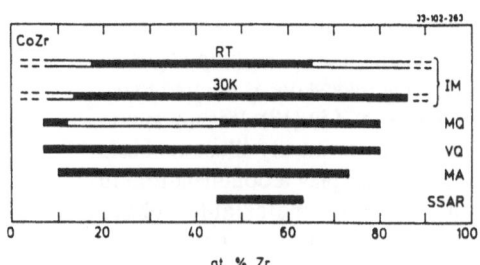

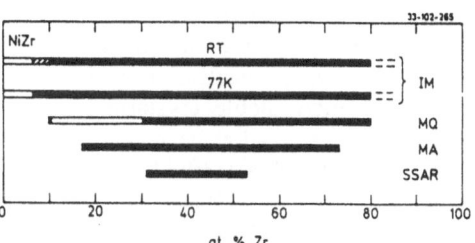

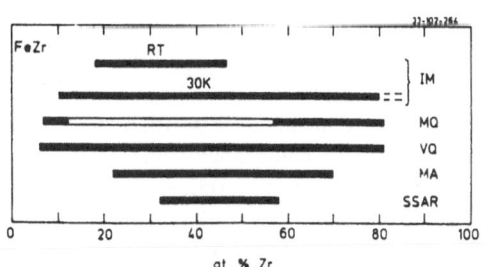

FIGURE 10. Glass forming range of CoZr, NiZr and FeZr with various techniques, IM: ion mixing, MQ: melt quenching, MA: mechanical alloying, VQ: vapour quenching, SSAR: solid state amorphization reaction.

114

tions. Indeed, one may speak of "ion implantation metallurgy" in modern materials science. In the previous sections it was discussed how the penetration of heavy energetic ions leads to alloying by ballistic mixing, cascade mixing and radiation enhanced diffusion. Ideally one would like to predict the phase formation from the essential properties of ion mixing and those of the phase diagrams of the alloyed species (17). Over the years a few attempts in this direction have been made and for a comprehensive and recent review one is referred to the work of B.X. Liu (18). It is my aim in this section to merely illustrate basic concepts that play a dominant role in ion induced metastable phase formation and especially amorphization.

Initially the synthesis of amorphous metal alloys was carried out by quenching from the liquid phase. To bypass nucleation and growth of more stable crystalline phases in undercooled melts, cooling rates in the range of 10^4 - 10^{10} K/s are required. This is easily obtained in ion mixing. If one considers the time evolution of collision cascades heating and cooling rates in excess of 10^{10} K/s should place ion mixing in a rather unique position, perhaps only surpassed by condensation from the vapour phase in sputter deposition or evaporation of amorphous films on low temperature substrates. Yet, amorphous phases do not only result from kinetic limitations. In solid state reactions, on a time scale of hours, one has demonstrated formation of amorphous phases driven by thermal diffusion and by mechanical alloying (19). So, under certain conditions the amorphous state may be more stable than the crystalline state. Due to irradiation, the final phases formed by ion

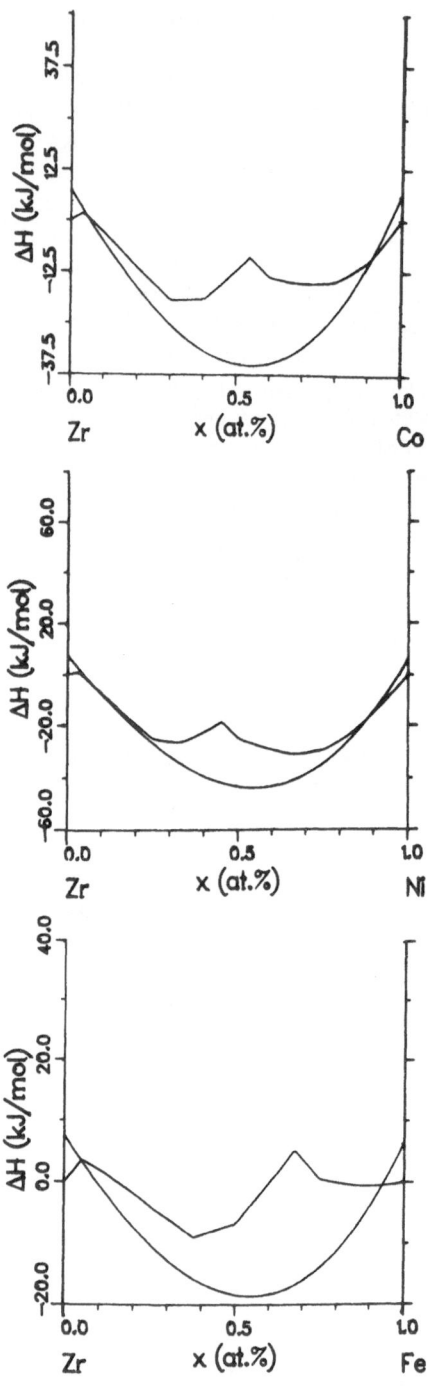

Figure 11. Comparison of the enthalpy differences of the amorphous phase and the solid solution of CoZr, NiZr and FeZr.

mixing have a high concentration of defects. Amorphization by irradiation, therefore, has been explained by the accumulation of stored energy, amorphous phases being formed if the free energy of a crystalline structure during irradiation becomes higher than the free energy of the amorphous phase.Although there is certainly some truth in the argument, it cannot be applied as a general rule, for irradiation of amorphous phases in most cases reduces the crystallization temperature. If one wants to arrive at metastable phases by ion mixing the ambient temperature must be low enough to avoid long range diffusion. In section 4 radiation enhanced diffusion was discussed and a correlation was shown between the critical temperature, T_c, for radiation enhanced diffusion and the systems cohesive energy. Here it may be added that metastable phases are more likely to result from ion mixing in the temperature independent regime, i.e. below T_c.

Attempts have been made to explain the phase formation during ion mixing by focussing on the cooling of the energy spikes. Hereby, the phase formation during ion mixing is thought to be similar to fast quenching from the melt. Estimated quench rates in spikes are of the order of 10^{12} K/s. This model is too simplified, not only because there is as yet no evidence for thermal equilibrium in the spike and for a molten phase (as was discussed in section 3), but also because the composition range over which amorphous alloys are obtained by ion mixing and melt quenching differ significantly. Fig. 10 shows a comparison by Bøttiger et al. (20) of the glass forming range in Zr based alloys. As a function of Zr concentration the regions are indicated over which amorphous alloys are obtained after ion mixing (at room temperature and at low T), after melt quenching, after mechanical alloying, after vapour quenching, and after solid state reaction. Clearly, different methods yield different glass forming ranges.

Metastable phase formation in ion mixing may be successfully explained with the aid of calculated phase diagrams comparing the enthalpy differences between the amorphous phase and the metastable solid solutions (21). Fig. 11 shows such a comparison for Ni-Zr. The smooth drawn line represents the enthalpy of the amorphous phase as function of Ni concentration in Zr. In addition to the left of the diagram the calculated enthalpy of the hcp phase as function of composition is shown. At about 45% Ni the crystalline phase peaks in energy well above the amorphous phase. In the region between 45-100% Ni the fcc crystal structure is lower in energy than the hcp structure but only above 90% Ni does the crystalline line come below the amorphous line. Clearly, the diagram indicates that over the composition range from 10-90% the amorphous phase has the lowest free energy. It is precisely in this region that amorphous phases are observed following ion mixing of NiZr at low temperature. Although at an early stage there will be enough energy in the ion indiced collision spike to form metastable solid solutions, the system relaxes to the lowest free energy state that it can reach without long range diffusion. Note that crystalline compound phases have not been included in the diagram of fig. 11. For compounds to form not only nucleation of the crystalline phase is needed but also long range diffusion. If ion mixing is done at low enough temperatures only polymorphous transitions in the phase diagram will occur and thus the system will go through the metastable solid solution down into the amorphous state in the composition range of 10-90%, outside this region on the one hand the hcp structure has the lowest free energy, on the other it is the fcc metastable solid solution that is expected to form. For the other Zr based alloys similar diagrams can be constructed and the above arguments may be applied, see fig. 11. The experimentally observed glass forming ranges are consistent with the picture where phases form during ion mixing that have the lowest free energy and that can be reached in polymorphous transitions, i.e. without long range diffusion.

116

This work is part of the research program of the stichting voor Fundamenteel Onderzoek der Materie (Foundation for Fundamental Research of Matter) and was made possible by financial support from the Nederlandse Organisatie voor Wetenschappelijk Onderzoek (Netherlands Organization for Advancement of Research).

REFERENCES

1. Van der Weg WF, Sigurd D, Mayer JW: Applications of Ion Beams to Metals. Edited by Picraux ST et al. Plenum Press, New York 1974, p. 209.
2. Averback RS: Nuclear Instr. & Methods **B15** (1986) 675.
3. Surface Modification and Alloying by Laser, Ion and Electron Beams. Edited by Poate JM et al. Plenus Press, New York 1983.
4. Surface Alloying by Ion, Electron and Laser Beams. Edited by Rehn LE et al., ASM, Ohio 1987.
5. Westendorp JFM, Saris FW, Koek B, Vliegers MPA, Fenn-Tyne I: Nucl. Instr. & Methods **B26** (1987) 539.
6. Paine BM and Nicolet MA: Nuclear Instr. & Mehods **209/210** (1983) 173.
7. Westendorp JFM, Littmark U, Saris FW: Nucl. Instr. & Methods **B18** (1986) 54.
8. Roush et al.: Nuclear Instr. & Methods **209/210** (1983) 67.
9. d'Heurle F, Baglin JEE, Clark GJ: J. Appl. Phys. **57** (1985) 1426.
10. Cheng YT, Van Rossum M, Nicolet MA, Johnson WL: Appl. Phys. Lett. **45** (1984) 185.
11. Cheng YT: Thesis Caltech 1987.
12. Johnson WL, Cheng YT, Van Rossum M, Nicolet MA: Nuclear Instr. & Methods **B7/8** (1985) 657.
13. Vineyard GH: Rad. Effects **29** (1976) 245.
14. Workman TW, Cheng YT, Johnson WL, Nicolet MA: Appl. Phys. Lett. **50** (1987) 1485.
15. Matteson S, Roth J, Nicolet MA, Rad. Effects **42** (1979) 217.
16. Cheng YT, Zhao XA, Banwell T, Workman TW, Nicolet MA, Johnson WL: J. Appl. Phys. **60** (1986) 2615.
17. Nastasi M: PhD thesis Cornell Univ. 1985.
18. Liu BX: Phys. Stat. Sol. A **94** (1986) 11.
19. Samwer K: Physics Reports **161** (1988) 1.
20. Bøttiger J, Dyrbye K, Pampus K, Poulsen R: to be published.
21. Cohesion in Metals, De Boer FR et al.: North Holland, Amsterdam 1988.

ION BEAM MIXING OF METALS AND CERAMICS — MATERIAL CONSIDERATIONS*

CARL J. McHARGUE*

Metals and Ceramics Division, Oak Ridge National Laboratory,
Oak Ridge, Tennessee 37831-6118

1. INTRODUCTION

Most studies on ion beam mixing have been concerned with metal-metal or metal-silicon systems. The great activity in this field is attested by the fact that a 1983 review tabulated data for 38 bilayer and 25 multilayer systems (1). The initial models for mixing focused on the dynamics of ion-solid interactions and produced models based on direct recoils, cascades, and enhanced diffusion due to radiation-produced defects.

An earlier chapter contains a description of ion beam mixing processes and the mechanisms that have been proposed to describe material transport across the layer/substrate interface. In this chapter, the issues raised by the application of ion beam mixing to a system where at least one component is an insulating compound, e.g., a ceramic, are discussed. In order to narrow the subject, this chapter will deal with some of the questions arising from ion bombardment of thin metal films and insulator substrate systems.

Much of the interest in ion beam mixing as applied to metal-insulator couples arises from the observation that bombardment with an ion beam often increases the adhesion of the film to the substrate. However, this chapter will not address that issue but will discuss the role of the materials properties that influence the microstructures so produced.

2. THERMODYNAMICS AND MIXING

Recent studies have suggested that the thermochemical properties of the participating species exert important effects. The compositions of layers formed at the interface between metal films and substrates (in the bilayer configuration) are often those that form during thermal annealing (2-5). In the case of transition metal-silicon systems, Lau et al. (4) have correlated the metal silicide formation during ion irradiation with the stable phases that lie near eutectic compositions in the respective equilibrium phase diagrams.

It has been proposed that the enthalpy of reaction between the metal and substrate is an important parameter in determining the nature of the inter-facial reactions that occur during ion irraditions. For metal-metal couples, Westendorp et al. (6) and Wang et al. (7) associated mixng rates with the solubility of the metals, which is determined in part by the enthalpy of chemical mixing. D'Heurle et al. (8) correlated the amount of mixing between metal bilayers with the square of the electronegativity difference between the constituent metals. The electronegativity is proportional to the enthalpy of mixing.

*Research sponsored by the Division of Materials Sciences, U.S. Department of Energy, under contract DE-AC05-840R21400 with the Martin Marietta Energy Systems, Inc.

C.J. McHargue et al. (eds.), Structure-Property Relationships in Surface-Modified Ceramics, 117–134.
© 1989 by Kluwer Academic Publishers.

Observations in differences in the amount of mixing induced in collisionally similar but chemically different systems again suggests that
cascade mixing is strongly influenced by local chemical processes. The
systems Cu-Au-Cu and Cu-W-Cu are collisionally similar (i.e., the relative
masses Cu/Au and Cu/W are similar) but are quite different chemically
(thermodynamically). The equilibrium phase diagrams show extensive solid
solubility for Cu and Au and the formation of ordered intermediate phases
whereas Cu and W are essentially immiscible. The amount of mixing in the
Cu-W-Cu samples was about that expected from calculations of recoil mixing
alone while that in the Cu-Au-Cu samples was an order of magnitude
greater (2).

Starting with the premise that the variations in collisionally similar
but chemically-dissimilar systems can be explained by thermodynamic characteristics, Johnson and co-workers (9) proposed a model in which the
final distribution of chemical species is determined by chemically biased
diffusion within a thermal spike. This model uses the differences in
enthalpies of mixing (ΔH_m) and cohesive energies (ΔH_c) to analyze the
variations in the amount of mixing between systems with similar collisional
properties. Their experiments involved both heavy ions and heavy metal
matrices ($Z > 20$) where dense collision cascades are likely to occur and low
temperature irradiation where radiation-enhanced-diffusion is suppressed.
The model predicts a mixing rate given as

$$\frac{d(4\ Dt)}{d\phi} = K_1 \frac{\epsilon^2}{\rho^{5/3}\ (\Delta H_c)^2}\ (1 + K_2 \frac{\Delta H_m}{\Delta H_c})\ , \tag{1}$$

where (4 Dt) is the variance of the spreading at the bilayer interface,
ϕ is the ion fluence, ϵ is the energy deposited per unit path length,
ρ is the average atomic density, and K_1 and K_2 are fitting constants.

Figure 1 shows the correlation between a normalized mixing parameter and
$\Delta H_m/\Delta H_c$ for bilayers of 4d-5d metals irradiated with 600 keV Xe^{++} at 77 K.
In this correlation, a regular solution model is assumed and the values of
ΔH_m, ΔH_c, and ρ used are those for a composition of $A_{50}B_{50}$.

In an attempt to rationalize many observations on metastable phase formation during ion mixing, Lilienfeld et al. (13) qualitatively introduced
kinetic factors in their guidelines. Crystalline phases form during mixing
for constituents that normally form: (a) solid solutions with extended
solubilities; (b) intermetallic compounds with simple crystal structures
(e.g. CsCl-type) with wide phase fields; or (c) complex crystal structures
with high atomic mobility. Amorphous alloys result during mixing if the
constituents normally form compounds with simple crystal structures but with
narrow phase fields or complex crystal structures.

The enthalpy of formation (ΔH_f) or the free energy of formation (ΔG_f) has
often been used to predict the behavior of metallization wherein thin metal
films are deposited on various substrates. Pretorius et al. (14) considered
the heats (or enthalpy) of formation for all reactions of the type

$$SiO_2 + M_x \rightarrow M_ySi + M_{x-y}\ O_2 \tag{2}$$

for a series of transition metals in contact with SiO_2 at 800°C. If the sum
of the enthalpies of the products was less than that of the reactants, it
was assumed that a chemical reaction should occur. In this experiment, the
calculations and observations agreed with the exception of Cr and Mn which
were borderline cases.

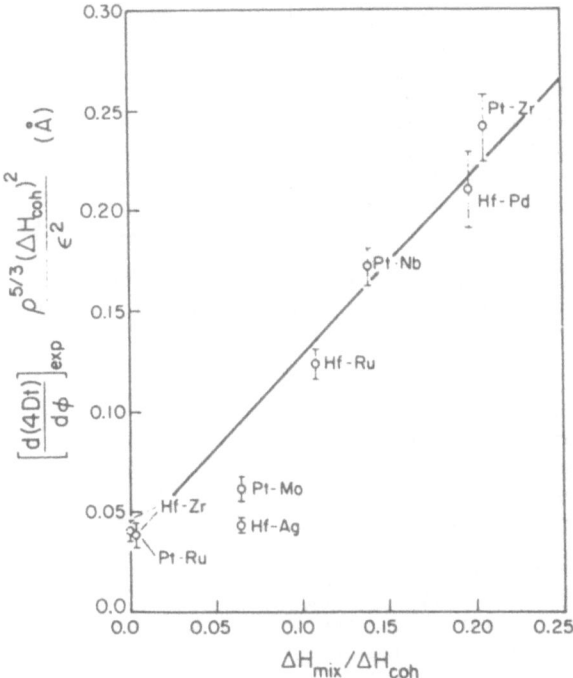

FIGURE 1. Correlation between the normalized mixing parameter and $\Delta H_{mix}/\Delta H_{coh}$ for bilayers irradiated with 600 keV Xe^{++} (ref. 12).

Banwell et al. (15) used the same approach to analyze ion beam mixed samples of Cr, Ni, and Ti on SiO_2 substrates. Farlow and co-workers (16) tested this enthalpy rule for beam mixing of 33 materials combinations and reported that mixing always occurred when predicted. However, mixing occasionally occurred when not predicted by the calculation. The use of the enthalpy rule to predict or rationalize results of ion mixing experiments has become common, although as will be discussed later, this approach can lead to incorrect conclusions.

3. ION-SOLID INTERACTIONS IN CERAMICS

Before proceeding to discuss the results of ion mixing experiments involving insulators or ceramics, some discussion of chemical bonding, crystal structure, and ternary multicomponent systems is required.

The interaction of energetic ions with ceramics differs from metals in a number of significant ways. Ceramics contain two or more chemical species distributed over at least two sublattices, usually in a very ordered manner. Hence atom (ion) A is unlikely to be found at a B sublattice site. The atoms have different atomic masses and the energy of displacement due to elastic collisions may be different. The influence of these factors on damage (defect) production was discussed in the paper by Parkin in these proceedings. There is a range of chemical bonding types in the materials of interest; ionic to covalent to "metallic-like." The type of bonding influences the kind of defects that are produced during irradiation which, in turn, may influence the mobility of various atomic species. In ionic crystals, the introduction of a defect (or impurity) must provide for local electrical neutrality and charged-point defects or impurity-point defect complexes often are produced. Catlow has discussed the charge compensation around impurities (17) in Al_2O_3 and Thevenard, in an earlier chapter, has discussed the types of microstructures produced by irradiation.

3.1. Charge states of implanted cations

Implantation of metallic ions (cations) into insulators must drive the system toward some nonequilibrium state since an excess positive charge is introduced unless the implanted ions aggregate to form metallic precipitates. Even though the ions occupy a well-defined charge state as they exit from the accelerator, it is the residual charge state of the implanted species in the solid that will determine the nature of the defect state created to compensate for the excess positive charge.

Among the methods that have been used to study the charge (valence) state of implanted cations in insulators are conversion electron Mössbauer spectroscopy (CEMS), x-ray photoemission spectroscopy (XPS), extended x-ray absorption fine structure (EXAFS), and x-ray absorption near-edge structure (XANES).

The most definitive results have come from CEMS studies of iron implanted into MgO (18), alkali halides (19), TiO_2 (20), and Al_2O_3 (21,22). Some less direct indications of the existence of a variety of charge states have been obtained from XPS, and XANES/EXAFS studies of titanium-implanted sapphire (23,24).

From a measurement of the Mössbauer spectra from an implanted ion, one can deduce information about its electronic structure, its position in the matrix, the local symmetry, and the local magnetic properties. Research groups at the Oak Ridge National Laboratory and the University of Claude Bernard have conducted a series of studies on the charge state of ^{57}Fe implanted into sapphire at room temperature (10^{16} to 10^{17} Fe/cm^2, 160 keV) (ref. 21,22). Conversion electron Mössbauer spectroscopy was used in conjunction with TEM and RBS-C.

Micrographs obtained by TEM show the as-implanted microstructure to consist of tangled arrays of dislocations extending from the surface to a depth of about 170 nm. The corresponding electron diffraction patterns showed that the implanted zones remained crystalline.

The CEMS spectra measured at room temperature are given in Fig. 2. The spectra consist of the superposition of several overlapping components. The following components were identified on the basis of consistent sets of computer fits for all the spectra: three quadrupole split doublets and one single line. These components can be assigned to a ferric ion (Fe^{3+}), two forms of a ferrous ion (Fe^{2+}_I and Fe^{2+}_{II}) and metallic iron (Fe^0). The fluence dependence of the relative amount of each component is given in Fig. 3.

The ferrous iron ions, Fe^{2+}, are described by two different quadrupole-split doublets. The parameters for the Fe^{2+}_I component are similar to those for the Fe-O bond in wustite (25). The Fe^{2+}_{II} component represents a more ionic state and is consistent with iron in $FeAl_2O_4$ (ref. 24). Since $FeAl_2O_4$ has a partly inverse spinel structure, the iron likely resides in octahedral sites, consistent with the relatively large value of the isomer shift (IS) observed. Essentially all the iron resides in the ferrous states at the lower concentrations (fluences) of iron. The relative amount of Fe^{2+} decreases as the implantation fluence increases (Fig. 3), although the total amount in each state continues to increase.

The ferric iron, Fe^{3+}, is represented by a doublet having the parameters: IS ≈ 0.22 $mm \cdot s^{-1}$, quadrupole splitting (QS) ≈ 1.0 $mm \cdot s^{-1}$. These values are comparable to those of iron substitutionally located in alumina (Al_{1-x} $Fe_x)_2O_3$ (ref. 12) and are indicative of covalent-distorted octahedral

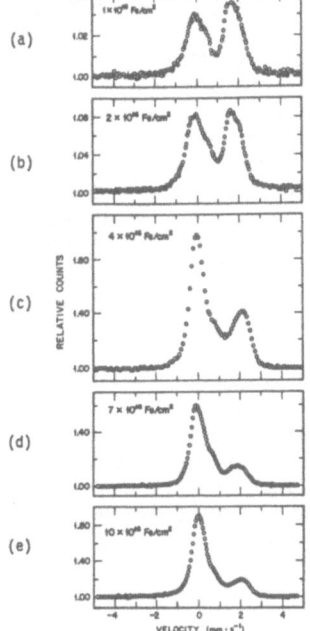

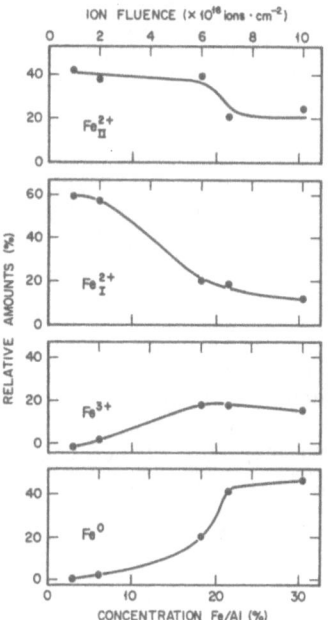

FIGURE 2. Conversion electron Mössbauer spectra measured at room temperature on α-Al₂O₃ implanted with ^{57}Fe:
(a) 1 x 10¹⁶ ions·cm⁻² (160 keV);
(b) 2 x 10¹⁶ ions·cm⁻² (160 keV);
(c) 4 x 10¹⁶ ions·cm⁻² (160 keV);
(d) 7 x 10¹⁶ ions·cm⁻² (160 keV);
(e) 1 x 10¹⁷ ions·cm⁻² (160 keV).

FIGURE 3. Fluence (i.e., concentration) dependence of relative amounts of components in Mössbauer spectra of iron-implanted Al₂O₃ (ref. 21).

122

surroundings. The relative amount of this Fe^{3+} component increases from
about zero at 1×10^{16} Fe/cm² to a maximum of 20% at 1×10^{17} Fe/cm².

The single line with IS ≈ 0 mm·s^{-1} is attributed to small metallic iron
clusters (precipitates) which behave superparamagnetically. The spectra
taken at 77 K were similar to the room temperature spectra (Fig. 2e) but
those taken at 4 K show magnetic splitting. The spectrum taken at 4 K indicates a superparamagnetic sextet and the Fe^{2+} doublet which distorts the
symmetry of the line shapes. Thus, this component arises from small
metallic clusters or precipitates with a size of ≈ 2 nm. The relative fraction of this component increases with fluence and represents about 48% of
the implanted iron at the highest fluence. The presence of these small
metallic iron precipitates has been confirmed recently by a TEM study (28).

The equilibrium phase diagram for the Fe-Al-O system at 1000°C is shown in
Fig. 4 of ref. 29. If thermodynamic equilibrium conditions were maintained
as Fe is added (implanted) to Al_2O_3, the composition will follow the tie-
line between Fe and Al_2O_3 and a two-phase mixture of these components would
be formed. Slight deviations would be expected to give mixtures of Fe-Al
solid solution and Al_2O_3 or iron + $Fe(Fe,Al)_2O_3$ spinel + Al_2O_3. The low-
fluence implants do not contain the implanted iron in such configurations.
The phase diagram suggests that the local regions are behaving more or less
independently, perhaps on a scale as small as the individual cascade or
thermal spike.

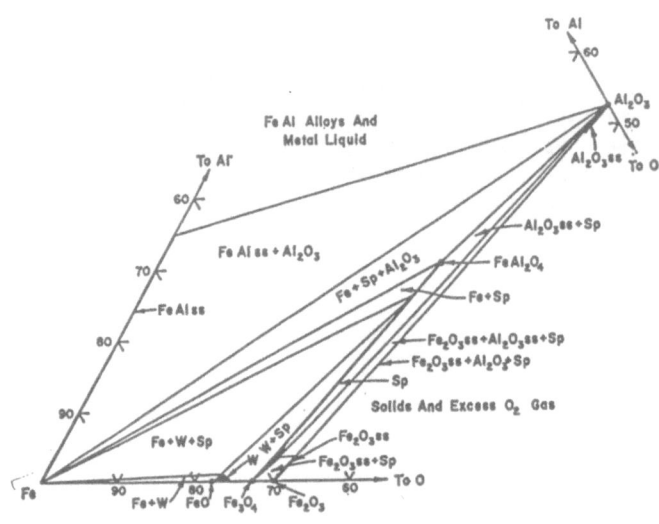

FIGURE 4. The Fe-Al-O phase diagram at 1000°C (ref. 29).

Similar studies have been conducted on iron-implanted alkali halides (19), MgO (18), and TiO_2 (3). The spectra from LiI, NaF, NaCl, KIF, KBr, KI, and RbCl are consistent with the implanted iron being in the Fe^{3+} state on cation sites and associated with two vacancies ($V^- - Fe^{3+} - V^-$). Additional defect complexes could not be ruled out. The CEMS spectra for LiF and KCl implanted with iron indicated that considerable fraction of Fe^{2+} and Fe^0 were also present, amounting to 40 and 30%, respectively, of the total iron content. The presence of Fe^{2+} might have been associated with second-phase precipitates or other Fe-V complexes.

The three charge states Fe^{3+}, Fe^{2+}, and Fe^0 were also found in MgO. In that material, most (>70%) of the iron occupied the Fe^{3+} state at low fluences and relative occupancy of this state decreased with an increase in the Fe^{2+} and Fe^0 states at higher fluences. About 3-5% of the iron was present as metallic clusters at a fluence of 8×10^{16} Fe/cm^2 (100 keV). The results suggest that Fe^{3+} in MgO exists only as an isolated impurity. The presence of another iron ion at a distance closer than four coordination spheres (cation) causes a change of the valence state, to either Fe^{2+} or Fe^0. The Fe^{2+}_{II} state represents the probability of finding one other iron as a closer neighbor (i.e. in the first cation coordination sphere), while the Fe^{2+}_I state represents a random distribution. The concentration of Fe^0 is proportional to the probability of finding three or more iron atoms in neighboring cation sites.

The studies of the charge state of titanium implanted into sapphire (23,24) indicate a range of charge states to exist in the implanted state but few conclusions can be drawn regarding their exact nature and relative amounts. Bull's AES data (24) suggested the presence of metallic titanium, but its presence was not confirmed by other measurements.

Annealing the iron-implanted sapphire in oxygen oxidizes the implanted iron to the ferric state (Fe^{3+}) and the final structure is a mixture of Al_2O_3 and a defect-spinel $(Fe_xAl_{1-x})_2O_3$. This spinel is similar in structure to $\gamma-Al_2O_3$ and contains all the iron in the 3+ state. In this annealing, oxygen is added to the system by diffusion from the surface. Annealing in hydrogen reduces the iron to the metallic state Fe^0 and TEM photographs confirm the presence of large α-Fe particles.

3.2. <u>Amorphization of sapphire by ion bombardment</u>

Sapphire can be rendered amorphous by displacement damage if the implantation temperature is low enough to suppress dynamic recovery events (30-32). Stoichiometric implants of Al(4×10^{16} Al/cm^2, 90 keV) plus O(6×10^{16} O/cm^2, 55 keV) were made using substrate temperatures of 77 K. Both TEM and RBS-C measurements confirmed the presence of an amorphous surface layer approximately 155 nm thick. This fluence corresponds to damage of 2 to 3 displacements per atom (dpa). Similar results were obtained for implantation of Cr (3×10^{15} Cr/cm^2, 150 keV) and Ti(10^{16} Ti/cm^2, 150 keV) at 77 K.

Implantation of Cr and Ti at room temperature produced a damaged crystalline surface layer containing large numbers of dislocation loops and tangles. The lattice disorder measured by RBS-C saturated at a value of $\chi = 0.67$ at about 10 dpa and remained at that level to a fluence corresponding to 110 dpa.

There are clear indications that in other instances, "chemical" effects may be involved in the amorphization of sapphire. Figure 5 shows the disorder (by RBS-C) in the Al-sublattice produced by several cations for 300 K substrate temperatures. The defect structures produced by Cr, Ti, and Nb are similar but more disorder is produced by Zr-implantation and the amorphous state is reached at about 90 dpa for Zr-implantation. Implantation with Nb to the same fluence and at the same energy produced the same results as Cr-implantation; i.e., no amorphization. Since Zr and Nb differ

124

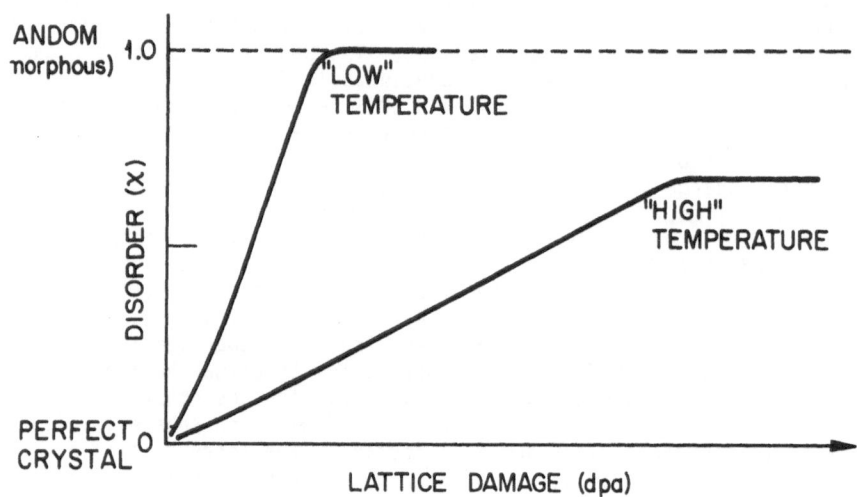

FIGURE 5.   Damage accumulation during implantation of ceramics.

by only one mass unit, the details of the cascades must be identical and the
amorphization by Zr must be chemical in nature.

We do not know if the zirconium inhibits the dynamic recovery of
irradition-produced defects and thus increases the rate of damage accumulation or acts during the quenching of the cascade to produce an amorphous
region. Recent TEM studies (33) show that the amorphous-crystalline
interfaces in buried amorphous layers is associated with a specific Zr concentration level rather than a specific damage level.

3.3. Stoichiometry in implanted ceramics

There have been several attempts to determine whether the disorder in the
two sublattices represents displacements in a stoichiometric ratio. The
experiments are usually based on RBS-C or nuclear reactions and have low
accuracies with respect to determining the number of displaced atoms (ions).

Naguib et al. (34), Turos et al. (35), and Drigo et al. (36) concluded
that Al- and O-ions were displaced in approximately a 3:2 ratio for implants
of 40 keV Kr, 30 keV Pt or Xe, and 100 keV Pb, respectively, into Al_2O_3.
The low fluence results of Naguib et al. and Drigo et al. seem to indicate a
greater number of oxygen displacements at low fluences but the ratio
approached 3:2 at higher fluences. Hart and co-workers (37) likewise
reported that implantation of Sb (9×10^{13} Sb/cm^2, 40 keV) into SiC
displaced equal numbers of Si and C atoms.

The requirements to maintain stoichiometry in Al_2O_3 are illustrated by the
results of Pells and Stathopoulos (38). Using electron bombardment in a high
voltage electron microscopy, these investigators determined that the displacement energies for Al and O are 18 and 76 eV, respectively. By proper
selection of the electron energy, they could study the form of damage when
only Al ions could be displaced by elastic collisions. The damage observed
by TEM after irradiation at about 800 K consisted of pure edge dislocation
loops of interstitial character and faulted with respect to the Al-
sublattice. A model for formation of stoichiometric loops when only Al-ions
were being displaced was proposed. It assumes that the displaced Al-ions

segregate to form a two-layer wide precipitate between an oxygen and an aluminum layer parallel to the (0001) plane. Oxygen diffuses from both sides of the loop to restore the electrostatic balance and creates a four-layer loop faulted on the cation lattice. Such a process would leave a stoichiometric ratio of Al- and O-vacancies in the lattice.

4. THE "ENTHALPY RULE"

In the past, heat of formation (ΔH_f) or Gibbs free energy of formation (ΔG_f) data have commonly been used to predict the interaction of thin metal films on ceramic substrates. For example, the heats of formation of metal oxides compared to SiO_2 has been used to predict the stability of various silicides in contact with SiO_2 substrates in oxidizing atmospheres (39). Pretorius et al. (40) used a similar approach to predict the reactions of metal films on SiO_2 substrates during vacuum annealing at 800°C.

The concept that mixing will occur during ion irradiation if the reaction enthalpy results in a net decrease in energy of the system has received considerable attention (15,16,41-43). The approach has been to write balanced chemical equations for all possible reactions between the metal film and the insulator substrate. The standard enthalpies of formation, $\Delta H°_f$, for each reaction is summed to give the enthalpy change (ΔH_r) under standard conditions. If ΔH_r is negative for any one reaction between metal and substrate, the enthalpy rule predicts mixing. If ΔH_r is positive for all possible reactions, the rule predicts no mixing and probably some sort of segregation will occur. Strictly, the free energy change should be considered rather than enthalpy change but in solid-solid systems, the entropy contribution is small.

Farlow and co-workers (16) published a survey of 33 metal film-substrate combinations that had been subjected to bombardment by Xe or Kr (1-2 x 10^{16} ions/cm²) and compared the occurrence or absence of mixing with the calculated enthalpy changes. The degree of mixing was determined from the RBS spectra. Ballistic mixing was not considered to violate the enthalpy rule and the possiblity of de-mixing was not considered.

A modified version of the table from Farlow et al. (16) is given as Table 1. It should be noted that in some cases the thermodynamic data from various sources gives significantly different results. In these material combinations, there appears to be four exceptions to the "enthalpy rule," Nb/Al_2O_3, V/Al_2O_3, Si/ZrO_2, and Cr/SiO_2 at 25°C. In the case of Si/ZrO_2, Farlow et al. (16) proposed that the compound $ZrSiO_4$ (whose enthalpy is lower than ZrO_2) formed and thus this system was not an exception to the rule. Notice that all the exceptions exhibit mixing where the rule predicts no mixing.

5. MIXING AND TERNARY PHASE DIAGRAMS

Beyers and co-workers (44,45) have proposed that ternary phase diagrams should be used to predict and underestand reactions between thin metal films and compound substrates. Singer and co-workers (46) have suggested that such diagrams would provide more useful guidelines for mixing and/or phase formation than the enthalpy rule. The enthalpy rule considers all possible reactions among the three elements (in the case of M/A_xB_y) and concludes that mixing will occur if any reaction enthalpy is favorable. The phase diagram, on the other hand, may indicate whether or not such a reaction occurs; however, again under equilibrium conditions.

It is instructive to examine the results for observations on mixing experiments involving Ti films on Si_3N_4 substrates. Bhattacharya et al. (43) studied the effects of 1.5 MeV Au+ on Ti/SiC and Ti/Si_3N_4 couples. Mixing occurred at the Ti/SiC interface but not at the Ti/Si_3N_4 interface.

TABLE 1. Mixing map for metal films – insulator substrates

Substrate / Film	Al_2O_3 (S.C.)	ZrO_2 S.C.	SiO_2	$Y\,PO_4$ (S.C.)	SiC (S.C.)	Si_3N_4
Ni					Y / SM	N / NM
Au	N / NM					
Cu	N / SA	N / SA	N / SA	N / SA	N / SA	
Cr	N / BM	N / NM	(N) / SM*	N / NM	Y / SM	
Ti	N / BM?	N / NM	Y / SM	N / NM		
Zr	N / BM		Y / SM			
Si	N / BM	(N) / SM		N / NM		
Nb	(N) / SM		Y / SM			
Pd	N / SA		N / SA			
V	(N) / SM		Y / SM			
W	N / NM		N / SA			
Al			Y / BM?	N / NM	N / SA	
Te			Y / SM			

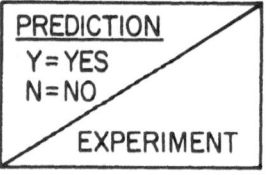

PREDICTION
Y = YES
N = NO
EXPERIMENT

BM = BALLISTIC MIXING
NM = NO MIXING
SA = SEGREGATION/AGGLOMERATION   SM = MIXING
*Cr/SiO_2 SM at 25°C, NM at 600°C

These authors note that there is a negative enthalpy for the reaction

$$Ti + SiC \rightarrow TiC + Si \tag{3}$$

and therefore mixing should occur.

The sign of the enthalpy of reaction for the Ti/Si_3N_4 couple depends upon the assumed products. The reactions

$$3\ Ti + Si_3N_4 \rightarrow 3\ TiSi + 4\ N \tag{4a}$$

and

$$3\ Ti + 2\ Si_3N_4 \rightarrow 3\ TiSi_2 + 8\ N \tag{4b}$$

both yield positive reaction enthalpies, indicating no mixing. However, the reaction

$$4\ Ti + Si_3N_4 \rightarrow 4\ TiN + 3\ Si \tag{5}$$

has a negative enthalpy, indicating mixing should occur.

Singer and co-workers (46) used the schematic ternary diagram for the system Ti-Si-N (shown in Fig. 6), to predict the phases that should form during alloying Ti with Si_3N_4. The compositions will fall on the dashed line drawn from the Ti corner to Si_3N_4. Until the concentration of Ti reaches a value of about 35 at. %, the phase composition should be Si_3N_4 + TiN + Si. Only if the Si_3N_4 were the dilute component would the formation of TiS_2 or TiS be expected.

Thus, it appears that the enthalpy rule may not be valid for this system. Singer (47) determined the phase composition of Si_3N_4 implanted with Ti (4×10^{17} Ti/cm^2, 190 keV) at ~900°C. The XPS and TEM examinations of the hot implants showed the presence of TiN and perhaps Si_3N_4, both as crystalline phases. Low temperature (40°C) implants apparently produced an amorphous Si_3N_4 "stuffed" with Ti and there was no evidence for any compound formation.

6. SOME RESULTS OF BOMBARDING METAL-INSULATOR COUPLES

Romana et al. (48) have attempted to mix Au or Ag into sapphire with 1.5 MeV Xe^+ at 77 and 300 K. Their RBS results suggests that only ballistic mixing occurred and optical absorption measurements suggest that at least some of the Au and Ag formed metal clusters or precipitates.

Mössbauer spectroscopy gives information on the "chemical" state and environment of implanted or ion beam mixed ions. This technique has been used to study the structures produced by ion beam mixing of Fe/SiO_2 (ref. 49), Fe/Al_2O_3 (ref. 50), and Nb/Fe_2O_3 (ref. 51). The results of these three studies have similar features. In each case the structure after high fluence mixing contained very small α-Fe clusters or precipitates. In the case of Fe/SiO_2, the clusters were reported to be less than 10 nm in diameter. In each system, an iron oxide phase and a ternary phase containing iron, oxygen, and the other component was present after mixing. The phases present are summarized in Table 2.

Also included in this table are the results of a Mössbauer study on iron implanted into Al_2O_3 at room temperature (21). At low fluences, i.e., very dilute concentrations of Fe in Al_2O_3, only the state Fe^{2+} was observed; part of it was in a FeO (wustite)-like bonding and part as in Fe Al_2O_4-bonds. bonds. At high fluences, the same Fe^{2+} states were present but also the states Fe^{3+} [as in $(Al,Fe)_2O_3$] and Fe^0 (metallic clusters of ~2 nm size). As much as 50% of the implanted iron was present as metallic clusters at a fluence of 10^{17} Fe/cm^2.

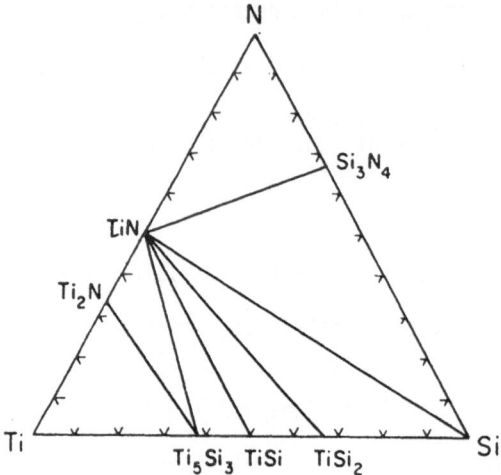

FIGURE 6. Ti-Si-N phase diagram. T = 700 – 1000°C (ref. 45).

TABLE 2. Phases identified by Mössbauer spectroscopy.

Ion beam mixing

    Nb/Fe_2O_3

        Fe^{2+}-Nb-O phase
        α-Fe precipitates

    Fe/SiO_2

        Fe-silicate
        Fe^{2+} in oxide phase
        Fe^0 clusters

    Fe/Al_2O_3

        Low fluence   High fluence
          Fe^{3+} in Al_2O_3 FeO-like bonds
          FeO-like bonds $FeAl_2O_4$

Implantation

    Fe into Al_2O_3

        Low fluence   High fluence
          FeO-like bonds   FeO-like
          $FeAl_2O_4$-like bonds $FeAl_2O_4$-like
                                    $(Al_{1-x}Fe_x)_2O_3$
                                      α-Fe clusters

The enthalpy rule predicts no reaction for the three (Au,Ag,Fe) metals with Al_2O_3 or for Fe with SiO_2. However, strong reactions are indicated for Nb with Fe_2O_3. Nevertheless metal precipitates formed in all instances and reactions did occur for both Fe and Nb.

Zhang et al. (49) suggested that the iron first injected into SiO_2 is quenched as a silicate or oxide and this is followed by radiation enhanced diffusion to allow the iron to cluster and precipitate. Thus, process thermodynamics and kinetics both may play roles in determining the final structure.

Farlow et al. (16) concluded that the RBS spectra for Zr/Al_2O_3 (Xe, 200 keV) could be described as ballistic mixing. However, Lewis and McHargue (52) analyzed the RBS data for Zr/Al_2O_3 bombarded with 1 MeV Fe^+ (2×10^{17} Fe/cm^2) in terms of recoil mixing plus cascade or enhanced diffusion. Figure 7 shows the concentration profile for Zr mixed into Al_2O_3 as well as the concentration profiles for diffusion and binary collision calculations.

Doi and co-workers (53) noted mixing occurred for Zr/Al_2O_3 when using N^+ (2×10^{17} N/cm^2, 200 keV) as the mixing ion. Figure 8 shows their concentration profiles. These data indicate that a small amount of Al was mixed into the Zr film but that a larger amount of Zr was transferred across the interface into the Al_2O_3. Cross-sectional TEM examination did not detect a distinct interface between the zirconium and sapphire after the ion beam treatment. There was no evidence for a second phase in their TEM micrographs.

Banwell and co-workers (15,54,55) conducted a series of experiments on the mixing of transition metals with SiO_2. The metals Ti, Cr, and Ni were chosen on the basis of being collisionally similar but significantly different with respect to chemical behavior with SiO_2. There are many thermodynamically favorable reactions between Ti and SiO_2, a few for Cr and SiO_2 and none for Ni and SiO_2. It was originally thought that such a series would clarify the role of thermodynamics in the mixing of metals-insulators as it had for metals with metals or semiconductors (6-10). However, at room temperature and below, there was essentially no difference in the composition profiles after mixing although cross-section TEM examination showed differences in the physical form of the metals. The Ti samples exhibited crystalline clusters and a band of amorphous or partly crystalline material. The lattice spacings of the crystalline clusters and XPS data indicated that the crystalline phase was a titanium oxide, but the presence of $TiSi_2$ could not be ruled out. There was no evidence for a crystalline phase or clustering in the Cr specimens. Most of the nickel was located within a 5-nm region centered 5 nm from the interface. There were Ni-containing crystalline clusters (2 nm diameter) that TEM and XPS indicated to be NiO. The transport of the metals into SiO_2 appeared to be the same in spite of their chemical differences. The chemical effects were expressed in the final morphology (i.e., phases). Banwell et al. (55) argue that kinetic constraints impede the influence of chemical driving forces in these systems at room temperatures and below.

7. TEMPERATURE EFFECTS

The temperature of the substrate during ion bombardment significantly affects the mixing processes but there have been few systematic studies to date. Whereas, Banwell and co-workers (15,54,55) found little difference in the concentration profiles for Ni, Ti, and Cr films on SiO_2 substrates bombarded with Xe at 77 or 300 K. They did find differences in the microstructure of the interface region. Using a substrate temperature of 750 K, however, gave strikingly different distributions of Ti and Cr compared

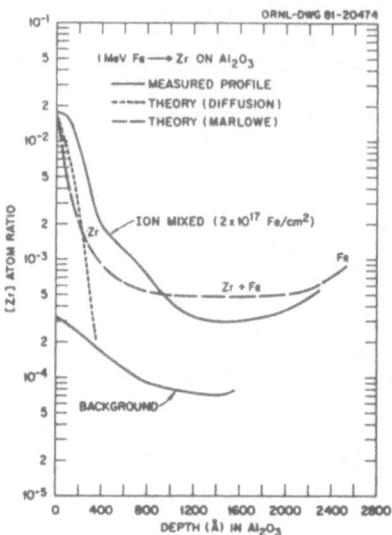

FIGURE 7.   Depth concentration profiles of a Zr coated Al_2O_3 sample following Fe^+ ion irradiation at $T \le 300°C$ (ref. 52).

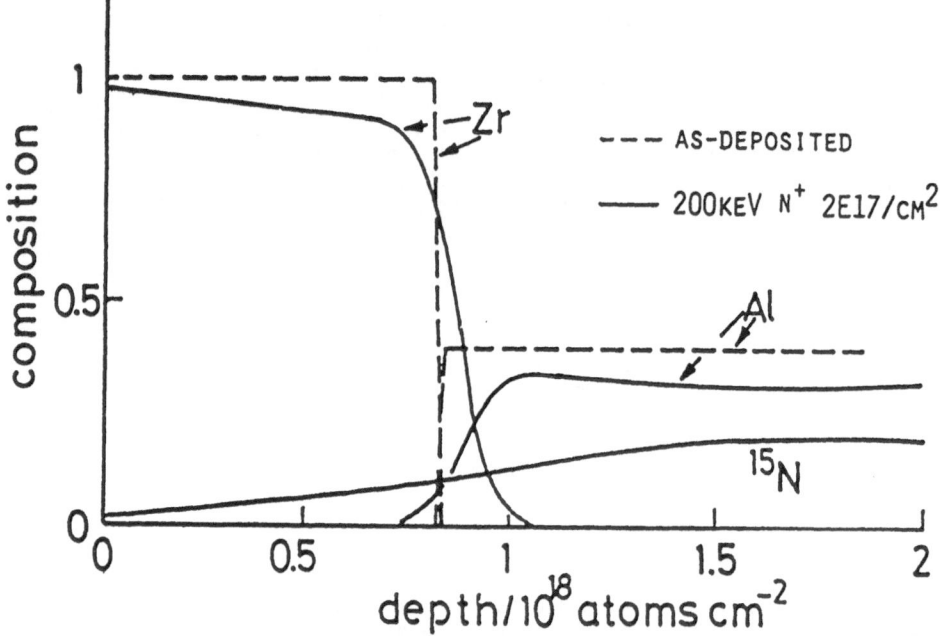

FIGURE 8.   Changes of concentrtion profiles at the interface of Zr/sapphire by $^{15}N^+$ ion irradiation (ref. 53).

to that for Ni. Figure 9 shows plots of areal density of metal, $(M)_S$, in the SiO_2 versus Xe fluence for 77, 300, and 750 K substrate temperatures. Enhanced mixing is evident for both Ti and Cr at 750 K while suppressed mixing is indicated for Ni. The RBS profiles indicate that only the interfacial region was modified and the deeper portion of the metal profiles (i.e., recoils) was unaltered. Unfortunately, this study did not provide information on the microstructural differences.

Farlow and co-workers (16) reported that Cr was mixed into SiO_2 at room temperature but not at 600°C under ion beam conditions similar to those used by Banwell et al. (15). The reason for the different results in these two studies is not known.

Lewis and McHargue (52,56) reported that raising the substrate temperature from 300 to 600°C caused a 50% increase in the amount of Cr transported across the interface into Al_2O_3 when bombarding with 4 MeV iron ions. A larger effect was evident for Zr films for which there was only slight penetration at 300°C but much more at 600°C. A further increase to 900°C markedly decreased the mixing in the case of Cr films. Excessive evaporation of the zirconium film in the vacuum chamber prevented that material from being studied at 900°C. The reason for the reversal in the effect of temperature on the mixing of Cr into Al_2O_3 is not known.

8. OPPORTUNITIES AND NEEDS FOR STUDY

It is clear that the field of ion beam mixing involving insulators (ceramics) is in an early stage of understanding and application. Relatively few material combinations have been studied and the range of experimental conditions has been limited. The nature of the materials must be a factor in selecting experimental conditions.

As discussed earlier, insulators include compounds that exhibit little or no deviation from stoichiometry, those that contain constitutional defects, and those with large deviations from stoichiometry. In the first instance, mixing experiments that keeps the system as close to compositional equilibrium may shed light on the mechanisms involved and the relative roles of kinetics and thermodynamics. For example, mixing of an oxide film with an oxide substrate, using combinations of soluble and insoluble oxides, might be a better experiment than mixing a metal into an oxide. Another approach might use oxygen as the mixing ion or inject oxygen with a second accelerator during mixing. A comparison of mixing into a compound that has varying stoichiometry versus one with a fixed stoichiometry would be useful. Many of the transition metal nitrides, oxides, and carbides exhibit wide phase fields and thus the ability to accommodate injected impurities. Differences in bonding between oxides and nitrides offers an additional parameter for study.

As previously discussed, few studies have explored the effect of temperature, although its influence has been shown to be large for some systems.

The cascade density can be varied by using ions of different masses and energies. Systematic variation of these parameters can help separte the contributions of various mixing mechanisms.

Future studies need to use a wider range of experimental techniques to characterize the effects of ion bombardment on the system. Progress will be much faster if information is obtained on the microstructure and the chemical environment of the various species.

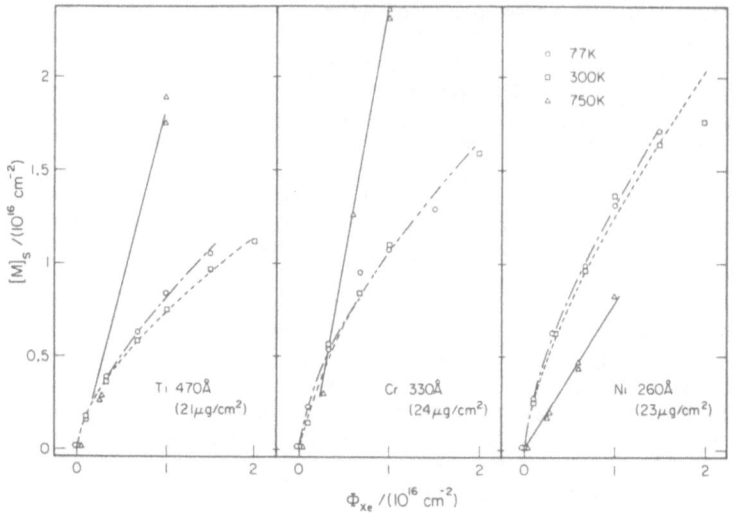

FIGURE 9. Plots of $(M)_S$ versus Φ_{Xe} with 290 keV Xe irradiation of 23 ± 1 μg/cm² Ti, Cr, and Ni layers on SiO_2 for different substrate temperatures during irradiation (ref. 15).

REFERENCES

1. Matteson S and M-A Nicolet: Ann. Rev. Mater. Sci. 13 (1983) 339.
2. Mayer JW, BY Tsaur, SS Lau, and L-S Hung: Nucl. Instr. Meth. 182/183 (1981) 1.
3. Wang ZL: Nucl. Instr. Meth. Phys. Res. B2 (1984) 784.
4. Lau SS, BX Liu, and M-A Nicolet: Nucl. Instr. Meth. Phys. Res. 209/210 (1983) 125.
5. White CW, G Farlow, J Narayan, GJ Clark, and JEE Baglin: Mater. Lett. 2 (1984) 367.
6. Westendorp H, ZL Wang, and FW Saris: Nucl. Instr. Meth. 194 (1982) 453.
7. Wang ZL, JFM Westendorp, S Doorn, and FW Saris: Metastable Materials Formation by Ion Implantation. ST Picraux and WJ Choyke (eds). Amsterdam: North-Holland, 1982, p. 59.
8. d'Heurle F, JEE Baglin, and GJ Clark: J. Appl. Phys. 57 (1985) 1426.
9. Johnson WL, YT Cheng, M vanRossum, and M-A Nicolet: Nucl. Instr. Meth. Phys. Res. B7/8 (1985) 657.
10. van Rossum M, YT Chen, M-A Nicolet, and WL Johnson: Appl. Phys. Lett. 46 (1985) 610.
11. Chen YT, M van Rossum, M-A Nicolet, and WL Johnson: Appl. Phys. Lett. 45 (1984) 185.
12. Workman TW, YT Chen, WL Johnson, and M-A Nicolet: Appl. Phys. Lett. 50 (1987) 1485.
13. Lilienfield DA, LS Hung, and JW Mayer: Nucl. Instr. Meth. Phys. Res. B19/20 (1987) 1.

14. Pretorius R, JM Harris, and M-A Nicolet: Solid State Electronics 21 (1978) 667.
15. Banwell T, BX Liu, I Golecki, and M-A Nicolet: Nucl. Instr. Meth. Phys. Res. 209/210 (1983) 125.
16. Farlow GC, BR Appleton, LA Boatner, CJ McHargue, and CW White: Ion Beam Processes in Advanced Electronic Materials and Device Technology. FH Eisen, TW Sigmon, and BR Appleton (eds). Pittsburgh: Materials Research Society, 1985, pp. 137-145.
17. Catlow CRA, R James, WC Mackrodt, and RF Smith: Phys. Rev. B 25 (1982) 1006.
18. Perez A, G Marest, BD Sawicka, JA Sawicki, and T Tyliszczak: Phys. Rev. B 28 (1983) 1227.
19. Kowalski J, G Marest, A Perez, BD Sawicka, J. Stanck, and T Tyliszczak: Nucl. Instr. Meth. Phys. Res. 209/210 (1983) 1145.
20. Guermazi M, G Marest, A Perez, BD Sawicka, JA Sawicki, P Thevenard, and T Tyliszczak: Mater. Res. Bull. 18 (1983) 529.
21. McHargue CJ, GC Farlow, PS Sklad, CW White, A Perez, N Kornilios, and G Marest: Nucl. Instr. Meth. Phys. Res. B19/20 (1987) 813.
22. McHargue CJ, GC Farlow, PS Sklad, CW White, A Perez, N. Kornilios, and G Marest: Mat. Res. Soc. Symp. Proc. 100 (1988) 119.
23. Bourdillon RJ, SJ Bull, PJ Burnett, and TF Page: J. Mater. Sci. 21 (1986) 1545.
24. Bull SJ: Ph.D. Dissertation, Cambridge University, 1988.
25. Greenwood NN and TC Gibb: Mössbauer Spectroscopy. London: Chapmann and Hall, 1971, p. 249.
26. Rossiter MJ: J. Phys. Chem. Solids 26 (1965) 775.
27. Janet C and H Gilbert: Bull. Soc. Fr Mineral Crystallogr. 93 (1970) 213.
28. Sklad PS, JD McCallum, SJ Pennycook, CJ McHargue, and CW White: Characterization of the Structure and Chemistry of Defects in Materials. Pittsburgh: Materials Research Society, in press.
29. Atlas LM and WK Sumida: J. Am. Ceram. Soc. 41 (1958) 157.
30. McHargue CJ, GC Farlow, GM Begun, JM Williams, CW White, BR Appleton, PS Sklad, and P Angelini: Nucl. Instr. Meth. Phys. Res. B16 (1986) 212.
31. White CW, GC Farlow, CJ McHargue, PS Sklad, P Angelini, and BR Appleton: Nucl. Instr. Meth. Phys. Res. B7/8 (1985) 473.
32. McHargue CJ, GC Farlow, CW White, JM Williams, BR Appleton, and H Naramoto: Mater. Sci. Engr. 69 (1985) 123.
33. Angelini P, PS Sklad, CJ McHargue, MB Lewis, and GC Farlow: Processing and Characterization of Materials Using Ion Beams. Pittsburgh: Materials Research Society, in press.
34. Naguib HM, JF Singleton, WA Grant, and G Carter: J. Mater. Sci. 8 (1973) 1633.
35. Turos A, HJ Matzke, and P. Rabette: Phys. Stat. Sol. 64 (1981) 565.
36. Drigo AV, SL Russo, P Mazzoldi, PD Goode, and NEW Hartley: Radiat. Eff. 33 (1977) 161.
37. Hart R, H Dunlap, and O Marsh: Radiat. Eff. 9 (1971) 261.
38. Pells GP and AY Stathopoulos: Radiat. Eff. 74 (1983) 181.
39. Murarka SP: J. Vac. Sci. Technol. 17 (1980) 775.
40. Pretorius R, JM Harris, and M-A Nicolet: Solid State Electronics 21 (1978) 667.
41. Appleton BR, H Naramoto, CW White, OW Holland, CJ McHargue, GC Farlow, J Narayan, and JM Williams: Nucl. Instr. Meth. Phys. Res. B1 (1984) 167.

134

42. Fathy D, OW Holland, J Narayan, and BR Appleton: Nucl. Instrum. Meth. Phys. Res. B7/8 (1985) 571.
43. Bhattacharya RS, AK Rai, and PP Pronko: J. Mater. Sci. 2 (1987) 211.
44. Beyers R: J. Appl. Phys. 56 (1984) 147.
45. Beyers R, R Sinclair, and ME Thomas: J. Vac. Sci. Technol. B2 (1984) 781.
46. Singer IL, RG Vardiman, and CR Gossett: Fundamentals of Beam-Solid Interactions and Transient Processing. MJ Aziz, LE Rehn, and B. Stritzker (eds). Pittsburgh: Materials Research Society, 1988, pp. 201-206.
47. Singer IL: Surface Coatings Technol. 33 (1987) 487.
48. Romana L, P Thevenard, R Brenier, G Fuchs, and G Massouras: Nucl. Instr. Meth. Phys. Res. B32 (1988) 96.
49. Zhang PQ, G Principi, A Paccagnella, S LoRusso, and G. Battaglin: Nucl. Instr. Meth. Phys. Res. B, 1987.
50. Ogale SB, DM Phase, SM Chaudhari, SV Ghaisas, SM Kanetkar, PP Patil, VG Bhide, and SK Date: Phys. Rev. B35 (1987) 1593.
51. Li J and BX Liu: Nucl. Instr. Meth. Phys. Res. B31 (1988) 407.
52. Lewis MB and CJ McHargue: Metastable Materials Formation by Ion Implantation. ST Picraux and WJ Choyke (eds). Amsterdam. North-Holland, 1982, pp. 85-91.
53. Doi H: Toyota Central R&D Laboratories, Private Communication, 1987.
54. Banwell T and M-A Nicolet: Nucl. Instr. Meth. Phys. Res. B19/20 (1987) 704.
55. Banwell T, M-A Nicolet, T Sands, and PJ Grunthaner: Appl. Phys. Lett. 50 (1987) 571.
56. Lewis MB and CJ McHargue: Ion Implantation and Ion Beam Processing of Materials. GK Hubler, OW Holland, CR Clayton, and CW White (eds). New York: North-Holland, 1984, pp. 771-776.

THE ROLE OF CHEMICAL DRIVING FORCES IN BOMBARDMENT-INDUCED
COMPOSITIONAL CHANGE

ROGER KELLY

IBM Research Division, T. J. Watson Research Center,
Yorktown Heights, NY 10598

1. INTRODUCTION
   This constitutes part of a discussion on bombardment-induced compositional
change with alloys, oxides, oxysalts, and halides. Already treated are the
role of the surface binding energy (1), the role of bombardment-induced
Gibbsian segregation ("BIS") (2), as well as an extended version of the present material (3). We here consider three aspects of what will be termed
chemical driving forces. We first consider how it is possible for BIS to
occur at all given the small numerical value of the driving force, 0.06 to
0.52 eV. The same is then done for bombardment-induced mixing, now
recognized to be chemically driven by the heat of mixing, now recognized to
be chemically driven by the heat of mixing, typically ≤1.3 eV (4,5). We
finally consider bombardment-induced decomposition, i.e. the characteristic
compositional changes found with many oxides and oxysalts. In early work
these changes were normally inferred from electron diffraction (e.g. 6) or
AES (e.g. 7), but more recently there has been an increasing use of XPS
("x-ray photoelectron spectroscopy"), thence the possibility of confirming
valence states which do not give stable bulk compounds (e.g. 8-11). A typical example of BIS is shown in Fig. 1 (12-14) and of decomposition in
Fig. 2 (10).
   As already pointed out in (1-3), compositional change has been the subject
of both experimental and theoretical study for the past two decades.
Concerning theoretical work, we emphasized that work written before 1980
tended to emphasize preferential sputtering as triggered by differences of
mass, chemical binding, or volatility.
   Mass is now regarded to be an important factor in preferential sputtering,
thence in compositional change, only under near-threshold conditions (7) and
in isotope sputtering (15). Chemical binding, reflectd explicitly in the
surface binding energy, was at one time the most popular framework for
explaining preferential sputtering, thence compositional change, with
alloys. It is easily shown (1,16), however, that chemical binding as manifested in the surface binding energy should, at least with alloys, normally
lead to significantly smaller compositional changes than are observed; only
when it is manifested in BIS can chemical binding lead to large effects. On
the other hand, there is a hint that with oxides and halides the surface
binding energy may play an unsuspected role (1,17).
   The role of volatility, i.e. of thermal sputtering, is more problematical
than that of mass or chemical binding. For example, differences in volatility were found to explain many examples of O loss from oxides, with both
the correlation ("yes" or "no") and the required magnitude of the volatility
($10^{2\pm1}$ atm at ~4000 K) being reasonably correct for the systems then known
(18). Since then, many additional oxides have been shown to lose oxygen,
including some like ZrO_2 (9) which have very low volatility. With alloys

135

C.J. McHargue et al. (eds.), Structure-Property Relationships in Surface-Modified Ceramics, 135–148.
© 1989 by Kluwer Academic Publishers.

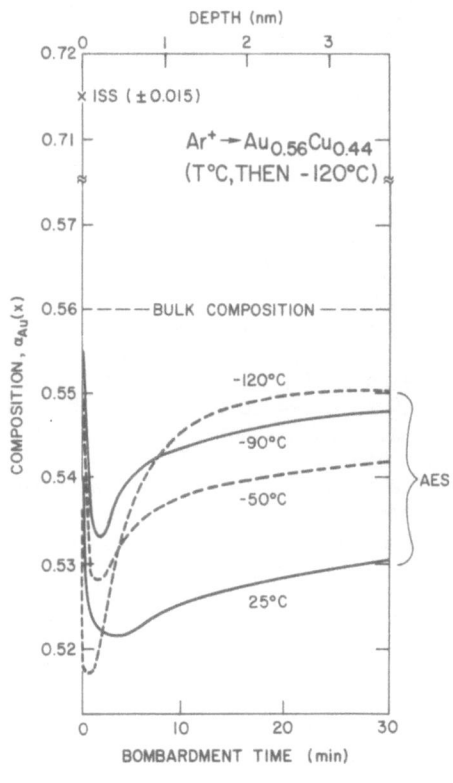

FIGURE 1. Composition profiles for $Au_{0.56}Cu_{0.44}$ showing BIS at various bombardment temperatures. The portions based on AES were obtained by first bombarding to steady state with a 40-μAcm^{-2} beam of 2 keV Ar$^+$ at temperatures in the interval -120 to 25°C, then cooling to -120°C, and finally profiling with a 0.4-μAcm^{-2} beam of 2 keV Ar$^+$. Compositions were obtained with low-energy AES, namely 60 eV for Cu and 69 eV for Au. The ISS point is an average of 11 results for various incident energies and current densities. Profiles similar to those shown have been identified with a total of seven different systems (2). Due to Li and Koshikawa (AES) (12) and to Kang et al. (ISS) (13,14).

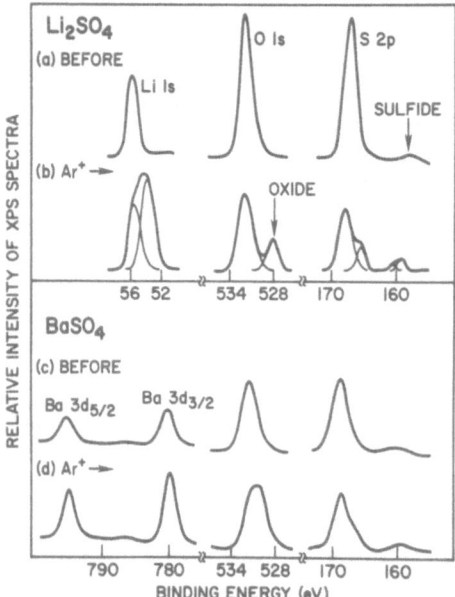

FIGURE 2. XPS spectra of metal, O, and S photoelectron peaks from Li_2SO_4 and $BaSO_4$: (a) and (c) are before Ar$^+$ bombardment; (b) and (d) are after 4 keV Ar$^+$ bombardment to a fluence of 2–8 x 10^{17} ions/cm^2. The targets were in the form of compressed powders. We agree with the authors that what is being observed is more nearly a chemical rearrangement of atoms displaced within each cascade than, for example, a process related to the thermal spike. Due to Contarini and Rabalais (10).

the best evidence for thermal sputtering relates to energy distributions, as for $Xe^+ \rightarrow Ag$ (19), which showed departures from the form expected for cascade sputtering as if a thermal distribution were superimposed. Temperatures of 20,000 to 30,000 K were implied, however, and this contradicts the idea that a condensed phase cannot exceed its critical temperature because of a "phase explosion" (20,21).

2. BOMBARDMENT-INDUCED SEGREGATION

We referred briefly in Section 1 to BIS, the most characteristic result of which is the development of profiles as in Fig. 1 (12-14). Such profiles are, we would emphasize, not exceptional but rather occur quite generally (2). The crucial detail in the present context is that, assuming BIS to resemble equilibrium segregation, then the driving force is only 0.06 to 0.52 eV, values which are very much less than the energies characterizing particle bombardment, ≥ 10 eV (columns 2 and 3 of Table 1).

Cheng et al. (4,5) have demonstrated that binary metallic systems showed ion-beam mixing rates which correlated with the heats of mixing, i.e. for a given fluence (dose)

$$(\text{distance of mixing})^2 \propto \Delta H_m .$$ [1]

Apparently a significant part of the driving force for the mixing is based on chemical energy differences and we note that the magnitude, ≤ 1.3 eV, is again small compared with incident particle or recoil energies (column 4 of Table 1.

It is clear that there is an extreme disparity between the chemical energy differences relevant to BIS or mixing (≤ 1.3 eV) and typical bombardment or recoil energies. We will therefore reconsider these effects and attempt to show a way out of the paradox.

In the case of BIS, it is plausible to argue in terms of the underlying process being ballistic but with some trajectories terminated by one or more low-energy steps. The steps in question could be identified variously with the interstitial migration postulated by Sigmund (22), the athermal rearrangement seen in simulations by King and Benedek (23), or the "crystal repair" seen in simulations by Harrison and Webb (24), where in each case the process of interest occurred in the cooling phase of the cascade.

Whatever the exact details, in the bulk the low-energy steps are necessarily random but in the outer 2 atom layers we suggest that they will be directed and therefore lead to segregation (Fig. 3). We will call the steps "chemically guided" and ask whether the jump rate is high enough to explain BIS.

Consider the system Cu-Ni. In eight different studies of BIS the dose to reach 80% of steady state lay between 5×10^{15} and 3×10^{16} ions/cm^2 (25-32). The corresponding number of chemically guided steps per target atom, $\Gamma^b t$, follows using the standard formalism for the bombardment-induced diffusion coefficient, D^b (e.g. 33):

$$D^b t = (1/6)\Gamma^b t m_1^2 \lambda^2 ;$$ [2]

$$\Gamma^b t = f \times 0.421 \times (dE/dx)2\lambda/E_d \times dose/N_s = 10 \text{ to } 60.$$

Here $m_1\lambda$ is the average length of a ballistic trajectory, λ is the mean atomic spacing (0.225 nm for Cu-Ni), f is the fraction of ballistic trajectories terminated in chemically guided steps (probably near unity), dE/dx is about 0.7×10^{10} eV/cm for keV energies (34), the factor 2 approximates the

TABLE 1. Examples of chemical energy differences which can act as driving forces for change sin bombarded solids. the heat of segregation, ΔH^{seq}, leads to BIS, while the heat of mixing, ΔH_m, contributes to bombardment-induced mixing. ΔH^{seg} is taken from ref. (2); ΔH_m is taken mainly from ref. (53) and applies to systems with 1:1 proportions. The entries AES, FIM, ISS, and XPS are the usual acronyms referring to the methods of surface analysis.

System	ΔH^{seg} (eV) (experimental from Arrhenius plots)	ΔH^{seg} (eV) (experimental from individual data points using Eq. (4) with $\Delta S^{seg} = 0$)	ΔH_m (eV)
Ag-Au	...	0.04-0.07 ISS	-0.048
Ag-Pd	0.09-0.13 AES	0.02-0.04; 0.098 AES	-0.052
Au-Cu	0.13 AES; 0.13 ISS	0.03-0.06 AES; 0.04-0.10 ISS	-0.053
Au-Ni	0.52 AES; 0.45 ISS	~0.4 AES[a]	+0.078
Au-Pd	...	0.05-0.13 AES; 0.09-0.11 ISS	-0.081
Cr-Mo	...	~0.02 AES[b]	+0.075
Cu-Ni	0.42 ISS	0.21 FIM; 0.2-0.3 ISS	+0.018
Cu-Pd	0.059 AES	0.02-0.07 AES	-0.111
Cu-Pt	...	0.1-0.2 AES, XPS; 0.09-0.2 ISS	-0.115
Mg-Al	...	0.1 AES	-0.034
Mo-W	...	~0.4 AES	+0.021
Ni-Co	0.069; 0.18 AES	0.04-0.09 AES	0.000
Ni-Mo	...	...	-0.011
Ni-Pt	0.15; 0.25 AES; 0.11-0.24 ISS	...	-0.096
Pd-Ni	0.31 AES	0.06 AES	-0.006

[a]This information is all for dilute Au. For concentrated Au, ΔH^{seg} is much lower, ~0.07 (54).

[b]The information on the surface composition is not fully self-consistent (55).

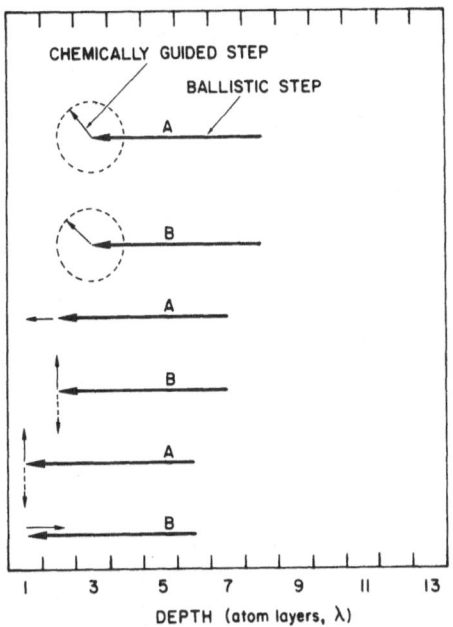

DEPTH (atom layers, λ)

FIGURE 3. Sketch of how chemical guidance could lead to BIS. Bombardment of a target is envisaged as leading to ballistic motion of target atoms with a fraction $\underline{f}$ of the trajectories terminated by one or more low-energy steps. In the bulk these steps are necessarily random (shown as dashed circles) but in the outer 2 atom layers we suggest that they will be directed ("chemically guided") and therefore lead to segregation. This description is preferred to one based on a thermal spike because the driving force for segregation is so weak that a thermal spike would often violate the condition of Eq. [6].

idea that segregation should occur for trajectories terminating in either of the outer two atom layers, E_d, the displacement energy, is about 32 eV (35), and N_s is the target surface number density ($\equiv \lambda^{-2}$). The kinetics of equilibrium segregation are described adequately by (3,36,37):

$$\alpha_{A(2)} = K\alpha_{A(2')} ; \tag{3a}$$

$$\alpha_{A(2')} = \alpha_{A(3)} - \alpha_{A(3)}(1 - 1/K) \exp(\xi + \tau) \times \mathrm{erfc}(\xi/2\tau^{1/2} + \tau^{1/2}); \tag{3b}$$

$$\xi = x/K\lambda ; \quad \tau = Dt/K^2\lambda^2,$$

where, in our usual notation, α_i is the atom fraction of component $\underline{i}$, "2" is an atom-layer one position, "2'" is a subsurface position, "3" is a bulk position, and $\overline{K}$ (which is intended to be time-invariant) is related to the equilibrium segregation ratio, K^{eq}. Thus we have

$$K^{eq} \equiv \exp(\Delta G^{seg}/kT) = \alpha_A^\infty(2)\alpha_{B(3)}/\alpha_B^\infty(2)\alpha_{A(3)}, \tag{4}$$

so that K follows as

$$K = \alpha_{A(2)}/\alpha_{A(2')} = \alpha_A^\infty(2)/\alpha_{A(3)}$$
$$= K^{eq}/(\alpha_{B(3)} + \alpha_{A(3)}K^{eq}).$$

140

Here "∞" indicates "steady state". Equation [3a] describes the situation
in which segregation equilibrium between atom-layer one and two is instantaneous, thence time-invariant. Equation [3b] describes a uniform initial
distribution with a surface boundary condition (x = t = 0),

$$\alpha_{A(2')} = \alpha_{A(3)}/K,$$

which triggers a diffusional outflow into a finite reservoir.

For a 1:1 target such as $Cu_{0.50}Ni_{0.50}$, K^{eq} is very large (2) so that K is
of order 2. The condition for 80% of equilibrium segregation finally
follows as the solution at x = 0 of

$$\alpha_{A(2)}/K\alpha_{A(3)} = 0.80,$$

namely, $\tau = Dt/K^2\lambda^2 \approx 1.0$. Since we can write $Dt = (1/6)\Gamma t\lambda^2$, the
corresponding number of jumps per target atom follows as

$$\Gamma t \approx 6K^2 \approx 20. \qquad [5]$$

The conclusion is that $\Gamma^b t$, Eq. [2], is similar to Γt, Eq. [5]. This by
itself would suggest that the extent of BIS at ambient temperature would be
similar to that of equilibrium segregation, but a second effect intervenes:
the composition spike due to BIS will tend to be sputtered away. The final
result is that chemical guidance adequately explains BIS but that the extent
of BIS at ambient temperature will always be small.

We finally note that it would not always be correct to attribute BIS to a
thermal-spike type of process such as that advocated by Cheng et al. for
mixing (characteristic energy 1-2 eV (4), by Oostra et al. for heavy-ion
sputtering (characteristic energy 2-3 eV (19), or by Urbassek for cluster
formation (characteristic energy 0.5-1 eV (21). This is because of the
problem that chemical guidance, in the broadest sense, is possible only when
the driving force obeys the following:

$$(\text{driving force}) \gtrsim kT. \qquad [6]$$

3. BOMBARDMENT-INDUCED MIXING
With mixing the correlation of Eq. [1] can be shown (3) to lead to the
conclusion that the purely ballistic effect, seen for $\Delta H_m = 0$, is less
important by a factor of about 3 than the purely chemical effect. This is
why Cheng et al. (4,5) argued that mixing takes place in an intermediate
phase of the cascade when the average particle energy simulates a temperature of 1-2 eV. Mixing was thus identified with a thermal-spike type of
process like thermal sputtering (18,19).

Although we cannot in general eliminate a thermal-spike interpretation,
there is the objection raised above: the ΔH_m values leading to Eq. [1] in
part violate the condition of Eq. [6]. We therefore regard it as
appropriate at this point to propose an alternative in which chemical energy
differences could be important to mixing in a more general sense. We accept
the traditional view that mixing is initiated by a ballistic event
(22,38,39). Consider a ballistic mixing process expressed in the "diffusion
limit" (39). Then if ϕ is the incident particle fluence (dose) and $d\sigma_i(z)$
is the relocation cross section of component i, we have

$$\partial \alpha_i / \partial \phi \;=\; -\; \frac{\partial \alpha_i}{\partial x} \int_{-\infty}^{\infty} d\sigma_i(z)\, z \;+\; \frac{\partial^2 \alpha_i}{2\partial x^2} \int_{-\infty}^{\infty} d\sigma_i(z)\, z^2$$

$$\equiv\; -\; v_i\, \frac{\partial \alpha_i}{\partial x} \;+\; D_i^b\, \frac{\partial^2 \alpha_i}{\partial x^2}\,,$$

where v_i describes the drift velocity of component i, D^b_i is the bombardment-induced diffusion coefficient of component i, and additional terms to conserve sites (39) have been omitted. Consider Fig. 4. We identify D^b_i with the ballistic part of each relocation and note that, if a given atom is displaced n times in a bombardment and each time executes random ballistic motion, then the total mean square distance moved <u>along a given axis</u> is

$$< z^2 > \;=\; nm_1^2 \lambda^2\, \frac{\int_0^{\pi} \cos^2\theta\, \sin\theta\, d\theta}{\int_0^{\pi} \sin\theta\, d\theta} \;=\; (1/3)nm_1^2 \lambda^2. \qquad [7]$$

Likewise, we identify v_i with the low-energy terminal steps which occur in a fraction f of the relocations and have an average length $m_2\lambda$. Because of this "v" the same atom executes an additional motion the mean square value of which along a given <u>axis</u> has a maximum $(fnm_2\lambda)^2$, but more generally can be written approximately as

$$< z^2 > \;\approx\; [fn(\Delta H_m/E^*)m_2\lambda]^2, \qquad [8]$$

where E^* is, judging from the results lying back of Eq. [1], of order 1-2 eV. (Equation [8] is not intended to be rigorous.) It follows that a minor chemical perturbation, i.e. what we call a chemically guided step, of a basically ballistic process could dominate as far as <u>mixing</u> is concerned. The condition for this to be so is just Eq. [8] > Eq. [7], an inequality which should be always valid if <u>n</u> or ΔH_m is large enough.

4. BOMBARDMENT-INDUCED DECOMPOSITION

Four separate groups of authors (8-11) have examined a long series of oxides and oxysalts by alternately bombarding them and studying the surfaces with XPS. The latter serves to detect different valence states, especially of cations and even when the states do not give stable bulk compounds. Typical results are reproduced in Fig. 2 (10), while Table 2 summarizes most of the results of Rabalais et al. (10,40-42) together with information on TiO_2 and V_2O_5.

The trends in Table 2 are clear. Fixed valence systems lose a volatile component (O, C, N, S) and at the same time evolve mainly to oxides rather than to carbides, nitrides, or sulfides. Variable valence systems lose oxygen and alkali metal or Ba or Ag (as is the case) and at the same time evolve towards lower valence states.

The results for various oxides, this time of a wider group of authors, are summarized in a different way in Fig. 5, which shows the enthalpy increases (eV/atom) for the observed changes versus the number of atoms in the formula

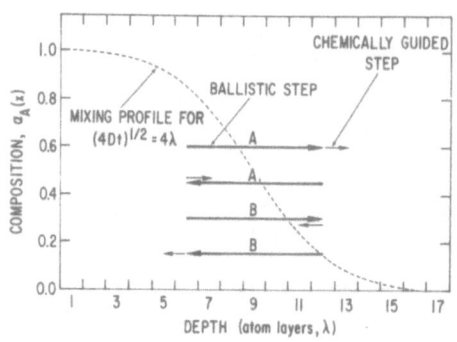

FIGURE 4. Sketch of how chemical guidance could lead to bombardment-induced mixing. As in Fig. 3 it is envisaged that the ballistic trajectories tend to be terminated by chemically guided steps. The ballistic steps are random and therefore lead to a diffusion coefficient, $\underline{D}$, whereas the guided steps are directed and lead to a drift velocity $\underline{v}$. For a sufficiently large number of jumps per atom, $\underline{n}$, or for a large enough driving force, ΔH_m, drift (Eq. [8]) will always dominate diffusion (Eq. [7]). Also shown is the mixing profile given by $(1/2)$erfc $[(x - x_0)/(4Dt)^{1/2}]$, with x_0 = 9λ and $(4Dt)^{1/2} = 4\lambda$.

of the starting substance, i.e. the "complexity" of the substance. The trend here is that a system will tolerate enthalpy increases up to about 0.7 eV/atom (marked with arrow), this result being independent of the "complexity"; the systems HfO_2, Nb_2O_5, SnO_2, Ta_2O_5, WO_3, and ZrO_2, however, constitute exceptions. We will overlook the stability claimed for Cu_2O (43), as this substance lies near the limit of 0.7 eV/atom.

There are at least four ways to rationalize results as in Table 2 and Fig. 5. The first concerns the surface binding energy, while the remaining three are different aspects of bombardment-induced decomposition.

4.1. Surface binding energy

We have already discussed in detail the possible role of the surface binding energy in governing compositional changes with oxides and oxysalts (1,17). In brief, for oxides of group III and beyond and which are reasonably ionic, the anion binding energy at an undisturbed surface should be distinctly lower than the cation binding energy, an effect which is tentatively supported by experiment (44). The binding energy argument would be particularly relevant to the exceptions noted above: HfO_2, Nb_2O_5, etc. The problem, as with the role of mass [Section 4.3(a) to follow], is the lack of universality: e.g. why do Al_2O_3, Cr_2O_3, and SiO_2 not show extensive O loss?

4.2. Stochastic rearrangement

Suppose that during the active phase of a cascade the atoms of, for example, $PbSO_4$ become interchanged or otherwise uncoordinated. Then in the cooling phase of the cascade the system would tend to re-establish local order and one possibility is that the new order, thence stoichiometry, would form randomly. To some extent this would be guided by a partial loss of a volatile species, by diffusional transport, or by BIS [Section 4.3(d), to follow]. One would then expect a wide range of products, with $PbSO_4$, for example, evolving to a mixture of PbO, Pb, and PbS. Since this is contrary to what is observed (8), stochastic rearrangement is evidently not the dominant process. Nevertheless, an element of such rearrangement is probably relevant with most systems, as follows from these examples. CoO is largely stable but evolves to a small extent to Co^0 either at ambient temperature (45) or at $\geq$550 K (46). NiO yields Ni^0 at $\geq$400 K (46). Fe_2O_3 evolves mainly to Fe^{II} but with some Fe^0 also present (45).

4.3. Energy-limited rearrangement

We again suppose that a bombarded system tends to first become uncoordinated but finally to re-establish local order. Then another possibility

TABLE 2. Summary of compositional changes as observed by Rabalais et al. (10,40-42) when oxides and oxysalts are bombarded with 4 keV Ar+. The surfaces were analyzed by XPS. Because of its importance, information on TiO_2 and V_2O_5 as studied mainly with XPS by other authors is included.

Substance	Loss deduced by assuming that one cation is fixed[a]	Fractional loss	Ref.
Li_2CO_3	$0.62CO_2 + 0.38CO$	0.82 (C)	10
$BaCO_3$	$0.78CO + 0.22CO_2$	0.84 (C)	10
TiO_2	$Ti^{IV} \rightarrow Ti^{II}, Ti^{III}$	. . .	9,45
$LiNO_3$	$0.72O_2 + 0.28N_2$	0.95 (N)	40
$NaNO_3$	$0.64O_2 + 0.36N_2$	0.82 (N)	40
$NaVO_3^{[b]}$	$V^V \rightarrow V^{IV}, V^{III}$	0.79 (Na), 0.46 (O)	41
$V_2O_5^{[c]}$	$V^V \rightarrow V^{III}$	. . .	. . .
$NaNbO_3$	$Nb^V \rightarrow Nb^{II}, Nb^{IV}$	0.84 (Na), 0.53 (O)	41
Nb_2O_5	$Nb^V \rightarrow Nb^{II}, Nb^{IV}$	0.43 (O)	9,41,45
$NaTaO_3$	$Ta^V \rightarrow Ta^0$ etc.	0.54 (Na), 0.48 (O)	41
Ta_2O_5	$Ta^V \rightarrow Ta^{II}, Ta^0$ or Ta^I, Ta^{IV}	0.43 (O)	9,41
Li_2SO_4	$0.74SO_3 + 0.26O_2$	0.67 (S)	10
$BaSO_4$	$0.66SO_2 + 0.34SO_3$	0.62 (S)	10
Li_2CrO_4	$Cr^{VI} \rightarrow Cr^{III}$	0.22 (Li), 0.51 (O)	42
Na_2CrO_4	$Cr^{VI} \rightarrow Cr^{III}$	0.61 (Na), 0.48 (O)	42
K_2CrO_4	$Cr^{VI}, Cr^{III} \rightarrow Cr^{III}$	0.69 (K), 0.45 (O)	42
$BaCrO_4$	$Cr^{VI} \rightarrow Cr^{III}$	0.13 (Ba), 0.32 (O)	42
$Cr_2O_3^{[d]}$	unestablished $\rightarrow Cr^{III}$	none	42
Na_2MoO_4	$Mo^{VI} \rightarrow Mo^{IV}, Mo^0$	0.35 (Na), 0.38 (O)	41
$MoO_3^{[e]}$	$Mo^{VI} \rightarrow Mo^{IV}$	0.56 (O)	41
Li_2WO_4	$W^{VI} \rightarrow W^0$ etc.	0.69 (Li), 0.49 (O)	41
Na_2WO_4	$W^{VI} \rightarrow W^0$ etc.	0.68 (Na), 0.53 (O)	41
Ag_2WO_4	$W^{VI} \rightarrow W^0$ etc.	0.75 (Ag), 0.54 (O)	41
WO_3	$W^{VI} \rightarrow W^0$ etc.	0.55 (O)	41

[a]For example, with Li_2CO_3 Li was taken as fixed, but with Li_2CrO_4 Cr was taken as fixed.
[b]As in Fig. 8 and not Table 3 of ref. (41).
[c]Electron diffraction (56).
[d]Described as "$(NH_4)_2CrO_4$" but in fact evolved to Cr_2O_3.
[e]Described as "MoO_2" but "MoO_3" is more accurate.

144

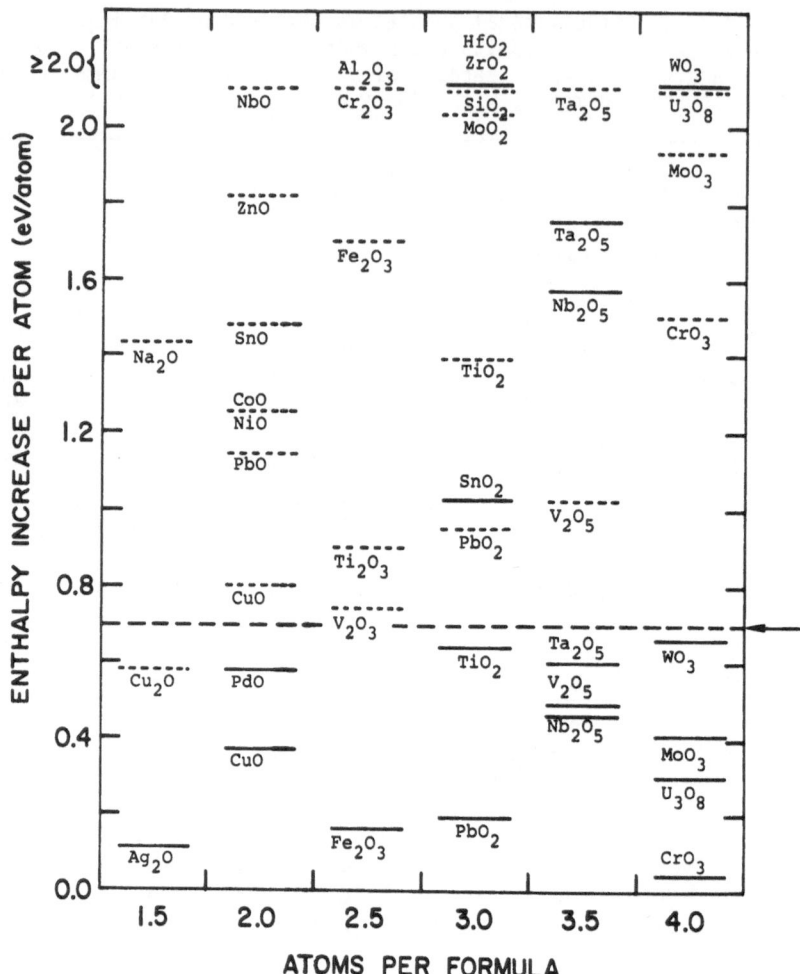

FIGURE 5. Diagram of enthalpy increase per atom (eV/atom) versus atoms in the formula, i.e. the "complexity", for various bombarded oxides. A solid line indicates that the oxide reduced, a dashed line that it did not, with the two categories separated at about 0.7 eV/atom (arrow). This figure constitutes a trend analysis appropriate to energy-limited rearrangement as discussed in Section 4.3. Details are as follows: column 1.5 (reduction to metal), column 2.0 (reduction to Cu, Pd, Cu_2O, Pb, Ni, Co, Sn, Zn, Nb), column 2.5 (reduction to Fe_3O_4, VO, TiO, Fe, Cr, Al), column 3.0 (reduction to PbO, Ti_2O_3, Pb, SnO, TiO, Mo, Si, Zr, Hf), column 3.5 (reduction to NbO_2, V_2O_3, interpolated TaO_2, VO, NbO, interpolated TaO, Ta), column 4.0 (reduction to Cr_2O_3, UO_2, MoO_2, WO_2, Cr, Mo, U, W).

is that this change is energy (enthalpy) limited, i.e. constitutes what is effectively a bombardment-induced phase change. For example, $PbSO_4$, known to evolve mainly to PbO rather than Pb or PbS (8), will be assumed to do so because the energy increase is 0.42 eV/atom in the first case, 0.80 in the second, and 1.42 in the third. One is thus again dealing with energy differences which, in units of eV/atom, are similar to those involved in BIS and mixing (Table 1).

The argument at this stage is imperfect, as it does not explain why the system did not return to $PbSO_4$, given that $PbSO_4$ constitutes the lowest energy state. By contrast, in BIS and mixing the final state that is observed is that of lowest energy. One can propose a number of reasons why the system might tend to avoid the state $PbSO_4$ (or whatever was the ground state), although, as will be clear, no one explanation appears to cover all systems.

(a) <u>Mass differences</u>: Mass is most clearly relevant under near-theshold conditions, including with Al_2O_3 (7), PbO (47), and Ta_2O_5 (7). Nevertheless, even at higher energies near-threshold behavior might be expected for large enough mass differences, and this would explain why WO_3 reduced to products including W^o but MoO_3 mainly to Mo^{IV}, or why Ta_2O_5 yielded products including Ta^o or Ta^I but Nb_2O_5 stopped at Nb^{II} (Table 2).

(b) <u>Bombardment-induced amorphization</u>: If the amorphization process is taken as being equivalent to fusion, then with $PbSO_4$ one can expect a ground-state energy increase of 0.07 eV/atom, whence a decrease in the energy needed for formation of PbO from 0.42 to 0.35 eV/atom. This is an unimportant change, though with some systems the change is larger, e.g. about 0.26 eV/atom [average from (48)] for oxides of the type MO_2. Indeed, as seen in Fig. 5, the latter change is sufficient to justify the recent claim that bombarded SnO_2 evolves to SnO (49).

(c) <u>Point-defect accumulation</u>: Fecht and Johnson (50) have argued that point-defect accumulation can cause a phase to become unstable. For example, they show that Al would amorphize if it acquired 5% vacancies, a number that is obtained by comparing the vacancy formation energy [0.7 eV (51)] with the heat of fusion [<0.11 eV for temperatures below the glass transition temperature (50)]. Considering now $PbSO_4$, then a possible (though simplistic) description of a high oxygen vacancy concentration is the formation of $PbSO_3$, leading to a ground-state energy increase of ≤0.43 eV/atom. With such an increase, together with amorphization, it is possible that $PbSO_4$ would become unstable.

(d) <u>Volatility, diffusional transport, BIS</u>: To some extent the evolution of $PbSO_4$ to PbO will be conditioned by the volatility of O_2, SO_2, and SO_3. This effect is clearly not as important as it might seem, however, as it would be expected to lead to a universal decomposition of all oxides and oxysalts, for example Al_2O_3, Cr_2O_3, and SiO_2. Ease of diffusional transport is somewhat different and in principal plays a role with selected systems. From a thermodynamic point of view these should either not lose O at all (HfO_2, SnO_2, ZrO_2) or should lose less O than is observed (Nb_2O_5, Ta_2O_5, WO_3). The key may lie in the fact that in most cases the diffusion coefficient for O transport in substoichiometric material is unusually large (3). Finally there is BIS. Very little is known about this effect with oxides or oxysalts except for the unique example of $Na_2 \cdot SiO_2$ (52). Here bombardment led to a profile like that of an alloy as in Fig. 1, thence to Na loss.

4.4. Equilibrium rearrangement

If the system underwent equilibrium decomposition at the transient high temperature ("thermal spike") which is conventionally assumed to exist during the cooling phase of a cascade, then changes essentially as observed could be expected. This was the point of view taken to explain O loss from oxides (18) and the loss of both alkali metals and other volatile species from oxysalts (41). We note, first of all, that this mechanism does not avoid the problematical cases such as HfO_2, Nb_2O_5, etc.: i.e. just as they lie at the top of Fig. 5, they also lack volatility.

The mechanism has never been properly tested, the current situation being that most tests are either a trend analysis (18,41) or are based on experiments (19) which are inconclusive. Our position is the same as with BIS and mixing. Although we cannot disprove a thermal-spike interpretation we can propose an alternative in which chemical energy differences are important (a) without regard to the inequality of Eq. [6], (b) without violating the problem of a phase explosion (20,21), and (c) without the more general objection that the mechanism requires chemical changes to occur in an assumed equilibrium during a highly proscribed time interval similar to 10^{-12} to 10^{-9} s. The alternative was given in Section 4.3.

5. DISCUSSION

Recent experiments suggest that weak chemical driving forces play a major role in bombarded targets even when, for reasons of extreme energy disparity, it might be difficult to understand why this should be so. With BIS the basic result is that bombarded alloys show segregation in the same sense, although to a factor of 10-100 lesser extent, than equilibrated alloys (2,3). The driving force is 0.06 to 0.52 eV/atom (Table 1) and we have suggested that the result becomes understandable if a fraction f of ballistic trajectories within the outer 2 atom layers of the target $\bar{1}$s terminated by one or more low-energy, chemically guided steps (Fig. 3). This mechanism always satisfies the inequality of Eq. [6] and is therefore more satisfactory than a thermal-spike model. With bombardment-induced mixing the prime result is that of Cheng et al. (4,5), who found that the extent of mixing scaled with the heat of mixing, Eq. [1]. This implies a driving force of, typically, ≤1.3 eV/atom (Table 1). A very similar model as with segregation is proposed (Fig. 4) and again there is the general caution that a thermal-spike model will often violate the condition of Eq. [6]. Finally we have considered what has been termed bombardment-induced decomposition. This is a multifaceted phenomenon in which bombarded oxides lose O, sulfates lose S and O, and other systems lose, as is appropriate, alkali metal, Ba, Ag, C, N, etc., the result being an energy increase of $\lesssim$0.7 eV/atom (Fig. 5). The idea of a chemical driving force is here at first sight unfavorable in that the "ground state" (e.g. $PbSO_4$) has a lower energy than the observed product (e.g. PbO). The point of view of Fecht and Johnson (50) is relevant, however, which is that if there are sufficient bombardment-induced vacancies (simplistically, conversion of $PbSO_4$ to $PbSO_3$) then the ground state will be unstable. This is how Fecht and Johnson described bombardment-induced amorphization and other phase changes, processes which we see now to be related to those that are discussed here. Besides point-defect accumulation, the changes will also be aided by amorphization, volatility, diffusional transport, and BIS.

REFERENCES
1. Kelly R: Materials Modification by High-Fluence Ion Beams. R Kelly and MF da Silva (eds). Dordrecht, Netherlands: Kluwer, p. 305, in press.
2. Kelly R: Nucl. Instr. Meth. B, in press.
3. Kelly R: Mater. Sci. Engr., in press.
4. Cheng Y-T, M van Rossum, M-A Nicolet, and WL Johnson: Appl. Phys. Lett. 45 (1984) 185.

5. Cheng Y-T, TW Workman, M-A Nicolet, and WL Johnson: Mater. Res. Soc. Symp. Proc. 74 (1987) 419.
6. Murti DK and R Kelly: Thin Solid Films 33 (1976) 149.
7. Taglauer E: Appl. Surf. Sci. 13 (1982) 80.
8. Christie AB, J Lee, I Sutherland, and JM Walls: Appl. Surf. Sci. 15 (1983) 224.
9. Hofmann S and JM Sanz: J. Trce Microprobe Tech. 1 (1982/83) 213.
10. Contarini S and JW Rabalais: J. Electron Spect. Related Phenom. 35 (1985) 191.
11. Marletta G: Nucl. Instr. Meth. B32 (1988) 204.
12. Li RS and T Koshikawa: Surf. Sci. 151 (1985) 459.
13. Kang HJ, R Shimizu, and T Okutani: Surf. sci. 116 (1982) L173.
14. Kang HJ, E Kawatoh, and R Shimizu: Surf. Sci. 144 (1984) 541.
15. Gnaser H and ID Hutcheon: Surf. Sci. 195 (1988) 499.
16. Kelly R and DE Harrison: Mater. Sci. Engr. 69 (1985) 449.
17. Kelly R: Nucl. Instr. Meth. B18 (1987) 388.
18. Good-Zamin CJ, MT Shehata, DB Squires, and R Kelly: Radiat. Eff. 35 (1978) 139.
19. Oostra DJ, RP van Ingen, A Haring, AE de Vries, and FW Saris: Phys. Rev. Lett. 61 (1988) 1392.
20. Martynyuk MM: Russ. J. Phys. Chem. 57 (1983) 494.
21. Urbassek HM: Nucl. Instr. Meth. B31 (1988) 541.
22. Sigmund P: Appl. Phys. A30 (1983) 43.
23. King WE and R Benedek: J. Nucl. Mater. 117 (1983) 26.
24. Harrison DE and RP Webb: Nucl. Instr. Meth. 218 (1983) 727.
25. Shimizu H, M Ono, and K Nakayama: Surf. Sci. 36 (1973) 817.
26. Ho PS, JE Lewis, HS Wildman, and JK Howard: Surf. Sci. 57 (1976) 393.
27. Shimizu R and N Saeki: Surf. Sci. 62 (1977) 751.
28. Saeki N and R Shimizu: Surf. Sci. 71 (1978) 479.
29. Yabumoto M, H Kakibayashi, M Mohri, K Watanabe, and T Yamashina: Thin Solid Films 63 (1979) 263.
30. Shikata M and R Shimizu: Surf. sci. 97 (1980) L363.
31. Berghaus T, C Lunau, H Neddermeyer, and V Rogge: Surf. Sci. 182 (1987) 13.
32. Jiang SL, A Oliva, A Amoddeo, and R Kelly: submitted to Surf. Sci.
33. Bøttiger J. SK Nielsen, and PT Thorsen: Nucl. Instr. Meth. B7/8 (1985) 707.
34. Gibbons JF, WS Johnson, and SW Mylroie: Projected Range Statistics, Semiconductors and Related Materials: 2nd ed. Stroudsberg, PA: Dowden, Hutchinson, and Ross, 1975.
35. Lucasson P: Fundamental Aspects of Radiation Damage in Metals. USERDA CONF-751006-P1, 1975, p. 42.
36. McLean D: Grain Boundaries in Metals. Oxford: Oxford UP, 1957, p. 136.
37. Lea C and MP Seah: Phil. Mag. 35 (1977) 213.
38. Sigmund P, A Oliva, and G Falcone: Nucl. Instr. Meth. 194 (1982) 541.
39. Oliva A, R Kelly, and G Falcone: Surf. Sci. 166 (1986) 403.
40. Aduru S, S Contarini, and JW Rabalais: J. Phys. Chem. 90 (1986) 1683.
41. Ho SF, S Contarini, and JW Rabalais: J. Phys Chem. 91 (1987) 4779.
42. Contarini S, S Aduru, and JW Rabalais: J. Phys. Chem. 90 (1986) 3202.
43. Herion J, G Scharlo, and M Tapiero: Appl. Surf. Sci. 14 (1982-83) 233.
44. Betz G and W Husinsky: Nucl. Instr. Meth. B13 (1986) 343.
45. Choudhury T, SO Saied, JL Sullivan, and A Abbot: J. Phys. D, in press.
46. Langell MA: Surf. Sci. 186 (1987) 323.
47. Kim KS, WE Baitinger, and N Winograd: Surf. Sci. 55 (1976) 285.
48. Chase MW, CA Davies, JR Downey, DJ Frurip, RA McDonald, and AN Syverud: JANAF Thermochemical Tables. 3rd ed. J. Phys. Chem. Ref. Data 14 (1985) Suppl. 1.

49. Marletta G: University di Catania, Catania, Italy, work in progress.
50. Fecht HJ and WL Johnson: Nature 334 (1988) 50.
51. Oliva A, R Kelly, and G Falcone: Nucl. Instr. Meth. B19/20 (1987) 101.
52. Torrisi A, G Marletta, A Licciardello, and O Puglishi: Nucl. Instr. Meth. B32 (1988) 283.
53. Hultgren R, PD Desai, DT Hawkins, M Gleiser, and KK Kelley: Selected Values of the Thermodynamic Properties of Binary Alloys. Metals Park, OH: American Society for Metals, 1973.
54. Biloen P, R Bouwman, RA van Santen, and HH Brongersma: Appl. Surf. Phys. 2 (1979) 532.
55. Dawson PT and SA Petrone: Surf. Sci. 152/153 (1985) 925.
56. Naguib HM and R Kelly: J. Phys. Chem. Sol. 33 (1972) 1751.

DEFECT CREATION IN ION BOMBARDED CERAMICS

P. THEVENARD

Universite Lyon I, Departement de Physique des Materiaux,
43 Boulevard du 11 Novembre 1918, 69622 Villeurbanne, Cedex, France

1. INTRODUCTION

The study of defect creation in ion bombarded ceramics is inherently complex due to the fact that first these materials are polyatomic solids having complicated structure and bonding, and secondly ion implantation is a non homogeneous process of defect creation which occurs out of thermodynamic equilibrium. In this lecture we have chosen to emphasize damage effects in ion implanted ceramics such as MgO and TiO_2. After a brief review on well characterized defects in simple ionic crystals, the intrinsic defect creation is described with regard to particle-matter interaction.

In the case of high energy particles, the track effect associated with electronic energy loss processes is interpreted by using a model of defect creation surrounding the particle trajectories. For low energy ion irradiation, atomic collision processes are preponderant and defects are created in small volumes associated with dense collision cascades. The kinetics of point or extended defect creation associated with the nuclear energy loss processes are explained by using a model which takes into account the recombination of defects.

An important part of this lecture deals with the extrinsic defects due to the implanted particles in the lattice: microstructural modifications in the implanted layer associated with bonding and charge transfer effects are pointed out in relation with critical parameters such as local concentration, chemical state and site location of the implanted species and temperature. Finally, the aggregation and precipitation or dissolution mechanisms responsible for new phase formation in ceramics are presented and discussed taking into account the high energy deposition and the high concentration of defects characteristic of ion bombarded solids. The consequences of the interactions of energetic ions with insulating materials span a wide range of aspects which can be divided into three major categories with regard to the interactions:

(1) Excitations and ionizations of the electrons of the solid which arise at all ion energies are preponderant at high velocities. The consequences of electronic collisions are important in insulating materials particularly those which present a high degree of ionicity: they can result in intrinsic defect production electron-hole pair generation, luminescence, change in the valence state of impurities, sputtering, new molecule formation, break in chemical bond, secondary electron emission, and radiation emission.

(2) Atomic collisions associated with direct momentum transfer from the incident ions to the atoms of the solid, are preponderant at low velocities. The consequences of nuclear collisions span intrinsic defect production, atomic mixing, sputtering, molecule formation or molecule decomposition.

(3) The ion implantation process occurs when the particles lose all their energy. The atomic addition in a solid can result in doping, new compound

C.J. McHargue et al. (eds.), Structure-Property Relationships in Surface-Modified Ceramics, 149–166.
© 1989 by Kluwer Academic Publishers.

formation, new phase formation and specifically in insulators a "new chemistry" may govern the process.

The defect production in insulators with ion beams is evidently a complex problem and a number of review papers and books on radiation damage and ion implantation effects in insulators are available: Crawford and Slifkin (1), Dupuy (2), Henderson and Hugues (3), Hobbs (4), Kelly (5), Matzke (6), Henderson (7), Mazzoldi (8), Arnold and Borders (9), Wilson and Webb (10), Clinard and Hobbs (11), Mazzoldi and Arnold (12), Thevenard and Perez (13).

The complexity of the topic is due to the fact that generally the bombarded insulators have complicated composition (two or multicomponent materials), complicated bonding (from highly ionic to partially covalent). Then they present a wide dispersion in damage response. As an example, Table 1 lists the crystal structures, the packing of anions and cations of simple refractory oxides and the variation of the ionicity of the chemical bond is indicated.

TABLE 1. Crystal structure of simple refractory oxides

Structure	Anion		Cation site occupancy	Ionicity
MgO	Rocksalt	ccp*	All octahedral	0.8
Al_2O_3	Corundum	hcp**	2/3 octahedral	0.7
TiO_2	Rutile	hcp	1/2 octahedral	0.5
Mg_2SiO_4	Olivine	hcp	1/2 octahedral 1/8 tetrahedral	0.4
$MgAl_2O_4$	Spinel	ccp	1/2 octahedral 1/8 tetrahedral	0.6

*ccp = cubic close packed.
**hcp = hexagonal close packed

The ion implantation consequences in insulating materials are strongly different from those observed in metallic compounds as the implanted particles may have in the lattice different charge-states or may react and establish chemical bonds with the host atoms.

The aim of this lecture is to review some aspects of defect creation and new compound formation in ion bombarded insulators and the presentation will be focused particularly in refractory oxides such as ceramics.

After a brief presentation of experimental techniques of characterization, the lecture begins with the correlation between the defect creation in ionic crystals and the electronic energy loss of high energetic particles. The second part of the topic concerns the intrinsic defect creation and the nuclear energy loss of the particles in refractory materials such as MgO. The third part treat on the recovery and new compound formation by ion implantation in MgO where the influence of different key parameters such as the electronegativity, size, charge state of the implanted particles, the dose, the temperature, the defects in the host lattice have been studied in detail. The last part of the topic deals with ion implantation effects in a partially covalent ceramics: TiO_2. The chemical bond formation between implanted particles and atoms of the host lattice is studied in relation with the associated intrinsic defect creation. The influence of the nature of implanted ions on the chemical effects and on the extended defect creation (planar precipitates, microtwins...) have been analyzed in detail. The present topic has not the pretention to unify all the results obtained in ion implanted insulators. In order to have an overview on this subject, the student has to see in addition the topic presented by C. J. McHargue on ion implantation effects in Al_2O_3 (this volume).

2. TECHNIQUES

The study of defect creation in materials is strongly dependent on the experimental techniques available. In insulating materials, optical measurements are useful to characterize point defects, aggregates or chemical bonding. The use of transmission electron microscopy TEM or HREM (high resolution electron microscopy) or x-ray diffraction is essential for the characterization of extended defects aggregates, precipitates and new compounds. The Rutherford backscattering analysis (RBS) in random or channeled position may give essential information on damage, depth distribution of implanted ions, and location of these ions.

Nuclear techniques such as Mössbauer spectroscopy give useful information on the electronic structure of implanted Mössbauer ions, the position in the host lattice, and the local symmetry.

2.1. Optical spectroscopy

Point defects in ionic crystals have been well characterized by absorption or emission optical spectroscopy. Point defects such as electrons trapped in anion vacancies have energy levels within the band gap. The optical absorption band shape of a defect depends on the electron-phonon interaction. The coupling between the electronic and the vibrational motion is defined by the Huang Rhys factor S. As a consequence of this coupling, there is an energy shift $Sh\omega/\pi$ between the absorption band peak and the emission band peak associated with a given defect, where $h\omega/2\pi$ is the phonon energy. For weak coupling S is lower than 1 and the energy is localized in the zero phonon line and for strong coupling S is larger than 10 and the optical bands are broad and structureless. The energy (or wavelength) position of the absorption band or emission band is typical of a defect. In the case of F-type centers in ionic crystals (LiF or MgO) the S factor is larger than 10 and the absorption band are situated near 5 eV (250 nm).

For absorption measurements the wavelength range is generally 200 to 2000 nm, and the optical density, OD = log (I_0/I_t) indicates the level of absorption. I_0 represents the intensity of the incident beam on the sample and I_t the intensity of the transmitted beam.

The absorption coefficient α is given by α = 2.3 (OD/e) where e is the thickness of the sample crossed by the optical beam. For a Gaussian band shape the concentration N of defects is given by Smakula's formula:

$$N \times f = 8.7 \times 10^{19} \left[\frac{n}{n^2 + 2} \right]^2 \alpha_{max} \times \Delta E \quad \text{defects} \times m^{-3}$$

where f is the oscillator strength of the transition, α_{max} the absorption coefficient at peak height, n the refractive index, and ΔE the half width of the band in eV.

For F-type centers, f is of the order of 1 and 10^{15} defects per cm^{-3} absorb approximately 10% of the intensity of the light in a sample of 1 mm thick. Optical measurements are then very useful to characterize defects at low concentration in insulating materials.

2.2. Electron microscopy

The observation of irradiated insulators by transmission electron microscopy is possible if the materials are insensible to ionizing radiation. The refractory oxides which are insensible to electron irradiation may be observed by this technique. The electrons that form the image must pass through the specimen which then must have a thickness lower than approximately 100 nm. TEM can be operated in different modes giving different information about the specimen: bright-field, dark-field, and selected-area diffraction. In bright-field the regions of the sample which scatter electrons appear black and diffracted electrons are blocked by the objective aperture. In dark-field only a selected diffracted beam is allowed to pass through the objective aperture, then diffracting regions appear bright and transmitting region black. Information can be obtained about the crystallite orientation and crystalline defects. The diffraction patterns give information on the crystal structures and orientations of different phases present in a sample. One inconvenience of the technique concerns the sample preparation, as this is destructive.

2.3. X-ray diffraction at glancing incidence

This nondestructive technique gives information on crystalline structures near the sample surface.

2.4. Rutherford backscattering spectroscopy

This technique uses light ion beams of energy in the MeV range. The measurement of the energy of ions scattered by coulombic interaction with nucleus of the target atoms gives information on:
--the concentration of impurities in a sample,
--the stoichiometry, and
--the depth distribution of implanted particles in a target.
The sensitivity of this analysis is about 10^{20} atoms$\cdot$cm^{-3}.

2.5. Mössbauer spectroscopy

This technique is based on recoil-free emission of γ-rays by a radioactive source and the resonant absorption of these γ-rays by Mössbauer nuclei in the sample.

The resonance between the source (^{57}Co for example) and the absorber (^{57}Fe for example) is achieved by modulating the energy of the emitted γ rays using the Doppler effect. The source is displaced at a velocity v (in mm$\cdot$s^{-1}); the velocity equal to zero is relative to metallic iron.

The Mössbauer parameters are (14):
--the isomer shift IS which depends on the electronic density at the nucleus givng the information on the oxidation states of iron and on the bonding;
--the quadrupole splitting QS as influenced by any modification of the local electric field gradient at the nucleus;
--the magnetic hyperfine field H which is the internal magnetic field at the nucleus.

The information given by this technique are the electronic structure of Mössbauer ions in the host, their positions in the host lattice and the local symmetry. In order to study implanted materials conversion electron Mössbauer spectroscopy (CEMS) is well adapted. After the absorption of γ ray by Mössbauer nucleus it decays by re-emitting a γ-ray or by emission of an atomic s-level electron. The ratio of emitted electrons to re-emitted γ is about 10. By measuring the emitted electrons, a near surface analysis (about 100 nm) can be performed. The sensitivity of this technique corresponds to 10^{15} ^{57}Fe$\cdot$cm^{-2}.

3. INTRINSIC DEFECT CREATION BY ELECTRONIC PROCESSES IN HIGHLY IONIC MATERIALS: LiF and $MgFe_2O_4$

The type of energy deposition (electronic excitations or nuclear colli- sions), the temperature, and the time scale are critical parameters for defect creation in highly ionic materials. The energy deposition is described in Fig. 1 which represents the two parts of the stopping power of an energetic ion as a function of the square root of Lindhard's energy para- meter ϵ (15) (S_e corresponds to electronic processes and S_n to nuclear processes).

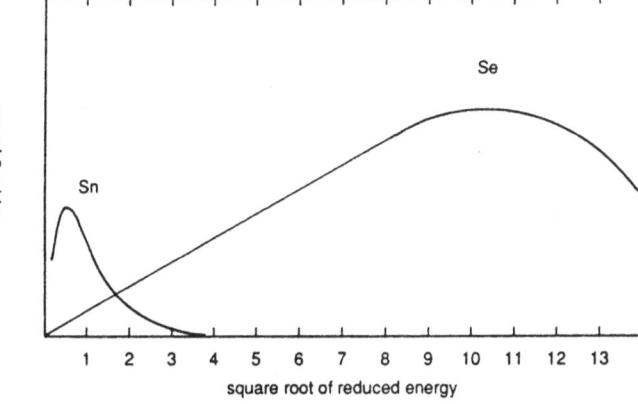

FIGURE 1. Stopping power of ions: Se electronic energy loss, Sn nuclear energy loss as a function of $\epsilon^{1/2}$ (ref. 1).

Relative importance at these two modes can be shifted by the choice of the ion mass and energy. The total energy deposition occurs within less than 10^{-13} sec and is strongly localized in a small volume surrounding the par- ticle trajectory.

In lithium fluoride, the calculations of Fain (16) have shown that 90% of the energy dissipated by electronic processes is deposed within a distance of 10 nm from the particle trajectories. Taking into account the spatial distribution of the energy deposited by the particle, the damage should be created very close to classical trajectory in a volume called the "ion track". The defect concentration is not homogeneous but there is a gradient of damage of the ion track. A model for defect creation has been proposed by Thevenard et al. (17) which takes into account this localization. Following this model the concentration of defects created per unit area should vary as an exponential law versus the ion dose ϕ:

$$C_{defects} \cdot cm^{-2} = C_{max} \exp (- \pi r^2 \phi)$$

where r is the track radius and C_{max} the defect concentration at saturation. Experimental results of defect creation (F centers) in LiF can be fitted using this model. The track radius r associated with different particles in LiF are reported on Table 2. These results suggest the following remarks:

(1) The track radius r deduced from the experimental results are mean values assuming in a first approximation that defects are created at satura- tion in a cylinder of radius r surrounding the particle trajectories.

(2) The track associated with a given particle has a geometry which depends on the particle energy. A conic form of the track appears to be more realistic than a cylinder one.

TABLE 2. Track radius r associated with different particles in LiF

Ion	Energy (MeV)	Range (μm)	Track radius (nm)
^{1}H	2	35	1,3
^{3}He	4	15,3	4,2
^{40}Ar	4	2,2	5,C
^{2}H	28	2000	7,C
^{7}Li	28	100	6,3
^{4}He	56	1000	16,5

Unfortunately the track effects in LiF cannot be revealed by TEM observations as lithium fluoride is sensitive to electron irradiation. The con-
firmation of the existence of track effects associated with energetic
particles have been obtained recently by Treilleux et al. (18) in
$MgFe_2O_4$ bombarded with 3.4 GeV xenon ions. From TEM observations the
experimental track radius is about 2.5 nm. Track effects have also been
observed in $Y_3Fe_5O_{12}$ by Houpert et al. (19).
The mechanisms which explain the track effect in such oxides are not very
clear as most of the oxides are insensitive to ionizing radiation.
Following Clinard and Hobbs (7) there are four criteria for efficient
coupling of the electronic energy losses to atomic displacement phenomena:
    (1) An electronic excitation must be localized to one or a few atoms;
    (2) The excitation must have a lifetime comparable to phonon period in
order to couple into a mechanical response;
    (3) The available excitation energy must be comparable to the atom displa-
cement energy in its excited state; and
    (4) An energy to momentum conversion mechanism must exist and compete
favorably with other excitation decay modes.
These four criteria imply that atomic displacement due to electronic excitation will occur in highly ionic solids with univalent anion. Many ionic
oxides fail the energy criteria as the binding energy E_b due to coulombic
interactions is proportional to the square of the common ion charge Ze in
the lattice: $E_b \sim 8Z^2$ (in eV). Thus radiolysis does not occur in most of
refractory oxides.

4. DEFECT CREATION BY NUCLEAR PROCESSES IN MgO
Magnesium oxide possesses a highly ionic binding and thus fails the energy
criteria for defect creation by electronic excitations. However, the
nuclear energy losses of energetic particles in MgO can result in defect
creation if the energy transfer to one atom of the lattice exceeds the
displacement energy. The two sublattices are affected by defect formation.
As the displacement energy for oxygen and magnesium is 60 and 64 eV, respectively the concentrations of displaced oxygen and magnesium atoms are
the same. The primary defects are F-type centers (electrons trapped in
anion vacancies), V-type centers associated with cation vacancies, and
interstitials which form perfect dislocation loops. F and F^+-centers give

an optical absorption band at 250 nm and V centers at 570 nm. Evans et al.
(20) has shown that the defect creation in MgO bombarded with energetic rare
gas ions is correlated with nuclear energy losses of the particles and that
a saturation in point defect concentration is reached near 10^{21} defects·cm^{-3}.
A typical growth curve for F-type center creation in MgO bombarded with 320
keV Rb$^+$ is represented on Fig. 2. For low doses, a yield of about one F$^+$
center per incident ion is observed. This is much lower than the displaced
oxygen atoms per incident ion given by LSS theory and indicates a high
degree of defect recombination.

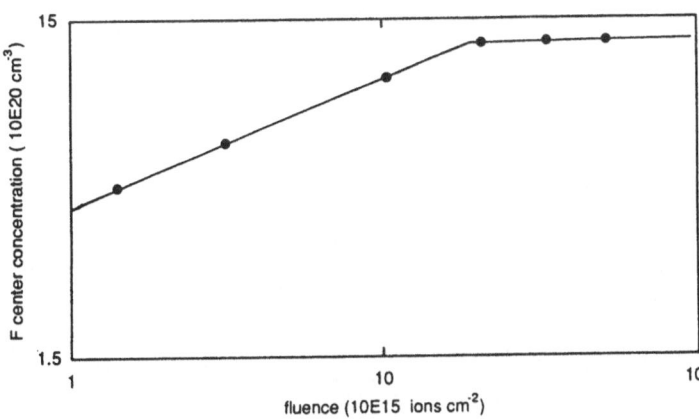

FIGURE 2. F$^+$
center growth
curve in MgO
bombarded at
300 K with 320
keV Rb$^+$ ions
(ref. 22).

The recombination volume for a vacancy can be deduced from the concentration at saturation. The value obtained is equal to 2 nm^3 which
corresponds to a radius of 0.8 nm = 2 a_0 (a_0 MgO lattice parameter). The
F$^+$ center creation varies as the logarithm of the dose. The kinetics
correspond to the evolution predicted by Hughes and Pooley (21). Extended
defect formation can occur in MgO bombarded with heavy ions at 300 K. Using
in-situ electron microscopy observations Treilleux (22) has shown that
dislocation loops are localized in {110} planes. They are of interstitial
type and perfect in nature. A correlation between F$^+$ type centers and
interstitials in the loops has been pointed out. Finally, it is not
possible to reach amorphization in MgO even for heavy ion bombardment and
very high doses > 10^{17} ions·cm^{-2}.

The intrinsic defect evolution with annealing has been studied with regard
to the influence of the implanted ions. The evolution of normalized point
defect concentration with temperature anneal is represented in Fig. 3 for
Na$^+$, Ne$^+$, Mg$^+$ implanted samples.

Vacancies are immobile in the temperature range 0 to 800°C as shown by
Chen et al. (23) in additive coloration samples and the decay of the F$^+$
absorption band is mainly due to mobile interstitials which recombine with
vacancies. From Fig. 3, it clearly appears that these recombinations depend
on the nature of the implanted species due to the coupling between interstitials and implanted particles. A correlation between the evolutions of the
intrinsic defects and the implanted particles seems to be evident. This is
confirmed by the study of the implanted particle evolution with temperature.

5. NEW PHASE FORMATION ASSOCIATED WITH THE IMPLANTED PARTICLES IN MgO

The metastable phases produced by a high dose implantation of metallic ions
into MgO have been studied in details with various ions in large doses and

156

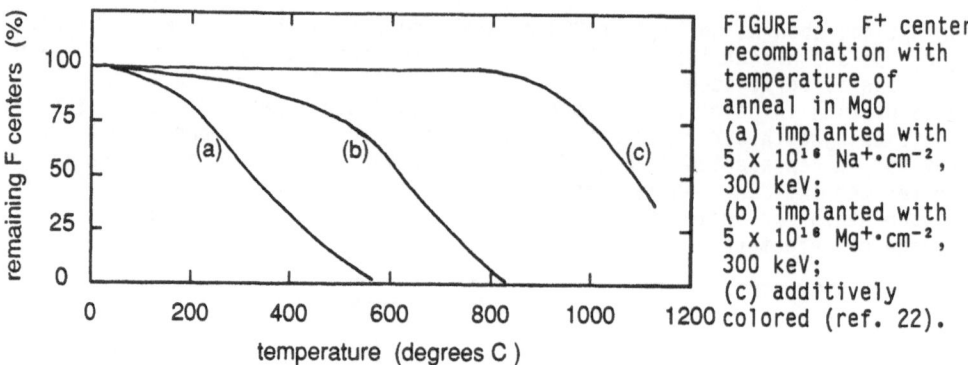

FIGURE 3. F⁺ center
recombination with
temperature of
anneal in MgO
(a) implanted with
5×10^{16} Na⁺·cm⁻²,
300 keV;
(b) implanted with
5×10^{16} Mg⁺·cm⁻²,
300 keV;
(c) additively
colored (ref. 22).

energy ranges. Two precipitation processes occur in MgO depending on the
implanted species and on temperature.
--The implanted particles can precipitate independently of the elements of
the matrix (Mg and O) to form small metallic clusters embedded in the lattice: e.g. Li, Na, K, Rb, Cs, Mg, and Fe.
--The implanted impurities can precipitate with elements of the matrix to
form new compounds: e.g. Fe, In, and Au.

5.1. Precipitate formation of extrinsic atoms

After implantation with alkali ions or magnesium ions and generally after
annealing at a temperature where intrinsic defects are mobile, the formation of metallic precipitates of implanted particles occurs in MgO (24-27).
The growth of the clusters with time t at a given temperature T is described
by a power law $t^{1/6}$ which is consistent with low strain energy and a
dislocation- assisted diffusional process. The small clusters have been
identified by optical absorption (OA), transmission electron microscopy
(TEM) and x-ray diffraction. Figure 4 shows the typical optical absorption
evolution of annealing with the temperature of annealing for Li-, Na-, K-,
and Mg-implanted samples. The characteristics of these samples are listed
in Table 3: ion energy E, dose ϕ, projected range R_p, range straggling ΔR_p,
atom concentration. In addition, the Fermi velocities ν_F, the plasma energy
E_p and the cutoff wavelength λ_c of each metal have been reported. The position λ of the optical absorption band of small spherical metallic precipitates embedded in MgO depends on λ_c of the metal and on the refractive index
n_0 of MgO (28) by the relation:

$$\lambda = \lambda_c (1 + 2n_0^2)^{1/2} .$$

The half width $\Delta E_{1/2}$ of this absorption band depends on the radius R of the
metallic aggregate as:

$$\Delta E_{1/2} = h \frac{\nu_F}{R} ,$$

h is the Planck constant and ν_F the Fermi velocity). The calculated position λ^* of the absorption band and the experimental value λ_{exp} for metallic
clusters dispersed in MgO are reported in Table 2. A good agreement is
observed between the values λ^* and λ_{exp} for the different implanted metals.
The displacement in the position of the absorption band with increasing temperature of anneal is typical of the growth in size of the metallic
clusters.

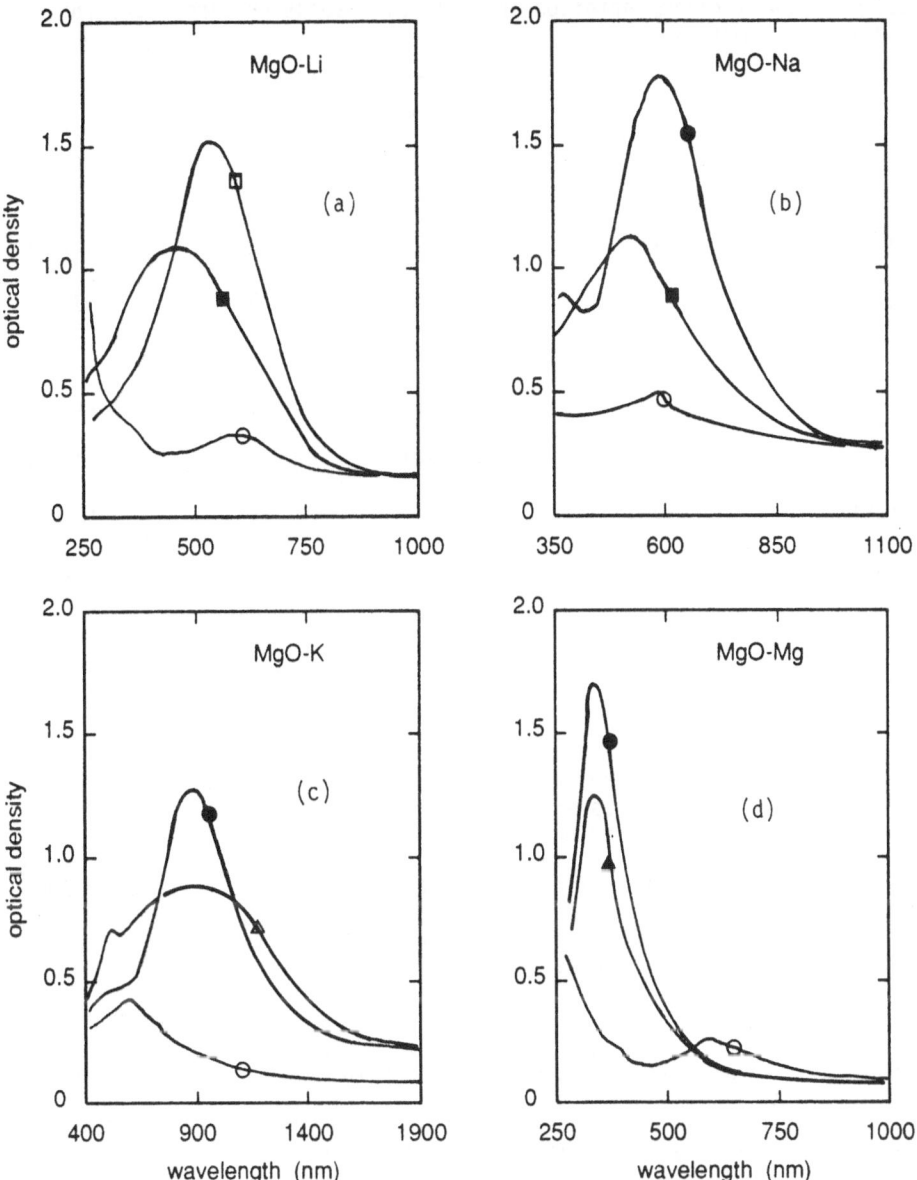

FIGURE 4. Optical absorption spectra evolution with the temperature of anneal of MgO samples implanted with: (a) 5 x 10^{16} Li$^+\cdot$cm^{-2}, 250 keV; (b) 5 x 10^{16} Na$^+\cdot$cm^{-2}, 600 keV; (c) 10^{17} K$^+\cdot$cm^{-2}, 450 keV; and (d) 5 x 10^{16} Mg$^+\cdot$cm^{-2}, 300 keV (ref. 22,24); O = as implanted; $\triangle$ = 550°C; ■ = 650°C; $\triangle$ = 750°C, □ = 800°C; ● = 850°C.

TABLE 3. Summary of implantation and optical absorption parameters for MgO
implanted with metallic ions.

Ion	Li	Na	K	Mg
E (keV)	250	600	450	300
ϕ (ion·cm^{-2})	$5\cdot10^{16}$	$5\cdot10^{16}$	10^{17}	$5\cdot10^{16}$
Rp (nm)	650	610	300	320
ΔRp (nm)	80	76	67	60
C_{max} (cm^{-3})	$2,5\cdot10^{21}$	$2,5\cdot10^{21}$	$6\cdot10^{21}$	$3\cdot10^{21}$
v_F (m·s^{-1})	$1,3\cdot10^{6}$	$1,1\cdot10^{6}$	$0,9\cdot10^{6}$	$1,6\cdot10^{6}$
E_p (eV)	7,1	5,7	3,7	10,6
λ_c (nm)	174	217	333	117
λ* (nm)	460	570	880	310
λ_{exp} (nm)	450	500	900	340

The orientation relationship and structure of these clusters are strongly
influenced by the lattice as shown by TEM observations. An unusual face
centered cubic structure has been observed for lithium precipitates (size
smaller than 20 nm) which are in epitaxy with the matrix. For sizes larger
than 30 nm, the structure is bcc and the orientation with respect to the
matrix is (011) Li // (111) MgO and <111> Li // <011> MgO. The fcc structure is imposed by the MgO lattice in order to reduce the surface energy of
the precipitates. If a precipitate is in total coherency with the matrix
one metal atom replaces one MgO molecule and the interface energy is equal
to zero. More generally the interface energy is proportional to the surface
of the precipitate. The strain energy of a precipitate depends on both the
bulk modulus of the precipitate and the shear modulus of the matrix, and is
proportional to the precipitate volume. Then the "coherent to incoherent"
transition occurs when the strain energy of a precipitate becomes higher
than the interfacial energy. The loss of coherency for a lithium precipitate in MgO is obtained when the radius of a precipitate is of the order of
12 nm in good agreement with the experimental observations. The same calculations for Na, K, or Mg precipitates in MgO give the limits for coherency
of these precipitates: 0.5 nm (Na), 0.3 nm (K), and 1.4 nm (Mg), respectively. These results explain why unusual phases of sodium, potassium or
magnesium have never been observed in small precipitates of these metals in
MgO.

5.2. Metastability of metallic clusters under ion bombardment: ion mixing
defects

The behavior of small metallic precipitates embedded in MgO when submitted
to a bombardment with energetic ions has been analyzed by optical absorption
and transmission electron microscopy. Strong modifications in the cluster

concentration arise when collision cascades take place in the precipitates. As shown, for example in Fig. 5, the optical absorption of metallic precipitates (Na) decreases when they are bombarded with energetic ions (Ar) (29).

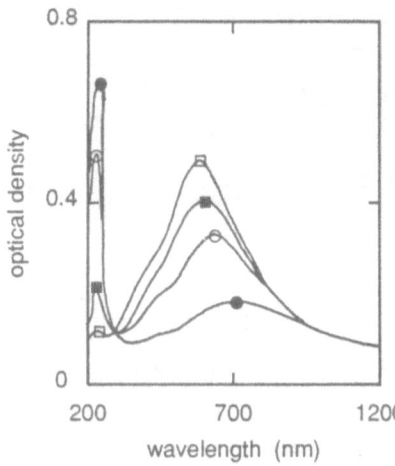

FIGURE 5. Optical absorption spectra of MgO crystal implanted with 2.7 x 10^{16} Na$^+$·cm^{-2}, 53 keV; □ = annealed 30 min at 973 K and then implanted with 107 keV Ar$^+$ ions; ■ = 2 x 10^{13} Ar$^+$·cm^{-2}; O = 3.2 x 10^{14} Ar$^+$·cm^{-2}; ● = 2 x 10^{15} Ar$^+$·cm^{-2} (ref. 31).

This effect is not due to electronic energy loss processes, it depends only on the nuclear energy losses of the bombarding particles. The area under the absorption band at 540 nm due to Na metal is proportional to the quantity of Na in metallic precipitate form. The reduction in band amplitude can be ascribed to either the dispersion (re-solution) of Na atoms in the MgO host, presumably due to the progressive break-up of the metallic precipitates under the impact of the incident particles, or to a change in the optical properties of the precipitates (from metallic to insulating). However, as shown on Fig. 6, this progressive break-up mechanism is in complete contradiction with in situ TEM observations (29-33) which show that as opposed to the decrease of the absolute number of visible clusters their average size is rather weakly affected by the ion bombardment.

In addition a "one-shot" effect is observed, i.e. a cluster that had not been affected by the ion bombardment at a given fluence (10^{14} ions·cm^{-2}) could become invisible after a further implanted fluence of only 10^{13} particles·cm^{-2}. This result precludes an interpretation in terms of progressive Na precipitate dissolution under the kinematical effect of incoming ions since the latter process would lead to a significant reduction in average precipitate size. Furthermore, though a precipitate becomes invisible in a dark-field micrograph a careful study of the corresponding bright field, micrograph reveals that a contrast remains at the place of the precipitate. So, the particle bombardment does not lead to precipitate disruption nor to a change in its size but does lead to a modification of its structure and chemical nature. It is inferred that collision cascades produced by the incident particles at the host matrix-precipitate interface lead to the incorporation of host (Mg or O) atoms into the metallic precipitates. Because of their small size, the metallic clusters are at very high pressure (several thousand atmospheres) and the mobility under irradiation of oxygen is then considerably enhanced inside the metal. Thus, the collision cascade at the interface essentially serves to trigger the inward diffusion of the host constituents. These effects are reminiscent of ion beam mixing processes and they strongly depend on the density of displacement

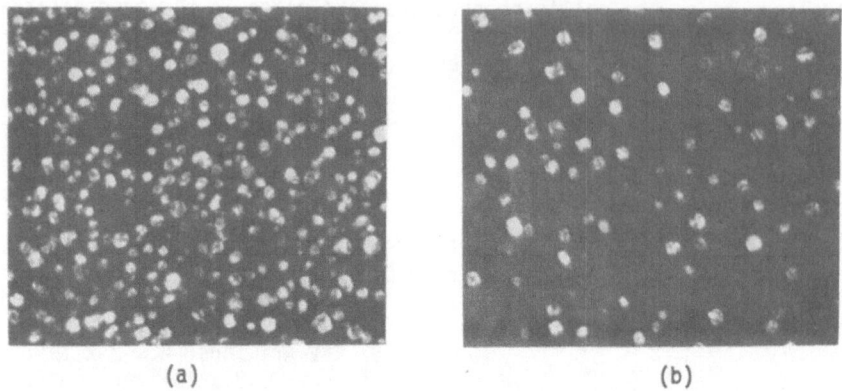

(a)   (b)

FIGURE 6. Dark field TEM images of Na precipitates in MgO crystal implanted with 2.7 x 10^{16} $Na^+ \cdot cm^{-2}$, 53 keV: (a) subsequently annealed for 30 min at 973 K and then (b) bombarded iwth 8 x 10^{14} $Na^+ \cdot cm^{-2}$, 120 keV (ref. 29).

cascades as shown by Treilleux et al. (31). As shown on Fig. 7, 2 x 10^{14} $Xe \cdot cm^{-2}$, 2 x 10^{15} $Kr \cdot cm^{-2}$, 10^{16} $Ar \cdot cm^{-2}$, and 2 x 10^{16} $Ne \cdot cm^{-2}$ have the same efficiency for eliminating precipitates.

In all the experiments, mixing effects also affect the intrinsic defect creation. The yield for F-center creation is enhanced when the metallic clusters are "re-dissolved" in the host MgO, by comparison with the kinetics in virgin MgO bombarded with the same ions at the same dose. This effect points out that metallic clusters are associated with intrinsic defects: their precipitation can take place in aggregate of intrinsic defects and then their dissolution generates vacancies in MgO which increases the yield for intrinsic defect creation.

5.3. Compound formation with host atoms

5.3.1. Oxide formation. Such a precipitation process is observed with iron implanted in MgO. This study has been performed by Perez et al. (34,35) by using Mössbauer spectroscopy, TEM, and optical absorption. They have shown that in as-implanted MgO samples iron is present in the lattice in different charge-states: Fe^{3+}, Fe^{2+}, and Fe^0. The relative fraction of these charge-states depends on the atomic concentration of implanted iron: Fe^{3+} is dominant at low concentration (<10^{15} ions $\cdot cm^{-2}$), Fe^{2+} with two different locations in the lattice (Fe_I^{2+} in magnesio-wustite and Fe_{II}^{2+} in dimers) dominant at intermediate local concentration and finally Fe^0 dominant at large concentration (>10^{17} ions $\cdot cm^{-2}$) in the form of small metallic precipitates (<2 nm).

The effect of thermal annealing at 700°C in air is to convert the iron mainly into Fe^{3+} ions, which is accompanied by some changes in the local symmetry. In the high temperature range (800–1000°C) annealing processes in air and in vacuum are drastically different. The formation of magnesio-ferrite precipitates coherent with the MgO structure was obtained by annealing in air, whereas annealing in vacuum converted all iron to Fe^{2+}. This result points out the key role of the oxygen partial pressure during the annealing process.

5.3.2. Metallic alloy formation. The precipitation of a binary alloy has been observed in MgO heavily implanted with indium and gold (34). After annealing at 700°C, MgO samples implanted with about 10^{17} $In^+ \cdot cm^{-2}$ contained

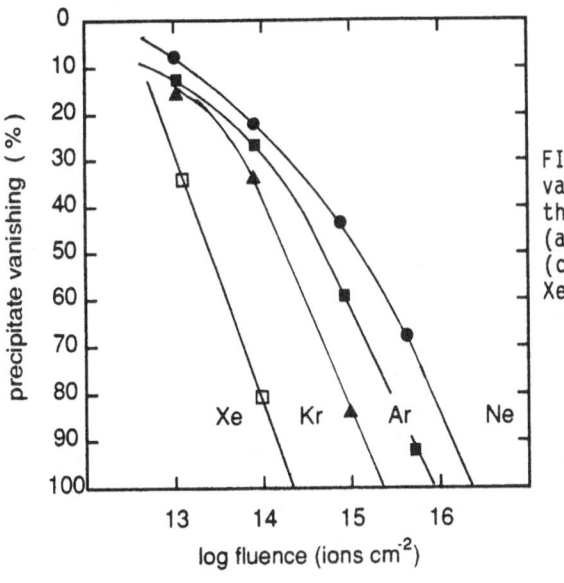

FIGURE 7. Na precipitate vanishing (in %) in MgO versus the dose of different ions: (a) 53 keV Ne$^+$; (b) 107 keV Ar$^+$; (c) 200 keV Kr$^+$; and (d) 500 keV Xe$^+$ (ref. 31).

small precipitates of a metallic alloy. Mg_3In was identified both by TEM and optical absorption. The alloy is crystallized in the hexagonal system and it has a preferred orientation relationship with the MgO matrix.

In the case of gold implanted in MgO, gold precipitates were observed both by TEM optical absorption and x-ray diffraction after annealing (<1000°C). They have the normal fcc structure and the size distribution is in the range 13 to 25 nm. For annealing at temperatures higher than 1000°C two different phases were identified: the metallic gold precipitates and a binary alloy Au_3Mg in the form of thin platelets crystallized in the hexagonal structure.

6. NEW PHASE FORMATION IN IMPLANTED TiO_2

When investigating the properties of implanted rutile, one should take into account the following characteristics of rutile:
--TiO_2 has the rhombohedric structure.
--Octahedrally or tetrahedrally coordinated interstitial sites are possible.
--Titanium can exist in rutile in a variety of states as oxidized Ti^{4+} or reduced Ti^{3+}.
--Nonstoichiometry can occur in TiO_2 which can be accommodated by extended defects, crystallographic shear structures or by the formation of Magneli phases.
--The phase relations for the systems metal-titanium-oxygen solutions are rather complicated.

There are only a few results on defect creation with high energy particles in covalent oxides like TiO_2 (36–42). Nevertheless, the ion implantation effects in TiO_2 are strongly different from those obtained in MgO. Chemical defects can occur in implanted TiO_2 which is not the case in a highly ionic oxide as MgO. The evidence of chemical effects has been shown in TiO_2 implanted with protons, deuterons (38,41) and alkali ions (43). At high doses precipitation of mixed oxides is observed together with the formation of extended planar defects.

Iron ions implanted in rutile give rise to small platelets of pseudo-brookite phase (44) and gold or silver ions form small metallic aggregates (39).

6.1. Evidence of chemical effects in TiO_2 implanted with protons or deuterons

In order to separate the effects of the electronic and the nuclear part of the energy losses from the implanted effects, TiO_2 single crystals have been implanted with 28 MeV deuterons and 56 MeV alpha particles. Optical absorption and conductivity measurements have been performed on the samples. The irradiation of TiO_2 samples with alpha particles does not lead to any coloration along the particles paths and in the implanted area. If defects are created they do not absorb light in the range 0.4-2 µm. This result is surprising as nonstoichiometric rutile TiO_{2-x} has a blue coloration due to an optical absorption in the infrared region.

When irradiated with deuterons, TiO_2 rutile exhibits a blue coloration in the implanted area. The optical absorption is located at 1.35 µm and is attributed to forbidden transition of Ti^{3+}.

The Ti^{3+} defect formation is associated with the formation of chemical bonds between implanted deuterons and oxygen of the host matrix. The O-D chemical bonds have been characterized by their optical absorption in the infrared near 2510 cm^{-1} (Fig. 8). This absorption band increases linearly with the ion dose for local concentration of deuterons lower than $3 \cdot 10^{22}$ $ions \cdot cm^{-3}$. It has to be noted that Ti^{3+} ions created concomitantly with O-D bonds induce a strong modification in the electrical properties of TiO_2 in addition to the optical ones.

High resolution transmission electron microscopy observations on heavily implanted TiO_2 with 50 keV deuterons (or protons) have shown that the high density of defects induced by the ion implantation process leads to the formation of dislocations and of microtwins with {101} twin plane. Formation of microtwins is obviously induced by the chemical reaction with deuterons (or protons) since such twins are not observed in the implanted area of TiO_2 crystals implanted with rare gas ions which do not react with oxygen of the host matrix.

Since doses of about 10^{18} deuterons$\cdot cm^{-2}$ (or protons) correspond to local concentration of deuterons in TiO_2 nearly equal to the concentration of oxygens, the formation of a TiO(OD) [or TiO(OH)] phase can be expected when chemical reactions lead to the formation of O-D bonds (O-H bonds) (41,42). The TiO(OD) or TiO(OH) phase has a structure almost identical to rutile and then it can be expected to precipitate in twinned orientation with respect to TiO_2 matrix.

6.2. Implantation effects with alkali ions

The formation of Ti^{3+} defects is observed when TiO_2 single crystals are bombarded with alkali ions (40). The optical absorption band of these defects is located near 1 µm. This absorption cannot be due to plasma oscillations of free electrons in metallic precipitates formed with alkali metal atoms.

As in implanted deuteron samples, the Ti^{3+} defect creation should be correlated with the formation of chemical bonds between implanted alkali atoms and oxygens of the host lattice. The observations by TEM of implanted rutile show that extended planar defects are formed. The electron diffraction patterns obtained on the planar defects point out additional spots, Fig. 9. The interplanar distances deduced from the extra spots do not correspond to those of known alkali metal, or alkali oxides or titanium oxides. The formation of a mixed compound $Ti_xO_yM_z$ where M is the alkali metal is a realistic hypothesis, which is confirmed by x-ray diffraction at glancing incidence. The synthesis of this new phase is certainly favored by the existence of planar extended defects. The behavior of implanted alkali particles in TiO_2 is strongly different from those observed in MgO.

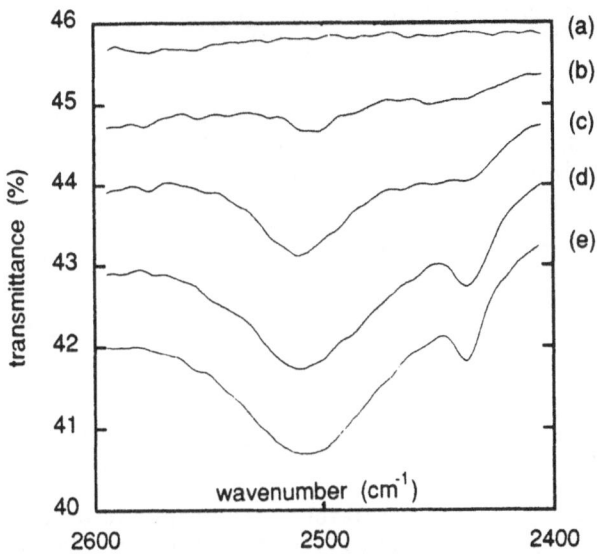

FIGURE 8. Infrared transmission spectra (FTIR) of rutile implanted with 50 keV deuterons at various doses: (a) before implantation; (b) 4 x 10¹⁶ D·cm⁻²; (c) 1.6 x 10¹⁷ D·cm⁻²; (d) 5 x 10¹⁷ D·cm⁻²; and (e) 10¹⁸ D·cm⁻².

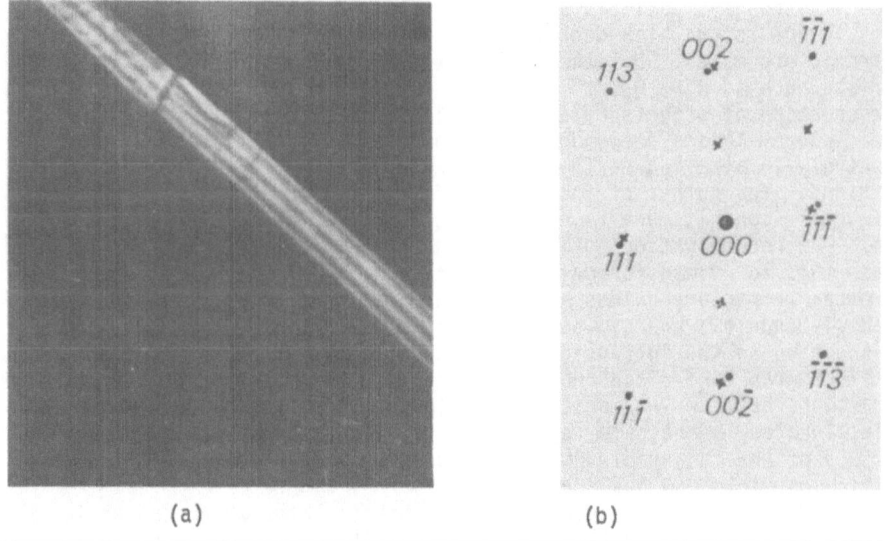

FIGURE 9. (a) Dark-field image (TEM) of TiO₂ implanted iwth 450 keV, 10¹⁷ K⁺·cm⁻²; (b) diffraction pattern with TiO₂ spots (•) and additional spots (x) attributed ot K₂TiO₃.

6.3. Iron implantation effects in titanium dioxide

The influence of bonding in the host matrix is not of great importance for defining the charge states of implanted iron: Fe^{2+} state is found predominantly as in MgO.

The existence of planar extended defects in TiO_2 favors the precipitation of new phases in the form of platelets. The metastable precipitated phase formed during the iron implantation has been identified as ferrous pseudobrookite $FeTi_2O_5$. These precipitates disappear upon annealing in air at relatively low temperatures which is correlated with both the oxidation of iron into the Fe^{3+} state and the iron out-diffusion (44).

6.4. Noble ion implantation effects

Gold or silver ions implanted in TiO_2 with high fluences form small metallic clusters. The existence of planar extended defects (as in Fe implantation) favors the precipitation of metallic aggregates of gold and silver.

The electronegativity of gold or silver is higher than titanium one which can explain the formation of these metallic clusters of implanted atoms.

7. CONCLUSION

In ion bombarded crystalline insulators, the intrinsic defect creation is strongly influenced by the energy deposition mechanisms. Defects can be created in both sublattices by atomic collision processes, the yield for defect creation depending on the specific anions and cations present in the matrix. In materials susceptible to ionization processes such as alkali halides, defects can be created by electronic process in the halogen sublattice in addition to the cation sublattice. In LiF it has been clearly shown that defect creation by electronic excitations induced by high energy particles is strongly localized around the particle trajectories. This track effect is due to the high densities of the localized deposited energy by electronic processes. Evidence of track effects have been shown in other materials such as $MgFe_2O_4$.

The creation of defects by atomic collisions is correlated with the nuclear energy losses preponderant at low energy. It has been studied extensively in alkaline earth oxides as they are not sensitive to ionizing radiations. The number of intrinsic defects created in MgO for example by atomic collisions is much lower than the number of calculated displaced atoms. The recombinations of defects in the displacement cascades can explain this low yield for defect retention.

In other refractory oxides as TiO_2, the intrinsic defect creation associated with nuclear energy losses is not so well known.

In addition to the intrinsic defects created by nuclear processes, chemical effects can take place due to the presence of implanted particles. The interactions between intrinsic defects (point or extended) and implanted species located in the same region lead to effects of a strong complexity. The study of the metastable phases produced by a high dose implantation of metallic ions into MgO has shown that:

--Alkali ions tend to form small metallic precipitates which can be in coherency with the host matrix. These precipitates can be re-dissolved by a subsequent ion bombardment.

--Magnesium ions form small metallic precipitates.

--Gold or indium ions tend to form binary alloys with magnesium.

--Iron ions present different charge states in MgO: Fe^0, Fe^{2+}, and Fe^{3+} depending on the local concentration; they tend to form spinal ferrite particles after annealing.

In TiO_2, chemical bond formations can occur between implanted particles and target atoms:

--Protons or deuterons form O-H or O-D chemical bonds correlated with the formation of intrinsic defects Ti^{3+}. The formation of TiO(OH) or TiO(OD) phase is observed, the planar precipitates are in the twinned orientation with respect to the TiO_2 matrix.

--Alkali ions tend to form a bond with oxygen leading to the formation of the intrinsic defects Ti^{3+}. Extended precipitates of mixed oxides $M_xTi_yO_z$ (M = alkali) are observed together with extended intrinsic defects.

--Gold ions form small metallic precipitates.

--Iron ions are present in various charge state: Fe^0, Fe^{2+}, and Fe^{3+}. They tend to form $FeTi_2O_5$ platelets after annealing.

From all these results it clearly appears that some specific parameters such as point or extended defect structures, bond ionicity, electronegativity, and/or crystallographic structure play key roles in this complex nonequilibrium phenomena but, up to now, general rules applicable to all ceramics have not been established.

REFERENCES

1. Crawford Jr. JH and LM Slifkin (eds): Point Defects in Solids. Vol. 1. New York: Plenum Press, 1975.
2. Dupuy CHS (ed): Radiation Damage Processes in Materials. Leiden: Noordhorf, 1975.
3. Henderson B and AE Hugues (eds): Defects and Their Structure in Non-Metallic Solids. New York: Plenum Press, 1976.
4. Hobbs LW: J. Am. Ceram. Soc. 62 (1979) 267.
5. Kelly R: Nucl. Instrum. Meth. 182-183 (1981) 351.
6. Matzke Hj: Radiat. Eff. 64 (1982) 3.
7. Henderson B: Radiat. Eff. 64 (1982) 35.
8. Mazzoldi P (ed): Radiat. Eff. 64 (1982) (Radiation Effects in Insulators-1).
9. Arnold G and J Borders (eds): Radiation Effects in Insulators-2. Amsterdam: North Holland, 1984.
10. Wilson I and R Webb (eds): Radiation Effects in Insulators-3. New York: Gordon and Breach, 1986.
11. Clinard F and LW Hobbs: Radiation Effects in Non-Metals. Chapter 7. Physics of Radiation Effects in Crystals. RA Johnson and AN Orlov (eds). Amsterdam: Elsevier Science Publishers, 1986.
12. Mazzoldi P and G Arnold (eds): Ion Beam Modification of Insulators. Amsterdam: Elsevier Science Publishers, 1987.
13. Thevenard P and A Perez (eds): Radiation Effects in Insulators-4. Amsterdam: North Holland Publishers, 1988.
14. Gibb TC: Principles of Mössbauer Spectroscopy. London: Chapman and Hall, 1976.
15. Lindhard J, M Scharff, and HE Schiott: K. Dan Vidensk. Selsk. mat. Fys. Medd. 33 (1963) 14.
16. Fain J, M Monnin, and M Montret: Rad. Res. 57 (1974) 379.
17. Thevenard P, G Guiraud, CHS Dupuy, and B Delaunay: Radiat. Eff. 32 (1977) 83.
18. Treilleux M, G Fuchs, A Perez, E Balanzat, and J Dural: Nucl. Instr. Meth. Phys. Res. B32 (1988) 397.
19. Houpert C, M Hervieu, D Groult, F Studer, and M Toulemonde: Nucl., Instr. Meth. Phys. Res. B32 (1988) 393.
20. Evans BD, J Comas, and PR Malmberg: Phys. Rev. B6 (1972) 2453.
21. Hughes AE and D Pooley: J. Phys. C4 (1971) 1963.
22. Treilleux M: Thesis, Lyon, 1982.

23. Chen Y, RT Williams, and WA Sibley: Phys. Rev. 182 (1969) 960.
24. Thevenard P: J. Physique C7 37 (1976) 526.
25. Treilleux M, P Thevenard, G Chassagne, and LW Hobbs: Phys. Stat. Sol. (a) 48 (1978) 425.
26. Treilleux M and G Chassagne: J. Phys. Lett. 40 (1979) 161, 283.
27. Treilleux M, P Thevenard, G Chassagne, and M Quermazi: Surf. Sci. 106 (1981) 165.
28. Doyle WT: Phys. Rev. 111 (1958) 1067.
29. Thevenard P, M Treilleux, MO Ruault, J Claumont, and H Bernas: Nucl. Instr. Meth. Phys. Res. B1 (1984) 235.
30. Treilleux M, P Thevenard, MO Ruault, J Claumont, and H Bernas: Nucl. Instr. Meth. Phys. Res. B12 (1985) 375.
31. Treilleux M and P Thevenard: Nucl. Instr. Meth. Phys. Res. B7/8 (1985) 601.
32. Treilleux M and P Thevenard: J de Phys. Lett. 46 (1985) L157.
33. Treilleux M, JP Dupin, G Fuchs, and P Thevenard: Nucl. Instr. Meth. Phys. Res. B19/20 (1987) 713.
34. Perez A, M Treilleux, P Thevenard, G Abouchacra, G Marest, L Fritsch, and J Serughetti: Metastable Materials Formation by Ion Implantation. ST Picraux and NJ Choyke (eds). New York: Elsevier Science Publishing Company, 1982, p. 159.
35. Perez A, G Marest, BD Sawicka, JA Sawicki, and T Tyliszczak: Phys. Rev. B28 (1983) 1227.
36. Parker TB and R Kelly: J. Phys. Chem. Solids 36 (1975) 377.
37. Siskind B, DM Gruen, and R Varma: J. Vac. Sci. Technol. 14 (1977) 537.
38. Guermazi M, P Thevenard, P Faisant, MG Blanchin, and CHS Dupuy: Radiat. Eff. 37 (1978) 99.
39. Thevenard P and M Guermazi: J. de Phys. C3 41 (1981) 113.
40. Guermazi M, P Thevenard, JP Dupin, and CHS Dupuy: Nucl. Instr. Meth. 182 (1981) 397.
41. Guermazi M, P Thevenard, and MG Blanchin: Radiat. Eff. 91 (1985) 125.
42. Guermazi M, P Thevenard, R Brenier, JP Thomas, and JM Mackowski: Nucl. Instr. Meth. 19 (1987) 912.
43. Guermazi M, P Thevenard, JP Dupin, and CHS Dupuy: Radiat. Eff. 49 (1980) 61.
44. Guermazi M, G Marest, A Perez, BD SAwicka, JA Sawicki, P Thevenard, and T Tyliszczak: Mater. Res. Bull. 18 (1983) 529.

ION BEAM - INDUCED CRYSTALLINE TO AMORPHOUS TRANSFORMATIONS
IN CERAMIC MATERIALS

J. L. WHITTON

Department of Physics, Queen's University, Kingston, Ontario, Canada K7L 3N6

1. INTRODUCTION

One of the first reports of a radiation-induced phase change from the crystalline to amorphous state was observed in naturally occurring minerals containing α-emitting radioactivity (1). The term "metamictisation" from the Greek meta (mixed or between) and miktos (compounded) was coined to describe the amorphization of the crystal structure of a mineral due to radiation from contained or nearby radioactive atoms. This was observed soon after von Laue's discovery of x-rays when mineralogists noted the loss of crystalline structure in minerals exposed to radioactivity.

Many neutron-irradiation-induced phase changes from one crystalline form to another have been observed. An early survey (2) reviewed these changes in metals and in the ceramics SiO_2, ZrO_2, and $BaTiO_2$.

Evidence of radiation damage or disordering following ion bombardment was reported some three decades ago. Trillat (3) observed a change from single crystal gold to polycrystalline after bombardment with oxygen ions, Gianolo (4) showed that 30 keV helium bombardment of the semiconductor silicon caused a change from single crystal to a quasi-amorphous state and others (5,6) observed similar changes in germanium by bombardment with various ions.

Similar damage effects were observed in the compound semiconductors GaAs after Kr^+ bombardment (7) and GaSb after 12 MeV deuteron bombardment (8). Disordered regions were also observed along fission tracks in natural mica (9).

It is clear, therefore, even from these early reports, that almost any kind of irradiation into most types of material results in disorder of one sort or another in the crystal lattice.

The impetus behind these early investigations was, of course, spurred by the development of fission reactors and the concomitant need to know about the effects of radiation on reactor structural materials. At the same time it was recognized that ceramic materials deserved special attention both from the point of view of fission reactor fuel (e.g., uranium dioxide and uranium carbide) and as radioactive waste containers. Now, some three to four decades later, the search for an understanding of radiation effects goes on, albeit for different reasons.

Ceramics have been used by the human race since time immemorial. Naturally occurring stone, because of its hardness, strength and resistance to heat, provided man with tools, containers, and dwellings. The creation of ceramics of pottery, bricks, etc., was a natural move away from forming shapes by chipping at stone since the new material of clay was plentiful and firing was by then possible. This early venture into manufacture of pottery shapes has developed into one of the major industries of our time and has

C.J. McHargue et al. (eds.), Structure-Property Relationships in Surface-Modified Ceramics, 167–179.
© *1989 by Kluwer Academic Publishers.*

expanded from the production of structural materials to becoming essential elements of our modern high-tech world. A few of the modern-day applications of ceramics are: refractory linings of steel-making furnaces, piezo-electric devices, light guides, coatings for highly efficient automobile engines, high temperature resistant tiles which allow re-entry to space vehicles into the atmosphere surrounding our planet Earth, radiation hard capacitors and, last but not least, the new high T_C superconductors. The recent advances in ceramic technology make possible the tailoring of compounds designed to cater to chemical, thermal, electrical and mechanical requirements which are impossible to satisfy with other materials.

What gives ceramics these special qualities? First, one could ask: What is a ceramic? The word ceramic stems from the Greek "keramos" meaning loosely, burnt stuff. This definition is pertinent when one thinks of bricks or pottery but in our modern age one must consider more carefully the physical and chemical properties.

Oxygen and metal (or semimetal) atoms form the basis of most ceramics. In the simplest cases equal numbers of oxygen and metal atoms are packed together in a manner dictated by the relative sizes of the ionized atoms. For BeO, the coordination number is four, each metal atom surrounded by four oxygen atoms and each oxygen by four beryllium atoms. In MgO each atom has six nearest neighbors. The comparatively large oxygen atoms act as a matrix, with the small metal atoms fitting comfortably in the spaces between the oxygen atoms. The prime characteristic of ceramics, however, is that these atoms are connected by bonds primarily ionic but also covalent. In this combination of metal atoms and oxygen atoms, the ionic bonds are especially strong because each oxygen, having two electron vacancies in the outer shell, takes two electrons from its metal neighbor. Both kinds of atoms thus become strongly ionized, one negatively, one positively and are bound by a correspondingly strong electrostatic attraction. In covalent bonding, electrons are shared between neighboring atoms and the bonding is highly directional, thus resisting the sliding of planes of atoms past one another.

We now consider the effect of these bondings on some particularly virtuous properties of ceramics, vis., (1) high temperature stability and chemical inertness, and (b) hardness.

High temperature stability and chemical inertness are due to several structural reasons. The dense packing of ceramics with the small cations fitting between the anions results in a very high concentration of bonds, so much thermal energy is required to separate atoms from each other. This results in a high melting point. The close spacing of the atoms provides a solid defense against chemical attack and, where there is an abundance of oxygen, e.g., in Al_2O_3, Ta_2O_3, since these already highly oxidized, the possibility of further oxidation, corrosion, etc., is remote.

The hardness of ceramics can be explained by several factors and may best be described by a comparison with metals. Metals are malleable because of non-directional interactions within the matrix. Positive metallic ions are held together by the sea of conduction electrons and the packing is mainly geometrical. Deformation occurs easily by the sliding of one plane of atoms over another, initiated by dislocation movement. The interatomic bonds in ceramics are partially covalent and these lead to directionality of movement. The slipping of one plane over another is extremely unfavorable so ceramics cannot be readily deformed. This accounts for one of the undesirable properties of ceramics: brittleness. Two examples of directional dislocations in ceramics are shown in reference (10): (a) a low-angle boundary in Al_2O_3, and (b) a two-dimensional twist boundary network in nickel oxide. Since the interatomic forces are directional, there is a tendency for dislocations to be in groups where adjacent individual defects partially cancel out each other's distorting influence.

This short introduction, based heavily on references (10-12) to which a newcomer to the field of ceramics is recommended, serves to show the complexity of ceramics. The directional effects, described in the previous paragraph, suggests that a crystalline to amorphous transition on ion bombardment of ceramics would require far fewer bombarding ions than would metals and such is the case. This then leads us naturally to a discussion of crystalline to amorphous transitions in ceramics — the subject of this paper.

2. THE CRYSTALLINE-AMORPHOUS TRANSFORMATION

As mentioned earlier, the impetus for a study of radiation damage came from the nuclear industry so the disorder processes and resultant microstructures in metals and alloys are reasonably well understood. The same is true for the elemental semiconductors silicon and germanium but not so well for the compound III-V and II-VI semiconductors. Ceramics, partly because of their mixed ionic and covalent bondings, pose a much more difficult-to-solve problem in an understanding of the ion bombardment induced crystalline to amorphous transition.

The interest is studying crystalline to amorphous transitions, or even crystalline to semi-crystalline is most clearly understood by consideration of a single crystal ceramic, say Al_2O_3. In a perfectly ordered lattice, the two types of atom are in their correct substitutional positions and the bonding is as expected from theoretical considerations. The instant, however, that ion bombardment commences, the radiation damage, due initially to displacement spikes where, for each incoming ion, hundreds to thousands of lattice atoms are moved from their original positions results in, after movement, the atoms locating either in new substitutional positions or in interstitial positions. Since bonds are broken, chemical, optical, and electrical properties are changed, these changes becoming greater the closer the lattice approaches amorphicity where only short-range order of a few atomic spacings is possible. This, along with the interest in basic studies of radiation damage is more than sufficient incentive to make measurement of these crystalline to amorphous changes.

The early studies, which will be described in detail here, used, for that period, relatively simple methods of assessing departures from crystal perfection as a result of ion bombardment. However, first note that the essential elements of a study of ion beam-induced crystalline to amorphous transitions in ceramics are: (a) an ion bombardment machine, and (b) equipment and expertise to measure the effects caused by the ion bombardment.

The first requirement poses no problem since ion implantation equipment is no longer rare, particularly since the traditional method of doping semiconductors by diffusion was supplanted by ion implantation. An exceptionally good account of developments in this field is given by Freeman (13) in "Canal Rays to Ion Implantation 1886-1986" and, for practical applications, by Sioshansi (14).

The second requirement can be accomplished in a number of ways, some of which are listed in Table 1.

These methods are listed in the chronological order of investigation and will be considered in turn. They can also be sub-divided into nondestructive and destructive techniques as will be shown.

Changes in refractive index and lattice parameter are, of course, nondestructive and quite simple to measure. By contrast, the measurement of changes in range profile (ion penetration) by use of radioactive tracers and

TABLE 1. Experimental methods of detecting radiation damage

Change in refractive index

Change in lattice parameter

Change in range profile (ion penetration) by use of radioactive
  tracers and sectioning techniques

Change in rate and temperature of gas release

Change in reflection electron diffraction pattern

Change in hardness

Changes observed by transmission electron microscopy

Changes observed by Rutherford backscattering of energetic light ions

---

sectioning techniques is destructive. Again, in no small part due to the interest of the nuclear industry in ion penetration, energy loss, etc., the use of radiotracers was quite commonplace, a great advantage being the wide choice of radiotracer half-life, enabling in some cases measurement of radiotracer implants as low as 10^{11} ions·cm^{-2}, i.e. 10^{-4} of an atomic layer.

The technique is simple: one implants a radiotracer then removes thin slices (in some cases 20 Å or 2 nm, equivalent to 5 atom layers) and sequentially measures the remaining radioactivity. When these measurements of remaining activity are plotted against thickness of layers removed, one obtains an integral distribution of activity, or ion distribution, with depth. Changes in range, or depth profile, reflect change in crystal structure (lattice disorder) and can be shown to be due to ion bombardment, or any other kind of damage.

The sectioning techniques available for use are shown in Table 2.

These sectioning techniques and the measurement of (a) change in rate and temperature of gas release, of change in diffraction pattern and (b) change in range profile as a function of ion bombardment dose were first described in references 21 and 22. The materials used in the former are listed in Table 3 along with the experimental results.

These materials have been subdivided into A, cubic and B, anisotropic substances. After suitable surface treatments of either cleaving, fracturing (with or without further polishing), cleaving and etching and chemical saw-cutting and etching, the specimens were bombarded.

Bombardment by xenon or krypton isotopes (^{84}Kr, ^{125}Xe, ^{131}Xe, ^{133}Xe) was used to achieve the required fluxes of 8×10^{10} to 2×10^{16} ions·cm^{-2}. After implantation, the specimens were heated for 5 min at each of several increasing temperatures and the gas release was monitored by measuring the remaining radioactivity after each anneal. The specimens were also examined before and after ion bombardment and at different temperature intervals by reflection electron diffraction. The results are shown in Figs. 1 and 2, respectively, electron diffraction patterns from MgO and gas release from the same MgO specimen.

TABLE 2. Sectioning techniques

Technique	Applicable to:	Layer thickness (minimum) (Å)
Anodic oxidation/dissolution	W,Si,Al,Au (15) Mo,V,Ta,Ta$_2$O$_5$	20-30
Chemical dissolution	Cu,Ag,Au (16)	
Sputtering	Any material (17,18,19) but with many possible side effects	20-30
Vibratory polishing	Any material (20) best for hard materials	20-100

TABLE 3. Summary of ion bombardment, diffraction, and gas release data

Material	Ion dose† (ions/cm²)	Change in diffraction pattern	Gas release (% at $0.4T_m$)	
			At low dose (8×10^{10} ions/cm²)	At dose listed in column 2
A. Cubic substances				
NaCl, KCl, KBr, KI	Up to 2×10^{16}	No change	$\leqslant 5$	~ 0
CaO, NiO	Up to 2×10^{16}	No change	$\leqslant 10$	$\leqslant 5$
UC	Up to 2×10^{16}	No change	Not measured	
CaF$_2$, BaF$_2$, UO$_2$, ThO$_2$	Up to 2×10^{16}	No change	$\leqslant 10$	$\leqslant 5$
MgO	$8 \times 10^{13(a)}$	Kikuchi lines weaker	$\leqslant 5$	6
	4×10^{14}	Kikuchi lines disappeared		32
	2×10^{16}	Faint single crystal +indication of polycrystal		60
B. Anisotropic substances				
TiO$_2$	Up to 8×10^{13}	No change	$\leqslant 5$	$\leqslant 5$
	2×10^{16}	Pattern disappeared		20-40
Al$_2$O$_3$	8×10^{10}	No change	0	0
	4×10^{13}	Kikuchi lines disappeared		~ 5
	2×10^{16}	Pattern disappeared		60-90
U$_3$O$_8$*	Above 8×10^{10}	Pattern disappeared	~ 5	60

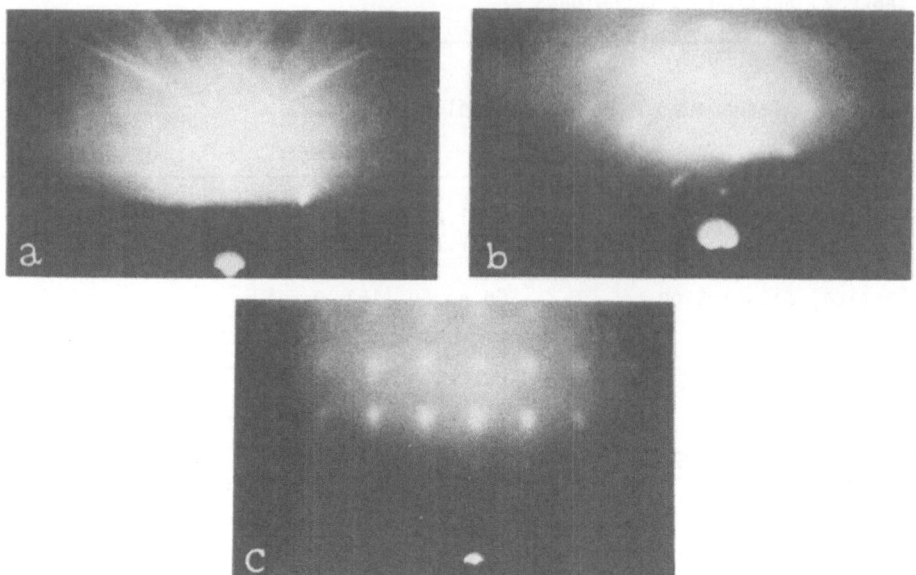

FIGURE 1. Reflection electron diffraction patterns of MgO; (a) unbombarded, (b) bombarded with 2 x 10¹⁸ ions·cm⁻² and (c) as (b) but heated 5 min to 500°C.

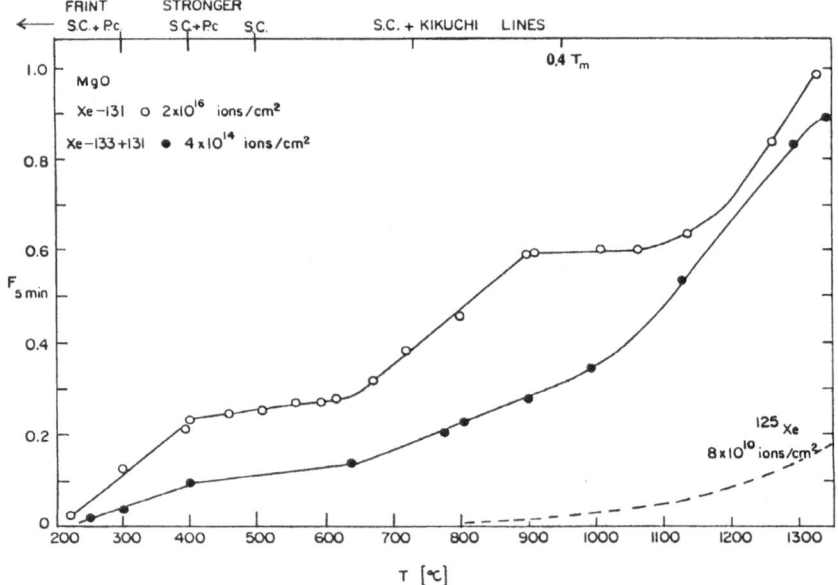

FIGURE 2. Fractional release of xenon from MgO. The broken line shows release following bombardment to a low dose (8 x 10¹⁰ ions·cm⁻²). The annealing of damage in MgO following bombardment to the high dose (2 x 10¹⁸ ions·cm⁻²) is indicated at the upper abscissa (S.C. = single crystal, P.c. = polycrystalline).

These patterns are for (a) unbombarded, showing clear diffraction spots and Kikuchi lines, (b) after bombardment to 2×10^{16} Xe^{131} ions·cm^{-2} showing still crystal structure, and (c) annealed for 5 min at 500°C when the single crystal pattern is nearly restored but without the perfection of the initial Kikuchi lines. These patterns should be compared to the gas release curves of Fig. 2 where the various stages of gas release may be equated with stages of lattice re-ordering. The pattern after annealing at 500°C is reasonably related to the second slow stage of gas release but it should be observed that neither in the diffraction pattern nor in the gas release curves is there a sign of amorphous to crystalline abrupt transition. It is also worth noting that MgO is unique in the cubic substances of showing any bombardment damage and significant gas release (c.f. Table 3).

It should be noted that fractional release at any temperature increases with increasing bombardment dose. This is a good indication that annealing of disorder is accompanied by a sweeping out of gas as the lattice re-orders.

As may be seen from Table 3 the cubic substances, with the exception of MgO, released almost no gas even at the high dose of 2×10^{16} ions·cm^{-2} and showed no change in diffraction pattern. We may now compare the MgO results of Figs. 1 and 2 with those of the anisotropic Al_2O_3, shown in Figs. 3 and 4.

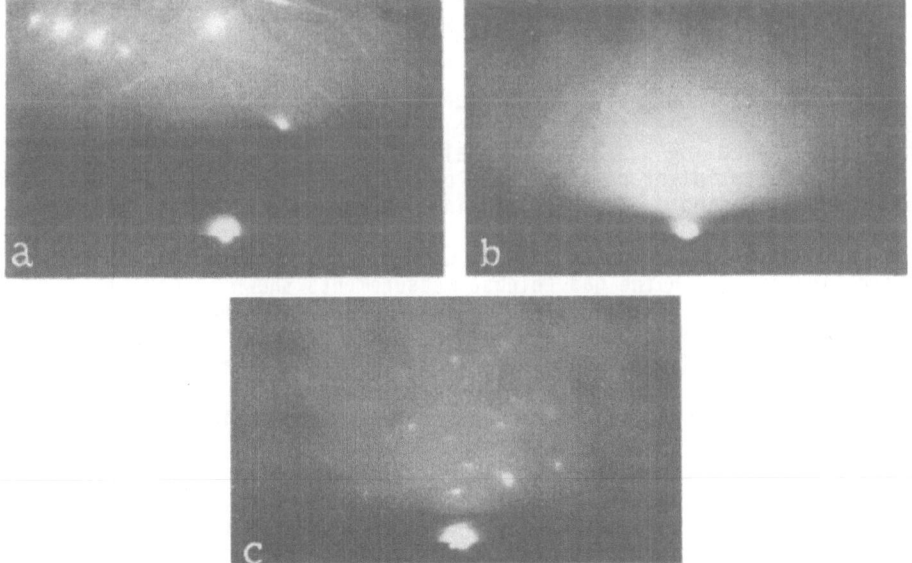

FIGURE 3. Reflection electron diffraction patterns of Al_2O_3; (a) unbombarded, (b) bombarded with 2×10^{16} ions·cm^{-2}, and (c) as (b) but annealed for 5 min at 730°C.

174

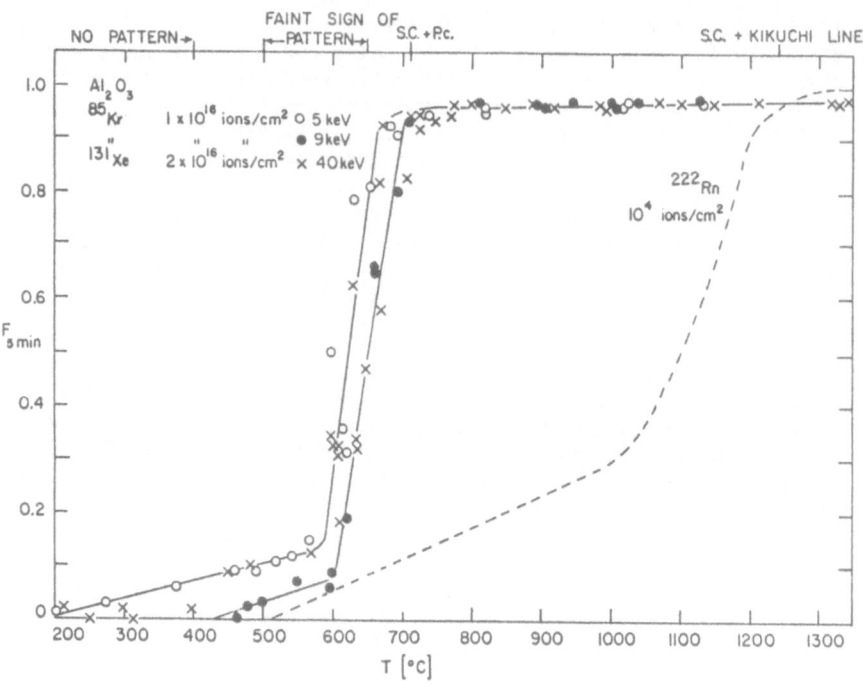

FIGURE 4. Fractional release of xenon, krypton and radon from Al_2O_3. The annealing of damage to the high dose of xenon is indicated at the upper abscissa (S.C. = single crystal, P.c. = polycrystalline.

These patterns show that the unbombarded material is good single crystal (spots and Kikuchi lines) which, on bombardment becomes completely amorphous. Annealing to 730°C, when all but a few percent of the gas is released results in the re-appearance of single crystal spots but still with polycrystalline rings. The Kikuchi lines appear only at about 1200°C and higher.

We may now follow the route we took with MgO and compare this electron diffraction data with the gas release measurements shown in Fig. 4.

It is quite clear that we see a one-to-one correspondence between the gas release measurements and the change in electron diffraction patterns. The latter are described on the upper part of Fig. 4 and it may be seen that slow release of gas from 200 to 600°C takes place within the amorphous (no pattern) phase. The sudden release at 600°C is accompanied by an indication of non-amorphous diffraction pattern and at 700°C almost all of the gas is released and the diffraction pattern shows polycrystalline and single crystal characteristics — Fig. 3(c). It is not, however, until a temperature of 1100-1200°C that a single crystal pattern with Kikuchi lines appears. It is at that temperature, interestingly enough, that complete release of a very low dose implant of ^{222}Rn is achieved. This one-to-one correspondence provides strong evidence that release of gas on annealing is accompanied by a re-ordering of the damaged lattice, or vice versa. However, since the ^{222}Rn is of such low concentration, the re-ordering of the lattice must be the deciding factor in the movement of gas to the surface with subsequent release.

This behavior may now be correlated with the change in depth penetration of radioactive energetic ions as a function of implantation dose. In MgO, shown in Fig. 5, a relatively light dose of 4×10^{13}, 40 keV Xe ions·cm⁻², allows the ions to penetrate deeply along the <100> direction. A similar dose on the same crystal tilted 25° from the <100>, to present a more random direction to the ion beam, results in a much reduced range but still with a penetrating tail. A high dose implant of 2×10^{16}, 40 keV Xe ions·cm⁻², which according to the reflection electron diffraction picture in Fig. 1, results in a transition from single crystal to near-amorphous/polycrystalline pattern, shows a reduced range similar to that in the 25° tilted crystal but with a much reduced tail. This is as expected from a near-amorphous material; a truly amorphous material would result in a Gaussian distribution of implanted ions as shown in Fig. 6 for xenon implanted into fused (amorphous) silica.

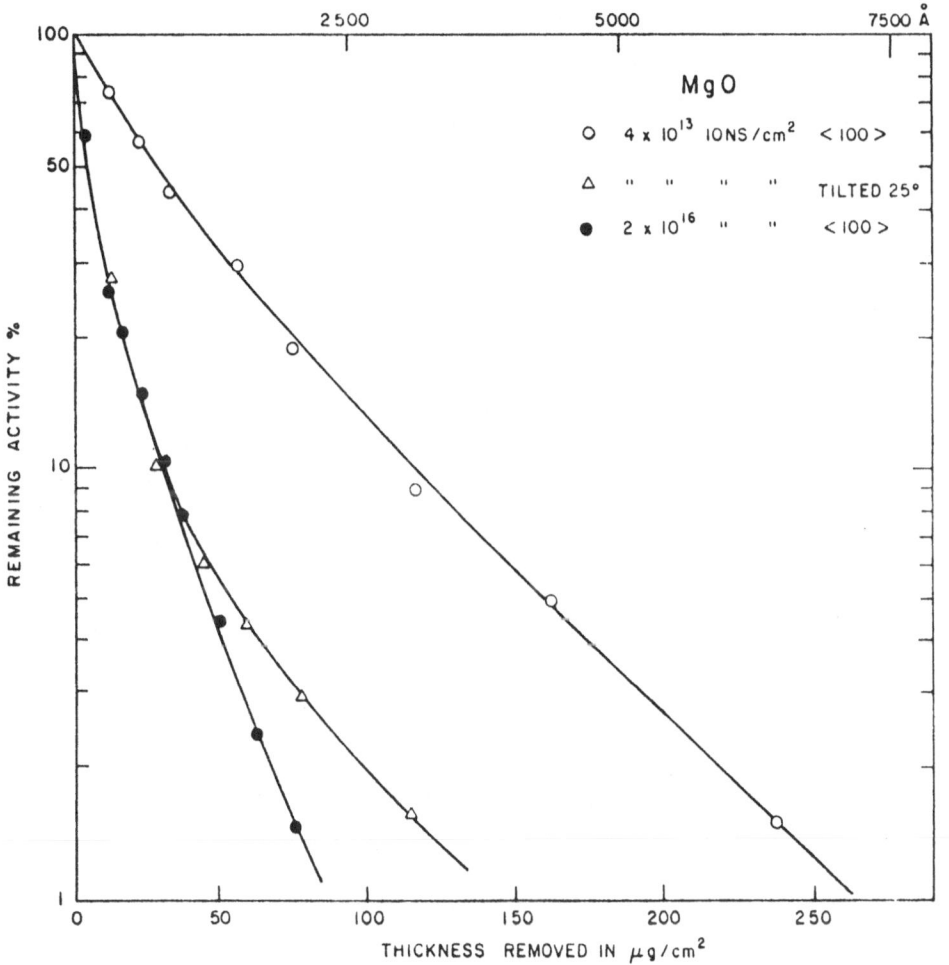

FIGURE 5. Effect of bombardment dose and crystal orientation on the range of 40 keV Xe ions in MgO.

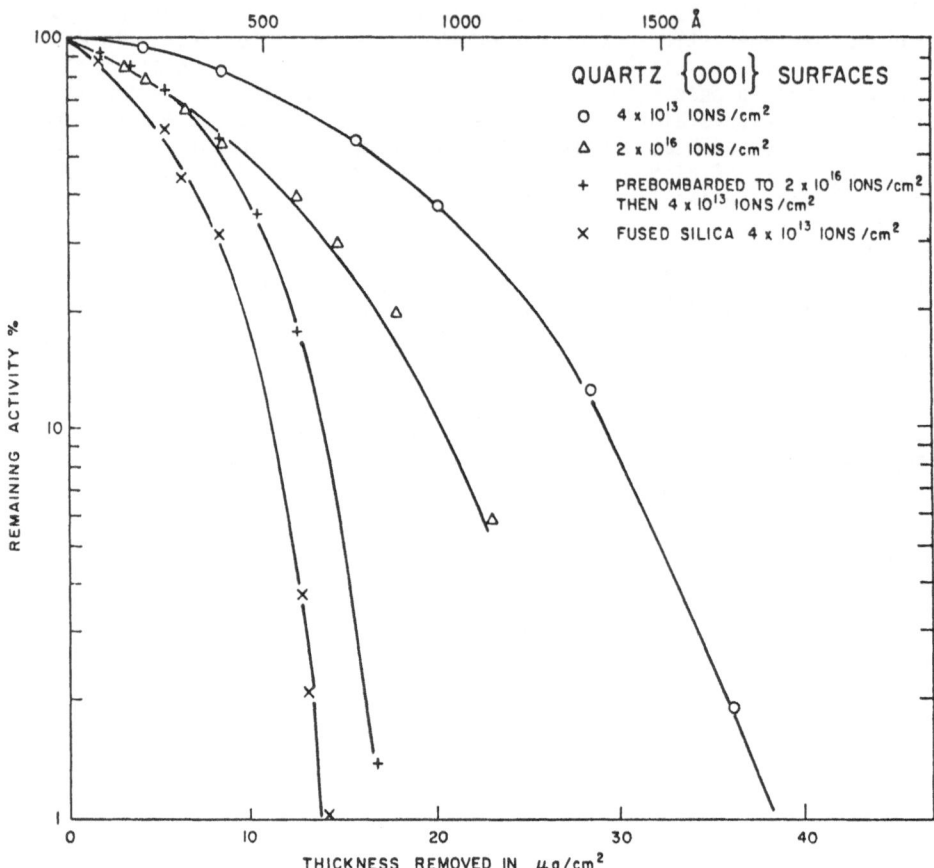

FIGURE 6. Effect of bombardment dose on the range of 40 keV Xe ions in quartz and in fused (amorphous) silica.

As stated earlier, MgO is the one cubic substance studied which shows radiation damage but which never exhibited a complete crystalline to amorphous transition. Of the anisotropic substances studied, Al_2O_3 transformed to the amorphous state and this is analogous with SiO_2 (unfortunately the range data in Al_2O_3 is no longer available) and a clear difference between those and of the range profiles in MgO is immediately obvious. The ranges in SiO_2 - along the <0001> axis - are successively shortened as the dose is increased to the point that a 2×10^{16} ions·cm^{-2} dose followed by 4×10^{13} ions·cm^{-2} has a range profile shape similar to a 4×10^{13} ions·cm^{-2} implant into fused (amorphous) SiO_2. The median ranges differ only by about 25%. Note that there is no penetrating tail in these two profiles. So, once again the one-to-one relationship between electron diffraction, gas release measurements and range profiles is established.

We may now turn to the final three experimental methods of detecting radiation damage; namely, changes in hardness, changes observed by transmission electron microscopy, and changes observed by Rutherford backscattering of energetic light ions.

Changes in hardness can be extremely difficult to measure due, in no small part, to the penetration depth of most hardness indenters being much greater than the depth of the implanted layer. This problem has been largely circumvented by the recent development of an ultra-sensitive indenter at the Oak Ridge National Laboratory (23). This new indenter can perhaps be used to investigate the apparently anomalous result of roughly an order of magnitude higher dose being required to make Al_2O_3 amorphous if the irradiation is directed near the $\langle 1\bar{2}10 \rangle$ axis rather than along the $\langle 0001 \rangle$, the latter being a much more open direction (24). This work by McHargue raises many intriguing possibilities for the mechanism of this anomalous behavior which could well lead to significant practical applications.

The early transmission electron microscopy experiments offered no in-depth structural information about the amorphous zones, showing, in fact only the linear dimensions of the disordered regions (5,6). The later studies (25) using cross-sectioned specimens have shown a fine correlation between the micrographs and channeling results on Al_2O_3 specimens implanted with 175 keV, 4×10^{16} Zr ions·cm^{-2}. The structure from the surface inwards is seen as (a) the surface region damaged but crystalline, followed by (b) an amorphous region followed by (c) a damaged but crystalline region and finally (d) the undamaged crystalline substrate.

In an attempt to relate the crystalline to amorphous transitions obtained by gas-release and diffraction measurements to Rutherford backscattering (RBS) channeling results, a comparison was made with the early work of Naguib et al. (26). This gave an indication of agreement although not completely conclusive since the maximum dose used was 1×10^{16} ions·cm^{-2} of the larger ion krypton which resulted in saturation disorder at that dose but not, according to the RBS signal, complete amorphicity. Since our work used 40 keV Xe ions at 2×10^{-6}·cm^{-2} it was felt that the mass and dose difference could explain the lack of amorphicity in the RBS/channeling study.

The more recent RBS/channeling studies (25) with 45 keV - 4×10^{16} Xe·cm^{-2} implanted into Al_2O_3 should allow for a better comparison than that of reference (26). Using the same ion mass but double the dose would lead one to expect complete amorphization of the implanted layer. This was not the case. The disorder peaks of both the Al and O sub-lattices fail to reach the random level - the measure of amorphicity. Figure 7 shows these spectra from reference (25). Perhaps only at the immediate surface, where the aligned and random spectra coincide, may one assume a thin layer of amorphous material.

This is, in fact, compatible with the glancing angle electron diffraction results which, because of the only few degrees of glancing angle, arise from the first 20 to 30 Å of the surface. The RBS/channeling, on the other hand, samples a much greater depth and with a depth resolution of some 200 to 250 Å. The two types of result should, therefore, not be directly compared but should be used in a complementary fashion. The sudden burst of gas release seen from Al_2O_3 at 600°C (Fig. 4) can equally be explained by the rapid re-crystallization of a very thin amorphous layer - a thicker layer would presumably take much longer to grow epitaxially from the underlying crystalline Al_2O_3 and would have the effect of slowing down this rapid release of gas.

3. CONCLUSION

An account has been given of many various ways of detecting crystalline to amorphous transitions in ceramic materials. An attempt has been made to relate the results from the different techniques to each other. The apparent anomaly between the earlier results from gas release measurements, reflection electron diffraction patterns and range profiles with later

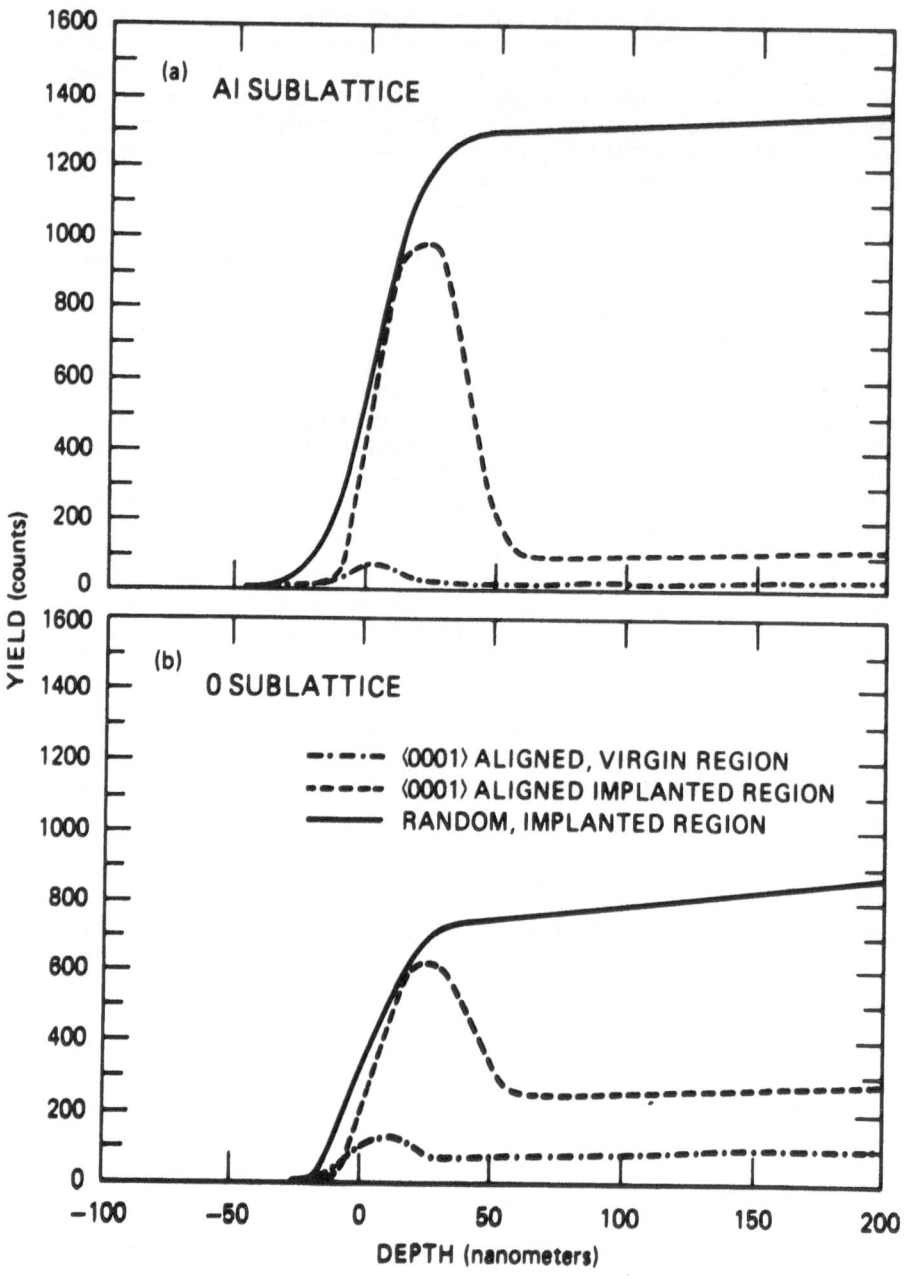

^{131}Xe (45 keV, 4 X 10^{16}/cm^2, RT) IN Al$_2$O$_3$

FIGURE 7. Backscattering spectra from 2 MeV He$^+$ incident on ^{131}Xe implanted Al$_2$O$_3$ showing the signals from (a) the aluminum sub-lattice and (b) the oxygen sub-lattice.

Rutherford backscattering/channeling results has been cleared up as due to different measurement resolutions (sampling depths) in these techniques.

4. ACKNOWLEDGEMENTS

I wish to express my thanks to Carl McHargue for inviting me to take part in this NATO-ASI which he organized in a superb fashion such that all participants benefitted scientifically, socially, and even aesthetically. Much collaborative work was seeded at Il Ciocco and will undoubtedly flourish. The financial assistance from NATO is much appreciated.

REFERENCES
1. Fassler A: Z. Krist. 104 (1942) 81.
2. Hauser O and M Schenk: Kernenergie 6 (1963) 655; Phys. Stat. Sol. 6 (1964) 83.
3. Trillat JJ: J. Chem. Phys. 53 (1956) 510.
4. Gianolo U: J. Appl. Phys. 28 (1957) 868.
5. Izvi K and H Suzuki: J. Phys. Soc. Japan 18(Suppl. III) (1963) 210.
6. Parson JR: Phil. Mag. 12 (1965) 1159.
7. Pohlau C, H Lutz, and R Sizman: Z. Angew Physik 17 (1964) 404.
8. Gonser U and B Okkerse: J. Phys. Chem. Solids. 7 (1958) 55.
9. Price PB and RM Walker: J. Appl. Phys. 33 (1962) 2625, 3400.
10. Cotterill, Rodney: The Cambridge Guide to the Material World. Cambridge: Cambridge University Press, 1985, 119.
11. Gilman JJ: Scient. American 55 (Sept. 1967).
12. Bowen H Kent: Scient. American 55 (Oct. 1986).
13. Freeman JH: Radiat. Eff. 100(3-4) (1986) 161.
14. Sioshansi Piran: Mater. Engr. (Feb. 1987).
15. Whitton JL: Channeling. DV Morgan (ed). New York: John Wiley & Sons, 1973) 225 and references within.
16. Andersen T and G Sorensen: Radiat. Eff. 2 (1969) 111.
17. Lutz H and R Sizman: Phys. Lett. 5 (1963) 113.
18. Pöhlau C, H Lutz, and R Sizman: Z. Angew. Physik 17 (1964) 404.
19. Behrisch R (ed): Sputtering by Particle Bombardment I. Topics Appl. Phys. 47. Berlin/New York: Springer-Verlag, 1981.
20. Whitton JL: J. Appl. Phys. 36 (1965) 3917.
21. Matzke Hj and JL Whitton: Can. J. Phys. 44 (1966) 995.
22. Whitton JL and Hj Matzke: Can. J. Phys. 44 (1966) 2905.
23. Joslin DL: this volume.
24. McHargue, CJ: this volume.
25. McHargue CJ, GC Farlow, CW White, BR Appleton, P Angelini and H Naramoto: Nucl. Instr. Methods Phys. Res. B10/11 (1985) 569.
26. Naguib HM, JR Singleton, WA Grant, and G Carter: J. Mater. Sci. 8 (1973) 1633.

ION BEAM MIXING AT METAL/Al$_2$O$_3$ INTERFACES

L. Romana, P. Thevenard, G. Massouras, G. Fuchs, R. Brenier, and B. Canut

Departement de Physique des Materiaux, Universite Claude Bernard-Lyon I
69622 Villeurbanne, France

1. INTRODUCTION
   Metal films on insulator substrates such as Al$_2$O$_3$ or SiO$_2$ have applications in industrial fields such as electronics, connectics, biotechnology,
etc. One of the most important problems arises from the poor adhesion of
the metallic layer to the insulator. This poor adhesion is generally attributed to a lack of chemical affinity, Van der Waals or electrostatic forces
between the two materials (1,2). Metals having a large negative value of
free energy for oxide formation show a strong adhesion to the oxide
substrate (3,4), whereas metals with low oxygen affinity are weakly
bonded (5-7).
   Ion beam mixing has been proved to be a useful low temperature technique
to improve adhesion between a thin film and an oxide substrate (8-10). This
improvement has been proposed to be due to the formation of an intermediate
layer at the interface (even at a thickness of one or two monolayers). Some
attempts have been made to correlate the high temperature thermal processing
of metal films deposited on SiO$_2$ (11,12) or on Al$_2$O$_3$ (13,14) to ion beam
mixing treatment. These studies show that the reactivities of thin metallic
films with oxide substrate can generally be predicted by thermodynamical
considerations.
   In this work we have investigated the modifications of
metal-Al$_2$O$_3$ interfaces by ion beam mixing. Using the heats of formation of
metal oxides, we chose two metals: silver and niobium, which have a marked
difference in the heats of formation of metal oxides — -7.3 kg·cal·g^{-1}·mol^{-1}
(Ag$_2$O) and -463 kg·cal·g^{-1} (Nb$_2$O$_5$)(-399 kg·cal·g^{-1} for Al$_2$O$_3$) (ref. 15). In
addition, Nb-Al$_2$O$_3$ interfaces appear to be an ideal case for understanding
the basic mechanisms of ion beam mixing of metal-insulator interfaces since
epitaxial growth of niobium on sapphire has been observed (16-18).
   Thermal annealing treatments were performed in order to study the interfacial energy modifications of the metal/Al$_2$O$_3$ structures.
   These modifications were characterized by different techniques such as
Rutherford Backscattering Spectrometry (RBS), Transmission Electron
Microscopy (TEM), x-ray diffraction at glancing incidence and electrical
resistivity measurements. The influence of the alumina structure on the
interface modification by ion beam mixing was studied by using either single
crystal (α-Al$_2$O$_3$) or evaporated alumina film substrates.

2. EXPERIMENTAL PROCEDURE
   Alpha-Al$_2$O$_3$ single crystals of (0001) orientation with optically polished
surfaces were cleaned at room temperature in the following sequential ultra-
sonic baths: trichlorethylene, distilled water, acetone. They were then
annealed at 1273 K in air for 48 h and immediately introduced into the evaporation chamber. The evaporated alumina layers were deposited by electron

181

C.J. McHargue et al. (eds.), Structure-Property Relationships in Surface-Modified Ceramics, 181–187.
© 1989 by Kluwer Academic Publishers.

beam evaporation at room temperature (RT) on microscope glass plates or NaCl substrates at a pressure of 4×10^{-8} Torr and with a deposition rate of 2 Å·s^{-1}. The film thicknesses were about 1000 Å. The metallic films were deposited on alumina in the same chamber, also by means of the electron beam evaporation system. The depositions were performed either in a high vacuum (HV) chamber or in an ultra high vacuum (UHV) system online with the accelerator. Table 1 summarizes the evaporation conditions.

TABLE 1. Evaporation conditions of metallic films

Element	Pressure (Torr)	Deposition Rate (Å·s$^-$)	Thickness (Å)	Deposition Temperature (K)
Ag	7×10^{-7}	1	200	300
Nb	4×10^{-8}	0.2	200	300
Nb	7×10^{-8}	0.2	560	773

The samples were irradiated with Xe$^+$ ions (1.5 MeV) at 300 and 77 K, using a 2-MeV Van de Graaff accelerator. Samples prepared in the UHV chamber were irradiated in situ under 10^{-8} Torr pressure, whereas for those evaporated in the HV system, the irradiation was performed at 10^{-6} Torr.

The irradiation fluences ranged from $3 \cdot 10^{15}$ to 10^{16} ions·cm^{-2} and the flux was kept less than 3×10^{12} ions·cm^{-2}·s^{-1}. The beam was electrostatically scanned across the samples to provide homogeneous irradiation. Rutherford Backscattering Spectroscopy (RBS) analyses were conducted with 2 MeV helium ions. Thermal annealing treatments of the samples were carried out in air at 723 K for 1 h.

Transmission electron microscope observations using a Philips EM 300 TEM were made on metal-evaporated alumina samples placed on microscope grids by dissolving the NaCl substrate in distilled water. X-ray diffraction patterns were obtained with CuKα radiation and with a fixed glancing incidence of 2°. Electrical resistivities were measured with a four-point probe apparatus.

3. RESULTS AND DISCUSSION
3.1. Ag-Al$_2$O$_3$ interfaces
The modification of the Ag/Al$_2$O$_3$ interfaces by ion beam mixing was investigated by following the change in the interface energy, qualitatively obtained by the wetting test described by Sood and Baglin (6).

The samples were irradiated with $3 \cdot 10^{15}$ Xe$^+$·cm^{-2} at two temperatures: 300 and 77 K and thermally annealed at 723 K in air for 1 h.

Figure 1 shows TEM micrographs of silver on evaporated alumina film as evaporated, after irradiation and after annealing treatments.

The as-evaporated silver layer is smooth, continuous and featureless [Fig. 1 (a)]. The silver films beak up into islands (300 nm mean size) after thermal annealing (Fig. 1 (b)] indicating a high Ag/Al$_2$O$_3$ interfacial energy (i.e., poor adhesion).

The samples irradiated at 300 K show a strong modification of the surface topography [Fig. 1 (c)]. Lateral segregation of the silver layer causes the formation of isolated metallic islands with average size of 150 nm, in addition to bubble formation into the evaporated alumina layer previously reported in ref. (19). After annealing, the TEM micrograph [Fig. 1(d)] shows that silver undergoes the same island formation as that shown in Fig. 1 (b). These results indicate that the interfacial energy has not been modified by irradiation at 300 K, or that such a modification is not stable up to 723 K.

The surface topography of the sample irradiated at 77 K is shown in Fig. 1 (e). A percolation of clusters is observed which suggests that a lesser amount of surface diffusion of silver occurred. After the anneal, the island formation is limited and connected silver clusters still exist [Fig. 1 (f)]. In this case, it appears that the adhesion between the silver and alumina substrate has been improved by xenon bombardment.

In agreement with Farlow et al. (14), we believe that the lateral motion of metal atoms at the alumina surface during irradiation is induced by atomic collisions and not by thermal activation since xenon bombardment causes lateral diffusion at both irradiation temperatures (300 and 77 K).

The improvement of the silver-alumina adhesion observed after irradiation at 77 K, could be explained as follows: undoubtedly, ballistic mixing occurs at the interface whatever the irradiation temperature. If chemical driving forces are effective, a demixing effect may take place, particularly for non-reactive systems as such is the case of Ag/Al_2O_3. As chemical driving forces are less effective at low temperature (20), the silver atoms introduced into the alumina layer by ballistic processes, may be retained into the alumina matrices, leading to a metastable interface configuration with a lower interfacial energy. The thermal stability of a similar metastable interface (up to 723 K) has been observed for the Cu/Al_2O_3 system irradiated with Ne ions (21).

Silver deposited on Al_2O_3 single crystals undergoes similar lateral segregation after Xe^+ irradiation but no bubbles were observed in this case.

3.2. Nb/Al_2O_3 interface

Niobium thin films are difficult to prepare because of its high affinity for oxygen which may induce reactions with active gases during the deposition (22). Table 2 reports the principal characteristics of the evaporated niobium films. The metallic oxidation was determined from RBS data.

We note that the HV-deposited films were oxidized and had a high electrical resistivity (one order of magnitude higher than the niobium bulk resistivity). The amorphous structure of the as-evaporated "niobium" layer is indicated out by TEM diffraction [Fig. 2 (a)]. Xenon irradiation of these samples (up to 10^{16} $Xe^+ \cdot cm^{-2}$) did not induce modification of the interface, according to the RBS analysis. The electrical resistivity and the amorphous network of the layer are the same as those before irradiation. The surface of the films before and after irradiation [Fig. 2 (b) and (c), respectively) is smooth without lateral segregation of the film.

A vacuum of $7 \cdot 10^{-8}$ Torr during niobium evaporation seems to be sufficient to prevent oxidation of the film. This is confirmed by the metallic electrical resistivity of the niobium layer. As the niobium evaporation was carried out by holding the substrate temperature at 773 K, we could expect, from previous studies, to obtain epitaxial growth of the niobium film on the sapphire substrate (18). In our case, x-ray diffraction analyses show a polycrystalline niobium film without preferential orientation. This difference might be due to the lower vacuum compared to the one of ref. (18).

Figure 3 shows the RBS spectra of Nb/Al_2O_3 before before and after irradiation with xenon fluence of 10^{16} ions $\cdot cm^{-2}$ at 1.5 MeV energy. These RBS spectra were obtained by tilting the sample with respect to the direction of the incident beam to an angle of 50° in order to enhance the depth resolution. After irradiation, there is a decrease in the niobium scattering

184

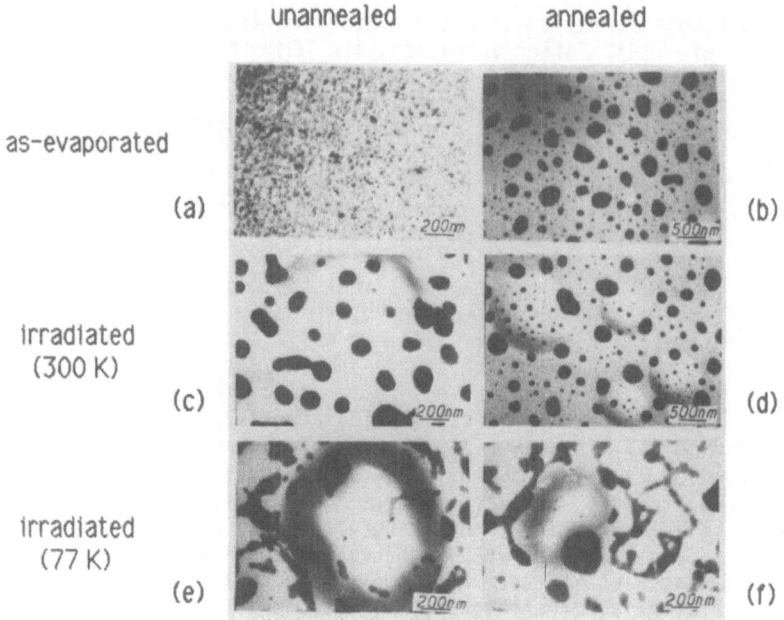

unannealed   annealed

as-evaporated

(a)   (b)

irradiated
(300 K)

(c)   (d)

irradiated
(77 K)

(e)   (f)

FIGURE 1. TEM micrographs of silver-alumina samples before and after heat treatment (first and second column, respectively). Irradiation temperatures are indicated at the left side of each row.

TABLE 2. Properties of evaporated niobium films

Niobium Pressure Deposition (Torr)	Electrical Resistivity (10^{-4} $\Omega \cdot$cm)	Microstructure	Oxidation
10^{-6}	4	amorphous	Yes
$7 \cdot 10^{-8}$	0.4	polycrystalline	No

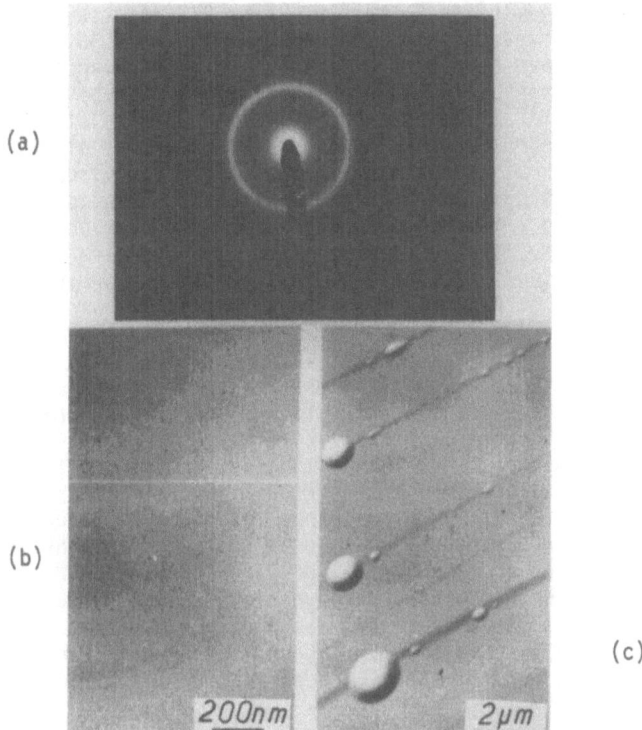

(a)

(b)

(c)

200nm

2μm

FIGURE 2. (a) Electron diffraction pattern of the as-deposited niobium-alumina sample, (b) and (c) TEM micrographs of these samples before and after irradiation, respectively.

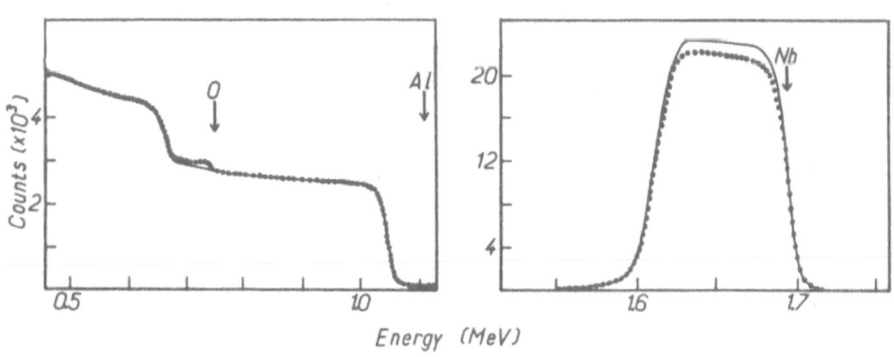

FIGURE 3. RBS spectra of niobium-alumina samples before irradiation (full line) and after irradiation with 10^{16} Xe$^+$ cm^{-2} at 300 K (dotted line). The arrows indicate the surface energy position of each element.

yield, whereas oxygen is distributed throughout the metallic layer, suggesting a niobium oxide formation. From RBS results the mixed layer was approximately an average composition NbO_x with x = 0.33. Since RBS analysis and irradiation have been performed in situ in the evaporation chamber under 10^{-8} Torr, the oxygen into the niobium layer is probably due to some reaction between the metallic layer and the anions of the alumina substrate. Ballistic mixing should cause oxygen penetration into niobium up to only 30 Å range (23) and, thus, cannot explain the oxygen diffusion throughout the niobium layer (560 Å). Some radiation enhanced diffusion of the oxygen might be suspected (24) due to the high affinity of niobium to oxygen. The electrical resistivity value of the mixed layer confirms that oxidation occurred. The value increases from 44 x 10^{-6} $\Omega \cdot cm$ for the as-evaporated sample to 120 x 10^{-6} $\Omega \cdot cm$ for the irradiated (10^{16} ion$\cdot cm^{-2}$) sample.

4. CONCLUSIONS

Metall-Al_2O_3 interface modifications induced by xenon irradiation have been studied in the present work. Niobium and silver were selected because of their different chemical reactivities with Al_2O_3.

The wetting test data show that a modification of the interfacial energy of Ag/Al_2O_3 samples occurred after irradiation performed at 77 K, whereas for those irradiated at 300 K no change was observed. Such a discrepancy between the two irradiation temperatures could be explained by the less effective influence of the chemical driving forces at low temperature.

Irradiation of Nb/Al_2O_3 induces oxygen diffusion from alumina substrate through the entire metallic layer. Such oxygen mobility could be due to the high oxygen affinity of niobium. Oxygen migration induced by irradiation could be enhanced by diffusional forces leaving the mixed layer in a metastable phase.

No differences were observed between substrates of single crystals or evaporated alumina films except for the formation of blisters in the latter material.

REFERENCES

1. Baglin JEE: Ion Beam Modification of Insulators. P Mazzoldi and G Arnold (eds). Amsterdam: Elsevier, 1987.
2. Kim YH, YS Chaug, NJ Chou, and J Kim: J. Vac. Sci. Technol. A5(5) (1987) 2890-2893.
3. Mattox DM: Thin Solid Film 18 (1973) 173.
4. Chapman BN: J. Vac. Sci. Technol. 11 (1974) 106.
5. Schmidt PH, JM Rowell, and WL Feldmann: Appl. Phys. Lett. 39(2) (1981) 177-179.
6. Sood DK and JEE Baglin: Nucl. Inst. Meth. B19/20 (1987) 954-958.
7. Vossen JL: Adhesion Measurement of Thin Films, Thick Films and Bulk Coating. KL Mittal (ed). Philadelphia, PA: ASTM, 1978.
8. Baglin JEE: Mat. Res. Symp. Proc. 47 (1985) 3-10.
9. Sood DK, PD Bond, and SPS Badwall: Mat. Res. Symp. Proc. 27 (1984) 565-570.
10. Griffith JE, Y Qiu, and TA Tombrello: Nucl. Inst. Meth. 198 (1982) 607.
11. Pretorius R, JM Harris, and M-A Nicolet: Solid State Electronic 27 (1978) 667-675.
12. White CW, G Farlow, J Narayan, GJ Clark, and JEE Baglin: Mat. Lett. 2(5A) (1984) 367-372.
13. Zhao XA, E Kolawa, and M-A Nicolet: J. Vac. Sci. Technol. A4(6) (1986) 3139-3141.

14. Farlow, GC, BR Appleton, LA Boatner, CJ McHargue, and CW White, Mat. Res. Soc. Symp. Proc. 45 (1985) 137-145.
15. Handbook of Chemistry and Physics: G Weast (ed). W. Palm Beach, FL: CRC Press, 1981-82.
16. Florjancic M, W Mader, M. Rühle, and M Truwitt: J. Phys. C4(46) (1985) 129-133.
17. Burger K, W Mader, and M Rühle: Ultramicroscopy 22 (1987) 1-14.
18. Wolf SA, S Qadri, JH Claassen, TL Francavilla, and BJ Dalrymple: J. Vac. Sci. Technol. A4(3) (1986) 524-527.
19. Romana L, P Thevenard, R Brenier, G Fuchs, and G Massouras: Nucl. Instr. Meth. B 32 (1988) 96-99.
20. Lau SS, BX Liu, and M-A Nicolet: Nucl. Instr. Meth. 209/210 (1983) 97-105.
21. Baglin JEE and GJ Clark: Nucl. Instr. Meth. B7/8 (1985) 881-885.
22. Wolf SA, JP Kennedy, and M Nisernoff: J. Vac. Sci. Technol. 12(1) (1976) 145-147.
23. Ziegler JF, JP Kennedy, and U Littmark: The Stopping and Range of Ions in Solids. JF Ziegler (ed). New York: Pergamon, 1985.
24. Rehn LE, RS Averback, and PR Okamoto: Nucl. Sci. Engr. 69 (1985) 1-11.

ION BEAM MIXING OF Fe IN SiO$_2$: STRUCTURAL AND OPTICAL PROPERTIES

G. BATTAGLIN,[1,2] S. LO RUSSO,[2] A. PACCAGNELLA,[2,3]
P. POLATO,[4] AND G. PRINCIPA[5]

[1]Dipartimento di Chimica Fisica, Calle Larga S. Marta 2137,
 30123 Venezia, Italy
[2]Unita GNSM-CISM, Dipartimento di Fisica, Via Marzolo 8,
 35131 Padova, Italy
[3]Dipartimento di Ingegneria, 38050 Mesiano di Povo (Trento), Italy
[4]Stazione Sperimentale del Vetro, Via Briati 10,
 30121 Murano (Venezia), Italy
[5]Dipartimento di Ingegneria Meccanica, SezioneMateriali,
 Vi8a Marzolo 9, 35131 Padova, Italy

1. INTRODUCTION

Ion beam mixing effects in metal/insulator systems have been investigated for a few metal/SiO$_2$ couples irradiated with different ions, with particular attention to mixing mechanisms (1-3). Because of applications of ion mixing to metal/insulator couples, most interest was devoted to modifications of the interfacial region, which may lead to film adhesion improvements useful for microelectronic device fabrication (4,5). In a recent paper (6) we presented some results on modifications of optical properties in ion beam mixed Fe/SiO$_2$ samples, which could be relevant for applications in optoelectronic devices. In this paper we focus our attention both on mixing mechanisms and on the microstructural and optical properties of the interfacial region of the Kr irradiated Fe/SiO$_2$ system before and after thermal annealings for 5 h at 500 and 700°C.

2. EXPERIMENTAL

Iron films 9 to 80 nm thick were e-beam evaporated onto high-purity fused silica sheets. Samples were irradiated at room temperature (RT) with a 200-KeV Kr^{++} beam with a current density of 1 µA cm^{-2} in the fluence range 3.6×10^{14} to 1.3×10^{17} ions cm^{-2}. Under these irradiation conditions the temperature of samples was lower than 100°C. The projected range of 200 keV Kr ions in Fe is 42 nm with a range straggling of 11 nm (ref. 7).

The surface unmixed Fe was chemically etched following a procedure reported in ref. (3). The mixing behavior was characterized by measuring the total amount of mixed Fe by Rutherford backscattering (at an angle of 160°) spectrometry (RBS) of ^{4}He$^+$ particles with energy of 2.2 or 2.3 MeV. Areal densities of Fe and Kr, as well as the energy-to-depth conversion were obtained following standard procedures (8).

All samples were also analyzed by scanning electron microscopy (SEM). Some samples were prepared by evaporating a 32-nm thick ^{57}Fe (95% enriched) film onto the silica substrates. They were irradiated with the 200 keV Kr beam to doses of 2.5×10^{15}, 8.2×10^{15}, 1.8×10^{16}, and 3.5×10^{16} ions cm^{-2}, etched and analyzed by conversion-electron Mössbauer spectroscopy (CEMS) with a standard Mössbauer spectrometer equipped with a ^{57}Co source and a gas flowing proportional counter to detect electrons emitted after resonance absorption. A least squares minimization routine was used to fit spectral profiles (9). Specular reflectance curves of the same samples

189

C.J. McHargue et al. (eds.), Structure-Property Relationships in Surface-Modified Ceramics, 189–198.
© *1989 by Kluwer Academic Publishers.*

were recorded in the wavelength range 350 to 850 nm using a Perkin-Elmer Mod. 330 spectrophotometer. Spectral transmittance was also recorded in order to calculate the samples absorbance. Finally, these samples were subjected to sequential thermal annealings at 550 and 700°C in flux of forming gas for 5 h. In order to reduce the possibility of oxidation, during annealing samples were wrapped in Ta foils. RBS, CEMS, and optical measurements were repeated after each thermal treatment.

3. RESULTS AND DISCUSSION
3.1. RBS analysis
In order to find optimum mixing conditions, the effect of the Fe film thickness on the mixing behavior was studied for thicknesses in the range 9-8 nm and Kr irradiation doses of 1×10^{16}, 2×10^{16}, and 5×10^{16} ions cm^{-2}. Figure 1(a) shows RBS spectra for samples with Fe film thickness: 41, 63, and 80 nm, irradiated to Kr doses close to 2×10^{16} ions cm^{-2} and etched. Figure 1(b) summarizes all results.

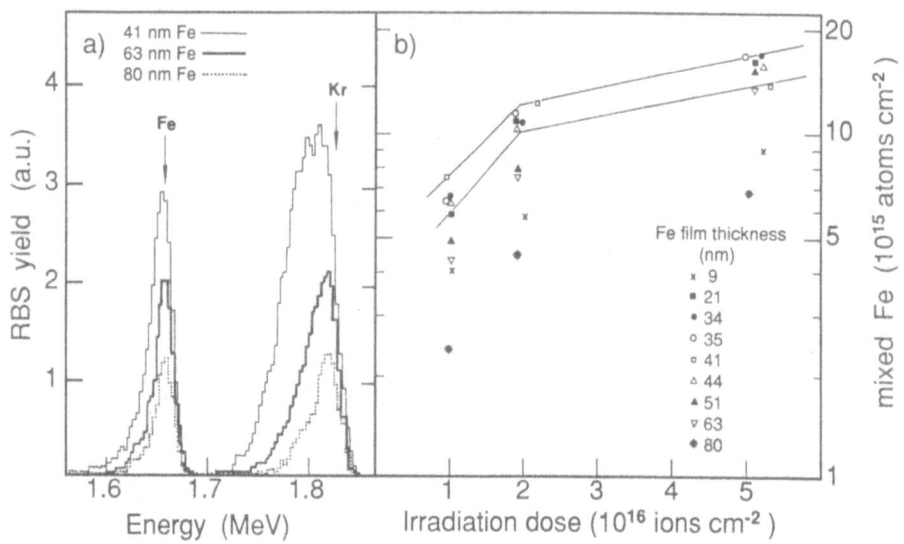

FIGURE 1. (a) Fe and Kr signals in RBS spectra (^{4}He$^+$, 2.2 MeV, 160°) of three Fe/SiO$_2$ samples (with different Fe film thicknesses) irradiated with 200 keV Kr ions to doses close to 2×10^{16} ions cm^{-2} and etched. Arrows indicate the energy of He particles backscattered from Fe and Kr atoms at the surface of samples. (b) Mixed Fe amount versus Kr irradiation dose for 200 keV irradiated Fe/SiO$_2$ samples with different Fe film thicknesses. The two lines were drawn to guide the eye.

From Fig. 1(a) we note that with the 80 nm thick Fe film about the 15% of the incident Kr ions crossed the Fe/SiO$_2$ interface and were retained in the silica matrix. Since (projected range) +3 (range straggling) = 75 nm (ref. 7), probably the stopping power of Fe for Kr particles is actually slightly lower than that evaluated using formulae reported in ref. (7).

In the thickness range 21-44 nm [Fig. 1(b)] the mixed Fe amount is maximum and, within 20%, independent of the film thickness. Points corresponding to this thickness interval lie within the two lines that we drew in Fig. 1(b) in order to guide the eye.

Detailed investigations of the mixing behavior were performed on samples with a Fe film thickness of 41 nm in the Kr dose range 3.6×10^{14} to 1.3×10^{17} ions cm^{-2}. Since SEM analyses showed that irradiations to doses higher than 3.6×10^{16} ions cm^{-2} caused the formation of holes in the Fe film and blistering in the SiO_2 substrate [in ref. (3) we showed that these features strongly affect the mixing behavior], we consider here only samples unaffected by such morphological modifications, i.e. samples irradiated up to the dose of 3.6×10^{16} ions cm^{-2}. The measured amount of mixed Fe versus the Kr irradiation dose is reported in Fig. 2. Even though experimental points could be reasonably fitted in the log-log plot by a single straight line with slope 0.85, a better interpolation can be achieved using a curve with decreasing slope with increasing Kr fluence (solid line in Fig. 2). Up to a dose close to 3×10^{15} ions cm^{-2} this curve lies very close to a straight line with slope 1 (dotted in the plot) from which experimental points significantly deviate at higher doses.

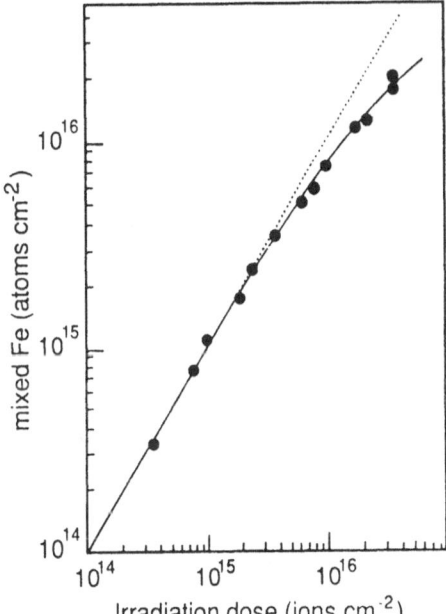

FIGURE 2. Measured amount of mixed Fe versus the Kr irradiation dose for samples with a 41-nm thick Fe film. The solid line connecting experimental points was drawn to guide the eye. The dotted line has slope 1.

It is generally accepted and verified in metal/metal and metal/semiconductor systems [see ref. (10) and references therein] that a linear dependence of the number of mixed atoms, N, on the irradiation dose, Q, indicates the occurrence of anisotropic mixing due to primary collisions. On the contrary a square root dependence of N on Q is connected to the isotropic diffusion in well developed cascades. In principle it should be possible to describe the trend of our results with the superposition of a linear and a square root term: $N = AQ + BQ^{1/2}$, where parameters A and B depend on irradiation conditions. In the past we used this formula to fit our mixing results in Ar-irradiated Fe/SiO_2 samples (3). Even if the obtained fits were (and can be) accurate enough, the physical interpretation is not straightforward. In fact if we consider low irradiation doses, for which we observe a linear behavior, we see that the ratio between mixed Fe atoms and incident Kr ions is very close to unity: this would imply an extremely high, non-physical, value for the cross section of recoil processes.

The behavior evidenced in Fig. 2 is similar to that observed by Banwell and Nicolet for the Xe irradiated Ni/SiO_2 system (11). These authors suggested that at lower doses each cascade causes, on the average, the same amount of mixed metal atoms, due to both collisional and diffusive processes within the single cascade. When the metal concentration in SiO_2 reaches a threshold value, isotropic diffusive processes contribute not only to the injection of new atoms, but also to the exit of already mixed ones from the silica matrix. The activation of such demixing process leads to the decrease in the mixing efficiency observed in Fig. 2.

The fundamental role played by cascades generated in the insulating substrate is evidenced by data shown in Fig. 3 where we reported the amount of mixed Fe versus the amount of Kr which crossed the Fe/SiO_2 interface and was retained in the substrate in all our samples. With the only exception of data of samples with the 9 mm thick Fe film (but RBS analyses showed that this very thin film contained about 25 at. % of O and this can really influence the mixing behavior of this set of samples) all experimental points lie with good approximation on the same curve: samples with different Fe film thicknesses, irradiated at different doses, display the same amount of mixed Fe atoms provided that the same amount of Kr ions crossed the Fe/SiO_2 interface. We believe that physical parameters connected with generation, characteristics and quenching of cascades in insulating materials must be quantitatively accounted for in order to develop a satis-factory description of the ion mixing phenomenon in metal/insulator systems.

In connection with CEMS and optical studies, selected samples were sequen-tially annealed for 5 h at 500 and 700°C. We report in Fig. 4, RBS spectra for a ^{57}Fe (32 nm)/SiO_2 samples 200 keV Kr irradiated to the dose of 3.5×10^{16} ions cm^{-2}, etched and annealed. The spectrum for the unannealed sample (RT) is reported for comparison.

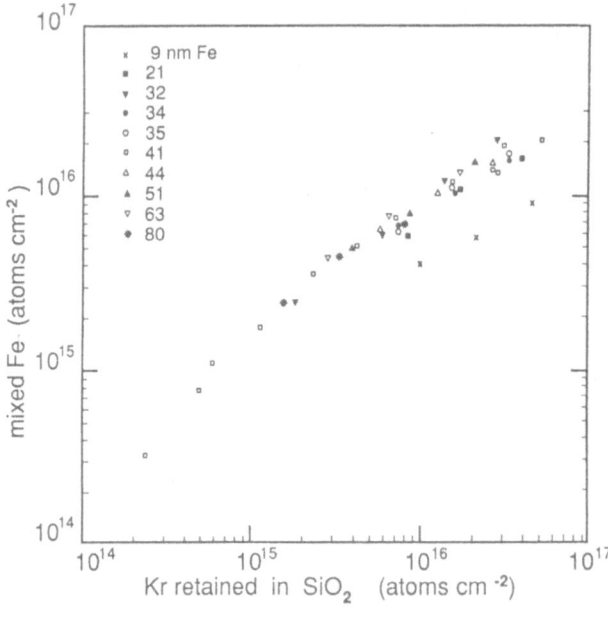

FIGURE 3. Mixed Fe amount versus the amount of Kr retained in SiO^2 for 200 keV irradiations, to dif-ferent doses, of samples with the reported Fe film thicknesses.

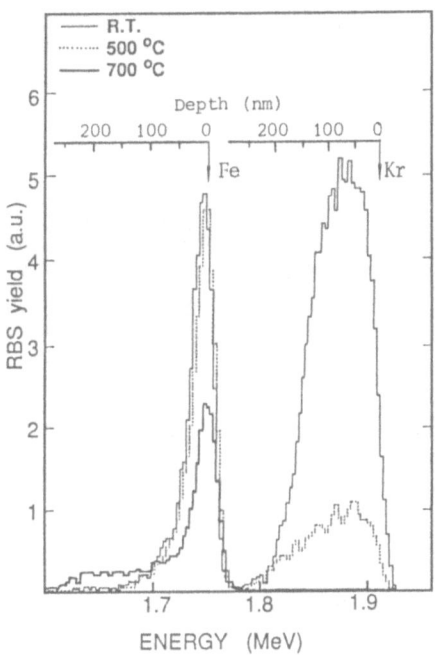

FIGURE 4. Fe and Kr signals in RBS spectra (^{4}He$^+$, 2.3 MeV, 160°) for a ^{57}Fe (32 nm)/SiO$_2$ sample 200 keV Kr irradiated to the dose of 3.5 x 10^{16} ions cm^{-2}, after etching (RT) and annealings.

The 500°C annealing did not significantly modify the amount and the distribution of mixed Fe atoms in SiO$_2$. On the contrary, the retained Kr amount underwent a reduction of about a factor of 4: the annealing caused the diffusion of Kr in the highly damaged region and its desorption at the surface.

The 700°C annealing caused a complete outdiffusion of Kr and a reduction of the Fe content to about one-half of its initial value. The missing Fe migrated to the surface where it segregated or oxidized forming a non adherent layer which we removed by rubbing the surface with a tissue paper. The increase of the backscattering yield of the Fe signal at energies between 1.6 and 1.7 MeV indicates that a small amount of the mixed Fe migrated also inside the SiO$_2$ substrate up to the depth of about 250 nm, the maximum depth at which Kr was originally present.

3.2. CEMS analysis

In order to obtain structural information concerning the mixed layer and to follow the thermal evolution of the ion-beam-induced Fe-phases in SiO$_2$, samples with the ^{57}Fe film were analyzed by CEMS. In Fig. 5 we reported CEM spectra of the sample irradiated to the Kr dose of 3.5 x 10^{16} ions cm^{-2}. The CEM spectrum of the as-evaporated sample (not reported) displayed the typical shape of magnetically ordered metallic Fe.

The spectrum of the Kr-irradiated unannealed sample (RT) was resolved in a set of spectral components attributed to the following phases: small clusters of metallic Fe with superparamagnetic behavior dispersed in the SiO$_2$ substrate (singlet S); a non-stoichiometric oxide Fe^{2+} phase (doublet D1); a silicate phase with Fe in two octahedral sites (doublets D2 and D3). Computed spectral parameters are reported in Table 1, where the percent amount of each phase is represented by the spectral relative area, RA. From RA values and the amount of mixed Fe measured by RBS (2.1 x 10^{16} atoms cm^{-2} for the considered sample) we calculated Fe areal densities in the different phases and reported these values in the column AD of Table 1.

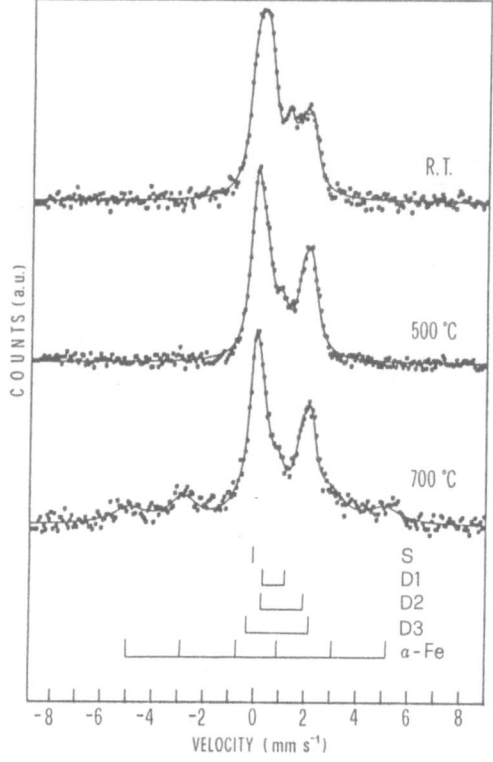

FIGURE 5. CEM spectra of the ^{57}Fe (32 nm)/SiO$_2$ sample Kr irradiated to the dose of 3.3×10^{16} ions cm^{-2}, etched (RT) and annealed. Sticky diagrams represent spectral components described in the text, the parameters of which are reported in Table 1.

TABLE 1. Mössbauer parameters of the ^{57}Fe/SiO$_2$ sample irradiated to the Kr dose of 3.5×10^{16} ions cm^{-2} and etched: isomer shift (IS) relative to α-Fe, quadrupole splitting (SQ) and full width at half maximum (Γ) in mm s^{-1}; relative subspectrum area (RA) in percent areal density, AD, in 10^{15} atoms cm^{-2}.

Fe-phases		IS	QS	Γ	RA	AD
Fe-clusters	(S)	0.08	---	0.64	28	5.9
Fe-oxide	(D1)	0.84	0.82	0.60	32	6.7
	(D2)	1.15	1.36	0.58	25	5.2
Fe-silicate						
	(D3)	1.02	2.38	0.50	15	3.2

CEM spectra after thermal annealings could still be resolved with the same set of spectral components (500°C) plus a broad sextet which was attributed to α-Fe precipitates (700°C). The percent amount of Fe-phases at the three different temperatures are reported in Table 2. As the temperature increases, the RA of both Fe-clusters and Fe-oxide decreases, whereas the silicate phase seems to be thermodynamically favored up to 500°C since its amount increases at the expense of the other two phases. The new spectral component which appears after the 700°C annealing is characterized by an internal hyperfine field lower than that of crystalline α-Fe (31 T instead of 33 T), and by a rather broad linewidth (larger than 1.0 mm s^{-1}). This indicates that at this stage of precipitation α-Fe particles are characterized by a broad grain-size distribution.

TABLE 2. Percent amount (RA) of the ion induced Fe-phases in the ^{57}Fe/SiO$_2$ sample irradiated to the Kr dose of 3.5 x 10^{16} ions cm^{-2}, etched (RT) and annealed (500 and 700°C).

Fe-phases		RT	500 °C	700 °C
Fe-clusters	(S)	28	15	9
Fe-oxide	(D1)	32	17	10
Fe-silicate	(D2+D3)	40	68	46
alpha-Fe	(sextet)	---	---	35

3.3. Optical analysis

Samples characterized by CEMS were also optically characterized by measuring the specular reflectance in the wavelength range 350 to 850 nm. In order to discriminate the contribution of the mixed Fe from that of the retained Kr and of the radiation damage, same measurements were performed on bare silica samples irradiated with 70 keV Kr ions at different doses. The measured reflectance is reported in Fig. 6 as a function of the wavelength both for irradiated bare silica and for Fe/SiO$_2$ ion mixed specimens. The curve of untreated silica is reported for reference.

The Kr irradiation of bare silica induced an overall increase of the reflectance [Fig. 6(b)], more evident at long wavelengths and independent of the Kr dose. This effect corresponds to an increase of the refractive index of the implanted layer and can be connected to the compaction of the irradiated region (12). A very small effect can be attribute also to the implanted Kr (13).

In Fe/SiO$_2$ mixed samples [Fig. 6(a)] the reflectance increases with the mixed Fe amount (which increases with the Kr irradiation dose), but is not proportional. In fact samples irradiated to 2.5 x 10^{15} and 8.2 x 10^{15} Kr ions cm^{-2} show nearly the same increment in reflectivity which is not significantly higher than that of implanted bare silica, indicating that at these irradiation doses the compaction of silica plays the major role. The mixed Fe has an effect mainly at short wavelengths. The effect of mixed Fe begins to be appreciable for the sample irradiated to the Kr dose of 1.8 x 10^{16} ions cm^{-2} (not reported) and is larger for the specimen Kr irradiated to 3.5 x 10^{16} ions cm^{-2}.

After the 500°C annealing, the reflectance of implanted bare SiO$_2$ [Fig. 6(d)] became lower than that of untreated silica. This antireflecting

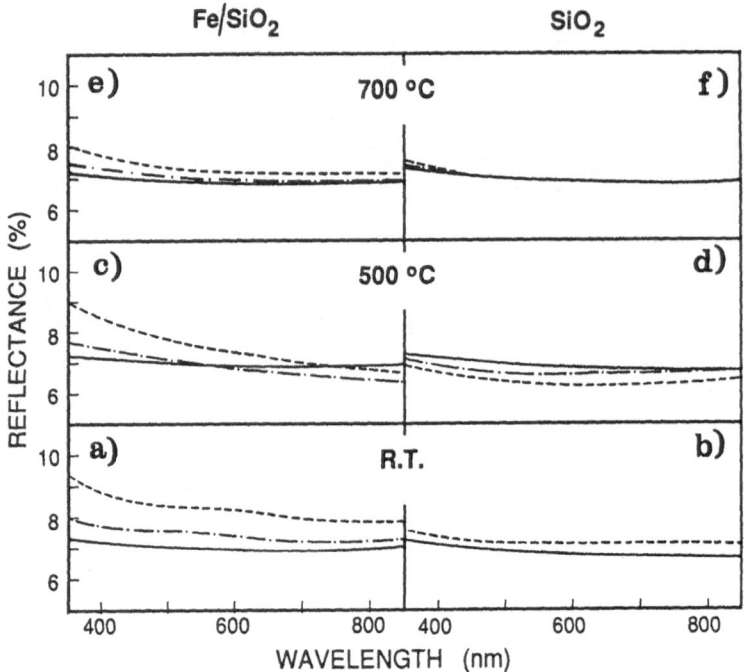

FIGURE 6. Reflectance curves of Fe (32 nm)/SiO$_2$ samples 200 keV Kr irradiated and etched (left) and of 70 keV Kr implanted bare SiO$_2$ (right) before
(RT) and after annealings. (——) untreated silica; (-·-·-) 2.5 x 10^{15} Kr
cm^{-2} irradiated samples; (- - -) 3.5 x 10^{16} Kr cm^{-2} irradiated specimens.

effect was larger the larger the Kr irradiation dose. A reflectance minimum
was present near the middle of the visible range. These features can be
explained with the formation of a perturbed surface layer with refractive
index lower than that of the bulk. This antireflecting layer probably forms
because of the formation of micropores due to the coalescence of the
implanted Kr in microblisters before its outdiffusion (revealed by RBS).
This hypothesis explains also the increasing of the antireflecting effect
with increasing the Kr dose: the larger is the number (or size) of blisters
formed upon annealing, the lower is the refractive index of the layer.

In Fe/SiO$_2$ samples the 500°C annealing [Fig. 6(c)] induced a decrease of
the reflectance with respect to unannealed samples. The effect was very
evident at long wavelengths where the reflectance became lower than that of
the untreated silica. A simulation of the reflectance curves (14) revealed
that the behavior of these samples was consistent with a bilayer configuration on the unperturbed SiO$_2$ substrate: the top layer has refractive
index higher than that of the substrate and the intermediate layer has
refractive index lower than that of untreated SiO$_2$. It is easy to identify
the top layer with the very near surface region where mixed Fe is present.
The formation of micropores is not effective enough to reduce the value of
the refractive index of the layer where Fe is present below that of
untreated silica, but causes the formation of an antireflecting layer at
larger depths where only Kr was present after ion mixing. Moreover RBS analyses showed that Kr was mainly distributed in a 75-nm-thick layer in the

case of implanted bare silica, and in a 140-nm-thick layer in the case of the Fe/SiO$_2$ mixed specimens. These values for the regions of modified refractive index are consistent with a reflectance minimum in the visible range for irradiated bare silica and in the near infrared for Fe/SiO$_2$ mixed samples.

Finally after the 700°C annealing, the reflectance in the visible range of the Kr irradiated bare SiO$_2$ was coincident with that of untreated silica, with the exception of small differences at short wavelengths [Fig. 6(f)]. This can be attributed to a nearly complete recovery of the radiation damage with structure reconstruction. The total disappearance of the implanted Kr is consistent with this hypothesis.

The reflectance of mixed Fe/SiO$_2$ [Fig. 6(e)] was still higher than that of untreated silica, but lower than that of unannealed specimens. The damage recovery caused the disappearance of the antireflecting intermediate layer while the large reduction in the Fe content can account for the lower reflecting effect.

In connection with reflectance measurements we recorded also the transmittance of our samples in the same wavelength range, in order to calculate their absorption. Using the untreated silica as a reference, differences greater than experimental errors were observed only for the Fe/SiO$_2$ sample irradiated to the highest Kr dose, i.e. 3.5×10^{16} ions cm^{-2}. Absorption (in %) was calculated as: A = 100-(reflectance %)-(transmittance %). The absorption of the unannealed sample increased monotonically with the decreasing of the wavelength, being about 2.5% at 700 nm and about 6% at 400 nm. After the annealing at 500°C, this absorption reduced of about a factor of 2, while after the 700°C annealing its values increased again up to about the 80% of those of the unannealed sample.

Perez and co-workers (14) studied the Fe ion implantation in silica and attributed the increase in the optical absorption of their samples to small metallic-Fe or Fe$_3$O$_4$ precipitates in the SiO$_2$ matrix. Probably this is true also in the present case: in fact CEMS analyses show that the amount of Fe in metallic precipitates decreased of a factor of 2 after the annealing at 500°C, while after the 700°C annealing the amount of Fe in small superparamagnetic and in larger α-Fe precipitates was only slightly lower than that present in the form of small superparamagnetic metallic-Fe precipitates in the unannealed sample.

The interpretation of the reflectance increase in ion mixed samples is still ambiguous and no direct connection can be safely established with Fe-phases detected by CEMS. The major difficulty stays on the lack of clear dose-dependent effects, probably because the total amount of mixed Fe is actually not very high. The use of Kr irradiation doses higher than 3.5×10^{16} ions cm^{-2} is not useful because the increase of the mixed Fe amount is counterbalanced by detrimental effects in the surface morphology. In the proceeding of this work we will use Xe ions to obtain a higher mixing efficiency.

REFERENCES
1. Banwell T, BX Liu, I Golecki, and M-A Nicolet: Nucl. Instr. Meth. 209/210 (1983) 125.
2. Banwell T, M-A Nicolet, T Sands, and PJ Grunthaner: Appl. Phys. Lett. 50 (1987) 571.
3. Battaglin G, S Lo Russo, A Paccagnella, G Principi, and PQ Zhang: Nucl. Instr. Meth. B27 (1987) 402.
4. White CW, G Farlow, J Narayan, GJ Clark, and JEE Baglin: Mater. Lett. 2 (1984) 367.

5. Baglin JEE, AG Schrott, RD Thompson, KN Tu, and A Segmüller: Nucl. Instr. Meth. B19/20 (1987) 782.
6. Battaglin G, S Lo Russo, A Paccagnella, P Polato, and G Principi: Nucl. Instr. Meth. B, in press.
7. Biersack JP and JF Ziegler: Ion Implantation Techniques. H Ryssel and H Glawisching (eds). Berlin: Springer Verlag, 1982, p. 122.
8. Chu WK, JW Mayer, and M-A Nicolet: Backscattering Spectrometry. New York: Academic Press, 1978.
9. Zhang PQ, G Principi, A Paccagnella, S Lo Russo, and G Battaglin: Nucl. Inst. Meth. B28 (1987) 561.
10. Cheng Y-T: Ph.D. Thesis, Caltech, 1987.
11. Banwell T and M-A Nicolet: Nucl. Instr. Meth. B19/20 (1987) 704.
12. Eernisse EP: J. Appl. Phys. 45 (1974) 167.
13. Polato P, P Mazzoldi, and A Boscolo-Boscoletto: J. Am. Ceram. Soc. 70 (1987) 775.
14. Perez A, M Treilleux, T Capra, and DL Griscom: J. Mater. Res. 2 (1987) 910.

ION-BEAM ALLOYING AND THERMOCHEMISTRY OF CERAMICS AT HIGH TEMPERATURES

I.L. Singer and J.H. Wandass[1] .

U.S. Naval Research Lab. Code 6170. Wash.D.C. 20375.

1. INTRODUCTION

High fluence ion implantation affords the possibilities of creating unique materials by overcoming chemical or physical processing limitations. Recently there has been considerable interest in high-fluence implantation into heated substrates in the semiconductor industry, where this "hot" implantation process has been shown to produce buried dielectric layers in Si [Ref. 1]. Hot implantation into metals [Ref. 2] and ceramics [Ref. 3] has also been investigated for producing graded interfaces with improved surface mechanical properties. Unlike "low" temperature implantation, which sustains only athermal mechanisms for solute redistribution and usually produces metastable phase formation [Ref. 4,5,6], hot implantation should be controlled by thermal processes (e.g., thermochemical reactions, diffusion, and defect annealing) leading to stable-phase formation. Hot implantation metallurgy, however, is expected to be more complicated in engineering materials than in Si because three or more elements are generally involved.

This paper has two goals. The first is to show the type of solute redistribution that can occur during high fluence implantation of Al^+ and Ti^+ into two ceramics, SiC and Si_3N_4, at high temperatures. The second is to demonstrate how simple calculated ternary phase diagrams can provide guidelines for interpreting the rather complicated compositions obtained during "hot" implantation and ion mixing of metals in ceramics.

2. EXPERIMENTAL

2.1 Substrate Preparation and Ion Implantation

Commercial SiC and Si_3N_4 substrates (SiC: ESK high-density sintered alpha; Si_3N_4: Norton NC132, MgO hot-pressed and Ceradyne 147, Y_2O_3 hot-pressed) were polished to a 3-μm diamond finish. Ion implantation was performed at NRL in a Varian/Extrion high current implanter. Base pressures before implantation were in the 10^{-7} Torr range. Ti^+ ions were implanted at and energy of 190 keV and Al^+ ions at 110 keV, both to a fluences of 4×10^{17} ions/cm^2, in order to achieve approximately 45 at. % peak concentration at the same depth in both SiC and Si_3N_4. The predicted range and range straggling values for the four combinations are about 116 nm and 36 nm, respectively [Ref. 7].

During implantation, substrates were either held at room temperature or implanted "hot." "Hot" implantation denotes direct heating of the substrates by the intense ion beam (up to 40 μA/cm^2) and was achieved by suspending the substrates in a Mo sheet basket during implantation. An optical pyrometer, calibrated by a thermocouple, was used to monitor the substrate temperature. A few minutes after implantation commenced, substrates reached temperatures of 900 °C. For both cold and hot implantation conditions, substrates were partially masked in order to retain a nonimplanted area of each surface.

---

[1] present address: AKZO Chemie America, Dobbs Ferry, NY

C.J. McHargue et al. (eds.), Structure-Property Relationships in Surface-Modified Ceramics, 199–208.
© 1989 by Kluwer Academic Publishers.

2.2. Surface Analysis

The near surface composition was investigated by X-ray Photoelectron Spectroscopy (XPS). XPS was performed with monochromatized Al X-rays in a Surface Science Instrument (SSI) small-spot analyzer. Sputter depth profiling was accomplished using 3 keV Ar^+ ion bombardment. Depths of the ion-milled craters were measured by Michelson interferometry. Data analysis was performed using SSI software routines. Composition vs depth profiles were quantified by integrating the photoelectron spectra and normalizing them using SSI's modified Scofield cross sections. Scanning electron microscopy (SEM) and energy-dispersive X-ray analysis (EDX) were performed in a commercial instrument.

3. RESULTS

3.1. Al-implantation. Scanning electron microscopy of implanted SiC showed submicron-size spherical particles distributed uniformly over the implanted surface. These particles can be seen in the SEM micrograph in Fig. 1, surrounding the Vickers' indentation in the hot-implanted surface. EDX analysis (V_e = 5 keV) on a sphere, between the spheres and over wide areas gave Al/Si ratios of 2.5, 0.4, and 0.5 respectively, indicating that the spheres were composed mainly of Al.

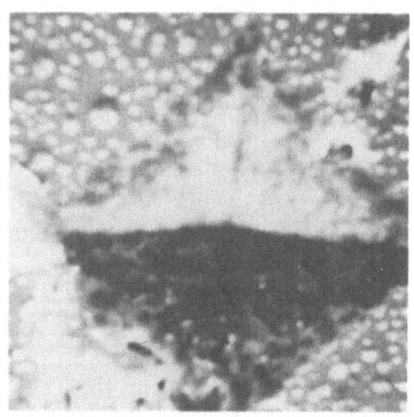

FIGURE 1. Scanning electron micrograph of Al-implanted (hot) SiC surface, showing Al-rich spheres.

—| 5μm |←—

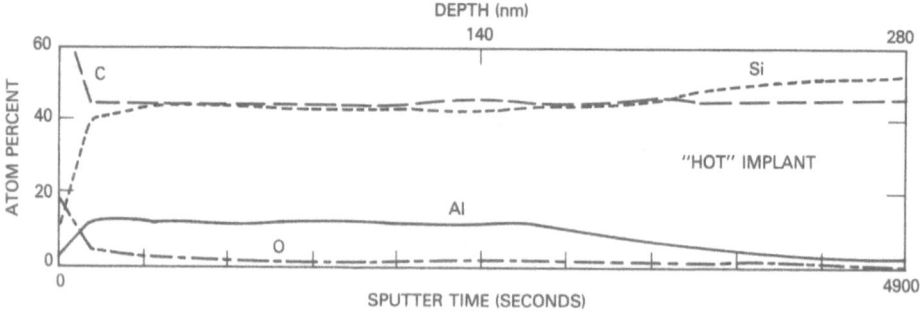

FIGURE 2. XPS sputter-depth profile of Al-implanted (hot) SiC.

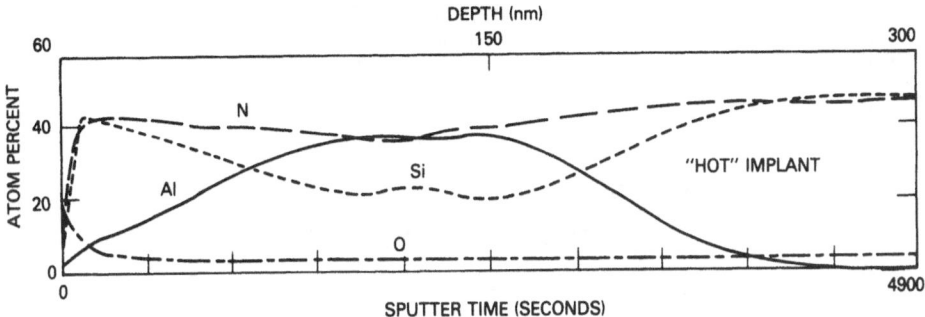

FIGURE 3. XPS sputter-depth profile of Al-implanted (hot) Si_3N_4.

In order to obtain reliable XPS depth profiles of the hot implanted SiC layer, the Al spheres were etched off the surface with a 20% aqueous sodium hydroxide solution. The XPS depth profile, displayed in Fig. 2, shows a flat Al concentration of 17% to a depth of about 200 nm.   Room temperature implantation profiles (not shown), by contrast, showed a Gaussian-like distribution, with a peak concentration at 110 nm of around 40 at. %.   Apparently, in the hot-implanted substrate, Al moved to the surface during implantation and, because the temperature was above the melting point of Al (ca. 660°C in bulk), it melted and formed droplets of Al.

An XPS depth profile of the hot implanted Si_3N_4 layer is shown in Fig. 3.   This profile was more nearly Gaussian, although it too displayed a somewhat flat top, leveling out at a concentration of about 40 at. %.   Room temperature Al implantation showed a more sharply peaked Gaussian profile (not shown).

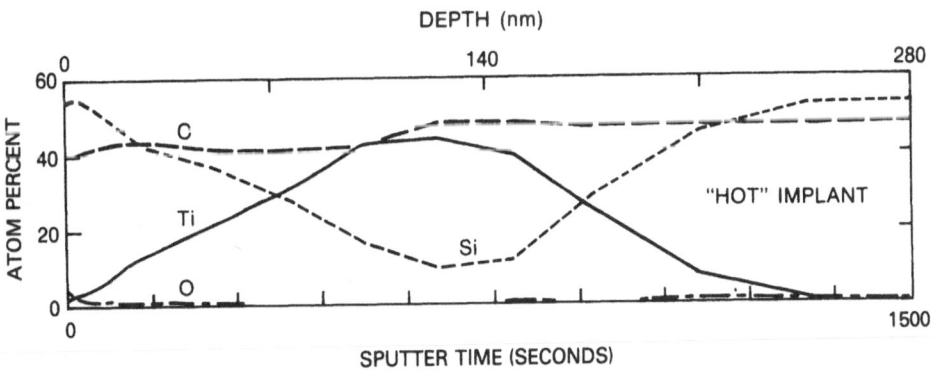

FIGURE 4. XPS sputter-depth profile of Ti-implanted (hot) SiC.

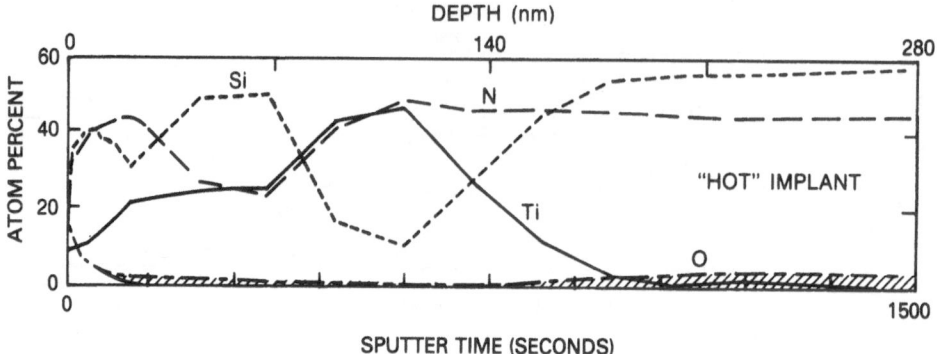

FIGURE 5.   XPS sputter-depth profile of Ti-implanted (hot) Si_3N_4.

3.2. Ti-implantation. XPS depth profiles of Ti implanted (hot) into SiC and Si_3N_4 are shown in Figs. 4 and 5, respectively. These profiles, reported and discussed elsewhere [Ref. 3], are presented here for comparison with the hot Al implantation profiles. The Ti depth profile in the SiC substrate shows a Gaussian-like distribution, similar to that found in the room temperature Ti-implanted SiC. By contrast, in the hot-implanted Si_3N_4 substrate, the profile shows considerable Si outmigration combined with TiN formation at two depths. These features have also been confirmed by TEM and RBS analysis [Ref.8], the latter a profiling technique which does not suffer from preferential sputtering and the resultant composition changes often found in XPS profiles.

4. DISCUSSION

In order to understand the compositions achieved by high fluence, hot implantation into compounds, both thermdynamic and kinetic factors should be considered. In this paper, however, we focus only on thermodynamics; in particular, on a simple method for predicting possible solute redistribution and phase formation. The method relies on calculated ternary phase diagrams, or more accurately, on isothermal sections of calculated ternary phase diagrams. Similar diagrams have been used recently to rationalize interfacial reactions in a number of metal-Si-O systems [Ref. 9] and ternary semiconductor systems [Ref. 10].

Ternary or higher order phase diagrams provide the correct description of competing phases where three or more components are in thermodynamic equilibrium [Ref. 11]. Unfortunately, complete ternary phase diagrams are not available for most materials at all temperatures of interest. For the purpose of assessing possible compositions in the highly nonequilibrium environment of the implanted layer, one may use readily available (albeit of variable quality) thermochemical data to calculate ternary phase diagrams for the elements of interest. In practice, one computes the stable "tie-lines" of the ternary system, i.e., lines joining binary compounds that do not react with one another. In the present treatment, we limit ourselves to ternary phase diagrams incorporating binary compounds. (see Appendix for more details.)

Calculated ternary phase diagrams for Al-Si-C and Al-Si-N are shown in Figs. 6 and 7. For Al-Si-C, two diagrams are shown because calculations showed that the tie lines may have switched from $Si-Al_4C_3$ to Al-SiC at 570°C or 900°C, depending on the thermochemical data used (see Appendix). The two tie lines for the Al-Si-N diagram, however, were found to be stable at all temperatures.

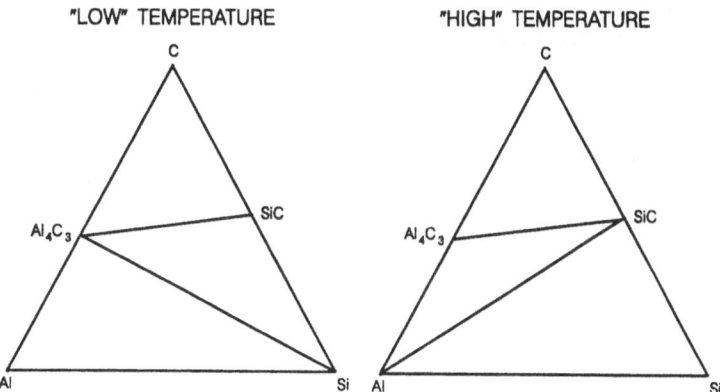

FIGURE 6. Calculated ternary phase diagrams for the Al-Si-C system at "low" and "high" temperatures. (see Appendix for details.)

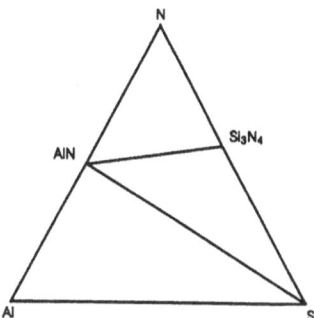

FIGURE 7. Calculated ternary phase diagrams for the Al-Si-N system.

According to the "low" temperature Al-Si-C diagram, Al initially reacts with SiC to form Al_4C_3, plus elemental Si. By constructing a reaction line from Al to SiC, one finds that no elemental Al is produced until the concentration of Al reaches 42 at. %. However, in the "high" temperature Al-Si-C system, elemental Al is the stable product at all concentrations. Considering the experimental data in Fig. 2, we suggest that at this temperature (900°C), implanted Al remained a solute species and was able to diffuse out of the implanted layer. On reaching the surface whose temperature far exceeded the melting point of Al (660°C), the Al agglomerated into the droplets seen in Fig. 1.

By contrast, Fig. 3 shows that very little Al migrated out of the hot-implanted Si_3N_4 layer because most of the Al, according to Fig. 7, tends towards compound AlN, not elemental Al, formation. At Al concentrations above 36 at. %, however, free Al should form, which could account for the flattening of the hot-implanted Al profile at peak concentrations about 40 at. %.

To more easily visualize and better quantify compound formation in implanted layers, we have constructed hypothetical phase distribution vs depth profiles for the two ternary systems. These profiles were calculated from the Al-Si-C and Al-Si-N ternary diagrams in Figs. 6 and 7, using the lever rule applied to the Gaussian depth distributions predicted for the energy/fluence conditions used. The phase profiles for Si_3N_4 implanted with Al to a fluence of $4 \times 10^{17}/cm^2$ at 110 keV are shown in Fig. 8. One observes the reaction products AlN and Si throughout the implanted layer and elemental Al from a depth of 100 to 140 nm. This is the depth range where the measured Al profile in Fig. 3 flattened out. The phase profiles for SiC implanted with Al to $4 \times 10^{17}/cm^2$ at 110 keV are shown in Fig. 9. As expected, the profiles give a Gaussian profile of elemental Al in a diluted SiC matrix. These profiles are particularly useful for a systems like Al-Si-N or even the more complex Ti-Si-N ternary [Ref. 8] where mixtures of compounds and elements are present.

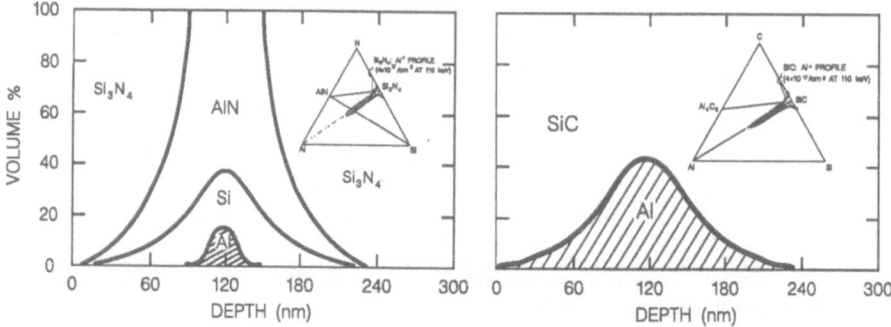

FIGURE 8. Calculated phase distributions vs depth distribution for Al implanted at 110 keV to a fluence of 4 x 10^{17}/cm^2 into Si$_3$N$_4$. (left)

FIGURE 9. Calculated phase distributions vs depth distribution for Al implanted at 110 keV to a fluence of 4 x 10^{17}/cm^2 into SiC. (right)

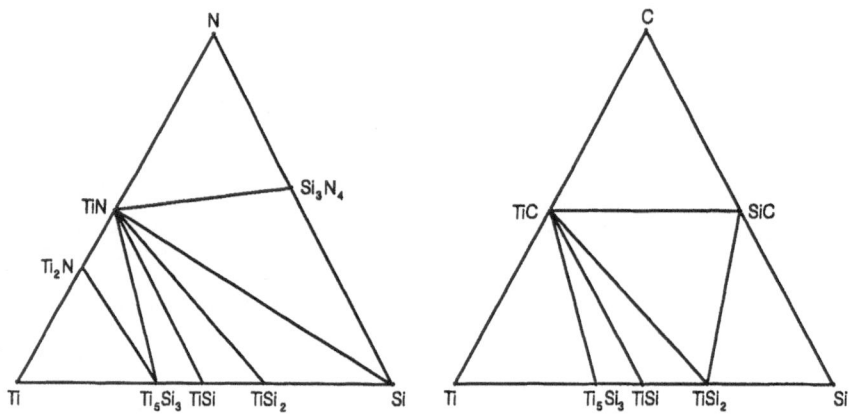

FIGURE 10. Calculated ternary phase diagrams for the Ti-Si-N system.

FIGURE 11. Ternary phase diagrams for the Ti-Si-C system, adapted from Ref. 13.

Calculated ternary phase diagrams have been used in a similar fashion to explain the concentration vs depth profiles of Ti implanted into Si$_3$N$_4$ and SiC (Figs. 4 and 5). The calculated Ti-Si-N diagram [Ref. 12] is shown in Fig. 10. In the reaction of Ti with Si$_3$N$_4$, elemental Si is produced at all concentrations of Ti up to about 44 at. % because of the tie line that joins TiN to Si. Again, we suggest that the redistribution of Si in Ti-implanted Si$_3$N$_4$, presented in Fig. 5, is possible because of the presence of free Si [Refs.3,8]. By contrast, there is no tie-line joining either Si or Ti in the Ti-Si-C ternary diagram (see Fig. 11) [Ref. 13], and the reaction of Ti with SiC up to elemental Ti concentrations of 45 at. % produces only compounds (carbides and silicides). A phase distribution vs depth profile for SiC implanted with Ti to 4 x 10^{17}/cm^2 at 190 keV is illustrated in Fig. 12. Hence, while many reactions should occur as the Ti implantation concentration increases, elemental Si is not expected to be one of the products therefore no solute redistribution of Si is predicted.

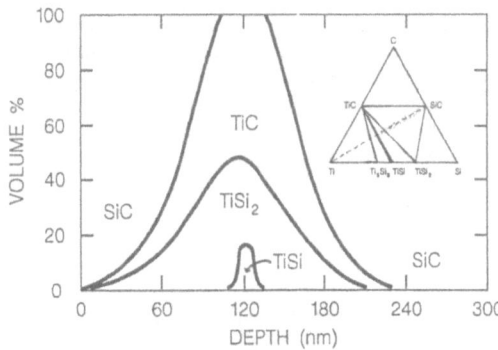

FIGURE 12. Calculated phase distributions vs depth distribution for Ti implanted at 190 keV to a fluence of 4 x 10^{17}/cm^2 into SiC.

Ternary phase diagrams may also be used to assess thermochemical factors influencing ion mixing of metals on insulators. Previous considerations of thermochemical factors in ion-mixing studies have been based on "the enthalpy rule of metal-insulator mixing," which holds that mixing occurs only if the reaction enthalpy is negative [Ref. 14,15,16]. The reaction enthalpy is determined by writing the balanced equation for reactants of a metal-insulator system against known reaction products. Reaction enthalpies are then calculated using standard heats of formation data as found in e.g. Ref. 17. For example, in the Zr-Al$_2$O$_3$ system, one calculates enthalpies for the equation

$$3Zr + 2Al_2O_3 = 3ZrO_2 + 4Al. \qquad (1)$$

Farlow et al [Ref. 15] recently ion mixed and analyzed more than 20 metal-insulator pairs and found the rule held for most, but could not account for mixing in the Zr-Al$_2$O$_3$ system, whose reaction enthalpy was nearly zero. This apparent ambiguity of the enthalpy rule, however, can be understood by examining the calculated Zr-Al-O ternary phase diagram, illustrated in Fig. 13. If one draws a reaction line (shown as a dashed line) between Zr and Al$_2$O$_3$, it is clear that Al$_2$O$_3$ is reduce (initially) to ZrO$_2$ and Al$_2$Zr, not to the reaction products given in eqn. 1. The problem with applying the enthalpy rule in this case was that the reaction enthalpy was calculated for the obvious product, but it was <u>the wrong product</u>. In calculating the ternary diagram in Fig. 13, all reactions products were considered.

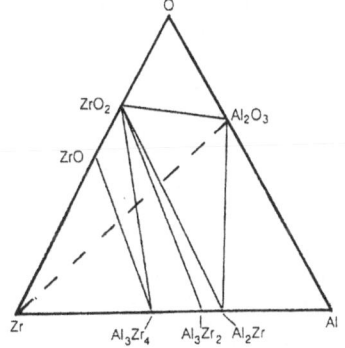

FIGURE 13. Calculated ternary phase diagrams for the Zr-Al-O system (T < 1400°C). The dashed line is a reaction line that depicts compositions obtained by reacting Zr with Al$_2$O$_3$.

A second example of the ambiguity inherent in the enthalpy rule is found in Ref. 16, where the authors attempted to use the rule to explain results of ion mixing Ti on Si_3N_4. They computed a positive enthalpy assuming TiSi or $TiSi_2$ products but a negative enthalpy assuming a TiN product and had to conclude that the reaction could go "...either way, depending on the assumed reaction product." The ternary phases diagram in Fig. 10 shows explicitly which reaction products are expected (e.g. TiN and Si) and predicts mixing if thermodynamics were the controlling factor. (In fact, although the authors in Ref. 16 did not observe mixing in the $Ti-Si_3N_4$ system, Noda et al [Ref. 18] have presented clear evidence for mixing at room temperature.) Ternary phase diagrams, therefore, should be used in place of the enthalpy rule of mixing because they perform the same function as the enthalpy rule but take into account all possible reactions. And, as discussed in the appendix, the calculated diagrams also include entropy contributions ignored by the enthalpy rule.

5. SUMMARY AND CONCLUSIONS

Surface analysis performed on Al^+ and Ti^+ implanted "hot" into SiC and Si_3N_4 indicated "anomalous" redistribution of Al in SiC and Ti in Si_3N_4. Thermochemical reactions expected in these ternary systems were explained with ternary phase diagrams calculated from thermochemical data for binary compounds. The calculated ternary phase diagrams indicated that elemental Al and elemental Si are favored in the two cases where these species were found to redistribute, but Al and Si compounds were favored in the two other cases where no solute redistribution was seen. Ternary phase diagrams were also shown to be more reliable than the enthalpy rule for predicting ion mixing of metal-insulator pairs. Thus, calculated ternary diagrams appear to be easy to use and useful tools for assessing and perhaps predicting compositions achieved in ion implanted and ion mixed ceramics.

ACKNOWLEDGEMENTS

ILS is indebted to M. Colette Einloth who, with the help of Ensign Tom Smith and Dr. Peter Ehni, developed the programs to calculate and plot the ternary diagrams; Mr. John Wegand for technical assistance; Mr. Dan Buntman, of Surface Science Laboratories, for several XPS analyses and DARPA for funding. JHW wishes to acknowledge the support of a National Research Council-Naval Research Laboratory Fellowship.

APPENDIX

Ternary phase diagrams provide a complete description of phase equilibria in three component systems. In such systems at equilibrium, Gibbs phase rule must be satisfied, i.e., $P = C - F + 2$, where P is the number of phases, C is the number of components, and F is the number of degrees of freedom. For a solid solution of 3 components at constant temperature and pressure, only 2 degrees of freedom are possible, i.e., the concentration of 2 of the 3 components; hence, the number of phases equals the number of components. In such an isobaric, isothermal section of a ternary phase diagram, a region of three phases in equilibrium is represented by a triangular area whose vertices are the three equilibrium phases; the sides of the triangle are tie lines and the vertices specify the compositions of the three co-existing phases. For simplicity (and, in part, out of ignorance) we ignore ternary phases (invariant points) and assume that binary compounds are stoichiometric (no bivariant lines).

An equilateral triangle provides a convenient framework for constructing this simplified ternary phase diagram and determining compositions. The three elements are placed at the vertices of the triangle. Binary compounds, known to exist at the specified temperature/pressure conditions, are placed according to some scale, e.g. at. %, at points on the sides of the triangle. The larger triangle is then subdivided into n + 1 tie triangles (where n is the number of binary compounds) whose vertices define, according to the phase rule, the three phases that determine the composition of any point within the triangle, or two phases if the point lies on a tie line. Tie lines, in other words, connect phases that are stable when brought in contact.

Tie lines are established by calculating the Gibbs free energy of reaction between competing reactions at the point where the two possible tie lines would cross. This is performed by writing balanced equations for all the possible reaction products, such as those given in eqns. (1) and (2) above and (A1) and (A2) below. Tie lines are then chosen by a process of elimination from those reactions which give the lowest negative free energy. Since free energies vary with temperature, tie lines may switch as the temperature changes. Once the ternary diagram is established, the amounts of each phase present at a point in the diagram can be computed by the lever rule [Ref. 11].

In this work, several methods for determining tie lines and their switching temperatures were used because the necessary thermochemical data were either inaccurate or contradictory. Calculations were performed for temperatures from 25°C to 1500°C. In the Si-N-Al case, the reaction

$$4Al + Si_3N_4 = 4AlN + 3Si \qquad (A1)$$

gave $\Delta G°_{298K} = -120$ kcal/mole, and a tie-line switch would not be expected. However, in the Si-C-Al case, $\Delta G°_{298K}$ values for the reaction

$$4Al + 3SiC = Al_4C_3 + 3Si \qquad (A2)$$

varied from -9 to +3 kcal/mole, depending on the method of calculation, so we had to consider that the tie lines might switch at some temperature between 25°C and the implantation temperature, 900°C.

The first method used room temperature $\Delta H°_{298K}$ and $S°_{298K}$ values to approximated $\Delta G_T = \Delta H°_{298K} - \Delta S°_{298K}(T - 298K)$ [Ref. 17]. The limitations of this approach are obvious, as it does not take into account the temperature dependence of $\Delta H°_{298K}$. This method predicted an Al_4C_3 - Si tie line up to about 570°C, then a switch to an Al - SiC tie line at higher temperatures. The second method used an algebraic representation for the Gibbs energies of reaction: $\Delta G°_T = A + BT \log T + CT$ [Ref.19]. The solution of these equations indicated the same tie lines but a switch temperature of about 900°C. A third method used tabulated values of the Gibbs function for the various reactants [Ref.20]. This method predicted that the Al - SiC tie line is stable over the temperature range 25 to 1500°C. We suggest, therefore, that the Al - SiC tie line should exist at temperatures above 600 to 900°C, or, perhaps, as low as 25°C.

REFERENCES

1. Wilson I.H., Nucl. Instrum and Meth. B1 (1984) 331; also, in "Ion Beam Modification of Insulators," Eds. P. Mazzoldi and G.W. Arnold (Elsevier, Amsterdam, 1987) p. 245.

2. Singer I.L., R.N. Bolster, J.A. Sprague, K. Kim, S. Ramalingam, R.A. Jeffries and G.O. Ramseyer, J. Appl. Phys. 58 (1985) 1255.

3. Singer I.L., Surf. Coat. and Technol. 33 (1987) 487.

4. "Metastable Materials Formation by Ion Implantation," edited by S.T. Picraux and W.J. Choyke (North Holland, N.Y. 1982).

5. Singer I.L., Vacuum 34 (1984) 853.

6. Follstaedt D.M., Nucl. Instrum. Meth., B7/8 (1985) 11.

7. Manning I., and G.P. Mueller, Computer Phys. Commun. 7 (1974) 85.

8. Singer I.L., R.G. Vardiman and C.R. Gossett, in "Fundamentals of Beam-Solid Interactions," ed. M.J. Aziz and L.E. Rehn (MRS, Pgh PA, 1988) 201-206.

9. Beyers R., J. Applied Physics 56 (1984) 147.

10. Lin J.-C., K.-C. Hsieh, K.J. Schulz, and Y.A. Chang, J. Mater. Res., 3 (1988) 148.

11. West D.R.F., "Ternary Equilibrium Diagrams," (Macmillan, New York, 1965).

208

12. Beyers R., R. Sinclair and M.E. Thomas, J. Vac. Sci. Technol. **B2** (1984) 781. [Note an error in the Ti-Si-N drawing in Fig. 4. The position of Si_3N_4 along the N-Si line was incorrectly placed at 33% Si and actually belongs at 43% Si.]; also, verified by the present authors.
13. Adapted from the Ti-Si-C ternary phase diagram in E. Rudy, "Ternary Phase Equilibria in Transition Metal-B-C-Si Systems, Part V," Technical Report AFML-TR-65-2 (Air Force Materials Lab, Wright-Patterson Air Force Base, OH, May 1969), p. 522.; the calculated diagram is similar, except that a tie line connects SiC to TiSi, not $TiSi_2$.
14. Banwell T., B.X. Liu, I. Golecki, M.-A. Nicolet, Nucl. Instrum. Methods **209/210** (1983) 125.
15. Farlow G.C., B.R. Appleton, L.A. Boatner, C.J. McHargue, C.W. White, G.J. Clark, and J.E.E. Baglin, Mater. Res. Soc. Symp. Proc. **45** (1985) 137.
16. Bhattacharya R.S., A.K. Rai, and P.P. Pronko, J. Mater. Res. **2** (1987) 211.
17. Kubaschewski O., and C.B. Alcock, "Metallurgical Thermochemistry," 5th edition (Pergamon Press, Oxford, 1979) Table A.
18. Noda, S., H. Doi, N. Yamamoto, T. Hioki, H. Kawamoto, O. Kamigaito, J. Mater. Sci. Letts. **5** (1986) 381.
19. Kubaschewski O., and C.B. Alcock, "Metallurgical Thermochemistry," 5th edition (Pergamon Press, Oxford, 1979) Table E.
20. Barin I., and O. Knacke, "Thermochemical Properties of Inorganic Substances," (Springer, Berlin, 1973 and 1977).

MÖSSBAUER SPECTROSCOPY OF IRON IMPLANTED IN ZIRCONIA

J. A. SAWICKI, G. MAREST* AND B. COX

Atomic Energy of Canada Limited, Chalk River Nuclear Laboratories,
Chalk River, KOJ1JO Ontario, Canada
*Institut de Physique Nucleaire, Universite Claude Bernard Lyon I,
69622 Villeurbanne, France

1. INTRODUCTION

Zirconium oxide ZrO_2, as an important representative of advanced ceramic
materials, has outstanding mechanic, refractory, and electrochemical properties. In addition to its wide use in ultra-strong stabilized zirconia tools
and as an ion conducting electrolyte in oxygen sensors, pumps, and fuel
cells, it has recently been examined for use in ceramic automotive engines
and as a support for high temperature catalysts (1,2). Zirconia plays also
a crucial role in nuclear reactors as a protective oxide layer on various
zirconium-alloy components, such as fuel cladding, fuel channels and
pressure tubes. It appears that iron impurities in zirconia may seriously
affect the properties of the zirconia-based ceramics as well as the oxidation characteristics of zirconium alloys (3-5). However, unlike the Zr-Fe
alloy system, which is rather well established (6), the knowledge of the
ternary Zr-Fe-O system is still in its infancy (7,8). Therefore, detailed
understanding of the state of iron in zirconia, and - particularly - its
role in oxygen and hydrogen transport, is of primary interest.

During the past several years it has been indicated that much can be
learned about the properties and structure of ceramic insulators from
Mössbauer spectroscopy on ion implanted specimens (9-16). In particular, it
was shown that the local charge states and environment of implanted Fe ions
in ceramic matrices can be most effectively studied by conversion electron
Mössbauer spectroscopy (CEMS). It was demonstrated previously (16), for the
ZrO_2 specimens implanted at room temperature to fluences of 5×10^{15} and
10^{16} Fe ions/cm^2 (~1 at. % Fe), that implanted iron exhibits very different
features from iron introduced into zirconia at thermal equilibrium by sintering, alloy oxidation or mineralization. Therefore, one can expect that
in rapidly quenched ceramics or in ceramics exposed to radiation environment
the chemical and physical properties of Fe impurities may dramatically
differ from those in equilibrium conditions. In order to establish the
variation of the various charge states versus concentration of implanted
iron in the present work the measurements were extended to the range of
fluences from 10^{15} to 10^{17} ions/cm^2. The work is still in progress and the
results reported below are preliminary and fragmentary.

2. EXPERIMENTAL RESULTS

The 0.2 μm zirconia films were prepared by anodic oxidation of very pure
(<20 ppm Fe) 50 μm thick zirconium foil in a 0.2-molar H_2SO_4 solution. The
resultant films consisted mostly of cubic ZrO_2 with small micropores of
~3 nm diameter. The 100 keV $^{57}Fe^+$ ions were implanted at ion currents of
~2-3 μA/cm^2 to fluences of 10^{15}, 2×10^{15}, 5×10^{15}, 10^{16}, 2×10^{16},
3.5×10^{16}, 5×10^{16}, 7.5×10^{16}, and 10^{17} ions/cm^2. In each case the
implanted area was 1 cm^2. During the implantations the targets were held at

C.J. McHargue et al. (eds.), Structure-Property Relationships in Surface-Modified Ceramics, 209–218.

room temperature in a vacuum of 10-4 Pa. The calculated mean projected range R_p and straggling in range ΔR_p for 100 keV ^{57}Fe ions in ZrO_2 are equal to 45 and 20 nm, respectively. For the above range of fluences, this corresponds to mean concentrations from 0.15 to 15 at. % Fe.

The conversion electron Mössbauer spectra were measured in the backscattering geometry - energy integral mode, using a helium flow proportional counter. The Mössbauer source consisted of 50 mCi ^{57}Co diffused in a Rh matrix. During the measurements both the source and the specimens were held at room temperature. Some representative spectra measured for as-implanted specimens are presented in Fig. 1 and the spectra obtained for thermally-annealed specimens are given in Fig. 2. The spectra were deconvoluted into sets of Lorentzian lines using a least-squares fitting program. The obtained positions, separations and relative intensities of individual components in the spectra are represented in the figures by the bar diagrams. The velocity scale is always referred to metallic α-Fe. The variation of the spectral parameters with the concentration of implanted iron ions is shown in Fig. 3.

The analysis of the Mössbauer spectra indicated that as-implanted iron occurs in zirconia in several different oxidation forms, with the relative abundances strongly dependent on the concentration (fluence) of implanted ions, as shown in Fig. 4.

The first fraction of iron, represented by the doublet with the isomer shift IS = 0.30-0.45 mm/s and the quadrupole splitting QS = 0.95-1.3 mm/s, was ascribed to high-spin ferric ions Fe^{3+} ($3d^5$). The values of the isomer shift and the quadrupole splitting suggest that Fe^{3+} ions are located substitutionally in distorted oxygen octahedra (17). The Fe^{3+} fraction predominates at the lowest fluences but quickly decreases to zero at a fluence of 10^{16} ions/cm² (1 at. % Fe), and thereafter it recovers slowly to about 10% relative abundance at high fluences. The maximum concentration of Fe^{3+} that can be substitutionally incorporated on Zr^{4+} sites in an as-implanted ZrO_2 matrix is about 5×10^{20} Fe/cm³, which is comparable with the solubility level of Fe in yttria stabilized zirconia, as determined by Burggraaf et al. (17) from Rutherford backscattering data. The predominant occurrence of iron in ZrO^2 in the form of Fe^{3+}, as observed at concentrations of ~0.1 at. % Fe, is in agreement with the results of EPR measurements (18) and with the conclusions based on lattice energy calculations (19).

Two other fractions of iron are represented by two doublets with large isomer shifts and quadrupole splittings, IS = 1.2-1.3 mm/s and QS = 1.5-1.8 mm/s, and IS = 1.1 mm/s and QS = 2.6-2.8 mm/s, respectively. The values of these parameters correspond to high-spin ferrous ions Fe^{2+} ($3d^6$). By analogy with MgO (10) and Al_2O_3 (12) the two different sets of parameters are thought to be representative of Fe^{2+} ions located at two crystallographically nonequivalent lattice sites with different electron densities and local symmetries. As seen in Fig. 4, the abundances of both Fe^{2+} fractions first rise rapidly with increasing Fe concentration and then, after reaching maxima of about 60 and 20% near 1 at. % Fe, they both decline slowly to about 10% magnitude with increasing fluence. Relatively large magnitudes for the IS and QS for both sites, as compared to the values of these parameters in other oxides (see Table 1), suggest that the bonds between Fe^{2+} and oxygen ions in ZrO_2 are characterized by a fairly low degree of covalence.

The last fraction of iron, represented by the single line with the isomer shift IS = -0.1 mm/s, should be ascribed to very small aggregates of metallic iron in neutral charge state, FeO. The relative abundance of this form of iron increases from nil for the fluence of 10^{15} ions/cm² to about 50% for fluences of 7.5×10^{16} and 10^{17} ions/cm². At these fluences one

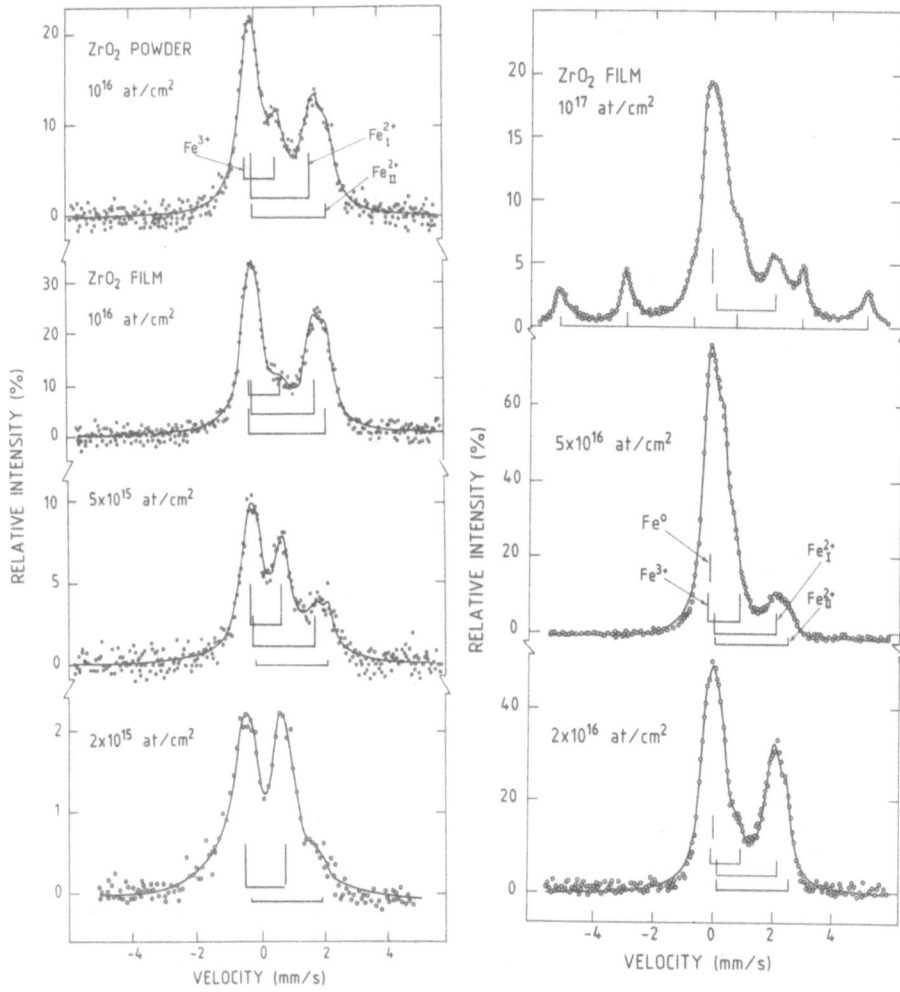

FIGURE 1. Conversion electron Mössbauer spectra of ^{57}Fe in as-implanted ZrO$_2$ obtained at several different fluences.

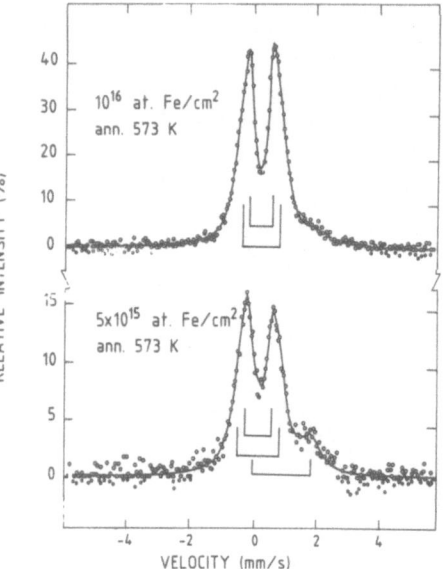

FIGURE 2. Conversion electron Mössbauer spectra of ^{57}Fe implanted in ZrO_2 and annealed in air at 573 K for 24 h.

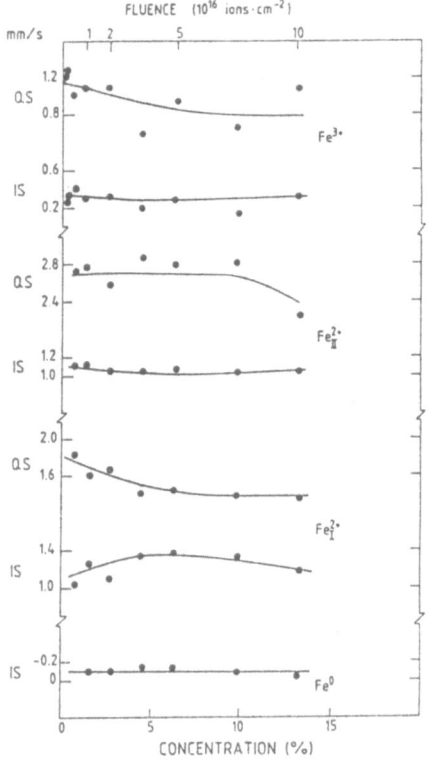

FIGURE 3. The parameters of the Mössbauer spectra, isomer shift - IS, quadrupole splitting - QS, and line width - W, plotted as a function of the effective concentration (fluence) of implanted iron ions.

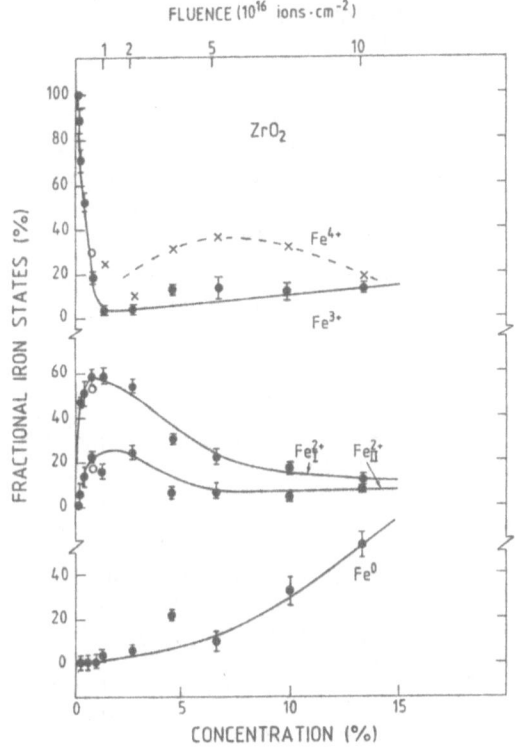

FIGURE 4. The relative populations of various charge states of iron in as-implanted ZrO_2 specimens, plotted as a function of the concentration (fluence) of implanted ions.

TABLE 1. Comparison of the Mössbauer spectra parameters for room temperature implantation of 100 keV Fe ions into various matrices up to the fluence of 10^{16} ions/cm², with the exceptions of Al_2O_3 (160 keV) and SiO_2 (100 keV, 4×10^{16} ions/cm²). Relative fractions F are given in percent and isomer shifts IS and quadrupole splittings QS in mm/s.

Matrix	d g/cm³	R_p nm	ΔR_p	x at%	Fe^0 IS	Fe^0 F	Fe^{2+}_I IS	Fe^{2+}_I QS	Fe^{2+}_I F	Fe^{2+}_{II} IS	Fe^{2+}_{II} QS	Fe^{2+}_{II} F	Fe^{3+} IS	Fe^{3+} QS	Fe^{3+} F
LiF	2.65	68	16	1.4	0.11	10	1.33	2.08	20	-			0.24	0.91	70
MgO	3.58	49	13	1.9	-		0.99	1.00	50	0.92	2.20	20	0.37	0.60	30
Al_2O_3	3.97	46	12	1.7	-		0.70	1.80	60	1.19	1.68	40	-		
SiO_2	2.27	78	21	1.6	-0.07	12	0.50	1.51	35	0.88	1.99	23	-		
TiO_2	4.26	48	16	1.6	0	5	0.95	2.20	95	-			-		
ZrO_2	5.65	45	20	1.5	-0.10	5	1.10	1.70	60	1.10	2.70	30	0.34	1.10	5

addition to the single line component (FeO) and three doublets Fe^{2+}_I, Fe^{2+}_{II}, and Fe^{3+}, a characteristic pattern of six lines due to ferromagnetic metallic iron aggregates with already fairly large mean diameter (>3 nm). To take into account the distribution of the sizes of the particles these spectra were best fitted with two sextets, corresponding to internal magnetic fields of H = 30 and 32 T. The relative intensities of the lines 2 and 5 compared to lines 1 and 6 ($A_{2/1}$ = $A_{5/6}$ = 1.3) in both sextets indicate that the metallic iron particles isolated in zirconia matrix are magnetically coupled and oriented in the plane parallel to the surface of the specimen. It is notable that the spectrum obtained by us for the ZrO_2 specimen implanted with 10^{17} Fe/cm² is very similar to that observed by Burggraaf et al. (17) for the ceramic specimen of yttria stabilized zirconia; 0.83 ZrO_2-0.17 $YO_{1.5}$, which has been implanted with 15 keV, 8 x 10^{16} Fe/cm².

Annealing in air of the specimens implanted with 5 x 10^{15} and 10^{16} ions/cm² resulted, even at fairly low temperature (573 K for 24 h), in almost complete transformation of Fe^{2+} to Fe^{3+} (see Fig. 2). We believe that this enhanced internal oxidation, which occurs at significantly lower temperatures than in any other oxide studied so far, is facilitated by high porosity of the specimens and the extraordinarily high mobility of oxygen in zirconia. The fact that the splitting of Fe^{3+} doublet in annealed specimens (QS = 0.9 mm/s) is smaller than in as implanted ones (QS = 1.15 mm/s) points to rearrangement of iron environment upon annealing to α-Fe_2O_3 like form.

The Mössbauer spectra obtained for room temperature Fe implants in ZrO_2 are clearly different from those obtained for single crystals of Al_2O_3 (sapphire) implanted at room temperature, but appeared to be very similar to those measured by McHargue et al. (12) for sapphire implanted at liquid nitrogen temperature with 150 keV Fe ions. In accordance with the interpretation of their data (20), this observation may suggest that zirconia can become amorphous at ion bombardment even at room temperature. Additional arguments in favor of this hypothesis will be presented in the discussion.

3. DISCUSSION

The comparison of the distribution of the valence states of iron in ZrO_2 with similar data in different crystalline insulators; LiF (9), MgO (10), Al_2O_3 (12), SiO_2 (13), and TiO_2 (14) (see, Table 1 and Fig. 5) shows that the various matrices are represented by characteristic occurrences of valence states of iron. It has been shown (9,10) that the occurrence of various valence states of iron as a function of the concentration of implanted iron ions provides valuable information about the nature of radiation damage and induced chemical effects in any given material. In particular, in the case of LiF and MgO it was demonstrated that the probabilities of finding various iron states (FeO, Fe^{2+} and Fe^{3+}) at an arbitrary iron concentration x can be described by a binomial distribution:

$$P_N(n,x) = \binom{N}{n} x^n (1-x)^{N-n} \qquad [1]$$

where N corresponds to the number of neighboring cationic sites which can be replaced by iron and which form complexes of n iron atoms with the iron probe atom. Whether such simple statistical model applies to ZrO_2 or not, is not clear so far. In particular, the observed evolution of the Fe^{3+} in ZrO_2 can be approximately described by a binomial distribution $P_{300}(1,x)$. Thus, in terms of the model it can be inferred that the existence of Fe^{3+} is only possible if another Fe ion is not present in a zone extending to as

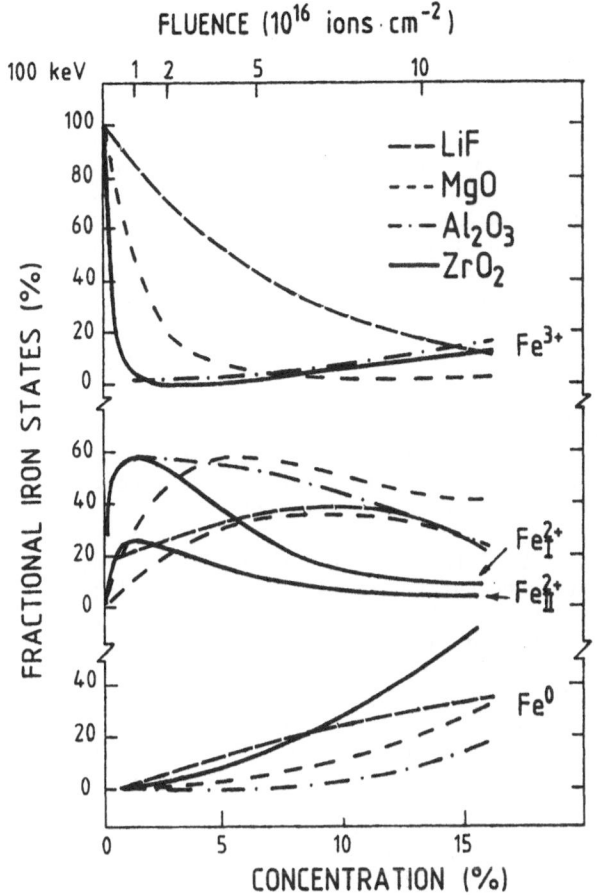

FIGURE 5. Comparison of the variation of the iron charge states as a function of the concentration of implanted iron ions in LiF (9), MgO (10), Al_2O_3 (12), and ZrO_2 (present work). The scale of fluences refers to ZrO_2.

many as 300 neighboring cationic positions. The reason for much faster evolution of charge states and aggregation of Fe in ZrO_2 compared to MgO (10) and Al_2O_3 (12) is not clear so far but it seems to be in line with the very low solubility of Fe in ZrO_2, the high density of oxygen vacancies and the high mobility of oxygen.

In comparison to other oxides studied so far and listed in Table 1, zirconia has a very different structure of distorted fluorite. It is based on the simple cubic packing of the oxygen ions with the Zr^{4+} in half the available sites with eightfold coordination and large voids in the center of the unit cell due to unoccupied positions in the simple cubic oxygen array. Iron is virtually insoluble in zirconia, whereas the large amount of vacant oxygen sites allows for high mobility of oxygen. If Fe^{2+} or Fe^{3+} is substituted for Zr^{4+}, the local charge defect must be compensated by oxygen vacancies (☐), possibly in a form of two types of complexes: Fe^{2+} - ☐ and Fe^{3+} - ☐ - Fe^{3+}. However, from an EPR study (18) and lattice energy calculations (19), it is known that Fe^{3+}, which cannot stabilize zirconia, does not modify the host lattice around it and that the oxygen vacancies should be dissociated from the substitutional Fe^{3+} ion, because the dissociated complex is more stable than the partly bound Fe^{3+} - ☐ complex by several electronvolts. On implantation, with increasing Fe concentration the system becomes more and more oxygen deficient (oxygen vacancy rich), so that the electroneutrality is more easily preserved by the creation of Fe^{2+} - ☐ complexes, in agreement with the experiment. Whether oxygen vacancies are sited adjacent to Fe^{2+} ions (substitution of a Fe^{2+} ion for a Zr^{4+} - O^{2-} complex) or are associated with the host cation sublattice (lowering the effective coordination number of Zr^{4+} ions) has yet to be resolved, for instance, on the basis of x-ray diffuse scattering, electron scattering or EXAFS analysis. By analogy with the case of Y^{3+} (0.91 Å) and Ca^{2+} (1.06 Å) substitution (21), it can be supposed that also in the case of Fe^{2+} (0.80 Å) oxygen vacancies are sited adjacent to zirconium, resulting in increased structural disorder around Zr^{4+} and lower coordination.

The question of whether or not ZrO_2 can become amorphous by ion bombardment has not been resolved in the literature to date. According to the criterion proposed by Naquib and Kelly (22), ZrO_2 is placed close to the boundary line between the oxides which do amorphize and those which do not. This boundary is defined by the ratio between the crystallization temperature and the melting temperature of the oxide, $T_C/T_m \sim 1/3$. (For ZrO_2, $T_C \sim 800$ K and $T_m \sim 3000$ K, or $T_C/T_m \sim 0.27$). The large mismatch between Pauling's electronegativities for Fe and Zr (1.83 and 1.33), as well as the preference of Fe for sixfold coordination and of Zr for eightfold coordination, seem to favor amorphization, whereas differences in ionic (crystal) radii ($R_{Fe}{}^{2+} = 0.80$ Å, $R_{Fe}{}^{3+} = 0.67$ Å, and $R_{Zr}{}^{4+} = 0.80$ Å) are not large enough to cause disorder on substitution. An additional fact favoring amorphization may be the relative structural instability of ZrO_2, since crystalline forms of zirconia can be modified by the presence of impurities, or can coexist and transform from one to another [monoclinic to face centered tetragonal (~1100°C) to cubic (~2300°C)]. The fact that the Mössbauer spectra of Fe implanted into anodic film (cubic) and polycrystalline powder (monoclinic) are very similar to each other (cf. spectra 1 and 2 in Fig. 1) may suggest that the local structure, which is reestablished upon implantation in the vicinity of the implanted Fe impurity, is identical in both cases.

Finally, we would like to comment on the possible existence of Fe^{4+} species in our specimens. Such species have recently been considered by McHargue et al. (23) in their interpretation of the data for Al_2O_3 implanted with Fe at liquid nitrogen temperature. It was postulated that the presence of high-spin Fe^{4+} ($t^3{}_{2g}d^1{}_{z^2}$) ions - situated in strongly elongated oxygen octahedra - helps to establish disordered amorphous regions in Al_2O_3. In crystalline compounds the Fe^{4+} oxidation state is characterized by an isomer shift, IS = -0.1 - 0.2 mm/s; considerably smaller than that observed for Fe^{2+} (1.2 mm/s) or Fe^{3+} (0.35 mm/s) with an oxygen sixfold coordination, and is consistent with a reduced-electron screening effect (24-26). As the analysis of our data showed, the inclusion of such an additional component considerably improved the fits of the spectra for all fluences equal to or higher than 5×10^{16} ions/cm². The abundances of the Fe^{4+} fraction, derived for various specimens, are depicted by the crossed points in Fig. 4.

Alternatively, this small additional component can be attributed to oxygen stabilized Zr^2Fe ($FeZr_2O_x$). This compound in its crystalline form has the parameters: IS = -0.12 mm/s and QS = 0.33 mm/s [Aubertin et al. (6)] or IS = -0.15 mm/s and QS = 0.24 mm/s [Vincze et al. (27)]. According to Harada et al. (7), $FeZr_2O_x$ (x ~ 0.6) forms at the Zr-Fe interface in the presence of oxygen and is stable at temperatures higher than 900°C. The implantation of additional specimens and measurements as a function of temperature should soon help to clarify this problem.

4. CONCLUSIONS

The results of this work showed that conversion electron Mössbauer spectroscopy permits the analysis of iron implanted in zirconia down to the level of 0.1 at. % Fe which is close to the concentrations of iron impurities in ceramics and reactor-grade zirconium alloys. The measurements indicated that the charge evolution of iron impurities in zirconia is qualitatively similar to that in MgO but different from that in Al_2O_3. At low fluences, corresponding to 0.1 at. %, iron tends to reside in zirconia mostly as substitutional Fe^{3+} ions dissociated from oxygen vacancies. Fe^{2+} - oxygen vacancy complexes are mostly formed at medium fluences, at iron concentrations in a range from 1 to 10 at. %. These complexes are formed in two different geometrical configurations. The concentration of metallic iron aggregates rises steadily with the total iron content and predominates above ~10 at. % Fe, where the sizes of iron aggregates become large enough to exhibit ferromagnetic properties. In addition, the measurements point to the possibility of amorphization of zirconia at room temperature as well as to the tendency for fast oxidation of Fe^{3+} at slightly elevated temperatures. The possibility of the formation of Fe^{4+} species, and their role in the stabilization of the amorphous structure is not clear as yet and has to be elucidated by additional studies.

The authors would like to thank Drs. B. D. Sawicka and J. H. Rolston of Chalk River Nuclear Laboratories, A. Perez from the University of Lyon, and C. J. McHargue from Oak Ridge National Laboratory for valuable discussions. The present work was supported by the CANDU Owners Group under the contract WPO35/6515.

REFERENCES

1. Ichinose N (ed): Introduction to Fine Ceramics, Applications in Engineering. J. Wiley, 1987.
2. Somiya, N Yamamoto, and H Yanagida (eds): Science and Technology of Zirconia III, Advances in Ceramics. Vol. 14, American Ceramic Society, 1988.
3. Franklin D and RG Adamson (eds): Zirconium in the Nuclear Industry, Proc. VIth Intern. Symposium, Vancouver-1986. Philadelphia: American Society for Testing and Materials. ASTM-STP-824, 1986.
4. Roberts JTA: Structural Materials in Nuclear Power Systems. New York: Plenum, 1981.
5. Cox B: Advances in Corrosion Science and Technology. MG Fontana and RN Staehle (eds). New York: Plenum, Vol. 5, 1976, p. 173-398.
6. Aubertin F, U Gonser, SJ Campbell, and H-G Wagner: Z. Metallkunde 76 (1985) 237.
7. Harada H, S Ishibe, R Konishi, and H Sasakura: Japan J. Appl. Phys. 25 (1986) 1842.
8. Babikova YF, AV Ivanov, VP Fillipov, and VN Abramtsev: Sov. Phys. Crystal 27 (1982) 236.
9. Kowalski J, G Marest, A Perez, BD Sawicka, JA Sawicki, J Stanek, and T Tyliszczak: Nucl. Instr. Meth. 209/210 (1983) 1145.
10. Perez A, G Marest, BD Sawicka, JA Sawicki, and T Tyliszczak: Phys. Rev. B28 (1983) 1227.
11. Guermazi M, G Marest, A Perez, BD Sawicka, JA Sawicki, P Thevenard, and T Tyliszczak: Mater. Res. Bull. 18 (1983) 529.
12. McHargue CJ, GC Farlow, PS Sklad, CW White, A Perez, N Kornilios, and G Marest: Nucl. Instr. Meth. B19/20 (1987) 813.
13. Perez, A M Treilleux, T Capra, and DL Griscom: J. Mater. Res. 2 (1987) 910.

14. McHargue CJ: Intern. Metals Rev. 31 (1986) 49.
15. Marest G: Surface and Interface Analysis 12 (1988) 509.
16. Sawicki JA, G Marest, B Cox, and SR Julian: Nucl. Instr. Meth. B32 (1988) 79.
17. Burggraff AJ, D Scholten, and BA von Hassel: Nucl. Instr. Meth. B32 (1988) 32.
18. Bacquet G, J Dugas, C Escribe, JM Gaite, and J Michoulier: J. Phys. C Solid State Phys. 7 (1974) 1551.
19. Escribe C and E van der Voort: Phys. Stat. Sol. (a) 36 (1976) 375.
20. McHargue, CJ, GC Farlow, GM Begun, JM Williams, CW White, BR Appleton, PS Sklad, and P Angelini: Nucl. Instr. Meth. B16 (1988) 212.
21. Catlow CRA, AV Chadwick, and GN Greaves: J. Am. Ceram. Soc. 69 (1986) 272.
22. Naguib HM and R Kelly: Radiat. Eff. 25 (1975) 1.
23. Marest G, CJ McHargue, and A Perez: private communication.
24. Demazeau G, B Buffet, M Pouchard, and P Hagenmüller: J. Solid State Chem. 54 (1984) 389.
25. Fournes L: Solid State Commun. 62 (1987) 239.
26. Demazeau G, Z Li-Ming, L Fournes, M Pouchard, and P Hagenmüller: J. Solid State Chem. 72 (1988) 31.
27. Vincze I, F van der Woude, and GM Scott: Solid State Commun. 37 (1981) 567.

ION IRRADIATION STUDIES OF OXIDE CERAMICS[*]

S. J. ZINKLE

METALS AND CERAMICS DIVISION, OAK RIDGE NATIONAL LABORATORY, P.O. BOX 2008,
OAK RIDGE, TN 37831-6376 USA

1. INTRODUCTION

Ion implantation is a useful tool for modifying the near-surface mechanical properties of ceramic materials. Depending on the implantation conditions, the near-surface region may become hardened or softened (amorphized) relative to the substrate (1). This may in turn exert a strong influence on the fracture toughness, strength, and wear resistance. It is now established that many ceramics can be made amorphous if the lattice damage exceeds some critical level (1–3). Burnett and Page (1) have shown that the near-surface hardness of ceramics typically increases with dose during the initial stages of implantation, then peaks and eventually shows softening due to the formation of an amorphous layer. The conditions for amorphization are material-dependent. For oxide ceramics such as MgO and Al_2O_3, amorphization has been observed in some cases when the damage energy level exceeds a value of ~5 keV per substrate atom for a room-temperature implantation (1,2). However, it is also established that chemical effects associated with the implanted ion species can have a large effect on the critical dose required for amorphization (2). The specimen temperature is also very important. The ion dose required for amorphization increases with increasing temperature (2). It is uncertain whether any ceramic can be amorphized if the implantation temperature is significantly higher than room temperature.

The fundamental processes responsible for amorphization are not well understood due to the complex interaction between chemical (implanted ion) effects and radiation damage effects. One approach that may be used to separate these effects involves the use of high-energy ions. In this case, the microstructure produced near the peak in the implanted ion distribution may be distinctly separated from the microstructure at shallower depths where only displacement damage effects have occurred. This paper presents the initial results of an investigation of the depth-dependent microstructures of three oxide ceramics following ion implantation to moderate doses. The implantations were performed using ion species that occur as cations in the target material; for example, Mg^+ ions were used for the MgO and $MgAl_2O_4$ (spinel) irradiations. This minimized chemical effects associated with the implantation and allowed a more direct evaluation to be made of the effects of implanted ions on the microstructure.

[*]Research sponsored by the Office of Fusion Energy, U.S. Department of Energy under contract DE-AC05-84OR21400 with Martin Marietta Energy Systems, Inc.

C.J. McHargue et al. (eds.), Structure-Property Relationships in Surface-Modified Ceramics, 219–229.
© 1989 by Kluwer Academic Publishers.

2. EXPERIMENTAL PROCEDURE

The starting materials for the implantation study consisted of poly-crystalline $MgAl_2O_4$ (grain size $\cong 30$ μm), two types of polycrystalline α-Al_2O_3 (grain sizes of 0.8 and 30 μm), and single crystal MgO. The 5 by 10 by 2 mm mechanically polished MgO specimens were implanted at an orien-tation ~7° from (100) in order to minimize ion channeling effects. The $MgAl_2O_4$ and Al_2O_3 specimens were irradiated as mechanically polished trans-mission electron microscope specimens (3 mm diam by 0.5 mm thickness) in a 9-specimen array. The irradiations were performed at room temperature and 650°C using the Van de Graaff facility at Oak Ridge National Laboratory (4). Table 1 summarizes the implantation conditions. The alumina implantations were performed with simultaneous beams of Al^+, O^+, and He^+ (the 0.2 to 0.4 MeV He^+ beam produced negligible displacement damage). The energies of the Al^+ and O^+ beams were chosen so that the calculated mean range of both ions in alumina was 1.2 μm. The spinel and MgO irradiations were performed using 2.4 MeV Mg^+ ions, which have a calculated mean ion range of 1.6 μm in both materials. In the following sections, the term "damage peak" refers to the calculated peak in the displacement damage profile. In the present experiment, this peak occurs at a depth that is ~0.1 μm closer to the sur-face than the mean ion ranges given above.

Table I. Ion implantation parameters

Material	Implanted Ion	Fluence	Peak Damage (keV/atom)	Irradiation Temperature (°C)
$MgAl_2O_4$	2.4 MeV Mg^+	$1.4 \times 10^{21}/m^2$	3.5	25, 650
Al_2O_3	2.0 MeV Al^+	$6.9 \times 10^{20}/m^2$	3.1	25, 650
	1.44 MeV O^+	$8.3 \times 10^{20}/m^2$		
MgO	2.4 MeV Mg^+	$4.1 \times 10^{19}/m^2$	0.10	25
MgO	2.4 MeV Mg^+	$2.1 \times 10^{19}/m^2$	0.05	650
MgO	2.4 MeV Mg^+	$8.1 \times 10^{20}/m^2$	2.0	650

Transmission electron microscopy (TEM) was performed on cross-sectioned specimens that were prepared by standard techniques (5). The depth-dependent hardness and elastic modulus of the implanted $MgAl_2O_4$ and Al_2O_3 specimens were measured at room temperature with a specialized micro-indentation ("nanoindenter") hardness tester (6). The indenter load and displacement were continuously measured during the indentation process so that depth-dependent hardness measurements could be obtained on as-irradiated specimens (6,7). Nine to fifteen indentation measurements were performed in the control and irradiated regions of each specimen. The max-imum load for each indent was ~0.12 N (12 g).

3. RESULTS

3.1 Spinel

Ion irradiation of spinel at 25 and 650°C produced faulted interstitial dislocation loops at intermediate depths (~1 μm) along the ion range. The microstructure near the implanted ion region and damage peak consisted of a high density of dislocation tangles for both temperatures. All regions of the irradiated specimens remained crystalline following both room temperature and 650°C implantation. A detailed description of the irradiated microstructure observed in these specimens is given elsewhere (7).

The cross-section microstructure of spinel following irradiation at 650°C is shown in Fig. 1. The near-surface region was devoid of any observable defects for depths up to ~0.9 µm. The damage microstructure at intermediate depths of the 650°C specimen is shown in more detail in Fig. 2. The loops exhibit fault contrast for g = <220> and no fault contrast for g = <440>. An analysis of the nature of the loops (7) revealed a mixture of interstitial loops of the type a/4<110>{110} and a/4<110>{111} which, as explained by Clinard et al. (8), are faulted on the cation sublattice.

The depth-dependent microstructure of spinel following irradiation at 25°C was very similar to that observed for the 650°C case (7), except that interstitial loop formation was also observed in the near-surface region of the specimen implanted at 25°C. The microstructures at intermediate depths and near the peak damage region of the 25°C specimen are compared in Fig. 3. Isolated dislocation loops are visible at intermediate depths whereas the peak damage region is a complex mixture of dislocations with no observable loops. A further inspection of the peak damage region reveals the presence of dislocation tangles along with other unidentified defects that exhibit contrast similar to antiphase boundaries (APBs) in ordered crystals (Fig. 4).

Nanoindenter measurements made on specimens irradiated at the two temperatures showed no significant change (<5%) in the elastic modulus. The hardness at an indent depth of 0.5 µm increased from 8.2 GPa (unimplanted) to ~8.6 GPa (implanted) for both irradiation temperatures as a result of the irradiation (7). The hardness change (implanted vs. unimplanted) of the 650°C specimen showed a gradual increase with increasing indentation depth that could be attributed to the loop-denuded region near the surface and the high defect density at deeper subsurface regions (7). The hardness increase of the 25°C specimen was essentially independent of depth for indent depths between 100 and 600 nm (7).

3.2 Alumina

The microstructure of irradiated alumina consisted of dense arrays of dislocation loops and network dislocations, with no evidence of amorphization at either irradiation temperature. The loops were observed on (0001) and {10$\bar{1}$0} habit planes. There were no obvious microstructural differences between the fine-grained and coarse-grained alumina specimens. Some preliminary observations of the microstructure of irradiated alumina are described below.

As shown in Fig. 5, the dominant microstructural feature in the midrange region of alumina irradiated at 650°C is a network of dislocations. A low density of dislocation loops with habit planes near {01$\bar{1}$0} were also observed under different diffracting conditions. An example of the loop microstructure observed near the peak damage region of a 650°C specimen is shown in Fig. 6. Edge-on loops lying on the basal plane and on the (01$\bar{1}$0) plane are visible in Fig. 6, with a loop diameter of ~20 nm. The midrange microstructure of the 650°C specimen suggests that the loop density is significantly less than that found near the damage peak (implanted ion region). The network dislocations observed at intermediate depths have yet to be observed in the peak damage region, which suggests that either dose or implanted ion effects may be significant at this temperature.

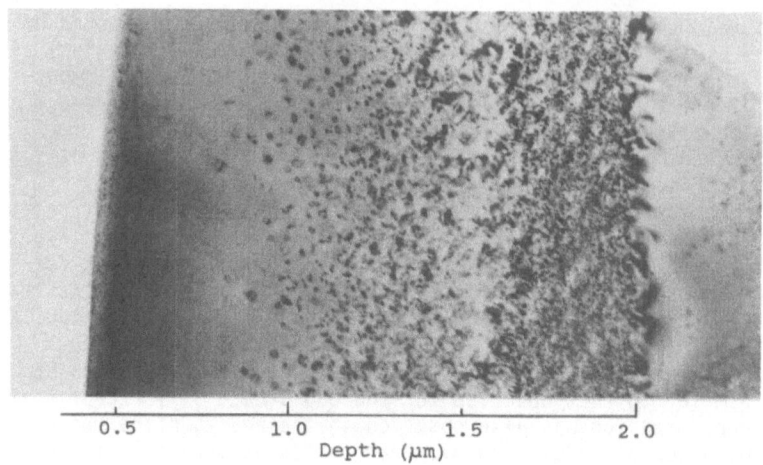

FIGURE 1. Depth-dependent microstructure of spinel implanted with
2.4 MeV Mg⁺ ions at 650°C.

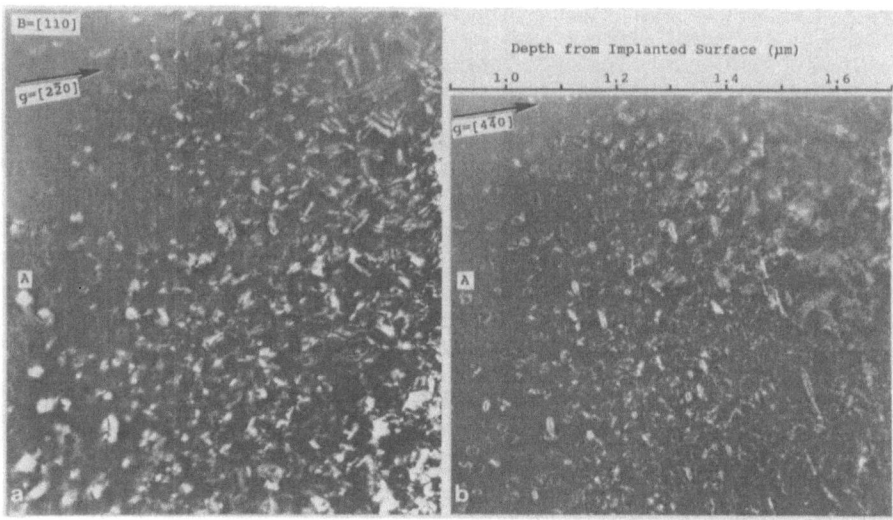

FIGURE 2. Comparison of the midrange microstructure of spinel implanted
at 650°C for two different diffraction vectors. a) weak beam (g,4g),
g = 2$\bar{2}$0. b) weak beam (g,2g), g = 4$\bar{4}$0. Note the absence of stacking
fault contrast in (b). The same area of the foil is shown for the two
different diffracting conditions.

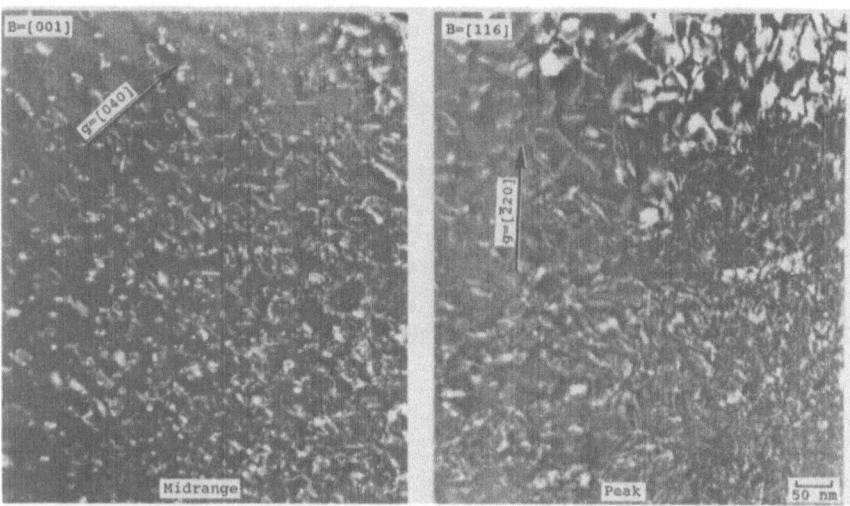

FIGURE 3. Typical irradiated microstructures of spinel implanted at 25°C. The left photograph shows the weak beam (g,3g) microstructure for damage depths of 1.0 to 1.5 μm. The right photograph shows the weak beam (g,4g) microstructure for damage depths of 1.4 to 1.9 μm. The original irradiated surface lies to the left of both photographs.

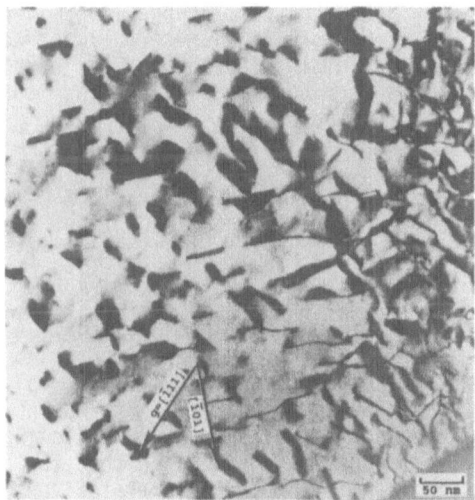

FIGURE 4. Weak beam (g,4g) microstructure of spinel implanted at 25°C at damage depths of 1.3 to 1.8 μm. The irradiated interface lies to the left of the micrograph.

224

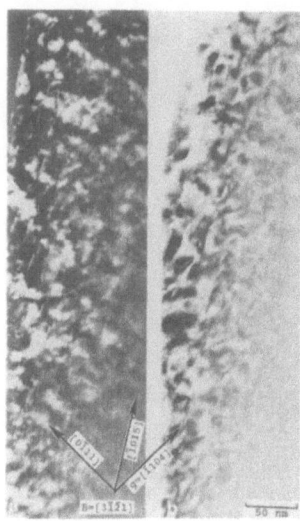 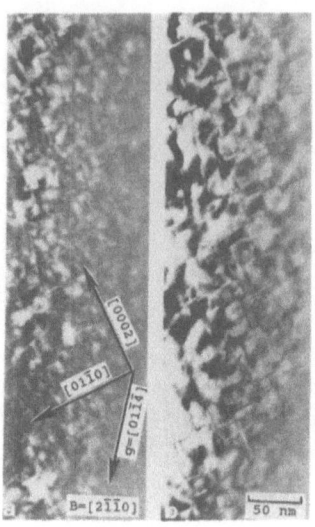

FIGURE 5. a) Weak beam (g,3g) and
b) bright field microstructure of
alumina at an intermediate depth
(~0.8 μm depth) following irradiation at 650°C.

FIGURE 6. a) Weak beam (g,3g) and
b) bright field microstructure of
alumina near the peak damage region
(~1.3 μm depth) following irradiation at 650°C.

The midrange microstructure of Al_2O_3 irradiated at 25°C was similar to
the 650°C case. Figure 7 shows the typical midrange loop microstructure
for alumina irradiated at 25°C. Loops with diameters ~10 nm were observed
on (0001) and {10$\bar{1}$0} planes. An example of a loop on the basal plane is
arrowed in Fig. 7. Dislocations were observed in both the midrange and
peak damage regions along with the loops. An example of the dislocation
microstructure observed near the peak displacement region of a 25°C specimen is shown in Fig. 8. A Burgers vector analysis of dislocations suggests
that b = 1/3<01$\bar{1}$1>. A second set of aligned dislocations were observed
with their line vectors nearly perpendicular to [0001], but their Burgers
vector has not yet been determined. Contrast similar to that associated
with APBs was observed under suitable diffraction conditions (Fig. 9).
Nanoindenter hardness measurements were performed at room temperature on
the irradiated fine-grained (d ~ 0.8 μm) alumina specimens. As shown in
Fig. 10, the elastic modulus of Al_2O_3 (determined from nanoindenter measurements) was unaffected by ion irradiation for indent depths up to 250 nm.
The nanoindenter measurements also indicated that there was no significant
(<5%) change in the depth-dependent hardness of Al_2O_3 following ion irradiation at 25°C (Fig. 11). The measurements suggested that the near-surface
hardness of alumina following irradiation at 650°C (relative to the
nonirradiated value) was increased by 10 to 20%. For the 650°C specimen,
the unimplanted hardness was about 10% less than the 25°C unimplanted

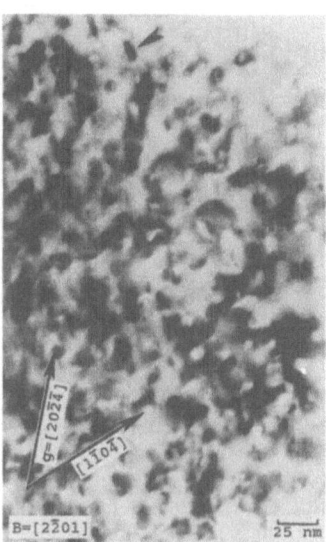

FIGURE 7. Microstructure of alumina at an intermediate depth (~0.8 μm depth) following irradiation at 25°C. The arrow points to a loop on the basal plane.

FIGURE 8. a) Weak beam (g,3g) and b) bright field microstructure of alumina near the peak damage region following irradiation at 25°C.

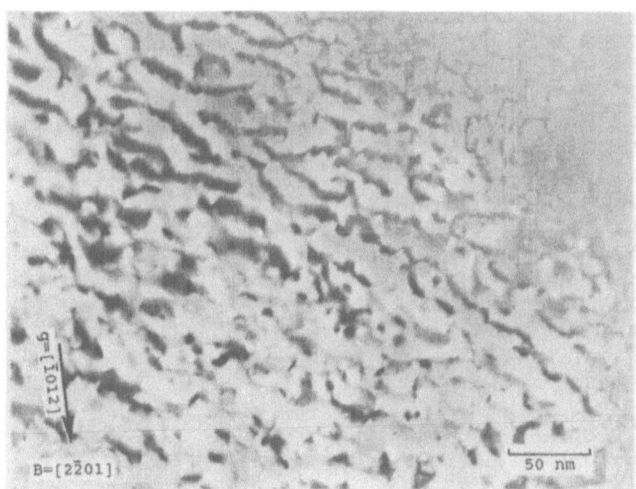

FIGURE 9. Weak beam (g,2g) microstructure of alumina irradiated at 25°C.

226

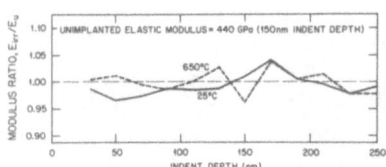

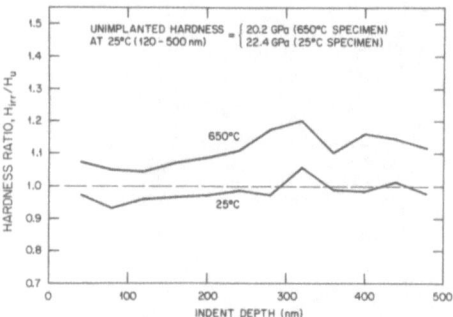

FIGURE 10. Depth-dependent elastic modulus ratio for ion-implanted Al₂O₃.

FIGURE 11. Depth-dependent hardness ratio for ion-implanted Al_2O_3.

hardness; the irradiated hardness for the 25 and 650°C specimens was essentially equal. For comparative purposes, Knoop indentation measurements were made at a load of 0.25 N and the hardness was calculated from the long diagonal of the residual indentation. The indentation depth was ~0.5 μm. The Knoop hardness measurements were in reasonable agreement with the nanoindenter results.

3.3 Magnesia

High densities of small dislocation loops were formed in MgO during irradiation at a variety of experimental conditions. Irradiation at 650°C to a fluence of 2.1×10^{19} Mg⁺/m² produced small (~10 nm diam) loops near the peak damage region. The loop density was lower and the mean loop size was larger in the near-surface region compared to the peak damage region. The damage microstructure following irradiation at 650°C to a higher fluence of 8.1×10^{20} Mg⁺/m² was qualitatively similar to that observed following the low dose irradiation. The loop density was lower in the near-surface region and the mean loop diameter changed from ~40 nm at a depth of 0.5 μm to ~10 nm at the peak damage region (1.6 μm depth). Figure 12 shows some loops that were present at a depth of ~1 μm from the irradiated surface.

There was not a large depth-dependence to the irradiated microstructure for the 25°C implantation. Figure 13 shows a cross-section view of MgO following irradiation at 25°C to a fluence of 4.1×10^{19} Mg⁺/m² (~0.7 dpa peak damage). An interesting feature observed for this moderate irradiation condition was the appearance of heterogeneous "patches" of small voids (Fig. 14). Small loops with diameters ~10 nm were also resolvable in the

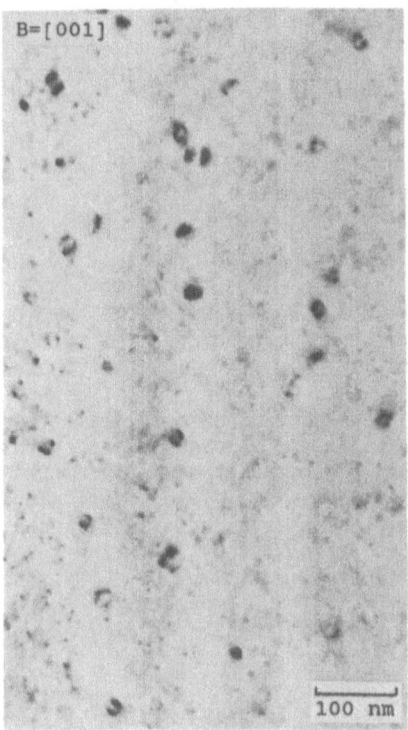

FIGURE 12. Dislocation loops observed in MgO at a damage depth of ~1 μm following irradiation at 650°C.

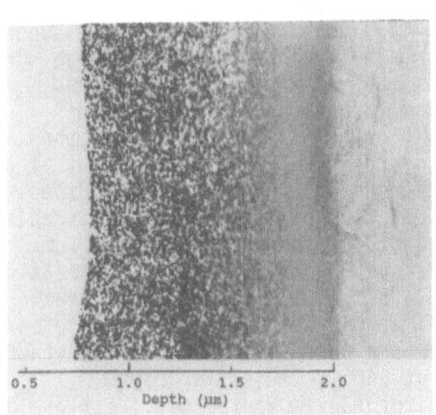

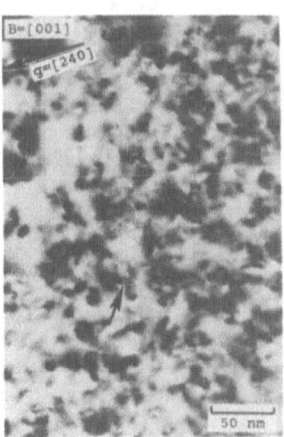

FIGURE 13. General cross-section microstructure of MgO following implantation at 25°C.

FIGURE 14. Heterogeneous formation of small voids (arrowed) at a depth of ~1 μm in MgO irradiated at 25°C.

microstructure. Knoop hardness measurements showed a large increase in
hardness for the 25°C case, and slight hardness increases for the two 650°C
cases. Complete details of the irradiation microstructure and hardness
measurements (including hardness anisotropy measurements) are given elsewhere (9).

4. DISCUSSION

4.1 General Microstructural Features

Amorphization was not detected in any of the implanted specimens. This
indicates that the critical damage level required to amorphize spinel or
alumina at room temperature by implantation with ions that minimize chemical effects is greater than 3.1 to 3.5 keV/atom. Implantation at damage
levels below 3 keV/atom with certain ion species such as zirconium has been
found to induce amorphization in alumina (2), which shows that chemical
effects can be important.

The fraction of defects retained from displacement processes in the form
of visible clusters is much lower in spinel compared to MgO and Al_2O_3. In
particular, only ~0.05% of the calculated displacements in spinel at an
irradiation depth of 1 µm result in visible defect clusters (7). Other
researchers have made similar observations (8,10) and have attributed the
resistance to damage accumulation in spinel to the structural complexity of
forming stoichiometric clusters (a minimum of 7 point defects are
required). This implies that higher implantation doses may be needed to
amorphize spinel compared to simpler oxide ceramics such as MgO or Al_2O_3.

4.2 Depth Dependence

The peak in the displacement damage profile occurs at a depth that is
close to the location of the implanted ions. It is therefore difficult to
attribute any depth-dependent microstructural changes solely to implanted
ion effects unless additional tests have been made at lower or higher doses.
On the other hand, if the observed microstructural features do not show a
strong depth dependence then it may be argued that implanted ion and displacement dose effects are of secondary importance.

The microstructure of implanted spinel exhibited a strong depth dependence
at both irradiation temperatures (Figs. 1–4). Some of the depth dependence
may be attributed to differences in the damage level — with increasing
dose, the density of loops increases until the close proximity of loops
induces interactions that could lead to the formation of a dislocation
network. However, the mechanism for this process is uncertain since the
loops in spinel contain a cation fault and are therefore not glissile. The
microstructure of alumina did not show a strong depth dependence for specimens implanted at room temperature (Figs. 7, 8). Alumina specimens irradiated
at 650°C showed a slight microstructural depth dependence, changing from
primarily aligned dislocations in the midrange region to aligned dislocations
and an array of dislocation loops on (0001) and $\{10\bar{1}0\}$ planes in the peak
damage region (Figs. 5, 6). The MgO specimens showed a weak dependence on
irradiation depth. The main observation was a decrease in loop size with
increasing depth for specimens implanted at 650°C. A similar depth
dependence of loop size has been reported for ion-implanted Al_2O_3 (11).

4.3 Temperature Effects

Irradiation at 650°C produced microstructures that were qualitatively similar to the room temperature irradiations. A general observation is that high implantation temperatures appeared to enhance the depth dependence of microstructural features. For example, spinel showed a loop-denuded zone within ~0.9 μm of the implanted surface following irradiation at 650°C whereas the 25°C specimen contained loops in the near-surface and midrange regions. Most of the loops (~70%) in spinel had {110} habit planes for the 650°C case, whereas the loops were nearly evenly divided between {110} and {111} for the 25°C implantation (7). Aligned dislocations were observed in the midrange region of Al_2O_3 following implantation at both 25 and 650°C. However, at 650°C there was a greater density of arrays of loops on (0001) and $\{10\bar{1}0\}$ in the peak damage region compared to the 25°C case. The loop density in MgO was lower for the high-temperature irradiations compared to the 25°C irradiation.

5. SUMMARY AND CONCLUSIONS

Initial microstructural observations have been made on three ion-implanted oxide ceramics. The main microstructural features associated with the implantations are dislocation loops and network dislocations. The effect on the resultant microstructure of implantation with ions that minimize chemical effects is uncertain, but the effect does not appear to be large for room temperature irradiations. Implantations at 650°C produced microstructural changes that were in general qualitatively similar to the 25°C implantation case. The high-temperature implantations appeared to enhance the depth dependence of the microstructure, although the effect was often slight.

ACKNOWLEDGMENTS

The author would like to thank C. P. Haltom, M. Williams, J. W. Jones, and A. T. Fisher for preparation of the TEM specimens, S. W. Cook and M. B. Lewis for performing the ion implantations, F. Scarboro for manuscript preparation, and P. Angelini and P. S. Sklad for manuscript review.

REFERENCES

1. BURNETT, PJ and PAGE, TF, *Radiation Effects,* 97 (1986) 123.
2. McHARGUE, CJ et al., *Materials Science and Engineering,* 69 (1985) 123.
3. OLIVER, WC et al., in *Defect Properties and Processing of High-Technology Nonmetallic Materials,* Y. Chen et al. (Eds.), Materials Research Society Symposium Proceedings, Vol. 60, Pittsburgh: Materials Research Society (1986), p. 515.
4. LEWIS, MB et al., *Nuclear Instruments and Methods,* 167 (1979) 233.
5. HORTON, LL, BENTLEY, J, and LEWIS, MB, *Nuclear Instruments and Methods, B* 16 (1986) 221.
6. OLIVER, WC, McHARGUE, CJ, and ZINKLE, SJ, *Thin Solid Films,* 153 (1987) 185.
7. ZINKLE, SJ, submitted to *Journal of the American Ceramic Society,* (July 1988).
8. CLINARD, FW, Jr., HURLEY, GF, and HOBBS, LW, *J. Nucl. Mater.* 108&109 (1982) 655.
9. LARAMIE, H, ZINKLE, SJ, and BRADT, RC, manuscript in preparation.
10. MITCHELL, TE, PASCUCCI, MR, and YOUNGMAN, RA, *Proc. 40th Annual EMSA Meeting,* (1982), p. 600.
11. KOBAYASHI, S., HIOKI, T, and KIAMIGAITO, O, *Proc. XI Int. Cong. on Electron Microscopy,* Kyoto (1986), p. 1303.

FURNACE ANNEALING OF SAPPHIRE SURFACES AMORPHISED BY ION IMPLANTATION

D.K. Sood, D.X. Cao*, L.A. Bunn and A.P. Pogany

Microelectronics and Materials Technology Centre, Royal Melbourne Institute of Technology, 124 La Trobe Street, Melbourne, Victoria 3001, Australia.

*Permanent Address: Institute of Nuclear Research, Academia Sinica, Shanghai, P.O. Box 8204, China.

1. INTRODUCTION

Surface modification of ceramics by ion implantation is being used to create novel mechanical [1,2], optical [3], chemical [4] and electrical [5] properties. The ion beam induced alterations in surface properties are brought about by one or more of the following factors – microstructure of implanted layer, residual (compressive) stress, lattice damage and defect–impurity interactions. Implantation parameters such as energy, ion dose, ion species and substrate temperature can be varied to produce surface microstructures which can be crystalline or amorphous. Heat treatment can be applied after implantation to produce further changes in the microstructure and the phases present in the implanted layer, in order to optimise its beneficial properties. Thus it is important to understand the fundamental atomic transport processes occuring during post implantation annealing.

Previous studies on ion implanted Al_2O_3 have shown, a) lattice damage accumulates to amorphise Al_2O_3 [6,7] at an optimum damage energy density, b) during annealing, amorphous Al_2O_3, prepared by stoichiometric implantation into c–axis Al_2O_3 crystals, converts first to the γ–phase which then transforms to the α–phase [6] by the outward motion of a well defined planar interface, c) implanted species either undergo a recrystallization – driven migration outwards to the surface or do not diffuse appreciably [7,8] and d) under certain conditions, a rapid isotropic diffusion of In within the amorphous Al_2O_3 layer can proceed at effective diffusion coefficients up to about 8 orders of magnitude larger than in crystalline Al_2O_3 [9].

In this work, we report on a study of annealing of amorphous surface layers produced by indium or zinc ion implantation to high concentrations (up to 45 mol % In) in a–axis or c–axis Al_2O_3. Damage free single crystal sapphire slices were implanted at liquid nitrogen temperature with 100keV In or Zn ions at doses between 0.8×10^{16} and 6×10^{16} ions/cm². The implanted surfaces were studied using Rutherford backscattering, channeling, reflection electron diffraction and scanning electron microscopy techniques. All implants produced an amorphous surface layer (60–90 nm thickness) having an abrupt interface with the underlying single crystal. The implanted crystals were annealed in a muffle furnace between 600 and 1190°C for up to 24 hours in flowing argon. Crystallization of the amorphous surface layer, diffusion and phase separation of implanted species in both crystalline and amorphous sapphire were observed. The dependence of these phenomena on ion species, ion dose, substrate orientation (a or c–axis) and annealing temperature is examined in detail, and has been shown to be quite complex.

C.J. McHargue et al. (eds.), Structure-Property Relationships in Surface-Modified Ceramics, 231–252.
© *1989 by Kluwer Academic Publishers.*

At lower doses ($<2x10^{16}$ ions/cm^2), the mode of crystallization depends strongly on the ion species, and is remarkably different from that reported reviously for intrinsic amorphous Al$_2$O$_3$ produced by stoichiometric implantation. Amorphous phase with implanted In, transforms directly to α–Al$_2$O$_3$ without any evidence of an intermediary γ–phase. Thus we also provide, to the best of our knowledge for the first time, evidence for truly epitaxial regrowth of amorphous Al$_2$O$_3$ to α–Al$_2$O$_3$ without any occurence of an intermediary γ–phase as reported previously [6] for c–axis samples. At higher doses ($>2x10^{16}$ ions/cm^2), crystallization is substantially retarded and rapid diffusion of In within the amorphous phase dominates. In is found to partially segregate as In$_2$O$_3$ particles. In contrast, Zn implantation shows quite different behaviour. These results are compared with previous published work on furnace annealing and a tentative unified model is proposed.

2. EXPERIMENTAL

Optically flat single crystals of α–Al$_2$O$_3$ (sapphire) were used after pre–annealing at 1400^{0}C in an oxygen environment for 5 days, so that these sapphire slices were damage free. The single crystals were either c–axis <0001> oriented or a–axis <1$\bar{2}$10> oriented. Indium or Zinc ion implantation was performed at 100keV energy, about 7^0 off the surface normal of the sample held at about 77^{0}K. Ion current density was less than 2μA/cm^2 and the chamber vacuum was about 5x10^{-7} Torr. Ion doses ranged from 0.8x10^{16} ions/cm^2 to 6x10^{16} ions/cm^2. A typical sapphire substrate, 13mm x 13mm square, was implanted over about two thirds of its area with the rest masked off, to serve as virgin control. After implantation, the sample was cut into three pieces, using a diamond saw. Each piece was subjected to an isothermal anneal sequence at a selected temperature. After a fixed anneal time, Rutherford Back–Scattering and Channeling (RBSC) analysis was performed and the same sample was used for further anneal times, to be followed by further RBSC analyses. The isothermal annealing was performed in flowing high purity Ar gas ambient at 600, 700, 800, 900 and 1190^{0}C. Most of the anneal times were varied as 10m, 30m, 1, 2, 3, 4, 7, 10 and 24h. The samples were analysed using the techniques of a) RBSC (2MeV incident He beam; and two detectors at scattering angles of 170^0 and 110^0 were used to get a depth resolution of 2.8nm per channel at 110^0), b) Reflection High Energy Electron Diffraction (RHEED) using a 100keV transmission electron microscope, and c) Scanning Electron Microscopy (SEM).

3. RESULTS AND DISCUSSION

The experimental data is first presented in section 3.1 for amorphous layers implanted in c–axis samples. This will be followed by results for a–axis crytals in section 3.2.

3.1 Results for c–axis Al$_2$O$_3$ crystals.

In order to appreciate the effect of implanted impurities on crystallization of amorphous Al$_2$O$_3$, it is appropriate to recall the annealing behaviour of intrinsic amorphous Al$_2$O$_3$ layers studied in detail by the Oak Ridge group [6,10].

3.1.1 The Oak Ridge model for recrystallization of intrinsic amorphous Al$_2$O$_3$.

White et al [6] employed a stoichiometric implant (two parts Al to three parts oxygen with the ion energies adjusted to give the same projected range) at liquid nitrogen temperature into c–axis Al$_2$O$_3$ single crystals to produce a 160nm thick surface layer of the amorphous phase of Al$_2$O$_3$ free from any unwanted impurities. Fig. 1 taken from their work [6] illustrates schematically the crystallization behaviour of such an intrinsic amorphous layer of Al$_2$O$_3$ deduced from a detailed RBSC and TEM study. During annealing in Ar ambient, the amorphous Al$_2$O$_3$ converts first to the γ–phase. The crystallized γ, then transforms to the α–phase by the motion of a well defined planar interface towards the free surface. Annealing at 800^{0}C for time periods as

RECRYSTALLIZATION OF AMORPHOUS Al_2O_3

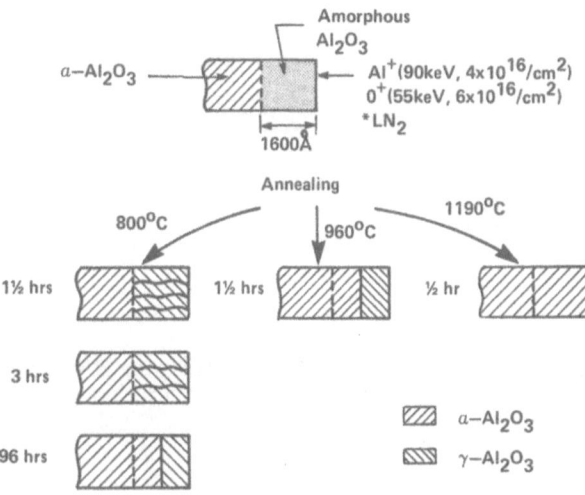

Fig. 1   Schematic representation of the crystallization behaviour of Al_2O_3 made amorphous by stoichiometric implantation into c–axis Al_2O_3 single crystal (after reference 6).

short as 1.5h converts the entire amorphous region into columnar crystallites of the γ–phase of Al_2O_3. Further heating at 800°C results in coarsening of crystallites which finally merge to produce a continuous film of γ–Al_2O_3. Fig. 2a shows the RBSC data of Farlow et al [10] for Al sublattice after annealing such an intrinsic amorphous layer. After annealing at 800°C for 3h, the aligned yield in the near surface region does not reach the random value confirming the presence of (highly oriented) columnar crystallites of γ–phase extending right through to the original amorphous–crystalline interface (as shown by similar widths of the surface peaks at 800°C and as implanted). The RBSC results of White et al [6] after annealing at 960°C for 45m are reproduced in Fig. 2b. In the Al sublattice, the amorphous–crystal interface (a–c interface) lies at a depth of 160nm for as implanted specimen. After annealing the interface has moved toward the surface, and the aligned yield in the surface region does not reach the random value. From a comparison with TEM results, they identified the near–surface region down to 110nm as γ–Al_2O_3 and the region from 110–160nm as α–Al_2O_3 shown schematically in Fig. 1.

   Their further measurements showed that the position of γ/α interface varied linearly with anneal time yielding a velocity of the interface. The measured interface velocity showed Arrhenius behaviour with an activation energy of 3.6 eV, over the temperature range of 800–1190°C. It may be noted that these results deal with one crystal orientation (c–axis) only, and the activation energy determined represents that for growth alone since nucleation sites are provided by the underlying substrate. This review of previous work will provide the necessary background for our work in the following sections.

234

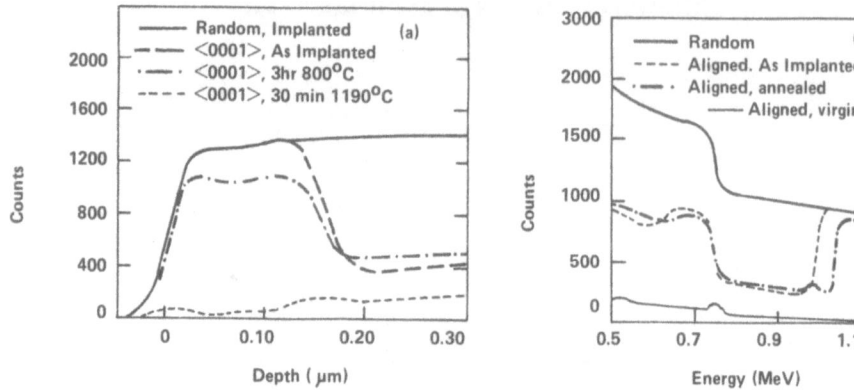

Fig. 2   RBSC results (after references 6 and 10) showing crystallization of α–Al_2O_3 made amorphous by stoichiometric implantation at 77^{0}K (a) anneal conditions, 800^{0}C, 3h; 1190^{0}C, 30m, Al sublattice only (b) annealing at 960^{0}C, for 45m.

3.1.2   The influence of implanted species on recrystallization of amorphous Al_2O_3.
This section contains our results on lower dose ($<2 \times 10^{16}$ ions/cm^2) implantation of Zn and In.

3.1.2a)   Zn implantation into c–axis Al_2O_3.   Fig. 3 shows a sequence of our RBSC data on a c–axis oriented sample implanted with 100keV Zn ions at about 77^{0}K, to a dose of 1×10^{16} Zn/cm^2; and annealed at 800^{0}C for 1, 3 and 4.5h.   Fig. 4 shows RHEED data taken from the same sample.   The as–implanted surface layer is amorphous as indicated by the absence of any structure in the RHEED pattern. RBSC data shows that the a–c interface lies at a depth of 76 nm, and zinc is randomly distributed in this layer since the aligned and random yields of Zn coincide. After annealing for 1h, the following observations can be made from Fig. 3 – a) in the Al sublattice (Al surface edge is at channel 812), the aligned yield in the near surface region drops significantly below the random yield, b) the a–c interface does not move, c) aligned yield of Zn impurity becomes less than the random yield showing the onset of substitutionality and d) the full width at half maximum (FWHM) of the Zn peak increases indicating diffusion of Zn.   As the annealing time is increased further, the implanted layer shows progressively lower aligned yield in the Al sublattice; beyond the original a–c interface, the crystal is perfect as seen by a sudden drop in the aligned yield.   While the Al_2O_3 regrows, Zn shows progressive increase in substitutionality and the FWHM of Zn distribution also increases.

After 800^{0}C 1h anneal, RHEED patterns show the presence of epitaxially regrown α–Al_2O_3 (Fig. 4c) and also of γ–Al_2O_3 with (111) planes parallel to the substrate (0001) surface (Fig, 4d).   The different patterns appear on varying the azimuth (rotating the crystal about the surface normal).   The size of the diffraction spots indicates small ($\sim$2–5nm) crystallites, suggesting a columnar or block morphology of the surface layer.   Further annealing to 4.5h produces no significant changes in the RHEED patterns

A detailed analysis of all Zn data (in addition to that shown in Figs. 3 and 4) leads to a tentative model (shown in Fig. 5) to describe the crystallization behaviour of amorphous Al_2O_3 produced by Zn implantation.   Annealing at 800^{0}C for times as

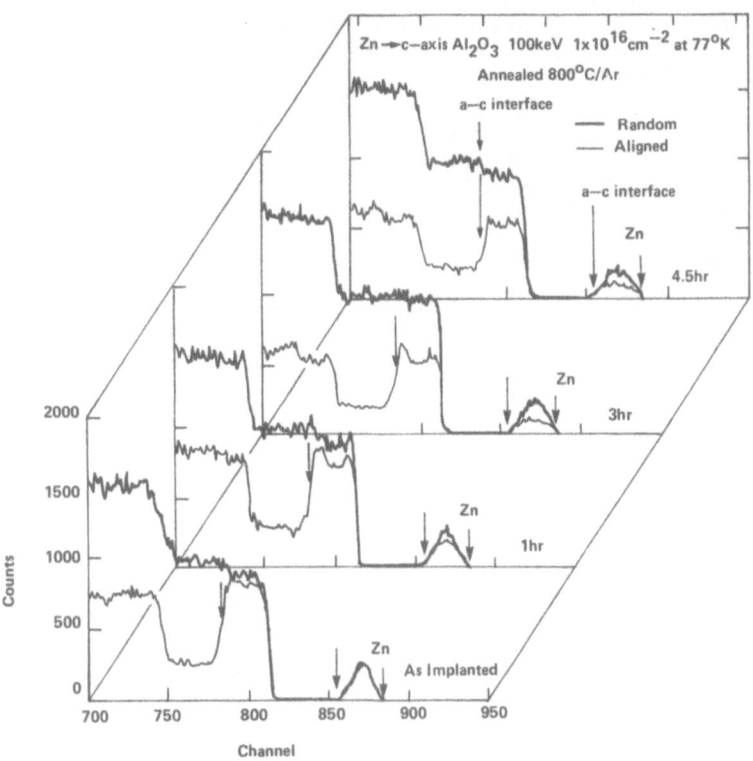

Fig. 3 RBSC spectra from c–axis sapphire implanted with Zn ions and then annealed isothermally at 800°C in Ar for indicated times

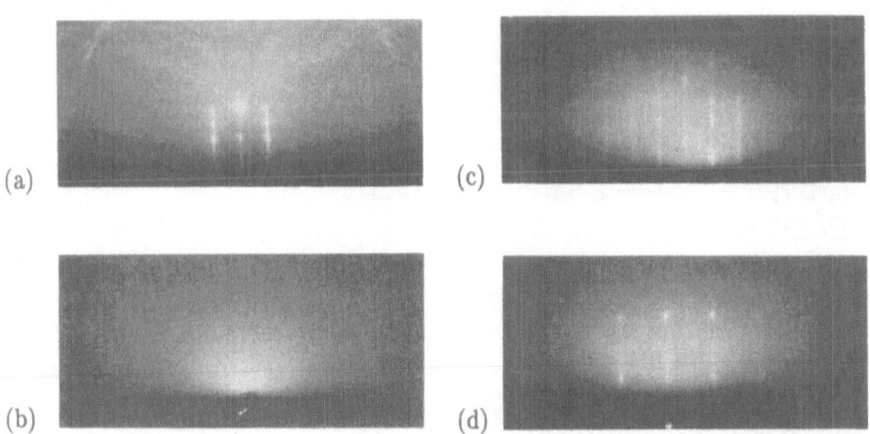

Fig. 4 RHEED patterns obtained from the sample of Fig. 3, a) virgin crystal, b) as–implanted showing amorphous pattern, c) after 800°C, 1h anneal, showing α–Al_2O_3 phase, d) same as c) but at different orientation showing γ–Al_2O_3 phase.

short as 1h converts the amorphous region into column crystallites of α–and γ–Al_2O_3. The α–Al_2O_3 is epitaxially aligned with the substrate, and the γ–Al_2O_3 is oriented with (111) planes parallel to the substrate (0001) planes. Further heating at 800^{0}C results in probable coarsening with progressively better single crystal regions. The Al sublattice part of the 3h RBSC data in Fig. 3 has a striking similarity to that in Fig. 2a for 800^{0}C 3h anneal. Thus it may be concluded that the addition of Zn to the amorphous Al_2O_3 phase leads to its conversion to columnar crystallites similar to those observed during crystallization of the intrinsic amorphous phase (Fig. 1). However these crystallites are now composed of a mixture of α and γ phases, rather than a single γ phase.

Implanted Zn profile undergoes gradual increase in FWHM from which an effective diffusion coefficient of 6×10^{-17} cm^2/s can be obtained. Zn becomes considerably sustitutional (aligned yield reduces by up to 45% of random value). There is no migration or pile up of Zn at the free surface. RHEED shows no evidence of any ZnO or Zn particles in the annealed samples. Isothermal anneal sequence (data not shown) performed on a similar sample at 900^{0}C gives very similar results consistent with the model of Fig. 5.

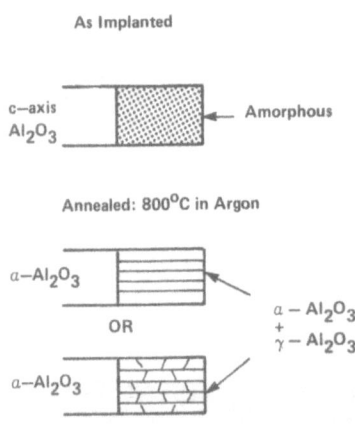

Fig. 5 Schematic representation of the crystallization behaviour of Al_2O_3 made amorphous by Zn implantation in c–axis Al_2O_3 single crystals to doses below 2×10^{16} Zn/cm^2.

Fig. 6 RBSC spectra from c–axis sapphire implanted with In ions and then annealed isothermally at 900^{0}C. The arrow marked In shows the channel corresponding to the energy for scattering from In atoms located at the free surface. Arrow also mark the depth (channel) at which a–c interface (as–implanted) is

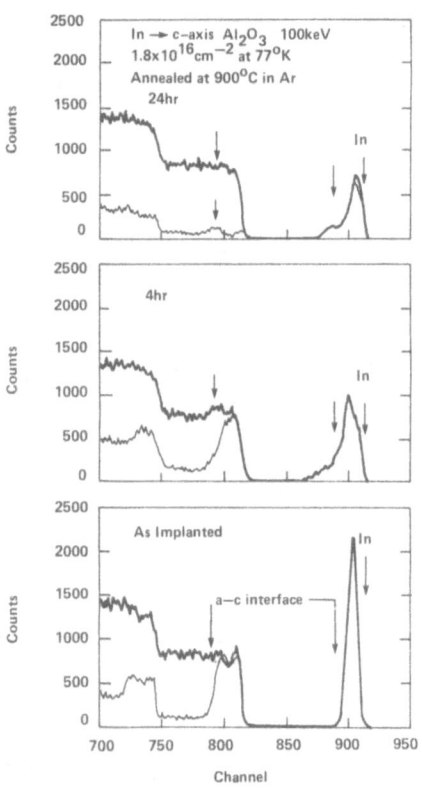

3.1.2b) **In implantation into c–axis Al$_2$O$_3$.** Fig. 6 shows a partial sequence of our RBSC data on a c–axis oriented sample implanted with 100keV In ions at about 770^{0}K, to a dose of 1.8x10^{16} In/cm^2; and annealed at 900^{0}C for selected times of 4h and 24h. The as–implanted surface layer is amorphous (confirmed by RHEED data not shown here) and the a–c interface lies at a depth of 59 nm in the as–implanted spectrum. In has nearly Gaussian distribution with its peak located at 14 nm below the surface. After annealing for 4h, the following observations can be made from Fig. 6 – a) in the Al sublattice (Al surface edge is at channel no. 812) there is clear evidence of the movement of planar a–c interface toward the surface. The aligned yield in the surface region nearly overlaps with the random yield and the RHEED data (not shown) confirms the surface layer is amorphous. The width of the surface damage peak has reduced. In redistributes with preferential migration towards the free surface and also inwards to the a–c interface. Similar trends continue as the anneal time is increased and after 24h anneal (Fig. 6), the implanted layer is regrown to a single crystal. There is a small damage peak at a depth close to the original a–c interface characteristic of 'end–of–range' extended defects. More In has migrated towards the surface and a buried peak develops at a depth corresponding to the a–c interface. This indicates that some In atoms are segregated around the 'end–of–range' extended defects. There is considerable loss of In (as shown by the reducing area under the In peak in Fig. 6) and the residual In is non substitutional.

Thus it may be concluded that the addition of In impurity to the amorphous Al$_2$O$_3$ phase dramatically alters the mode of its crystallization in that the amorphous phase converts directly to α–Al$_2$O$_3$ by the motion of a well defined planar interface towards the free surface. This behaviour is in contrast to that of Zn described in the previous section and perhaps emphasises the 'chemical' role of the implanted species.

3.1.2c) **Annealing at 1190^{0}C** for as short a time as 30m drives out the implanted Zn and In entirely as seen in Fig. 7 by the complete absence of any counts at surface edge positions of In and Zn. The implanted crystal regrows to near virgin perfection in the Al sublattice except for small damage peaks remaining at the surface and at the 'end–of–range' depth coinciding with the as–implanted a–c interface position for both In (Fig. 7a) and Zn (Fig. 7b) implanted samples. However, the oxygen sublattice does not regrow as well in either case. The inset in Fig. 7 contains RHEED patterns for the same samples. The RHEED patterns show good regrowth to single crystal epitaxial α–Al$_2$O$_3$. Detailed comparison of Kikuchi lines show a slight loss of the contrast in comparison with the virgin crystal indicating very slight residual damage in the epitaxial layer.

3.1.3 **Influence of Dose.** The impurity effects on crystallization are strongly dose dependent. Fig. 8 illustrates the case for Zn implants at doses of 1x10^{16} Zn/cm^2 and 6x10^{16} Zn/cm^2, after anneal at 900^{0}C in Ar for 0.5h. The as–implanted layers are amorphous at both the doses. After annealing, the lower dose sample exhibits conversion of the entire amorphous layer into non planar epitaxial growth of α–Al$_2$O$_3$ and probably γ–Al$_2$O$_3$ crystallites as described in Fig. 5. However, at the higher dose, crystallization is almost completely suppressed as shown by RHEED data (inset in Fig. 8) before and after annealing, and also confirmed by the RBSC data. The RHEED data show no detectable ZnO phase.

Zn diffuses within the implanted layer at both doses. From the observed increase in FWHM, we estimate a diffusion coefficient of 2.9x10^{-15}cm^2/s for diffusion in crystalline Al$_2$O$_3$ at 900^{0}C (for dose of 1x10^{16} Zn/cm^2); and a diffusion coefficient of 5.3x10^{-15} cm^2/s for diffusion in the amorphous phase of Al$_2$O$_3$ at 900^{0}C (for a dose of 6x10^{16} Zn/cm^2). More Zn diffusion data are discussed in section 3.3.

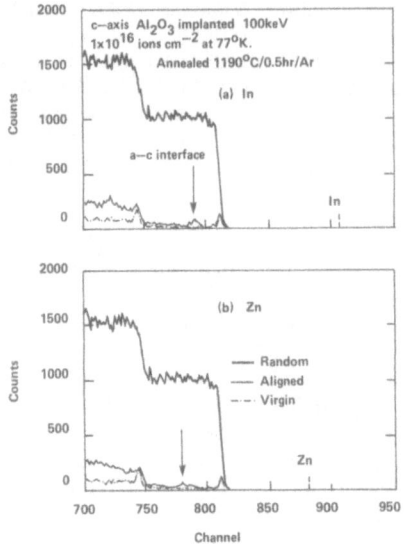

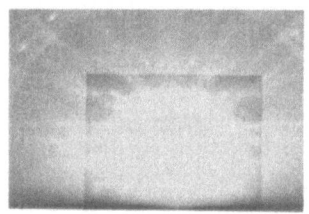

Fig. 7   Effect of annealing at 1190°C
shown by the RBSC spectra from
c–axis sapphire implanted with
1x10¹⁶ ions/cm² of a) In and b)
Zn. The inset shows RHEED
patterns taken from the same
samples and from a virgin region.

3.2   Results for a–axis Al$_2$O$_3$ crystals.

Indium implantation at 77°K into a–axis sapphire has been studied in detail.
Implantation to peak concentrations of 6–45 mol% In produces amorphous surface
layers. Isothermal annealing in Ar ambient between 600–900°C shows effects strongly
dependent on ion dose. At lower doses <2x10¹⁶ In/cm² (section 3.21), the amorphous
layer undergoes epitaxial regrowth as the amorphous to crystalline interface advances
out towards the surface. Regrowth velocity is high in about the first half hour of the
anneal; thereafter it shows a marked decrease. Regrowth obeys Arrhenius behaviour
with an activation energy of 0.7eV for initial faster growth and 1.28eV for further
anneal times.   The amorphous phase transforms directly to α–Al$_2$O$_3$ without any
evidence of an intermediary γ–phase.   At higher doses, epitaxial regrowth is
substantially retarded and rapid diffusion of In within the amorphous phase
dominates, as discussed in section 3.2.2.

3.2.1   In implantation in a–axis sapphire at doses <2x10¹⁶ In/cm².   Fig. 9 shows a
sequence of RBSC spectra on an a–axis Al$_2$O$_3$ sample implanted with 100keV In ions
at about 77°K to a dose of 8x10¹⁵ In/cm²; and annealed at 900°C for 10m, 1h and 2h.
The as–implanted surface layer (top spectrum in Fig. 9) is amorphous and the a–c

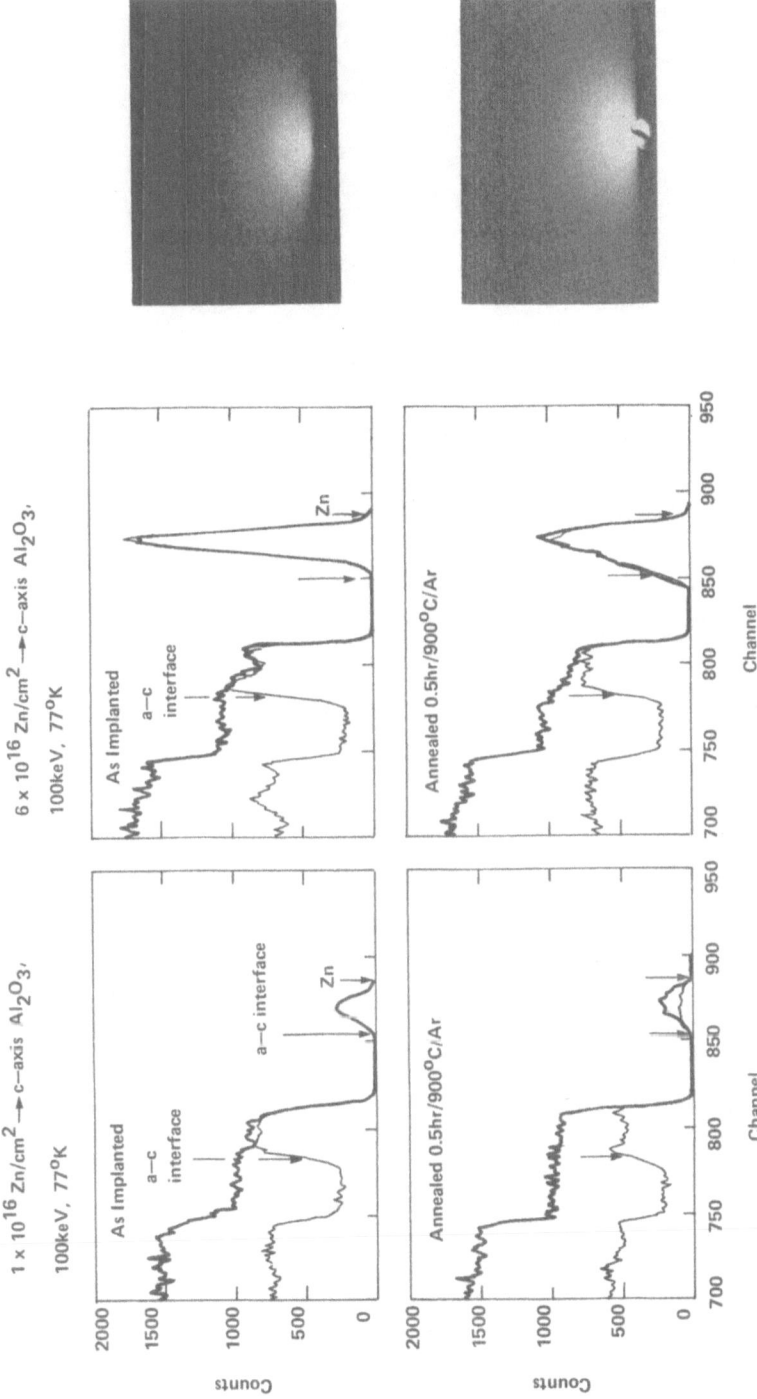

Fig. 8 Effect of dose on crystallization of Zn implanted into c–axis Al_2O_3. RBSC spectra before (first row) and after anneal at $900^{\circ}C$ for 0.5h (second row). The RHEED patterns are shown for 6×10^{16} Zn/cm^2 implanted sample.

240

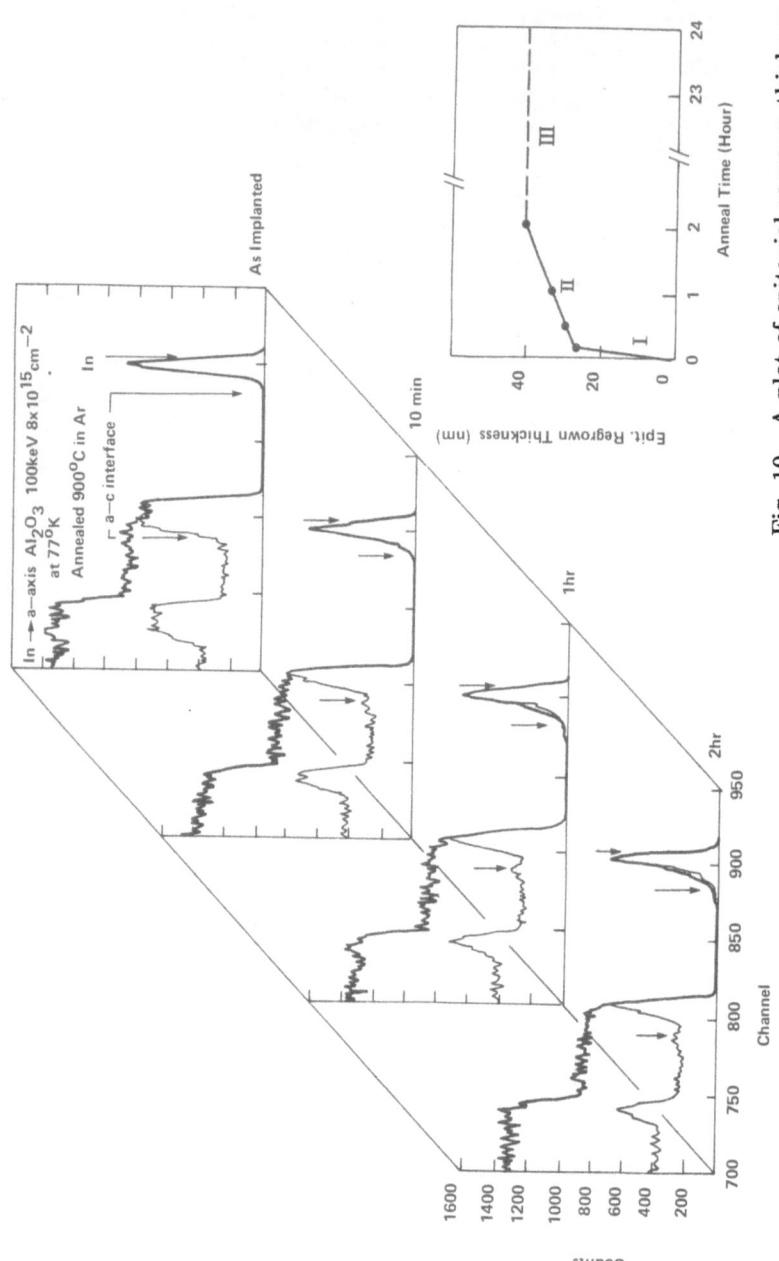

Fig. 9 RBSC spectra from an a–axis sapphire sample implanted with In ions and subsequently annealed isothermally at 900°C in Ar for indicated times.

Fig. 10 A plot of epitaxial regrown thickness vs. anneal time at 900°C for an a–axis Al₂O₃ sample implanted with In to a dose of 8x10¹⁵ In/cm² at 100keV, at about 77°K.

interface lies at a depth of 61nm. After 10m anneal, there is clear evidence of epitaxial regrowth of the as–implanted amorphous layer, as the a–c interface has moved toward the surface. As the anneal time is increased, the a–c interface moves progressively outwards to the free surface and the width of the amorphous layer reduces. The 2h anneal produces a change in crystallisation mode as dechanneling yield drops below random yield without any appreciable change in thickness of the layer. Fig. 10 shows a plot of the regrown layer thickness versus anneal time. Three regions can be clearly identified. Regions I and II show epitaxial regrowth which is faster in region I. Region III shows non–planar epitaxial transformation of amorphous phase into α–Al_2O_3. There is no evidence of any γ–phase formation from RHEED results. Region III persists up to 24h anneal as seen in Fig. 10. Very similar behaviour is observed for annealing at other temperatures 600, 700 and 800^{0}C with the extent of the three regions varying somewhat, e.g. 600^{0}C data shows no region III (see Fig. 14). Growth velocities (determined e.g. from slope of the curves of Region I and II in Fig. 10) are plotted in Fig. 11 as function of inverse of temperature T. They exhibit an Arrhenius behaviour and data points are fitted with activation energies of Q_I=0.7eV and Q_{II}=1.28eV for the two regions of epitaxial regrowth. The solid line is based on the Oak Ridge group data [6] on near epitaxial regrowth of intrinsic amorphous layers produced by stoichiometric implantation of O and Al ions. Thus the addition of In impurity ions to amorphous Al_2O_3 enhances growth velocity because of a clear reduction of the activation energy from 3.6eV for intrinsic amorphous Al_2O_3 to 0.7eV. From a detailed analysis of data at the four temperatures, we conclude that the effect of In induced enhancement in growth rate is dose dependent. As In dose is increased, the epitaxial growth is retarded. The transition in Fig. 10 from region I to II occurs as the c–a interface arrives at a depth where In concentration is high enough to appreciably slow down the epitaxial growth rate. Region III would commence as still higher In concentration is encountered (at depths closer to In range) wherein the epitaxial mechanism breaks down. For example, the depth at which region III commences in Fig. 10 is 25 nm from the surface, whereas the observed range of In is 24nm with straggling width of 11nm. The onset of region III occurs at a specific volume concentration of In and is temperature dependent, e.g. about $0.6x10^{21}$ In/cm^3 at 700^{0}C and about $1.0x10^{21}$ In/cm^3 at 900^{0}C. Similar implanted impurity concentration dependent effects have been reported for epitaxial regrowth of amorphous Si on underlying single crystal substrates [11]. The above mentioned observations are summarised in a schematic representation of recrystallization of Al_2O_3 amorphised by In implantation in the first column of Fig. 17.

3.2.2 Diffusion in Amorphous Al_2O_3 at doses >$2x10^{16}$ ions/cm^2.

In this high dose regime, epitaxial regrowth is substantially retarded and rapid diffusion of In within the amorphous phase dominates. Previously we published [9] a first report on such rapid diffusion observed during annealing at 600^{0}C of the amorphous surface layers produced by In implantation. We now present a more detailed study at several temperatures.

3.2.2a) Diffusion of In in amorphous Al_2O_3.

Fig. 12 shows a sequence of RBSC spectra on an a–axis Al_2O_3 sample implanted with 100keV In ions at about 77^{0}K to a dose of $6x10^{16}$ In/cm^2; and subsequently annealed at 600^{0}C for 1, 5, 7 and 24h. The as–implanted amorphous layer is 80nm thick as pointed by the arrow on the Al sublattice (in Fig. 12) marking the a–c interface. The RHEED pattern in the inset of Fig. 12 shows that the surface is amorphous. No significant change in the amorphous layer thickness is seen after a 1h anneal, but the In peak broadens significantly. As the anneal time is increased, the a–c interface moves out towards the surface with an

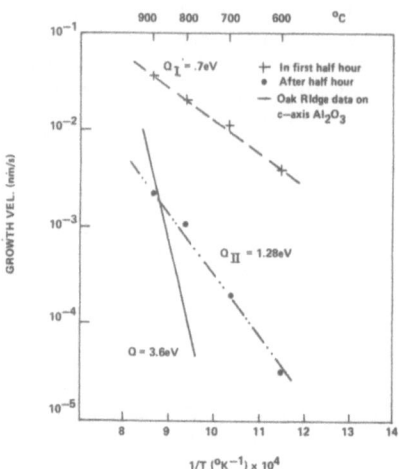

Fig. 11 An Arrhenius plot for a–axis Al_2O_3 samples implanted at 77^{0}K with 0.8–1.8x10^{16} In/cm^2 at 100keV. + for points in region I in first half hour of anneal, • Region II after half hour anneal, – Oak Ridge data for intrinsic amorphous Al_2O_3 produced by stoichiometric implantation (after ref. 6).

effective growth velocity of about 7x10^{-4}nm/s. This growth rate is about 4 orders of magnitude higher than that reported previously [6] for intrinsic amorphous Al_2O_3 on the c–axis sapphire. Thus the presence of In in the amorphous Al_2O_3 enhances the crystallization rate near the a–c interface. This behaviour is identical to that seen earlier for the lower dose samples discussed in section 3.2.1.

The corresponding In depth profiles undergo progressive broadening without any shift in the peak position. The square of the diffusion distance obtained from the change in FWHM of the In profile is plotted in Fig. 13 as a function of anneal time. This clearly shows rapid initial diffusion with $D_1 \sim 8.3 \times 10^{-16}$ cm^2/s for the first 5 hours of anneal, followed by a somewhat slower diffusion coefficient $D_2 \sim 1.3 \times 10^{-18}$ cm^2/s for longer anneal times. The initial rapid diffusion may result from radiation damage induced enhancement or from the chemical effect of In impurity atoms. These values are still about 8 orders of magnitude larger than those ($\sim$10^{-25} cm^2/s) obtained by an extrapolation to 600^{0}C of self (Al ion) diffusion coefficients measured by previous workers [12] in crystalline α–Al_2O_3. Thus we provide clear evidence of very rapid isotropic diffusion of an implanted species in amorphous Al_2O_3.

RHEED data (insets in Fig. 12) obtained on the same samples provides phase information as follows – a) the as–implanted layer is amorphous, b) 600^0, 24 hours anneal produces fine crystallites of In_2O_3 (all rings in inset of Fig. 12 index to In_2O_3), c) no evidence of polycrystalline Al_2O_3 and d) no evidence of precipitates of In or Al is seen.

For lower doses, no detectable In migration or lattice reordering is observed as shown in Fig. 14 for 1x10^{16} In/cm^2 (about 8 mol % In peak concentration), under an identical isothermal anneal sequence. However, formation of In_2O_3 proceeds in a similar manner (RHEED patterns in Fig. 14). Once again no evidence of formation of any other crystalline phase was found.

243

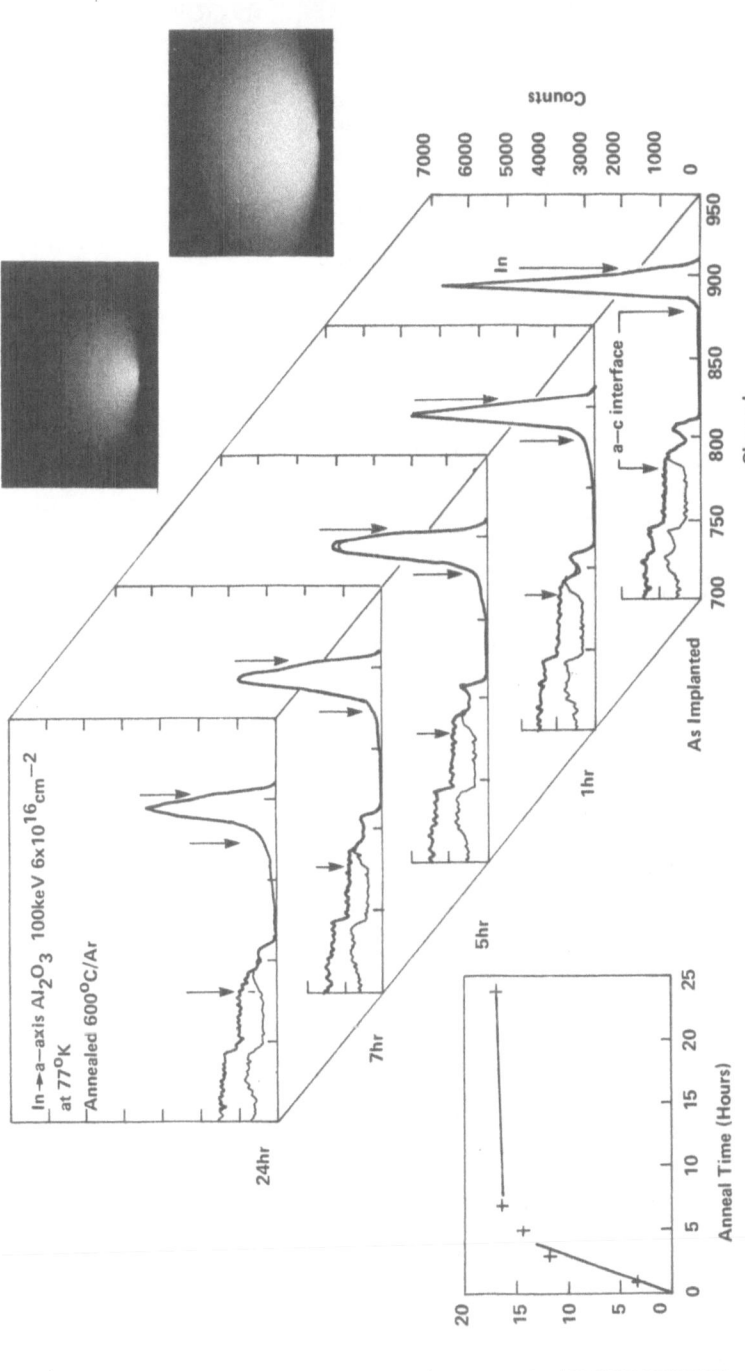

Fig. 12　A sequence of RBSC spectra from an a–axis Al_2O_3 sample implanted with In ions and subsequently annealed isothermally for indicated times. RHEED patterns are shown in the inset for as–implanted and 24h anneal.

Fig. 13　Plot of the square of the diffusion distance (obtained from the data in Fig. 12) vs. anneal time.

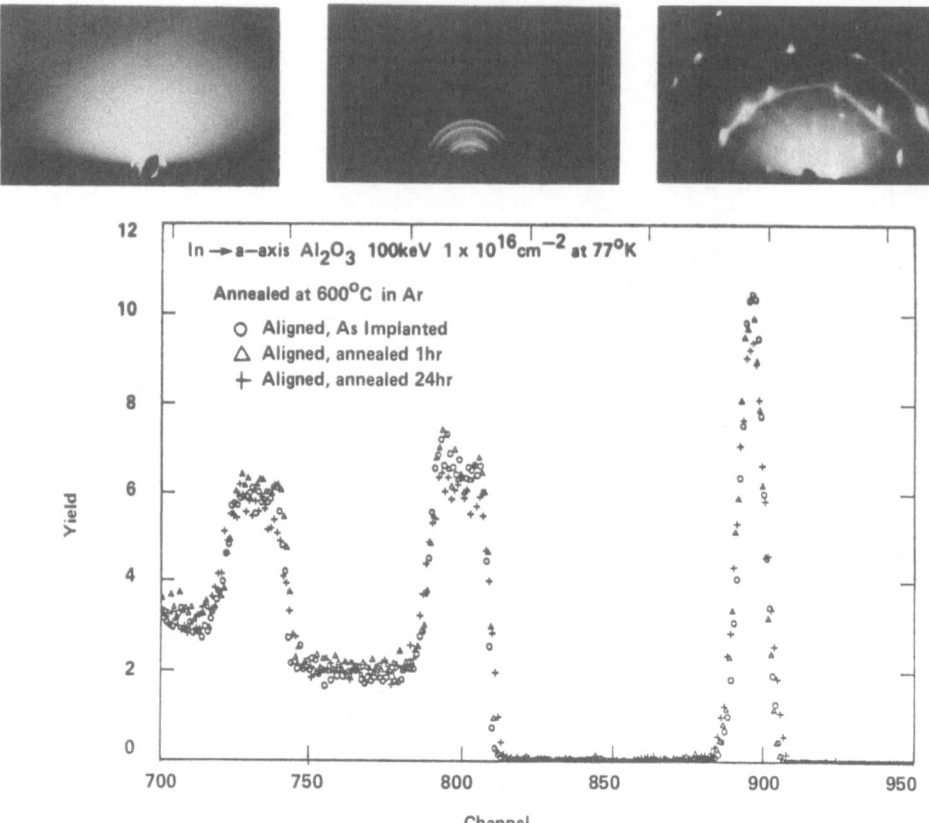

Fig. 14   RBSC spectra for an 1×10^{16} In/cm^2 implanted a–axis sapphire –
as–implanted, and after 1h and 24h anneals at 600^{0}C. RHEED data in the
inset is for (from the left) as implanted, 24h anneal and virgin crystal
respectively.

The effect of annealing a sample implanted with 6×10^{16} In/cm^2 for half hour at
700^{0}C is shown in Fig. 15. Epitaxial regrowth and rapid diffusion of In within the
amorphous layer occur simultaneously. The a–c interface moves out towards the
surface (Fig. 15a) and stops after advancing by about 28nm. The as–implanted profile
of In redistributes with two clear components (Fig. 15b) – I (shaded area)
corresponding to rapid diffusion within amorphous layer and – II a surface pile up
peak of In with a tail. Further annealing up to 24h produces no noticeable change in
the In distribution and the implanted layer remains amorphous. The RHEED data
(Fig. 15b) obtained on the sample after 24h anneal shows a family of well defined
rings which all index to In$_2$O$_3$ phase. Analysis of RHEED patterns indicates that
small (<10nm diameter) particles of In$_2$O$_3$ constitute the surface layer. There is no
evidence for any precipitates of In or crystalline Al$_2$O$_3$. The RHEED pattern in Fig.
15a confirms that the as–implanted layer is amorphous. The observed broadening
(shaded region in Fig. 15b) in component–I corresponds to an effective diffusion

coefficient of 1.04×10^{-14} cm^2/s for In in amorphous Al$_2$O$_3$, which is just over one order of magnitude higher than our previously reported [9] value of 8.3×10^{-16} cm^2/s at 600^{0}C.

At higher anneal temperatures, similar In redistribution behaviour continues, except that the relative amount of component–I diminishes as more and more In comes out to the surface to form In$_2$O$_3$. The migration of In gets faster, since all the observed migration is found to be completed within the first short anneal of 15 minutes at 800^{0}C and 10 minutes at 900^{0}C. Further annealing up to 24h produces no more change in the In profile. The underlying amorphous layer, however, regrows into an imperfect crystal. The details are presented elsewhere [13].

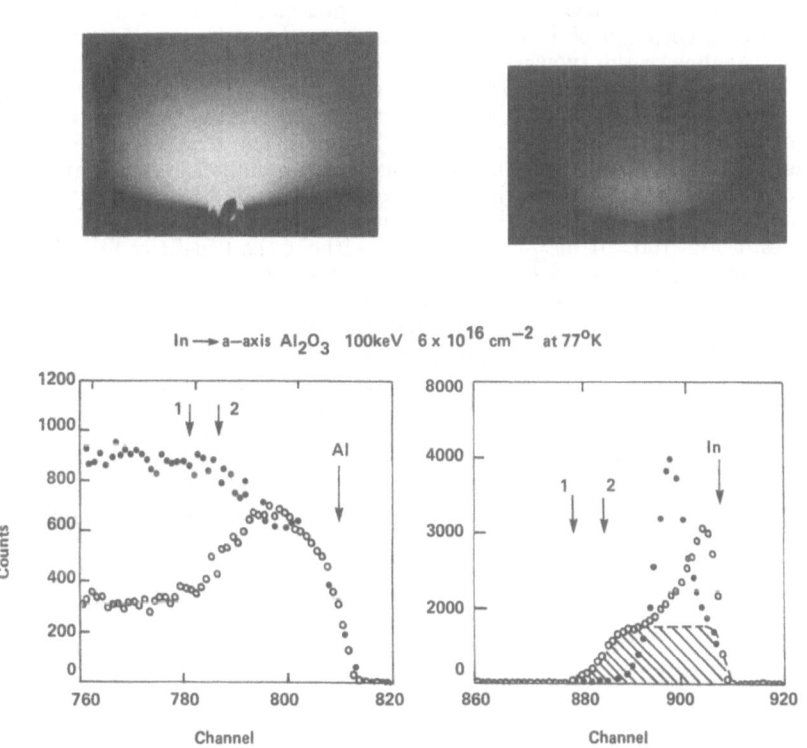

Fig. 15   RBSC spectra showing the effect of annealing a sample implanted with 6x10^{16} In/cm^2 for half hour at 700^{0}C. a) Al sublattice, random (•), aligned (o); b) In region, random (o), the as–implanted (•) distribution shown for comparison, In both figures, a–c interface positions are indicated – before annealing (1) after annealing (2).   Corresponding RHEED patterns shown inset are (a) as implanted (b) after annealing.

The surface topography of several annealed samples has been examined by SEM. Some selected results are shown in Fig. 16, for a–axis Al_2O_3 samples implanted with 6×10^{16} In/cm^2 (Fig. 16 a–c) and for a lower dose (1×10^{16} In/cm^2) sample (Fig. 16d). The surface of the as–implanted sample (Fig. 16a) is structureless. After annealing at 600^{0}C for 24h, the same surface develops (Fig. 16b) several bright islands. A detailed EDX analysis using a fine spot electron beam (~20nm diam), reveals that these bright islands are rich in In and the dark structureless regions contain much smaller amounts of In (similar to those present on the as–implanted surface in Fig. 16a). When annealing is performed at a higher temperature of 700^{0}C, the island structure is different (Fig. 16c) with much higher number density and smaller sizes of islands. The total surface coverage with the islands is now much greater. In contrast, the lower dose sample after annealing at 600^{0}C for 24h shows (Fig. 16d) a complete absence of In rich bright islands on the surface. By combining the complementary information obtained from RBSC, RHEED and SEM measurements on the same sample surface, we can identify these bright islands as In_2O_3 particles. For example, by comparing Fig. 16b with results of the 24h spectrum in Fig. 12 (both of which use the same sample), it can be seen that the presence of In at the surface and the presence of a deep tail of In extending well beyond the (as–implanted) a–c interface are consistent with the typical RBS spectrum expected from such an island structure [14]. When the impurity layer is laterally non–uniform, the tail in the RBS spectrum arises from a superposition of the yield from many islands of varying dimensions present within the area sampled by the helium beam used for RBS measurement (0.6mm diam in our experiment). Thus the maximum depth of penetration of the tail is a rough indication of the maximum thickness of islands. In Fig. 12 (24h spectrum), the tail extends to a depth of about 500nm which is of the order of the size of bright particles in Fig. 16b. It may be noted that without the benefit of SEM data, the tail in the RBS spectrum could be mistaken for 'anomalous' diffusion as occurred in our previous report [9]. However, the volume fraction of In trapped in the bright In_2O_3 particles may be quite small and therefore the diffusion measurements of Fig. 13 are representative of distribution of In in dark regions of Fig. 16b.

The as–implanted spectrum of Fig. 12 has no In at the surface and no tail on the In peak. The corresponding SEM micrograph (Fig. 16a) is completely devoid of any bright particles in agreement with the above discussion. Fig. 16c shows occurrence of a network of In_2O_3 particles and the corresponding RBSC spectrum (similar to that shown in Fig. 15b) does indeed exhibit a surface accumulation of In and a tail on the surface peak component II. Fig. 16d shows no In_2O_3 particles and the corresponding RBSC spectrum (Fig. 14) does not have any surface segregation of In or any tail on In peak. The RHEED data (Fig. 14 inset) however, show rings characteristic of In_2O_3 after annealing. Thus in this case, the In_2O_3 particles are embedded in the amorphous Al_2O_3 phase. In the other cases the In particles grow on the free surface and would be projecting out, (shown schematically in Fig. 17) so that RHEED patterns would show only characteristic In_2O_3 rings even at higher tilts, since the self shadowing of the particles would make them appear like a continuous layer.

The source of oxygen needed for formation of In_2O_3 is not clear, at present. The Ar gas used was of high purity grade. Oxygen could have come from one or more of the following – traces of moisture or oxygen in Ar gas used, an internal reduction reaction with Al_2O_3 or due to brief exposure to air when the sample was still hot. However, it may be noted that Zn implanted c–axis Al_2O_3 samples annealed under similar conditions do not form any detectable oxides of Zn, even though RBSC shows Zn migration to the surface (e.g. see Fig. 18).

6 x 10^{16} In / cm^2 1 x 10^{16} In/cm^2

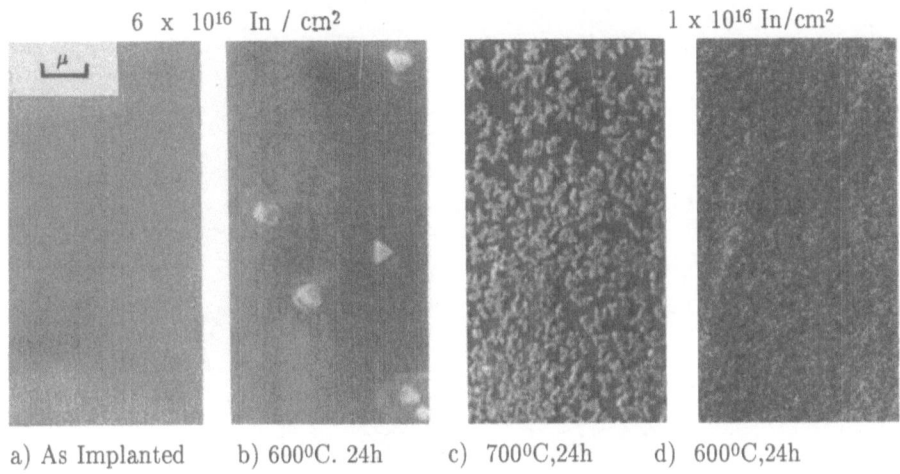

a) As Implanted   b) 600^{0}C. 24h   c) 700^{0}C,24h   d) 600^{0}C,24h

Fig. 16   SEM pictures showing surface topography of several a–axis Al$_2$O$_3$ samples

3.2.2b   <u>Temtative model for recrystallization of a–axis Al$_2$O$_3$ amorphised by In implantation.</u> A detailed analysis of all data (some in addition to that shown in this work) for In implantation into a–axis Al$_2$O$_3$ leads to a tentative model proposed in Fig. 17. The implantation of In at liquid nitrogen temperature produces an amorphous surface layer (about 60–80nm thickness) with an abrupt interface with the underlying a–axis single crystal. Ion range distribution is schematically shown by a curve within the amorphous layer in Fig. 17, which has two columns separating low dose ($<2 \times 10^{16}$ In/cm^2) and high dose ($>2 \times 10^{16}$ In/cm^2) regimes. The concentration of In at the a–c interface is very small. In the high dose regime, the thickness of amorphous layer is larger and so is the peak concentration.

Let us consider the low dose regime first. As isothermal annealing is started, the amorphous phase crystallizes epitaxially to α–Al$_2$O$_3$ at a well defined planar a–c interface which moves out towards the free surface. The growth velocity depends on the concentration of In in the amorphous phase. Since the concentration of In is depth dependent, three regions of growth velocity can be identified. At very low In concentrations, ('Region I'), the growth velocity is enhanced due to the 'chemical' effect of In in the amorphous phase. As the a–c interface moves towards the surface, it encounters progressively increasing concentration of In. The growth rate decreases, possibly due to In segregation or phase separation (within the amorphous phase) effects. This is called 'Region II'. As a–c interface advances still further, the In concentration reaches a value greater than a critical value above which the mode of crystallization itself alters from a planar to a non–planar epitaxial regrowth. This marks the beginning of 'Region III'. In this region there is no further movement of

RECRYSTALLIZATION OF Al_2O_3 AMORHISED

BY INDIUM IMPLANTATION

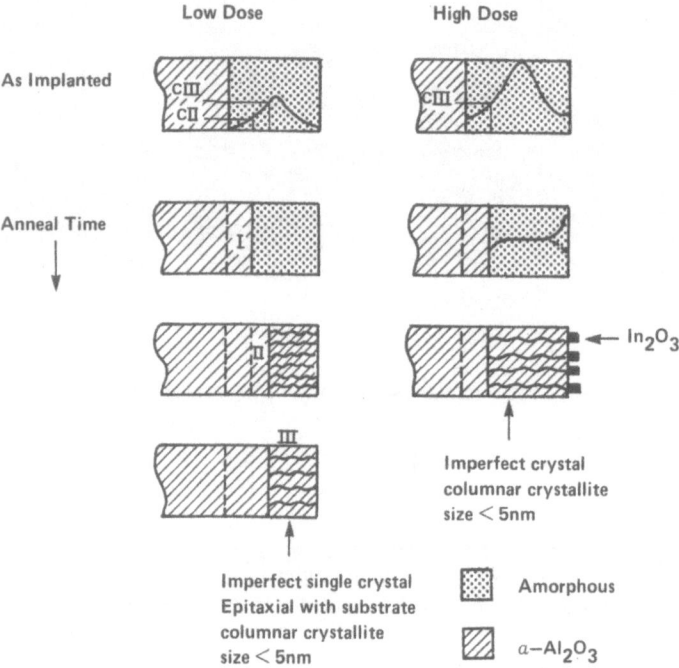

Fig. 17   Schematic representation of the recrystallization behaviour of a–axis Al_2O_3 made amorphous by In implantation.

the interface but perhaps columnar crystallites of $\alpha-Al_2O_3$ are produced which orient themselves during further annealing. Region III persists up to 24h annealing, and finally it becomes an imperfect single crystal epitaxial layer. The average columnar crystallite size is estimated <5nm.

This model operates within the temperature regime of 600° – 900°C. The extent of the three regions varies somewhat at different temperatures, e.g. 600°C data does not show any region III. The growth velocities show Arrhenius behaviour with widely different activation energies for the two regions, $Q_I = 0.7eV$ and $Q_{II} = 1.28eV$, for planar epitaxial regrowth. The onset of region III occurs at a specific concentration of In and is temperature dependent, e.g. about 0.6×10^{21} In/cm³ at 700°C and about 1×10^{21} In/cm³ at 900°C. This threshold concentration is presumed to be the same for both low and high dose regimes as shown in Fig. 17. However, the depth at which the transition takes place is larger for the higher dose implants.

In the high dose regime, rapid diffusion of In within the amorphous phase occurs simultaneously and competes with the epitaxial regrowth. As the a–c interface advances,In is redistributed within the amorphous layer ahead of the a–c interface. In diffusion within amorphous Al_2O_3 is up to about 8 orders of magnitude more rapid than in crystalline Al_2O_3. This causes rapid flattening of the In concentration profile and segregation of In at the free surface. Eventually region III sets in and the amorphous phase converts to α–Al_2O_3 by non–planar epitaxial mechanism. Rather large and numerous particles of In_2O_3 are formed at the free surface due to In segregation to high concentrations. Thus there are two clearly different In components – I which redistributes within the amorphous phase and II which segregates at the free surface.

This model operates well at all temperatures between 600–900°C. At higher anneal temperatures, the relative amount of component I diminishes as more In segregates to the surface.

The models for the effects of In (Fig. 17) and Zn (Fig. 5) are very different, and essentially emphasise the 'chemical' role of implanted impurities in affecting the transformation from amorphous to α–Al_2O_3 phase. Previous work [15] on transformtion of γ–Al_2O_3 to Θ–Al_2O_3 to α–Al_2O_3 suggested the critical role of impurities (or dopants). It was found that as cationic radius of the impurity increased, the transformation temperature decreased. Fink [16] reported that above 700°C, even 1% addition of V_2O could greatly accelerate the kinetics of $\gamma{\to}\alpha$ Al_2O_3 transformation, bypassing the Θ–Al_2O_3. Since the Oak Ridge model [Fig.1] suggests a${\to}\gamma{\to}\alpha$ as the transformation route bypassing the Θ phase for transformation of intrinsic amorphous Al_2O_3, it could very well be that the role of implanted impurities is to suppress the γ phase to a varying degree. If we pressume that Zn suppresses γ phase partly and In entirely, the three tentative models of crystallization in Fig. 1, Fig. 5 and Fig. 17 become quite consistent as far as the change of the mode of crystallization is concerned. Table I lists some useful data on the atomic species involved in the present experiments for assisting such comparisons.

TABLE 1. Some data on the atomic species involved in the present work

Element	Ionic Radius (charge state) nm	Valence	Electro–negativity	Melting Point °C	Boiling Point °C
Al	0.5(+3)	3	1.5	660	2450
In	0.8(+3)	3	1.7	156	2000
	1.3(+1)	1			
Zn	0.74(+2)	2	1.6	420	906
O	1.4(−2)	−2	3.5	−219	−183

3.3 <u>Diffusion of Zn in amorphous Al_2O_3.</u> Fig. 18 shows RBSC spectra on a c–axis Al_2O_3 sample implanted with 100keV Zn ions at about 77°K to a dose of 6×10^{16} Zn/cm^2; and subsequently annealed at 800°C for 3h in Ar. The as–implanted amorphous layer is about 80nm thick. After annealing, the layer remains amorphous as confirmed by the RHEED pattern. Zn has undergone rapid diffusion within the amorphous layer and has attained a nearly flat top distribution. Note the arrows marking a–c interface and Zn surface within which Zn is confined. There is no detectable loss of Zn. From the observed increase in FWHM of Zn, we estimate an effective diffusion coefficient of 1.6×10^{-15} cm^2/s at 800°C. This may be compared with the value of 5.3×10^{-15} cm^2/s at 900°C evaluated in section 3.1.3 from Fig. 8 at the same dose of Zn. Thus the diffusion scales with temperature qualitatively well. However, the diffusivities in crystallized Al_2O_3 as evaluated earlier in sections 3.1.2a and 3.1.3 are 6×10^{-17} cm^2/s at 800°C and 2.9×10^{-15} cm^2/s at 900°C. These values also scale well with temperature increase but their absolute value is much higher than those ($\sim 10^{-22}$ cm^2/s) obtained by an extrapolation to 800°C of self (Al ion) diffusion coefficients measured by previous workers [12]. This difference might be explained by the presence of columnar crystallites in the crystallized Al_2O_3 (Fig. 5), which may provide diffusion pathways along grain boundaries.

It is intriguing to note that RHEED (e.g. Fig. 18) shows no evidence of formation of any oxide of Zn even though RBSC shows Zn migration up to the surface.

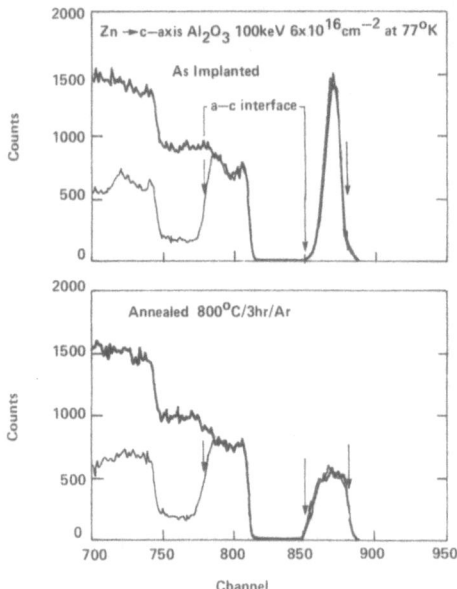

Fig. 18 RBSC spectra from a c–axis Al_2O_3 sample implanted with Zn ions, before and after anneal at 800°C for 3h. Inset shows a RHEED pattern taken after annealing.

4. CONCLUSIONS

In this work, we report on a study of annealing of amorphous surface layers produced by indium or zinc implantation to high concentrations (up to 45 mol % In) in a–axis or c–axis Al_2O_3. All implants produced an amorphous surface layer (60–90nm thickness) having an abrupt interface with the underlying single crystal. Crystallization of the amorphous surface layer, diffusion and phase separation of implanted species in both crystalline and amorphous sapphire were observed after annealing in argon at temperatures between 600°C and 1190°C. The dependence of these phenomena on ion species, ion dose, substrate orientation (a or c–axis) and annealing temperature is examined in detail, and has been shown to be quite complex. Implanted impurities in the amorphous phase not only alter the kinetics but also the mode of crystallization. These effects are strongly dependent on the impurity ion species and concentration.

At lower doses ($<2x10^{16}$ ions/cm^2), the mode of crystallization is remarkably different from that reported previously for intrinsic amorphous Al_2O_3 produced by stoichiometric implantation [6]. Our conclusions are presented separately for each species :

4.1 Zinc Implantation
4.1.1 Low dose regime ($<3x10^{16}$ Zn/cm^2):
1. The addition of Zn to the amorphous Al_2O_3 phase leads to its conversion to columnar crystallites similar to those observed during crystallization of the intrinsic amorphous phase (Fig. 1). However, these crystallites are now composed of a mixture of α and γ phases, rather than a single γ phase (Fig. 5).
2. Implanted Zn undergoes diffusion within the crystallised layer with anomalously large diffusion coefficients, perhaps due to the presence of columnar crystallites which may provide diffusion pathways along grain boundaries.
3. Zn becomes considerably substitutional (aligned yield reduces by up to 45% of random value).
4. There is no migration or pile up of Zn or formation of ZnO at the free surface.
4.1.2. High dose regime (~6x10^{16} Zn/cm^2):
1. Crystallization is inhibited
2. Zn diffuses rapidly within the amorphous surface layer and attains a nearly flat top depth distribution.
3. There is no loss or surface segregation of Zn.
4. Zn remains atomically dispersed within the amorphous layer and no phase segregation is observed either into Zn or ZnO.

4.2 Indium Implantation

4.2.1 Low dose regime ($<2x10^{16}$ In/cm^2)
1. The amorphous phase transforms directly to $\alpha-Al_2O_3$ without any evidence of an intermediary γ–phase.
2. Amorphous surface layer undergoes epitaxial regrowth at a well defined planar amorphous to crystalline interface which advances out towards the surface.
3. Regrowth velocity is retarded as the concentration of In increases.
4. Regrowth obeys Arrhenius behaviour with an activation energy of 0.7eV for initial faster growth and of 1.28eV for further anneal times.
5. Above a critical In concentration (about 0.6x10^{21} In/cm^3 at 700°C and about 1x10^{21} In/cm^3 at 900°C) epitaxial regrowth mechanism breaks down.

4.2.2 High dose regime ($>2x10^{16}$ In/cm^2):
1. Epitaxial regrowth is substantially retarded or even inhibited and rapid diffusion of In within and out of the amorphous phase dominates.

2. The effective diffusion coefficients for In in amorphous Al_2O_3 are at least 8 orders of magnitude larger than those obtained by an extrapolation of self (Al ion) diffusion coefficients reported in literature in crystalline $\alpha-Al_2O_3$.
3. In segregates at the surface to produce islands of In_2O_3.
4. In migration is completed within the shortest anneal times used (~10m). No further redistribution of In is observed up to 24h anneal times.
5. Tentative model is proposed to explain these observations.

5. ACKNOWLEDGEMENTS

This work was performed under an Australian Research Grants Scheme funding. We thank V. Sood for SEM measurements and Carl J. McHargue for stimulating discussions.

REFERENCES

1. C.J. McHargue, Nucl. Instrum. Meth. B19/20, 797 (1987), and references therein.
2. P.J. Burnett and T.F. Page, in 'Ceramic Surfaces and Surface Treatments', eds. R. Morrell and M.G. Nicholas (Brit. Cer. Soc., 1964) p.65.
3. H. Crawford, Nucl. Instrum. Meth. B1, 159 (1984).
4. G.K. Wolf and K. Roessler, in "Ion Beam Modification of Insulators', eds. P. Mazzoldi and G.W. Arnold (Elsevier, Amsterdam, 1987) p. 558.
5. R. Meaudre and A. Perez, Nucl. Instrum. Meth. B32, 75 (1988).
6. C.W. White, P.S. Sklad, L.A. Boatner, G.C. Farlow, C.J. McHargue, B.C. Sales and M.J. Aziz, Mat. Res. Soc. Symp. Proc. 60, 337 (1986).
7. A.P. Mouritz, D.K. Sood, D.H. St.John, M.V. Swain and J.S. Williams, Nucl. Instrum. Meth. B19/20, 805 (1987).
8. M. Ohkubo, T.Hioki and J. Kawamoto, J. Appl. Phys. 60, 1325 (1986).
9. D.X. Cao, D.K. Sood and A.P. Pogany, in Mat. Res. Soc. Symp. Proc. 100, 113 (1988).
10. G.C. Farlow, P.S. Sklad, C.W. White, C.J. McHargue and B.R. Appleton, Mat. Res. Soc. Symp. Proc. 60, 387 (1986).
11. J.S. Williams and J.M. Poate in 'Ion Implantation and Beam Processing' (Acad. Press, New York, 1984), Chapter 2, p.11.
12. A.E. Paladino and W.D. Kingery, J. Chem. Phys. 37, 957 (1962).
13. D.K. Sood and D.X. Cao, Mat. Res. Soc. Symp. Proc., December 1988, to be published.
14. 'Backscattering Spectrometry', W.K. Chu, J.W. Mayer and Marc–A. Nicolet (Academic Press, New York, 1978) Appendix E.
15. D.S. Tucker and J.J. Hren, Mat. Res. Soc. Proc. 31, 337 (1984).
16. G. Fink, J. Inorg. Nucl. Chem., 30, 59 (1968).

THE MECHANICAL PROPERTIES OF ION IMPLANTED CERAMICS

CARL J. MCHARGUE

Metals and Ceramics Division, Oak Ridge National Laboratory,
P.O. Box 2008, Oak Ridge, TN 37831-6118

1. INTRODUCTION

Ceramic materials generally fracture without appreciable plastic deformation under relatively low tensile stresses. Compressive strengths are
much higher and brittle materials can be often used in applications where
applied stresses are compressive in nature. The low strength in tension is
usually caused by the presence of surface flaws that grow into brittle fracture by the Griffith mechanism. The useful strength of these materials may
be raised by surface treatments which remove the flaws, reduce their
severity, increase the energy to propagate a crack, or generate a residual
compressive stress in the near-surface region.

Ion implantation introduces impurity or alloying atoms into the surface of
a sample in a controlled and reproducible manner. The injected ions lose
energy by elastic collisions and electronic interactions with the target
atoms. The elastic collisions can cause a cascade of atomic displacements
from the host lattice sites so that large numbers of point defects are
generated. As a result of the implantation, the strength of the host may be
increased by both alloying effects (solid solution or second-phase
strengthening) and radiation damage effects (defect strengthening). Both
processes will also contribute to formation of a residual compressive stress
state in the near-surface region.

There is a significant body of literature that describes recent attempts
to employ ion implantation to modify the near-surface structure of ceramics
and therefore the near-surface mechanical properties in a controlled manner.
Because the implanted layer is thin (a few tenths of a micron thick),
accurate determination of its properties is difficult and many of the data
describe the properties of a composite consisting of the implanted layer and
a portion of the substrate region. From such measurements, it is often
possible to deduce the direction of property changes, if not quantitative
values.

The changes in mechanical properties reflect changes in microstructure
and/or residual stresses. The effect of implantation on the microstructure
will first be summarized to form a basis for subsequent analysis of selected
properties.

2. MICROSTRUCTURE OF IMPLANTED CERAMICS

For a detailed discussion of the microstructural and phase changes induced
by ion beam treatments of ceramics, see the papers by Thevenard (1) and
Whitton (2) in this volume.

Implantation produces a wide range of non-equilibrium microstructure such
as high concentrations of point defects, supersaturated solid solutions, or
entirely new phases. The microstructure of implanted ceramics depends upon
the implantation parameters of fluence, ion species, substrate temperature,
and the material parameter of chemical bonding type (3-7). Many of the

C.J. McHargue et al. (eds.), Structure-Property Relationships in Surface-Modified Ceramics, 253–273.
© *1989 by Kluwer Academic Publishers.*

structural or property changes can be normalized with regard to the fluence-dependence by using deposited energy density (8,9) rather than ions per square centimeter.

The microstructure of as-implanted materials depends upon the damage retained after dynamic recovery processes annihilate most of the defects produced in the collision cascade and the remaining defects are rearranged into metastable configurations such as dislocation loops, stacking faults, three-dimensional voids, etc. At low fluences or low temperatures where recovery is suppressed, defects accumulate as the fluence is increased. If recovery is sufficiently inhibited, a concentration of defects may be reached where the long-range order of the crystal lattice is destroyed and an amorphous state is produced. There is evidence that some implanted species are more effective than others in stabilizing the disordering defects, leading to a "chemical" effect in addition to the damage energy effect (5). The temperature at which significant recovery occurs is governed primarily by the type of chemical bonding present; directional covalent bonds being more difficult to reform than ionic bonds.

The microstructure of oxide ceramics after implantation with relatively low fluences or in the temperature range where dynamic recovery prevents amorphization is characterized by point defect clusters (bounded by dislocation loops and dislocations). Figure 1 is a micrograph taken by transmission electron microscopy (TEM) of a back-thinned specimen of Al_2O_3 implanted with 2×10^{16} Cr/cm^2 (280 keV) (ref. 7). The micrograph contains a high density of "black spots" typical of point defect clusters. The large residual stress in the TEM foil prevented a determination of the character of the defects in this particular sample, but it is likely that they are dislocation loops bounding stoichiometric interstitial clusters of aluminum and oxygen faulted with respect to the cation (Al) lattice (10). Zinkle (11) identified dislocation loops on $\{0001\}$ and $\{10\overline{1}0\}$ habit planes after stoichiometric implants of Al + O.

Some oxides may become amorphous due to damage accumulation after high fluence implantation at room temperature or at lower fluences at 77 K (ref. 6,9). A fluence of chromium corresponding to 3 dpa (deposited energy density of about 2.25×10^{22} keV/cm^3) produces an amorphous surface if the implantation is carried out at 77 K (6), whereas a chromium fluence corresponding to 600 dpa (deposited energy density of $6-8 \times 10^{23}$ keV/cm^3) is required at 300 K (ref. 9).

The covalent-bonded ceramics SiC and Si_3N_4 are easily amorphized at 300 K (ref. 7,8,12,13). Figure 2 is a TEM photograph taken in cross-section of α-SiC implanted with 2×10^{16} Cr/cm^2 (280 keV) at room temperature. The surface is amorphous to a depth of 250 nm. The fluence corresponds to a damage production of 18 dpa in the Si-sublattice. The critical damage level is between 0.2 and 0.3 dpa (a damage energy density of $2-10 \times 10^{21}$ keV/cm^3) at this temperature, two orders of magnitude lower than for Al_2O_3. The critical damage energy density for Si_3N_4 is about a factor of two higher than that for SiC (ref. 9).

Implantation at elevated temperatures (~1050 K) where dynamic recovery is rapid does not produce the amorphous state in SiC for damage levels as high as 16 dpa (14-16).

In general, precipitates of second phases have not been observed in as-implanted oxide or carbide ceramics although there has been recently reported evidence for the formation of metallic iron clusters (~2 nm in size) in iron-implanted sapphire (17). Post-implantation annealing alters the microstructure as the non-equilibrium phases attempt to reach equilibrium. There may be several competing processes which may effect the mechanical properties in different ways. For example, the radiation-induced

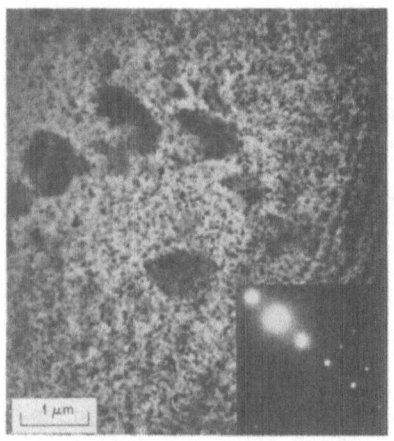

FIGURE 1. Transmission electron micrograph showing "black spot" damage characteristic of point defect clusters in Al_2O_3 implanted with 2×10^{16} Cr/cm^2 (280 keV). Insert contains the selected-area electron diffraction pattern.

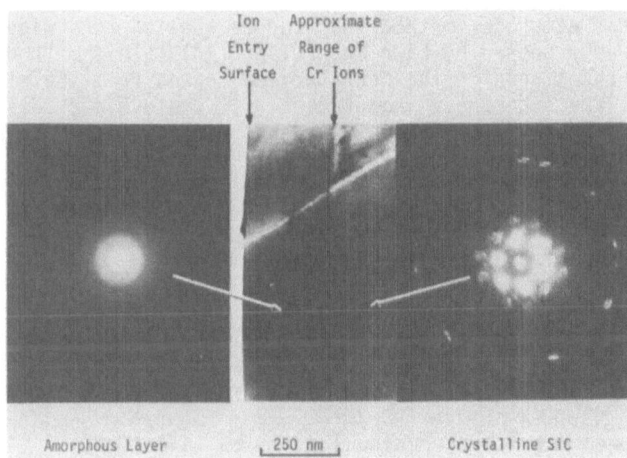

FIGURE 2. TEM of a cross section of an α-SiC crystal implanted with 2×10^{16} Cr/cm^2 (280 keV). The selected area electron diffraction patterns show the surface to be amorphous to a depth of 250 nm.

defects will be annihilated and thus remove the radiation or defect strengthening, but these changes may be countered by the precipitation hardening due to the formation of additional phases.

In summary, the implanted microstructure of ceramics may consist of a crystalline but highly disordered matrix containing individual charged point defects, defect clusters, and dislocations as loops or tangles. The phase structure may consist of a metastable solid solution (substitutional or interstitial) or have transformed to an amorphous state.

3. RESIDUAL STRESS

Both the injection of the implanted ions and the creation of point defects cause a volume increase in the implanted region. Since the material is free to expand only in one direction (normal to the free surface) the constraints of the substrate to hold the lateral dimensions constant produces a biaxial compressive stress in the implanted region. Because the major source of the volume expansion is the defect production, EerNisse proposed that the lateral stress varies with distance from the surface as the deposited damage energy (18). The integrated stress is then the force per unit width acting between the implanted layer and the substrate and results in bending of a sample implanted only on one side. The maximum compressive stress (T_{max}) is the integrated lateral stress (S) divided by the thickness of the damaged region (d).

If the implantation-produced residual compressive stress is large enough, it will affect the mechanical properties by increasing the applied stress necessary to place the stressed surface into tension, thus reducing the probability of propagating a pre-existing flaw or by affecting the crack opening stress field.

The residual stress can be measured by a number of techniques. X-ray diffraction techniques have been used with semiconducting materials (19). Attempts to employ these techniques with ceramics have been less successful due to the poor perfection of the starting material or the low absorption coefficient of materials such as Al_2O_3 and SiC which allow deep penetration of the x-ray beam and thus a high noise-to-noise ratio. EerNisse (18) developed a method to measure the bending of a cantilever beam during implantation and calculated the integrated stress from formulae for simple beam bending.

An alternate, though less accurate, method that can be applied to a wide range of brittle solids was proposed by Lawn and Fuller (20). This method deduces the magnitude of the stress in the near-surface region from the size of cracks produced by a Vickers hardness indentation. It employs a stress intensity formulation for a penny-like (radial) crack system subjected to a constant stress over a thin (relative to the crack depth) surface layer. The surface stress, σ_s, is given by

$$\sigma_s = \frac{K_c \left[1 - (c_0/c)^{3/2}\right]}{2 \, \phi \, d^{1/2}}$$

where K_c = fracture toughness of an unstressed specimen, c = crack length in the presence of the stress, c_0 = crack length in the unstressed surface, d = the thickness of the stressed layer, and (ϕ) is a crack geometry term of value about unity.

A summary of the published results on residual stresses in ceramic materials is given in Table 1 (revised from ref. 21). All the reported stresses are compressive in nature. In general, values determined by the indentation technique are lower than those given by the cantilever beam method. One of the difficulties of applying the indentation technique to ion implanted surfaces is the uncertainty in determining the thickness of the stressed layer. Burnett and Page (22) used a value of four times the calculated peak of deposited damage energy. McHargue (14) used the depth of visible damaged microstructure determined from cross-sectioned TEM samples.

The results listed in Table 1 were obtained for a wide range of implantation conditions. An attempt to normalize the data was made by examining the stresses as functions of the total deposited damage energy (per unit area) since it is this portion of the energy that produces defects (23). This term is designated as TDE or "energy density" (22). Total damage energy (TDE) is defined as (S_D x Φ) where

$$S_D = \int S_D(x) \, dx$$

as given by Eq. (19) in the EDEP-1 calculations (30) and Φ is fluence.

TABLE 1. Residual stress measured in ion implanted ceramics (after ref. 21)

Material	Implanted species	Fluence (ions·cm⁻²)	Energy (keV)	Temperature	Method	Integrated stress (S)	Maximum stress (T)
Al_2O_3							
c-axis	Cr	1×10^{16}	280	RT	Indent.	0.2×10^{-3} MPa·m	1 GPa
c-axis	Fe	$1\text{-}10 \times 10^{16}$	160	RT	Indent.	1.5×10^{-4} Mpa·m	1.16 GPa
c-axis	Ni	5×10^{15}	300	RT	CB	2.8×10^{5} dyne/cm	2 GPa
c-axis	Ni	5×10^{15}	300	100K	CB	13×10^{5} dyne/cm	9 GPa
c-axis	Ar	1×10^{15}	500	RT	CB	6×10^{-4} MPa·m	3.4 GPa
a-axis	Ar	1×10^{15}	500	RT	CB	1×10^{-3} MPa·m	5.8 GPa
c-axis	Ti	5.6×10^{16}	300	RT	Bowing	1.28×10^{-3} MPa·m	6.4 GPa
c-axis	Y	1.7×10^{16}	300	RT	Bowing	8.57×10^{-4} MPa·m	7.6 GPa
a-axis	Ti	5.6×10^{16}	300	RT	Indent.	9×10^{-4} MPa·m	4.5 GPa
a-axis	Y	1.7×10^{16}	300	RT	Indent.	6×10^{-5} MPa·m	0.5 GPa
SiC							
Sintered	Ar	3×10^{14}	800	RT	CB	9×10^{5} dyne/cm	
Sintered	N	1×10^{15}	400	RT	CB	6×10^{5} dyne/cm	
c-axis/ single crystal	N_2	1×10^{17}	90	RT	Indent.	5.6×10^{-5} MPa·m	0.4 GPa
Si_3N_4							
CVD film	Ar	2×10^{15}	280	RT	CB	5×10^{5} dyne/cm	3.5×10^{10} dyne/cm²
TiB_2							
Polycrystal	Ni	1×10^{17}	1000	RT	Indent.	$8\text{-}32 \times 10^{-2}$ MPa·m	1-4 GPa

Indent. = Indentation technique of Lawn and Fuller (20).
CB = Cantilever beam technique (18).
Bowing = Measured bowing of an initially flat sample.

Figure 3 shows the integrated stress (S) vs TDE for a series of Cr- and Fe-implantations into sapphire at several fluences and energies. The data for the Cr samples (numbers 1-3 in this figure) indicate a gradual increase in S with increasing TDE. The data for the Fe-implants fall into two groups, one group defined by the 160 keV beam energy and the other defined by the higher energies (1 - 4 MeV). The lower energy values show no dependence on TDE but the higher energy group suggest first a sharp rise then a sharp decrease in S with increasing TDE. Burnett and Page (22) associated a "peaking" of S with TDE with the onset of amorphization in sapphire implanted with Ti or Y. Examination by RBS-C and TEM shows that is not the case for the data contained in this figure.

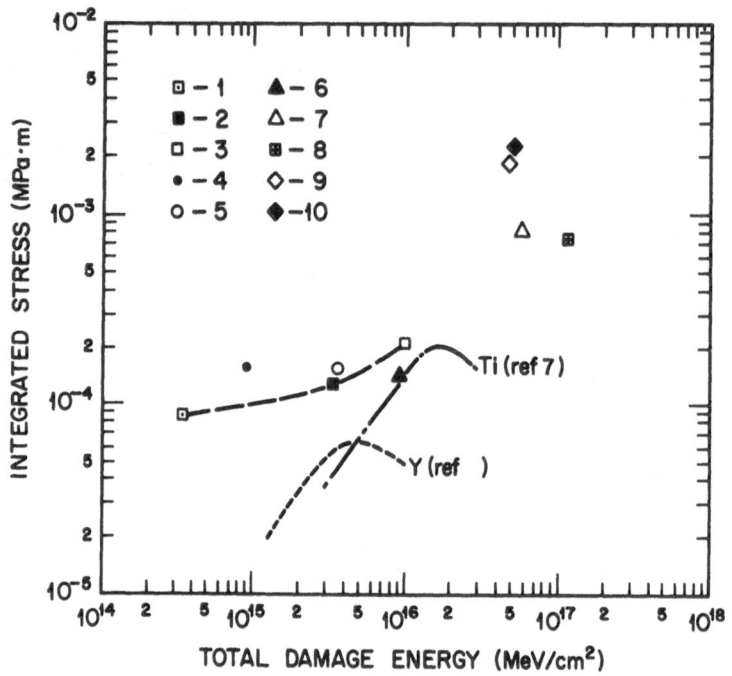

FIGURE 3. Integrated stress, S, vs total damage energy, TDE.

$1 = 4 \times 10^{15}$ Cr/cm^2, 150 keV;
$2 = 4 \times 10^{16}$ Cr/cm^2, 150 keV;
$3 = 1 \times 10^{17}$ Cr/cm^2, 150 keV;
$4 = 1 \times 10^{16}$ Fe/cm^2, 160 keV;
$5 = 4 \times 10^{16}$ Fe/cm^2, 160 keV;
$6 = 1 \times 10^{17}$ Fe/cm^2, 160 keV;
$7 = 2 \times 10^{17}$ Fe/cm^2, 1 MeV;
$8 = 4 \times 10^{17}$ Fe/cm^2, 1 MeV;
$9 = 1 \times 10^{17}$ Fe/cm^2, 3 MeV;
$10 = 1 \times 10^{17}$ Fe/cm^2, 4 MeV.

The use of total damage energy considers only the production of defects. It is, of course, the defects that are retained after various dynamic recovery processes have occurred which produce the observed residual stress state. McHargue et al. (3) observed that the peak damage in the Al-sublattice of Al_2O_3 (as measured by RBS-C) saturated at a value of $X_m \approx 0.67$ for a range of TDE = 1.36×10^{15} to 13.65×10^{15} MeV/cm².

Specimens 2 and 3 (Cr) and 4, 5, and 6 (Fe) lie in this range of TDE. The values of S shown in Fig. 3 are approximately constant in this range of TDE. It appears likely that both the Cr- and Fe-implants exhibit a saturation in S due to dynamic annealing effects. The values of X_m for the Fe-implanted samples are 0.55, 0.65, and 0.8 for specimen numbers 4, 5, and 6, respectively.

The values for S for the samples implanted with iron at the higher energies fall above the values for the other specimens. This grouping of S values may indicate a breakdown in the indentation technique for determining residual stress. A basic assumption in the derivation of T from the crack lengths is that the thickness of the stressed layer is much less than the crack length, c. The values of d for the lower energy implants (Cr and Fe) were in the range of 0.13 to 0.16 µm and the crack lengths were 8 - 10.5 µm, or d ≈ 1.5 - 2% of c. Corresponding values for the higher energy samples were 0.7 to 7.0 µm for d and 6 to 13 µm for c, or d ≈ 10 to 50 % c.

Since the total damage energy is deposited along a path length approximately equal to d, the average stress (T) might give a better description of the residual stress-damage energy relationship. Figure 4 shows the data plotted in this fashion (T vs TDE). These data suggest a saturation in stress at a value of 1125±57 MPa.

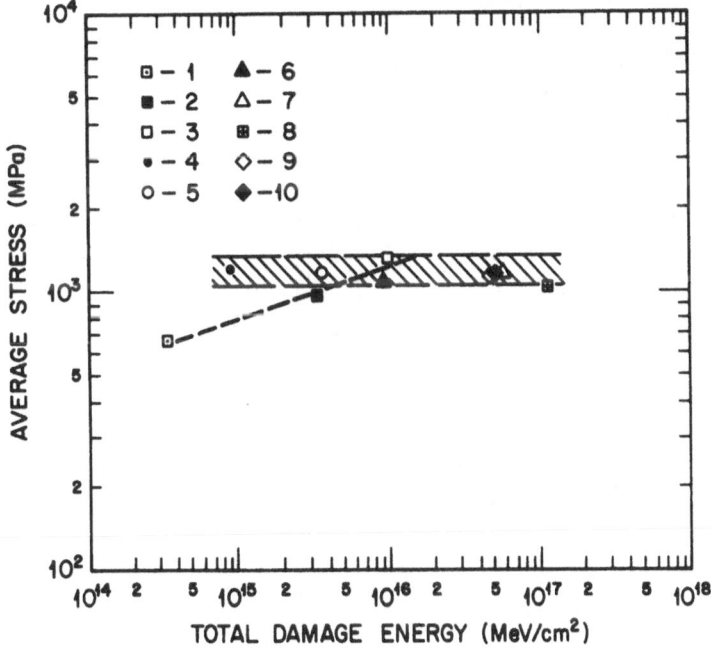

FIGURE 4. Average stress, T, versus total damage energy, TDE. Legend is same as in Fig. 3.

The residual compressive stress is very high and in fact equal to or greater than the rupture strength of all materials studied. For example, the rupture strength of TiB_2 is about 750 MPa whereas the average compressive stress in the implanted regions is 1 to 4 GPa. The stresses are in the range of 1-2% of the elastic modulus.

4. MECHANICAL PROPERTIES

Although ceramic materials are classified as "brittle", that is, the work to propagate a crack is less than the work of nucleation, plastic deformation at room temperature clearly can occur in some cases. The material under a hardness indenter undergoes (limited) plastic deformation by a range of deformation mechanisms--dislocation generation and motion, twinning, etc. The deformation is more localized than in metals due to the high stresses required to break co-valent bonds in materials such as SiC and Si_3N_4 or the necessity to maintain charge neutrality while moving dislocations through ionic materials. There can be substantial dislocation activity even at room temperature in the alkali halides and MgO. Dislocations have been observed beneath hardness indents in Al_2O_3 and SiC by Lawn et al. (31) and beneath wear tracks in Al_2O_3 Yust (32).

The deformation of amorphous materials occurs by very different mechanisms. Among the possible mechanisms are densification, shear band propagation, and viscous flow.

Ion implantation may alter the mechanical properties by defect strengthening, alloy or solid solution strengthening, or as a result of the crystalline to amorphous transformation. Second-phase formation during post-implantation annealing may further influence these properties.

4.1. Elastic Modulus

The development of ultra-low load micro-indentation hardness testers has made possible the measurement of the elastic modulus of implanted layers from the load-displacement curves for loading-unloading cycles. Figure 5 shows a plot of the elastic modulus as a function of depth (from the surface) of the indenter for crystalline (unimplanted) c-axis sapphire and a 150-nm-thick amorphous surface layer on the same substrate (33). The amorphous layer was produced by implantation of Al (4×10^{16} ions/cm^2, 90 keV) followed by O (6×10^{16} ions/cm^2, 55 keV) at 77 K. The value for the crystalline material is 539 GPa, which agrees well with the published values of C_{33} = 502 GPa. The value for the amorphous Al_2O_3 layer, taken at depth = 0 is 175 GPa or 32% of the crystalline value. Notice that the effects of the substrate on the measurement is discernable at indentation depths as low as 10 nm, only 6% of the amorphous layer thickness.

Similar measurements for single crystalline and amorphous (implanted with 10^{16} Cr/cm^2, 260 keV) SiC show that the elastic modulus of the amorphous material is about 50% of the crystalline material (34).

4.2. Hardness

A hardness value for a material is useful for comparative purposes but is not a fundamental property. In general, hardness implies a resistance to deformation but it is a manifestation of several related properties such as yield stress, tensile strength, ductility, work-hardening, elastic modulus, and residual stress states.

The apparent simplicity of the conventional microindentation hardness tests often belies the difficulties in obtaining the true properties of the subject material.

There are several factors that complicate the task of measuring the hardness of thin films, coatings, or surface-modified layers. The extent of the elastic-plastic zones relative to the penetration depth affects the indentation hardness measurements. For an ideal plastic material, the highest

MODULUS CHANGES DUE TO ION IMPLANTATION

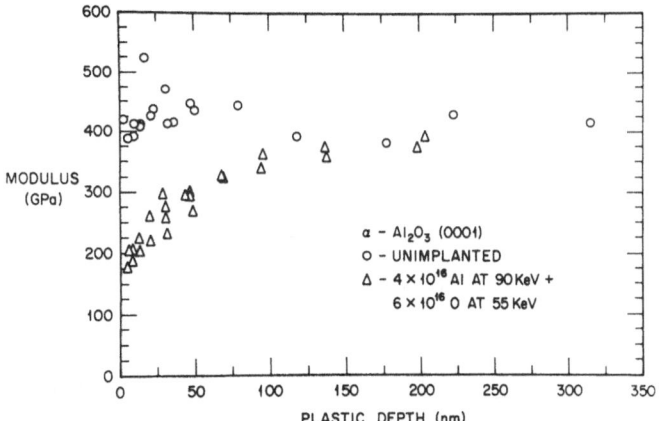

FIGURE 5. Modulus
versus plastic depth.

FIGURE 5. Modulus versus plastic depth.

pressure for a loaded sphere in contact with the surface occurs at a depth of about d/2, where d is the chord of the impression. As the indenter penetrates the surface, the deformation front continues to move inward and the deformed volume being approximately hemispherical with a radius of 7 to 14 times the plastic indent depth. Unless the layer thickness is large relative to the indent depth the indicated hardness will be that of the composite composed of the layer and the substrate.

In calculating a hardness number or index, one assumes that the size of the permanent impression is the same as when the loaded indenter was in contact with the sample. If there is a large elastic recovery, this assumption may be invalid. In materials with high hardness-to-elastic modulus ratios, that is, ceramics, such elastic recovery effects are significant (34). Figure 6 illustrates the displacement of a surface due to a loaded sharp indenter showing both the elastic displacement and the plastic displacement. Hardness is proportional to the contact pressure or the load divided by the projected contact area.

One approach to overcoming the problems caused by the depth of penetration relative to the coating or modified layer thickness is the development of ultra-low load microindentation techniques. These devices attempt to avoid the difficulties associated with imaging small indents by continuously measuring the depth of penetration during both loading and unloading. Both hardness and elastic modulus can be determined from a calibration of the shape of the indenter. However, there are still problems with assigning an absolute value to the hardness obtained by such instruments. Figure 7 shows the loading/unloading curves for a number of materials. Notice that in the case of a pure metal (copper) the assumption that the size of the residual indent represents the contact area under maximum load is reasonable. However, as shown by Fig. 7(b), this assumption is invalid for measurements made on a sapphire single crystal (35). The displacement AC is the total displacement under maximum load and includes both elastic and plastic contributions.

It has been shown that extrapolation of the stiffness, i.e., the slope of the unloading curve, to the x-axis gives the plastic depth of penetration (36). This is the displacement AB in Fig. 7(b). Thus, the elastic portion is the total displacement minus the plastic depth. The unloading curve

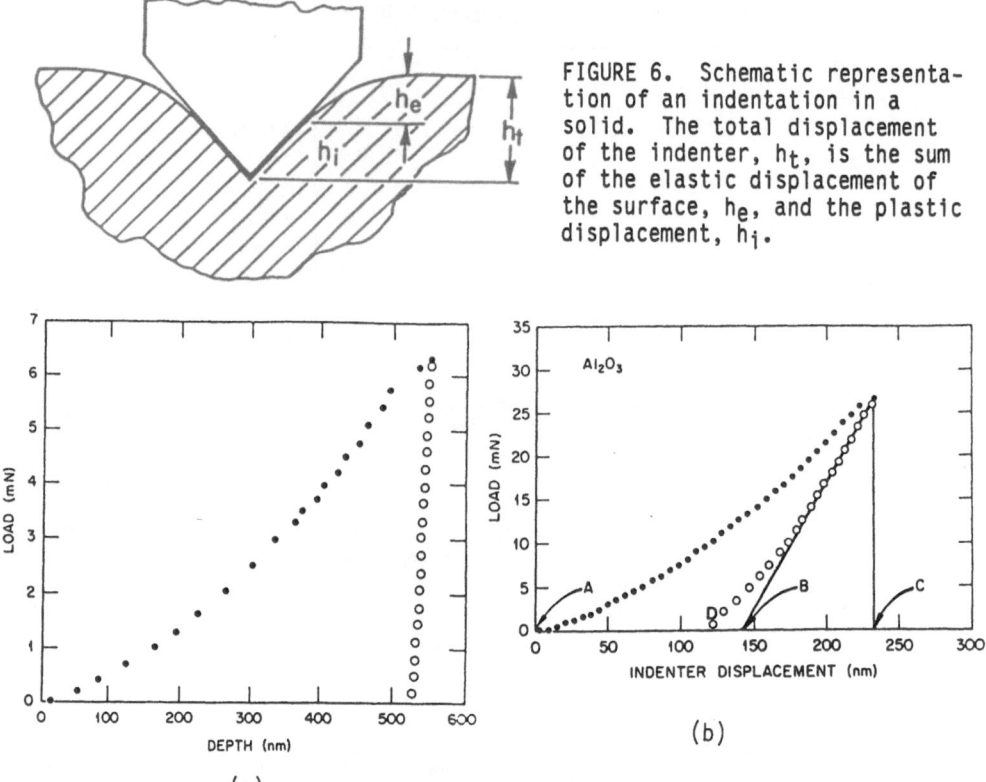

FIGURE 6. Schematic representation of an indentation in a solid. The total displacement of the indenter, h_t, is the sum of the elastic displacement of the surface, h_e, and the plastic displacement, h_i.

FIGURE 7. Load-displacement curves for (a) copper and (b) sapphire (c-axis normal to test plane) made with the Nanoindenter.

indicates that considerable elastic recovery will occur such that the residual depth (obtained by extrapolating the unloading curve to zero load) will be significantly smaller than the plastic depth. In Fig. 7(b), this final, recovered, indentation depth is AD. If the hardness were calculated from an image of this residual impression, its value would be too high.

Finally, the shallow indents made under low loads may exhibit the indentation size effect (ISE). The physical picture of the ISE is not understood but it is clear that the apparent hardness increases as the size of the indentation decreases. Thus the low load microindentation hardness values may sharply increase as the depth of the indentation approaches zero. Such an effect is apparent in the hardness vs indenter displacement curve for single crystalline Al_2O_3 shown in Fig. 8 (ref. 33).

Most of the published data on implanted ceramics comes from low load [10 g (0.09 N) to 50 g (0.5 N)] Knoop or Vickers microindentation tests. In all these cases, the depth of the indentation was equal to or exceeded the thickness of the implanted region. Thus, the reported changes in hardness clearly do not give the true values but do define the direction of changes.

Table 2 collects the information on the hardness of ion implanted ceramics. The hardness values are given as relative values, that is, hardness

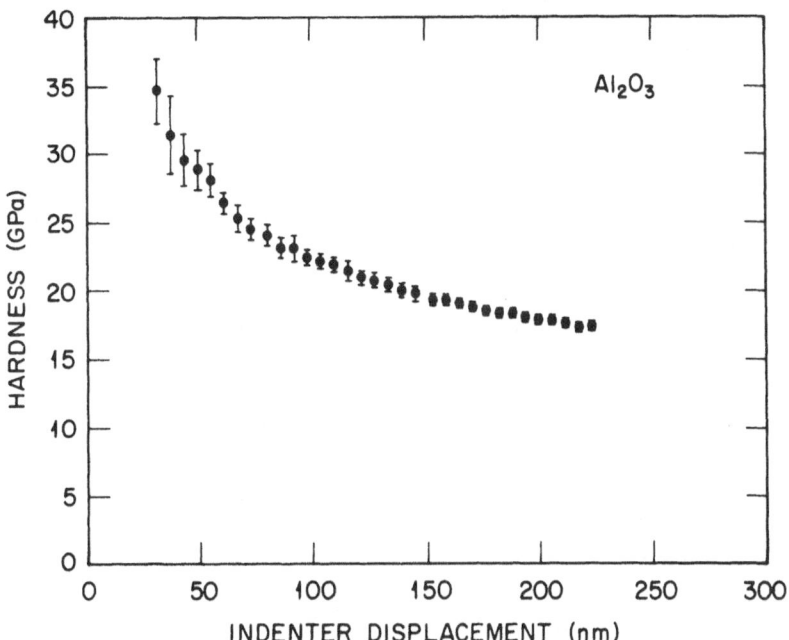

FIGURE 8. Apparent hardness as a function of indenter displacement for a high quality Al_2O_3 single crystal (c-axis orientation) tested on the Nanoindenter. The marked increase in the apparent hardness at low displacements is a manifestation of the indentation size effect (ISE).

of implanted sample relative to hardness of unimplanted sample, in order to minimize some of the problems associated with low-load hardness testing.

The hardness of the implanted material increased in each instance that the sample remained crystalline. Figure 9 shows the relative hardness as a function of fluence for Al_2O_3 (c-axis orientation) implanted with 280 keV Cr at room temperature. The amount of chromium in substitutional Al-sublattice sites (as viewed along the c-axis) and residual disorder in the Al-sublattice were determined from Rutherford backscattering-channeling spectra obtained with 2.0 MeV He^+. The amount of disorder was approximately constant for fluences from 10^{16} to 10^{17} ions.cm^{-2} whereas the substitutional fraction of chromium increased to 0.55 at the highest fluence. From such observations, it was estimated that defect hardening (radiation damage) accounted for essentially all the increase at fluences up to about 2×10^{16} Cr/cm^2 and more than half of it at 10^{17} Cr/cm^2. Bull has also concluded that radiation-induced defects are responsible for much of the hardening in Al_2O_3 by examining the hardness as a function of deposited damage energy, that is, defect production (36).

Implantation of Al_2O_3 at slightly elevated temperatures (~640 K) produced less disorder near the specimen's surface but about the same amount at the peak region as room temperature implantation (4). The hardness increase was less for the elevated temperature implant, reflecting the lesser damage (due to enhanced recovery). Such a behavior indicates the dominant role of defect strengthening.

All studies show that the crystalline to amorphous transformation results in a softening of the material. The data for Al_2O_3 (ref. 5,9,24,34),

TABLE 2. Hardness of Ion Implanted Ceramics (after ref. 21)

| Material | Species | Implantation Conditions | | | Hardness method* | Relative Hardness** implanted (unimplanted) | Comments | Reference |
		Fluence (ions·cm^{-2})	Energy (keV)	Temperature				
Al$_2$O$_3$								
c-axis	Cr	10^{16}–10^{17}	280	RT	K-15	1.27–1.55		7
	Cr	4 × 10^{16}	280	640K	K-15	1.1		4
	Cr	3 × 10^{15}	280	77K	K-15	0.6	amorphous	5
	Al+0	$\{\begin{matrix}4 \times 10^{16}Al\\6 \times 10^{16}O\end{matrix}\}$	90	77K	ULL	0.45	amorphous	34
	Fe,Cu,Ti, W,Mo	1.5–4×10^{16}	various	RT	K-15	1.1–1.4		37
	Ni	10^{17}	300	RT	K-25	1.3		24
	Ni	10^{15}	300	100K	K-	1.5		24
	Ni	10^{17}	300	100K	K-	0.6	amorphous	24
a-axis	Y	3 × 10^{16}	300	RT	K-25	1.57		9
	Y	6 × 10^{17}	300	RT	K-25	0.7	amorphous	9
	Ti	3.4 × 10^{16}	300	RT	K-25	1.3		38
	Cr	3.15 × 10^{16}	300	RT	K-25	1.11		38
MgO								
{100}SG	Ti	2 × 10^{16}	300	RT	K-10	2.3		39
	Ti	3.5 × 10^{17}	300	RT	K-10	0.8	amorphous	39
	Cr	6 × 10^{16}	300	RT	K-10	2.0		39
ZrO$_2$								
Y-FSZ	Al	1 × 10^{16}	190	RT	K-50	1.28		40
	Al	4 × 10^{17}	190	RT	K-50	0.83	amorphous	40
	Ti	3 × 10^{16}	400	RT	K-10	1.6		41
	Ti	1 × 10^{17}	400	RT	K-10	0.9	amorphous	41
TiB$_2$								
Sintered	Ni	1 × 10^{17}	1000	RT	K-15	1.7–2.1		42
SiC								
c-axis	Cr	4 × 10^{16}	280	RT	K-15	1.2		14
	Cr	2 × 10^{15}	280	RT	K-15	0.55	amorphous	7
	Cr	1 × 10^{16}	260	RT	ULL	0.4	amorphous	44
	N$_2$	8 × 10^{17}	80	RT	V-25	0.37	amorphous	43
Sintered	Ar	10^{16}	800	RT	V-100	0.5	amorphous	27

*K = Knoop; V = Vickers; ULL = Ultra-low load; SG = single crystal.
The number following K or V indicates the load used in the hardness test in values of grams (force).

**The value listed here is the maximum (or minimum) reported in the indicated reference.

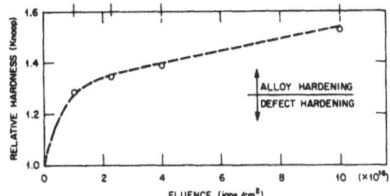

FIGURE 9. Relative hardness (implanted/unimplanted) for Al_2O_3 implanted with chromium (280 keV) at 300 K. Knoop indentations were made on the (0001) face with 15 g (f) (ref. 14).

MgO (ref. 39), ZrO_2 (ref. 40,41), and SiC (ref. 7,11,27,43,44) indicate that the hardness of the amorphous state is about 40 to 60% that of the crystalline counterpart. Figure 10 shows the behavior for Al_2O_3 and SiC implanted with 280 keV Cr ions where the hardness first increases with damage (or disorder) until a subsurface amorphous layer forms at the depth of peak damage and then decreases as the thickness of the amorphous layer increases.

The importance of indentation depth relative to layer thickness is shown by the hardness data obtained by the ultra-low load technique for an amorphous surface on Al_2O_3 (ref. 34). Figure 11 gives the hardness as determined for various indenter penetrations into a 155-nm-thick amorphous layer produced by 77 K implantation of 4×10^{16} Al cm^{-2} (90 keV) followed by 6×10^{16} O cm^{-2} (55 keV). Note that substrate effects are apparent for penetrations as low as 20 nm and that the surface layer affects the values of the apparent hardness of the harder substrate for considerable penetration depths. The true hardness versus depth curve should have the shape of a step function with a value of 7.5 GPa for the region from 0 to 155 nm and 27 GPa at depths greater than 155 nm.

4.3. Transverse Rupture Strength

Since a direct determination of the tensile strength of ceramics is difficult because of its sensitivity to the presence of any small surface flaws, the flexural or transverse rupture strength is generally determined from a bend test. A sample is loaded at the center and supported on the opposite side near the ends. The breaking stress in the outermost layer is calculated from a simple beam formula. There is usually a large amount of scatter in the data due to pre-existing flaws, and a large number of specimens are required to give the values statistical meaning. Data are often presented as plots of the frequency of failure versus applied load.

Hioki and co-workers were the first to report the effects of implantation on the flexure stress of Al_2O_3 (24,46). Implantation with gas ions caused the flexure strength (in 3-point bending) to increase with fluence to some peak and then suddenly decrease. Maximum increases of 65% at 6×10^{15} Ar/cm^2 (800 keV) and 55% at 10^{17} N/cm^2 (400 keV) were obtained for single crystal specimens. The decrease coincided with the onset of amorphization and bubble formation. The increases were much less for polycrystalline specimens, (about 10%).

The flexure strength was also determined for Al_2O_3 implanted with 300 keV Ni at 100, 300, and 523 K (ref. 24). Strength increases of about 10% were found at fluences of 10^{17} ions.cm^{-2} for the 300 and 523 K implants. The increase was about 30% in the 100 K specimens and was approximately the same for fluences of 10^{15} to 10^{17} ions.cm^{-2}. The 100 K implants produced amorphous surfaces. These results were interpreted as being due to the large (~9 GPa) compressive stress caused by the volume increase associated with the crystalline to amorphous transformation.

Figure 12 shows the effect of room temperature implantation of 1×10^{17} Cr/cm^2 (150 keV) on the flexure strength (in 4-point bend tests) of single

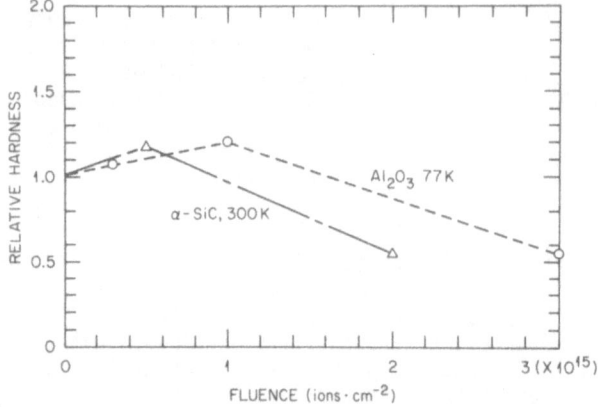

FIGURE 10. Relative hardness of Al_2O_3 and SiC implanted with 280 keV chromium ions (ref. 14).

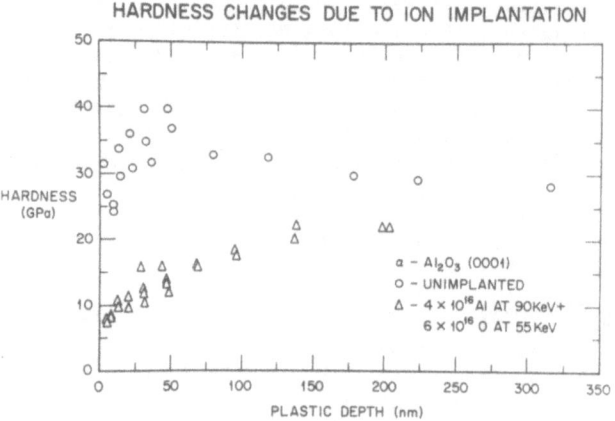

FIGURE 11. Hardness versus plastic depth for virgin single crystal and amorphous Al_2O_3.

crystal Al_2O_3 (47). These specimens remained crystalline. The data are shown as Weibull probability plots wherein the cumulative failure (fraction of specimen failed) is plotted versus applied stress. The curves show that implantation affected both the propagation of flaws (failures at lower stresses) and the intrinsic strength, the stress at which all specimens fail. The increase in stress for the characteristic life was 88%. Most of this effect appears to be due to the residual stresses although there is also some "intrinsic" strengthening. The slope or shape parameter was also affected by the implantation, indicating a change in the "effective" flaw size responsible for the failure.

Doi and co-workers studied the effects of implantation on the flexure strength of Y_2O_3-stabilized ZrO_2, sintered Si_3N_4 and sintered SiC (27). Implantation of 400 kV N in the fluence regime where the specimens remained crystalline increased the strength of single crystal specimens of (Y)-ZrO_2 (by 57% at 1×10^{16} N/cm^2). Likewise, implantations that did not amorphize SiC and Si_3N_4 also increased the flexure strength (each by about 25% at 5×10^{15} Ar/cm^2, 800 keV). However, in contrast to the behavior of Ni-implanted sapphire where amorphization also produced an increase in strength, it caused decreases in the flexure strength of SiC and Si_3N_4 (by about 20% at 5×10^{16} Ar/cm^2, 800 keV).

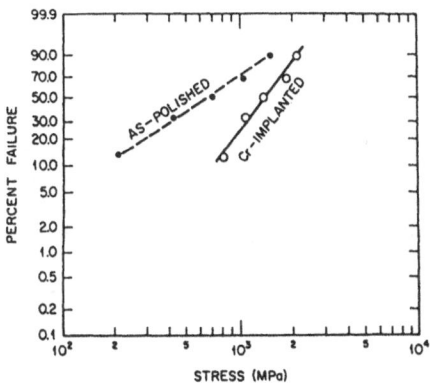

FIGURE 12. Weibull distribution
of cumulative failure of single
crystal sapphire tested in 4-point
bending (47).

4.4. Fracture Toughness

Failure in "brittle" materials often occurs by the extension of flaws
(cracks) that are present before the application of stress. A useful
measure of the properties of such materials is their toughness. A general
approach has been developed in engineering mechanics that indicates the
resistance to propagation of a crack under stress once it has been initiated. This measure of resistance is called fracture toughness, K_c.

The length of cracks associated with sharp-indenter impressions in ceramics reflect the toughness of the material (48). There are two types of
cracks: radial cracks that are the tensile extension of median cracks
formed during loading, and lateral cracks that are formed during unloading
due to the influence of residual tensile stresses in the near-surface
region. Radial cracks generally run perpendicular to the free surface
whereas lateral cracks have a component more or less parallel to the surface. There have been several attempts to formulate simple relationships
between the size of the radial cracks and the fracture toughness (e.g.,
49,50). Such relationships allow the apparent change in fracture toughness
for implanted ceramics to be determined from an indentation technique.

Increases of 15 to 100% in the apparent indentation fracture toughness
have been reported for ion implanted Al_2O_3 (7,22,24,51-53). Figure 13
(taken from ref. 24) illustrates the effects of fluence (of nickel ions into
Al_2O_3) and substrate temperature on the relative fracture toughness
(implanted/unimplanted). Samples implanted at the higher temperatures (300,
523 K) exhibit an increase of 80% at a fluence of 10^{17} ions/cm². These data
showed a hardness increase of 20 to 25% for the same samples. The presence
of an amorphous surface layer (on specimens implanted at 100 K) caused
significantly greater increases in the fracture toughness, the increase
being 110% for 10^{17} Ni/cm². Such changes in fracture toughness are
generally attributed to the residual compressive surface stress induced by
implantation. Detailed examination of the cracks around Vickers indentations made on the $\{10\bar{1}2\}$ surface of Al_2O_3 showed that implantation had
little effect upon the incidence of radial cracking (51). However, observations on sections broken along the radial cracks revealed that the semi-
circular crack trace became oblate as cracks were deflected by the implanted
surface layer and thus their trace on the free surface appeared shorter.
This change in the shape of the fracture path was observed both in samples
implanted with low fluences (crystalline surfaces) and with high fluences
(amorphous surfaces). Again, the change was attributed to the residual
surface compressive stress introduced by implantation.

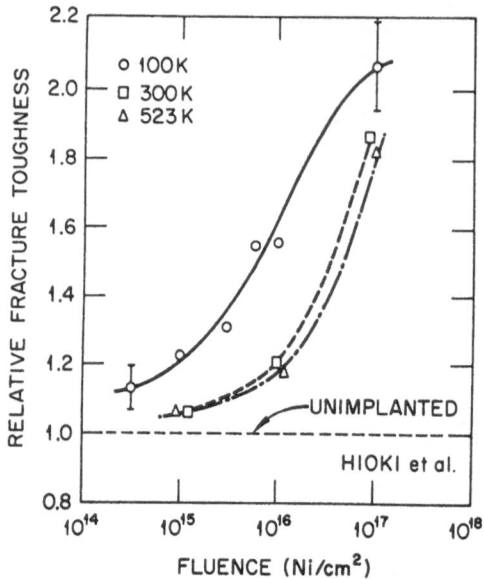

FIGURE 13. Indentation fracture toughness as unction of fluence for Al_2O_3 implanted with 300 keV Ni ions at 100, 300, or 523 K. The K_C value was evaluated for Vickers indentations at a load of 0.49 N. After Hioki et al. (24).

The apparent fracture toughness of SiC was increased by implantation, both for damaged but crystalline layers and for amorphous layers (13,22,54). The increases were in the range of 30 to 40%, depending upon the load used to initiate cracks and the computational method used. The incidence of lateral cracking was markedly reduced but there was little change in the radial cracking.

In contrast to Al_2O_3 and SiC, a decided change in the radial crack trajectories was found for implanted MgO (39). The usual four <110> crack traces for indents on {001} surfaces were replaced by an irregular array of cracks in other directions. The cracks in unimplanted MgO were consistent with a model of crack nucleation based on simple dislocation reactions. After implantation, this form of cracking was absent and cracks seemed to nucleate at the points of highest stress concentration. A large (80%) increase in indentation fracture toughness of the implanted MgO was attributed to both the surface compressive stress and the change in method of crack nucleation.

In the case of ZrO_2, the variation of apparent fracture toughness with implantation is different for partially stabilized and fully stabilized materials. The apparent fracture toughness of fully stabilized zirconia (with yttria) is increased by implantation of Ti (41), and Al (40,55) at fluences below that necessary to induce amorphization. Lateral cracking was again suppressed by implantation. On the other hand, in partially stabilized zirconia (with MgO or Y_2O_3) the apparent fracture toughness decreased with an increasing fluence of 400 keV Ti until the amorphous state was reached and then remained approximately constant (41). The amount of radial cracking was greater in the implanted PSZ samples than in the unimplanted controls.

Increases of 80 to 100% in the indentation fracture toughness have been reported for TiB_2 (7,42,52).

In the indentation fracture toughness method, the cracks are initiated in the unimplanted substrate and propagate toward the surface. The observations note the influence of the modified surface layer on the final stages of crack propagation. It is clear that the properties of the implanted zone affect the crack propagation process but whether or not it is accurate to call this an effect on fracture toughness is not yet resolved, but the method serves as a convenient and useful means of following changes induced by ion implantation.

4.5. Tribological Properties

This subject is considered in detail in the paper by Kossowsky in the present volume; however, his emphasis is on the properties at elevated temperatures. Some of the observations on the effect of ion implantation on the room temperature friction and wear of ceramics will be briefly reviewed in this section.

Friction is the important parameter that determines the manner in which stress is transferred from one member to another for contacting moving components. Ion implantations of single crystal sapphire to fluences less than those required for amorphization result in an increase in the coefficient of friction for metal pins under lubricated sliding (56) and metal, sapphire, and diamond pins under dry sliding conditions (27,57). The onset of amorphization is accompanied by a decrease in the coefficient of friction [e.g., from 0.24 to 0.04 for a diamond pin and a normal force of 0.49 N (ref. 14)]. At first, these effects were thought to be due to ploughing in the harder or softer implanted layers. However, Bull and Page (57) have shown that the effects are due to adhesion between the pin and disc and the effects of implantation on this property.

The scratch test or single pass pin-on-disk test gives an indication of resistance to abrasive wear where gouging or plowing may be involved. The scratch tests corresponds to a single pass pin-on-disk test. If the cross-sectional area of the scratch is measured and the corresponding tangential force is known, a work of material removal can be calculated for quantitative comparisons among samples. The single-pass test lacks the repeated loading and unloading of the pin-on-disk test and will not reflect any effect of fatigue or of abrasion by chips of material removed in previous passes.

Figure 14 shows a SEM photograph of scratches made by a loaded diamond stylus on a single crystal of Al_2O_3 (52). The implanted (crystalline) region is to the right-hand side and the implanted-unimplanted interface is indicated by the arrows. It is immediately obvious that the amount of lateral cracking and the resultant spalling of material is much less in the implanted region. The depth of each groove extends beyond the depth of the implanted zone by at least a factor of two. The tangential force (hence, dynamic coefficient of friction) increased by 20 to 30% upon going from the unimplanted to the implanted region.

The specific work of material removal is 50 to 140% greater for the implanted region of TiB_2 (1×10^{17} Ni/cm^2, 1 MeV) than for the unimplanted material (42). In these polycrystalline specimens, material removal was mostly by grain boundary cracking which was greatly suppressed by implantation. Transgranular cracks were also found in the wear path made in unimplanted regions but not in the implanted areas. As noted in Table 1, implantation produced a residual compressive stress of 1-4 GPa in these samples which apparently suppresses crack-formation and/or propagation.

The production of an amorphous surface on ceramics causes a surface compressive stress due to the volume change accompanying the transformation. In addition, deformation of amorphous phases occurs by viscous flow, shear band propagation, or densification, rather than by dislocation slip and cleavage fractures typical of brittle solids. Both effects could influence the wear properties. Figure 15 contains SEM photographs of scratches made in Al_2O_3 by a diamond stylus. The amorphous surface was produced by implantation of 2×10^{16} Cr/cm^2 (260 keV) at 77 K. The groove in the unimplanted (crystalline) region is accompanied by cracks that extend into

270

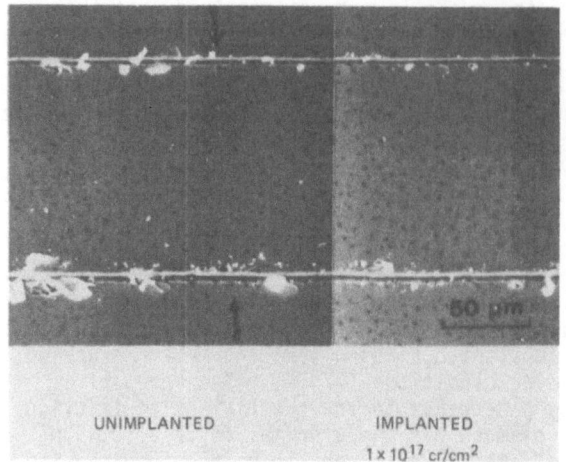

UNIMPLANTED   IMPLANTED
                          1×10^{17} cr/cm^2

FIGURE 14. SEM photograph of scratch made by diamond stylus with normal forces of 0.29 N (upper) and 0.49 N (lower) in Al_2O_3. The interface between the implanted (1×10^{17} Cr/cm^2, 280 keV) and unimplanted regions is marked by the arrows (implanted region to the right) (ref. 42).

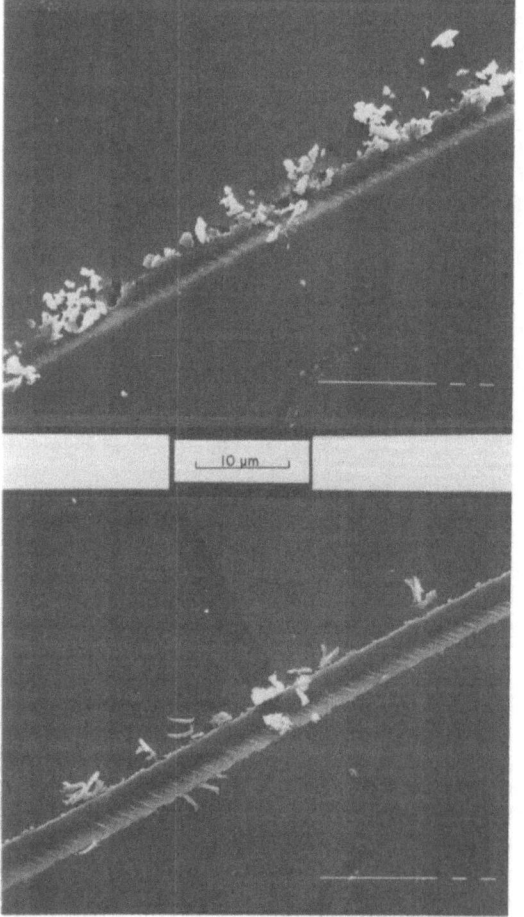

FIGURE 15. SEM photographs of scratch made by a diamond stylus in Al_2O_3. (a) Crystalline, unimplanted; (b) amorphous, implanted with 4×10^{16} Cr/cm^2 (150 keV) at 77 K (ref. 14).

the surrounding material for distances comparable to the groove width. There is also profuse cracking in the bottom of the groove. Material spalled due to the lateral cracks and much of the debris consists of angular pieces fractured from the wear track. On the other hand, the track in the implanted (amorphous) region is characterized by ductile-appearing chips and there are no visible cracks (radial or lateral). Profilometer traces across the track indicate extensive plastic deformation, with much of the material removed from the track present in ridges next to the groove. Even though the amorphous material is only 50% as hard as the crystalline material, the suppressing of cracking has greatly increased its resistance to this gouging or ploughing wear.

5. SUMMARY

The surface mechanical properties of ceramics are altered by ion implantation. The alteration results from the microstructural changes and the residual stresses produced in the implanted zone. Because of difficulties associated with accurate property determination of thin layers, data published to date are generally qualitative in nature and indicate only the direction of the changes. Implantation conditions which give a damaged but crystalline layer cause increases in hardness, fracture toughness, and flexural strength. The mechanisms responsible appear to involve both defect- and solid-solution strengthening, and the residual compressive stress.

If implantation produces an amorphous phase, the hardness decreases but the apparent fracture toughness and the flexural strength may be increased relative to the unimplanted value. These increases have been attributed to the residual compressive stress state. The amorphous surface layer suppresses the incidence of lateral cracking associated with indentation or scratching with a sharp stylus. Since much of the material that is removed from a brittle solid during a wear test is the result of lateral cracking, implanted ceramics with amorphous surfaces show less wear in laboratory tests than their crystalline counterparts.

6. ACKNOWLEDGEMENT

The author acknowledges the contributions of many colleagues to the studies conducted at ORNL: P. Angelini, B. R. Appleton, G. C. Farlow, M. B. Lewis, H. Naramoto, W. Oliver, P. S. Sklad, C. W. White, J. M. Williams, and C. S. Yust. Special thanks to Dr. P. Burnett, Prof. T. F. Page, Dr. S. J. Bull, Dr. H. Doi, and Dr. T. Hioki who made their results available before publication. This research is sponsored by the Division of Materials Sciences, U.S. Department of Energy, under contract DE-AC05-84OR21400 with the Martin Marietta Energy Systems, Inc.

REFERENCES
1. Thevenard P: this volume.
2. Whitton JL: this volume.
3. McHargue CJ, GC Farlow, GM Begun, JM Williams, CW White, BR Appleton, PS Sklad, and P Angelini: Nucl. Instr. Meth. Phys. Res. B16 (1986) 212.
4. McHargue CJ, GC Farlow, CW White, BR Appleton, JM Williams, PS Sklad, and P Angelini: Application of Ion Plating and Ion Implantation to Materials. R Hochman (ed). Metals Park, OH: American Society for Metals, 1986, 255-266.
5. McHargue CJ, GC Farlow, CW White, JM Williams, P Angelini, and GM Begun: Mater. Sci. Engr. 69 (1985) 123-127.
6. White CW, GC Farlow, CJ McHargue, P Angelini, PS Sklad, MB Lewis, and BR Appleton: Nucl. Instr. Meth. Phys. Res. B7/8 (1985) 473-478.

7. McHargue CJ, MB Lewis, BR Appleton, H Naramoto, CW White, and JM Williams: Science of Hard Materials. RK Viswanadham, DJ Rowcliffe, and J Gurland (eds). New York: Plenum Publishing Co., 1983, 451-465.
8. Williams JM, CJ McHargue, and BR Appleton: Nucl. Instr. Meth. 209/210 (1983) 317.
9. Burnett PJ and TF Page: J. Mater. Sci. 19 (1984) 845.
10. Pells GP and AY Stathopoulos: Radiat. Eff. 74 (1983) 181.
11. Zinkle SJ: this volume.
12. Spitznagel JA, S Wood, WJ Choyke, NJ Doyle, J Bradshaw, and SG Fishman: Nucl. Instr. Meth. Phys. Res. B16 (1986) 237.
13. Burnett PJ and TF Page: Science of Hard Materials. EA Almond, CA Brooks, and R Warren (eds). Institute of Physics Con. Ser. Vol. 75. Bristol: Adam Hilger, 1986, 789-802.
14. McHargue CJ: Nucl. Instr. Meth. Phys. Res. B19/20 (1987) 797.
15. Edmond JA and RF Davis: unpublished research at North Carolina State University.
16. McHargue CJ, PS Skad, P Angelini, and MB Lewis: Nucl. Inst. Meth. B1 (1984) 246.
17. McHargue CJ, GC Farlow, PS Sklad, CW White, A Perez, N Kornilios, and G Marest: Nucl. Instr. Meth. Phys. Res. B19/20 (1987) 813.
18. EerNisse EP: Appl. Phys. Lett. 18 (1971) 581.
19. Speriosu VS, HL Glass, and T Kobayashi: Appl. Phys. Lett. 34 (1979) 539.
20. Lawn BR and ER Fuller: J. Mater. Sci. 19 (1984) 4061.
21. McHargue CJ: Defect and Diffusion Forum 57/58 (1988) 359.
22. Burnett PJ and TF Page: J. Mater. Sci. 20 (1985) 4624.
23. McHargue CJ, ME O'Hern, CW White, and MB Lewis: to be published in Mater. Sci. Engr.
24. Hioki T, A Itoh, M Ohkubo, S Noda, H Doi, J Kawamoto, and O Kamigaito: J. Mater. Sci. 21 (1986) 1321.
25. Kreft GB and EP EerNise: J. Appl. PHys. 49 (1978) 2725.
26. Arnold GW, GB Kreft, and CB Norris: Appl. Phys. Lett. 25 (1974) 540.
27. Doi H: private communication, 1987.
28. Burnett PJ: Ph.D. dissertation, Cambridge University, 1984.
29. EerNisse EP: J. Appl. Phys. 48 (1977) 3337.
30. Davisson CM and I Manning: Naval Research Laboratory Report NRL 8859, Washington, DC, 1986.
31. Lawn BR, BJ Hockey, and SM Wiederhorn: J. Mater. Sci. 15 (1980) 1207.
32. Yust CS: unpublished work at ORNL.
33. Oliver WC and CJ McHargue: submitted to Thin Solid Films.
34. Oliver WC, CJ McHargue, GC Farlow, and CW White: Defect Properties and Processing of High-Technology Nonmetallic Materials. Y Chen, WD Kingerly, and RJ Stokes (eds). Pittsburgh: Materials Research Society, 1986, 515-523.
35. Page TF, CJ McHargue, and WC Oliver: submitted to J. Mater. Res.
36. Doerner M and WD Nix: J. Mater. Res. 1 (1984) 601.
37. McHargue CJ: Ion Beam Modification of Insulators. P Mazzoldi and G Arnold (eds). Chap. 6. Amsterdam: Elsevier, 1987.
38. Barnett PJ and TF Page: Plastic Deformations of Ceramic Materials. RC Bradt and RE Tressler (eds). New York: Plenum Press, 1984, 669-680.
39. Burnett PJ and TF Page: Ion Implantation and Ion Beam Processing of Materials. GK Hubler, OW Holland, CR Clayton, and CW White (eds). Amsterdam: North-Holland, 1984, 401-406.
40. Legg KO, JK Cochran, HF Solnick-Legg, and XL Mann: Nucl. Instr. Meth. Phys. Res. B7/8 (1985) 535.
41. Burnett PJ and TF Page: Ceram. Bull. 65 (1986) 1393.

42. McHargue CJ, CS Yust, P Angelini, PS Sklad, and MB Lewis: Science of Hard Materials. EA Almond, CA Brooks, and R Warren (eds). Inst. Phys. Conf. Series Vol. 75. Bristol: Adam Hilger, 1986, 803-812.
43. Roberts SG and TF Page: J. Mater. Sci. 21 (1986) 457.
44. Joslin DL and CJ McHargue: to be published in Proc. Radiation Effects in Insulators-5, 1989.
45. Bull SG: Cambridge University, private communication, 1987.
46. Hioki T, A Itoh, S Noda, M Doi, J Kawamoto, and O Kamigaito: Nucl. Instr. Meth. Phys. Res. B7/8 (1985) 521.
47. McHargue CJ: Unpublished work at ORNL.
48. Palmquist S: Arch. Eisenhuttenwes 33 (1962) 629.
49. Lawn BR, AG Evans, and DB Marshall: J. Am. Ceram. Soc. 63 (1980) 574.
50. Anstis GR, P Chantikul, BR Lawn, and DB Marshall: J. Am. Ceram. Soc. 64 (1981) 533.
51. Burnett PJ and TF Page: J. Mater. Sci. 19 (1984) 3524.
52. Yust CS and CJ McHargue: Emergent Process Methods for High Technology Ceramics. RF Davis, H Palmour III, and RL Porter (eds). New York: Plenum Press, 1984, 533-547.
53. Burnett PJ and TF Page: Proc. British Ceram. Soc. 34 (1984) 65-76.
54. Roberts SG and TF Page: Ion Implantation into Metals. V Ashworth, WA Grant, and RPM Proctor (eds). New York: Pergamon Press, 1982, 135-146.
55. Cochran JK, KO Legg, and HF Solnick-Legg: Defect Properties and Processing of High-Technology Nonmetallic Materials. JH Crawford Jr., Y Chen, and WA Sibley (eds). Amsterdam: North-Holland, 1984, 173-179.
56. Burnett PJ and TF Page: Wear 114 (1987) 85.
57. Bull SJ and TF Page: submitted to J. Mater. Sci.

TRIBOLOGICAL PROPERTIES OF ION BEAM MODIFIED CERAMICS AT ELEVATED
TEMPERATURES

RAM KOSSOWSKY

Applied Research Laboratory
The Pennsylvania State University
Post Office Box 30
State College, PA 16804

1. INTRODUCTION
1.1 About surfaces
Surfaces of technological materials are often quite different from
the bulk material they represent. Because of the essentially two-
dimensional bond structure, and the direct exposure to the ambients,
there are usually extensive variations in the chemistries, microstructures and morphologies of the surfaces in comparison to their
bulk constitution.

The now classical paper by Fireston and Abott [1] established that
no surface is truly flat but is made up of irregular arrays of hills
and valleys. This was the beginning of the statistical determination
and specification of surface topography known as "surface finish"
parameters. This work also set the stage for the introduction of the
concept of asperities to the studies of tribology and the early
attempts to quantify the real area of contact. Typical orders of
magnitude of typical metal surface features are shown in Fig. 1 [2].

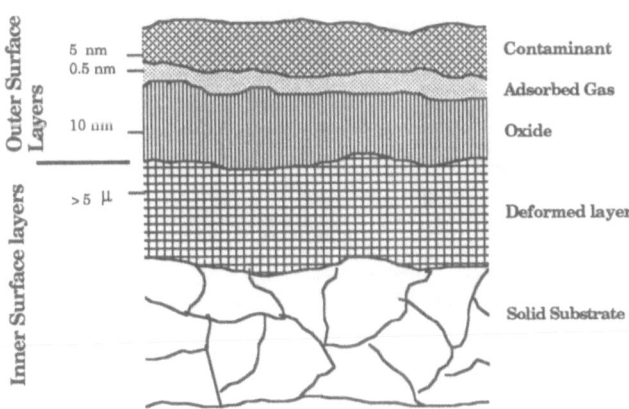

Figure 1. Typical surface features (after Ref. 2).

C.J. McHargue et al. (eds.), Structure-Property Relationships in Surface-Modified Ceramics, 275–294.
© 1989 by Kluwer Academic Publishers.

276

Although the existence and extent of deformed surface layers may
not be usually found on ceramic surfaces, the characteristics of the
surfaces of ceramic materials are determined by the same general
factors cited above. In addition, the characteristics of ceramic
surfaces may be affected by factors related to the specific ways by
which these materials are consolidated and fabricated, i.e., ball
milling, hot pressing, sintering, liquid phase sintering, and so on.
Furthermore, the enhanced sensitivities of ceramic materials to deformation and stress concentrations may be reflected in the structures and
morphologies of their surfaces. Figure 2 shows the change in the
response of the surface of titanium to mechanical deformation, with and

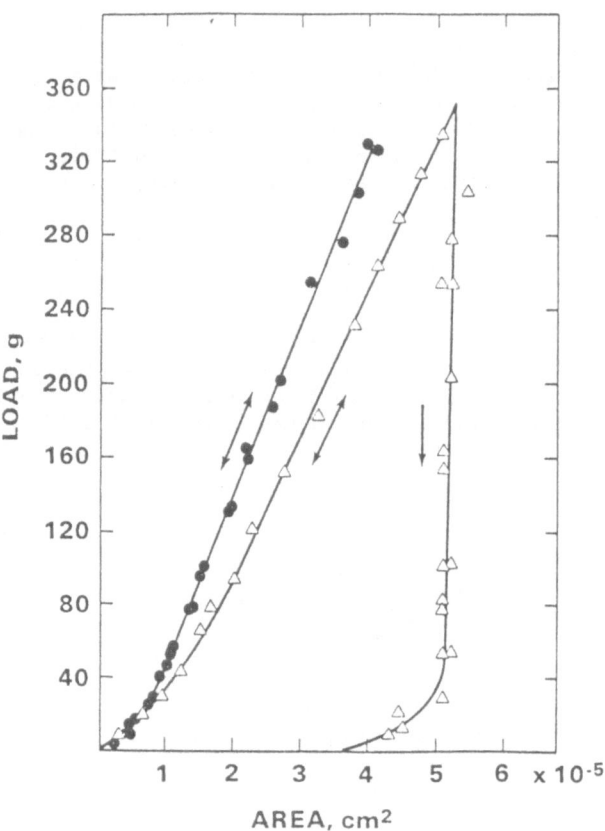

Figure 2. Hertz area/load curves showing the elastic behavior of
titanium covered with a virgin oxide (full symbols),
and after oxide removal (open symbols) (after Ref. 3).

without, an oxide cover. The oxide covered surface exhibits elastic behavior, while a plastic response is obtained under similar load conditions, after the oxide has been removed.

It is therefore not surprising that the friction and wear behaviors, which are specific surface phenomena, cannot be predicted reliably from first principles which are related to bulk structures and properties. Conversely, even a qualitative predictive approach to friction and wear could not be formulated without the definition of the characteristics of the wearing surfaces.

1.2 About Tribology

The science and technology of tribology, which encompasses friction, lubrication and wear, are among the most researched and most written about subjects. In spite of these activities, the science of tribology has not yet achieved the goals of quantitative formulation and parametric predictability. This is so because friction and wear are not materials properties. They are, rather, secondary to a multitude of parameters, intrinsic and extrinsic, which come together to dictate the behavior of a particular system composed of two materials at relative motion to one another across a common interface. Furthermore, the local conditions that fix the controlling parameters are not necessarily constant. Friction and wear are, actually, dynamic events where the controlling parameters change with time.

There are many ways to illustrate schematically the complexity of the genesis and evolution of friction and wear. In Fig. 3 we show the scheme proposed by Kossowsky and Wei [3].

The three basic elements that are involved in the frictional interactions of unlubricated surfaces [4] are listed at the top. These are the real area of contact, the interfacial bond strength, and the dynamics of materials interactions at the point of contact. These elements are not fully understood, let alone correctly quantified. Theoretical treatments of friction and wear resort, therefore, to some forms of approximation, guesses, or statistical descriptions.

Correct estimates of the real area of contact are difficult to attain. The contact surfaces of the wearing materials display complex topographies that are best described in terms of coverage by asperities. It is thus that the treatments of the interactions of the asperities provide the basis for some of the more advanced quantitative descriptions of wear [5-10].

The interfacial bond strength earns a place at the top of the friction and wear evolution line because the actual forces acting between the surfaces depend on the structures, crystal chemistries, and mitigating surface films. These are all parameters which do not lend themselves to unambiguous, time dependent, quantitative definitions.

The effect of velocity is usually not given a rigorous treatment when the tribological behaviors of ceramics are studied. While ceramics are likely to manifest insignificant shear rate effects, velocity will become important at elevated temperatures, where visco-elastic behavior might be involved.

278

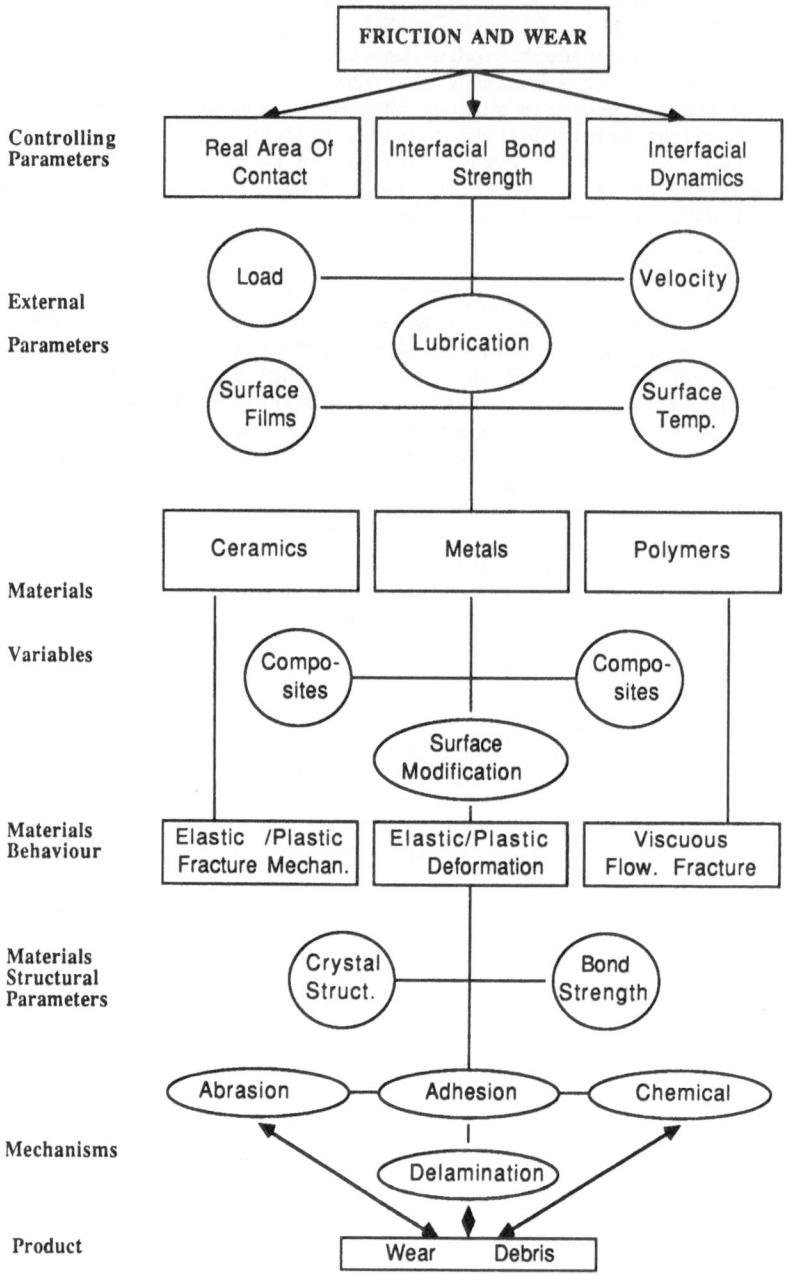

Figure 3. The genesis of friction and wear (after Ref. 3).

One of the early attempts to correlate a few of the parameters listed in Fig. 3 into a "unified" description of wear is due to Archard [11],

$$V = k \frac{S \times L}{H} \qquad (1)$$

where V, the wear volume, is proportional to the sliding distance S, the load L, and is inversely proportional to the hardness H. The difficulty with this equation is that to fit the data, the proportionality constant k may vary by as much as five orders of magnitude, from 10^{-3} to 10^{-8}. As pointed out by Samuels [12], this could mean that anywhere from 10^{-8} to 10^{-3} asperities are producing a wear particle, or that 10^3 and up to 10^8 events have to occur before an asperity becomes a wear particle. The reader may note that the above interpretation sets the stage for a cyclic, fatigue related description of wear. For ceramics, it may be tempting to involve, also, fracture dynamics principles.

The real area of contact depends on the average sizes, numbers, and geometries of the contact spots. For elastically loaded asperities, which is likely to be a dominant feature for wearing ceramic surfaces, the area of contact is given by [13-15] [see also Eq. (7)]

$$A = C(L/E)^{2/3} \qquad (2)$$

where C is a constant, L is the load and E is the elastic modulus.

2.   MECHANISMS OF FRICTION AND WEAR
2.1   Introduction

Most of the modern theories of friction and wear begin with the assumption of a dominating surface topography. Since the interactions among asperities are critical to the outcome of the friction and wear events, the geometrical factors of asperities should contribute to the events at the wearing interfaces.

The term adhesion is used, in its broad meaning, to define the medium of transfer of normal load across the interface between the two moving bodies which result in tangential, or frictional, forces. The specifics of load transfer, the events at the interfaces, and the responses of the materials to resulting stress fields, are then combined into a friction and wear mechanism.

Chemical reactions at the interface are likely to be important sources of adhesion for ceramics wearing at elevated temperatures, where diffusive processes across the interfaces are dominating. Elevated temperatures are usually considered extrinsic environmental factors. It has been also shown [16-18] that high loads, or high speeds, affect the wear behaviors of ceramics by inducing local high temperatures at the tips of asperities.

The origin of friction is thus generally defined as due to yet unspecified adhesion at the interface and, depending on materials and environmental parameters, friction may lead to wear. The purpose of a mechanistic model is then to define quantitatively the nature of adhesion at the interface, the conversion of adhesion to frictional, or tractive forces, and the evolution of wear through the interactions of the frictional forces with the materials involved.

2.2 Adhesive Wear

When ceramic materials are wearing under relatively low loads, that is, loads that do not cause local conditions to exceed the elastic limit, the real area of contact under an asperity is derived from the equations of Hertzian contact [17], as given by Eq. (2), or

$$A_0 = kL^{2/3} \tag{3}$$

where the constant k contains the elastic constants, ν and E, of both materials. [More rigorous derivations of the Hertzian related stresses are shown in Eqs. (6) and (7) below.] The frictional force is assumed to be equal to the asperity fracture stress which is scaled with the area of contact. Thus

$$\mu = \frac{\sigma_f A_0}{L} = k'L^{-1/3} \quad . \tag{4}$$

The results of expression (4) are interesting in that they predict that under low loads and in the pure elastic loading regime, the coefficient of friction will decrease with increasing load. This is in contradiction to the popular Amontons' Law which states that μ is independent of area of contact and load. This result is expected intuitively since the fracture stress for a ceramic material is a function of the fracture toughness parameter, K_c, and the critical crack size, a, or

$$\sigma_f = \frac{Q'K_c}{\sqrt{\pi a}} \tag{5}$$

all of which are independent of load. Since the true area of contact increases with load, the coefficient of friction will decrease [18-20].

The combination of normal Hertzian load and tangential frictional force introduces complex biaxial state of stress [21,22], as shown schematically in Fig. 4. The significant feature here is the occurrence of tensile stresses at the trailing edge of the moving asperity. Tensile stresses are responsible for crack opening. When the critical crack length, dictated by fracture mechanics principles, is reached, an unstable crack will develop leading to fracture and the eventual formation of a wear particle. Trailing edge cracks produced when a sapphire ball was sliding against a glass surface are shown in Fig. 5 [23]. Extensive cracking is seen in (a) where the coefficient of friction was of the order of 0.85 during sliding under water. Sliding in alcohol, (b) recorded μ of the order of 0.15 with concomitant decreases in the width of the wear track and the severity of cracking.

The critical load causing a crack to "pop in" under a pressing asperity may be approximated through the derivations of Lawn and Marshall [24,25], as shown in the schematic in Fig. 6. Following the equations of Hertzian contact, the spatial extent of the stress field is given by the dimension "a",

281

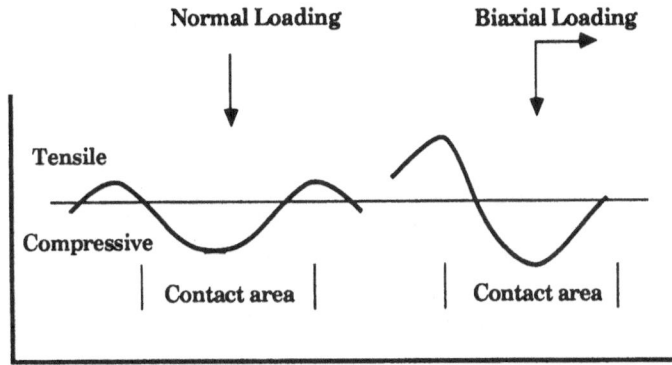

Figure 4. Schematic representation of stress distributions
under a sliding asperity in adhesive wear (after
Ref. 26).

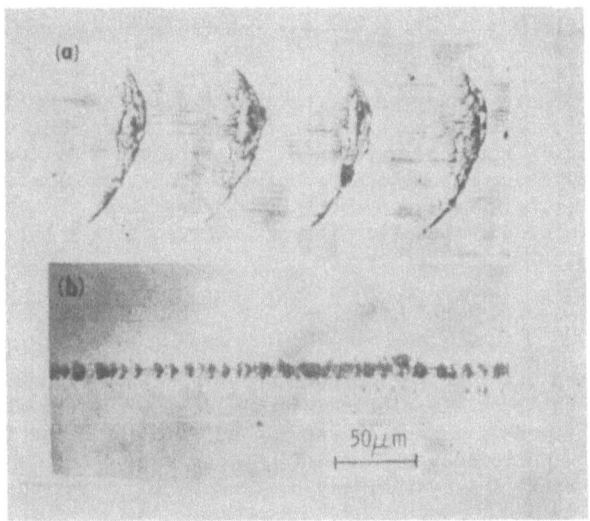

Figure 5. Trailing-edge cracks, sapphire ball sliding on glass.
(a) $\mu = 0.85$, water environment; (b) $\mu = 0.15$,
alcohol environment (with permission, Ref. 23).

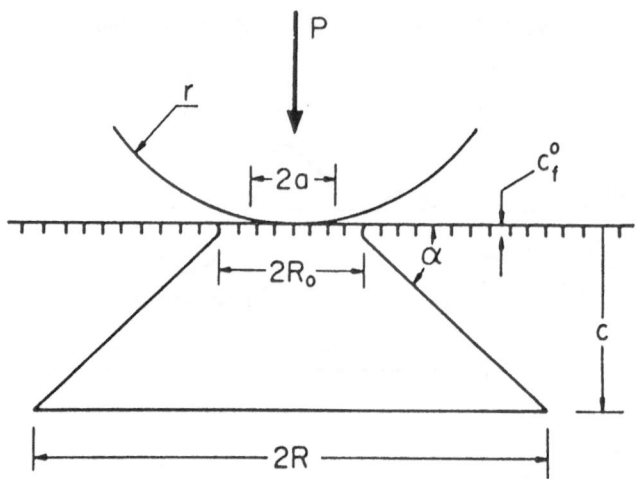

Figure 6. Schematic of the "Hertzian Cone" cracks system (after Ref. 24).

$$a = (4kr/3E)^{1/3} L^{1/3} \qquad (6)$$

where r is the sphere radius, L is the load (equivalent to P) and k is a complex dimensionless parameter that includes the elastic constants of both materials. For application to the tribology equation, we follow the simplifying assumptions adopted by many investigators that assign semi-spherical shapes to the tips of asperities. We can thus derive from Eq. (6) an expression for the measure of the Hertzian stress acting under the asperity, or

$$\sigma_0 = L/\pi a^2 = (3E/4\pi^{3/2} kr)^{2/3} L^{1/3} \qquad . \qquad (7)$$

This equation implies that for an unchanging asperity diameter and constant load, the stress field is uniquely defined by the elastic constants. For ceramic materials wearing under elastic loading conditions, the assumptions of contact geometries for the asperities, within a given operating system, may be acceptable as reasonable approximations.

Invoking elastic fracture mechanics relations between the critical load, crack length and fracture toughness, Lawn and Marshall wrote for the critical load to initiate crack propagation

$$L_c = \alpha r K_c^2/E \qquad (8)$$

where α is a dimensionless parameter, r is the tip radius of the asperity, as above, and K_c is the fracture toughness parameter. The critical fracture stress is given by Eq. (5). Since $\mu = \sigma_F/\sigma_c$, we obtain

$$\mu = \frac{QK_c}{(\pi a)^{1/2}} \frac{E}{\alpha r K_c^2} \tag{9}$$

or

$$\mu = \alpha' E / K_c \tag{10}$$

For a given class of materials, where the fracture toughness was found to vary with processing parameters, inverse dependencies of the coefficients of friction on fracture toughness have indeed been obtained [19,26].

2.3 Delamination Wear [27,28]

The ability of ceramic materials to deform plastically is quite limited. In terms of cracking induced delamination wear, the time element leading to the formation of a wear particle in metallic materials can be viewed as the time required for crack nucleation and propagation. In ceramic materials, however, cracks are assumed to be naturally present and the action of friction induced stress fields are to propagate cracks until the critical dimensions for fracture are reached. Wear of ceramic materials may thus be treated in terms of bulk fracture mechanics parameters, such as fracture toughness and rates of fatigue crack propagation.

Contract stresses play a dominant role in the evolution of friction and wear in ceramics [21,24-26] due to the combination of high elastic modulus and very low plastic deformation, as was shown above. The action of cyclic stresses that prevail during pure rolling, such as in bearings, or the combination of normal and tangential forces during sliding, result in local tensile stress fields that initiate and maintain the propagation of cracks. Wear rate is related to rates of crack propagation and the time element to reach critical crack size to form a wear particle. Adewye and Page [20] studied the wear behaviors of a few grades of hot pressed SiC and Si_3N_4 ceramics. They concluded that the wear of SiC was dominated by bulk cracking. That is, cracks that formed at the wear surface propagated into the bulk resulting in total material failure. Wear in Si_3N_4 was, however, characterized by local delamination. Delamination wear was also found to be the dominant mode in the wear of partially stabilized zirconia [29]. Common to the Si_3N_4 and ZrO_2 ceramics is their relatively high fracture toughness compared to the toughness of SiC. Wear of high toughness ceramics is controlled by micro-plastic deformation mechanisms at the front of the propagating crack [25]. Crack propagation is thus confined to the stress field that exists just below the surface. In the more brittle ceramics, such as SiC, fracture is effected by cleavage and intergranular cracking which move away from the localized stress fields into the bulk.

The dependence of wear on materials parameters is shown schematically in Fig. 7 with attention to the lower range of fracture toughness. As expressed also by Eqs. (8) and (10), the wear resistance of ceramics will improve with lower modulus. Thus, fracture toughness, hardness, strength, and corrosion resistance of the ceramic component should be maximized while elastic modulus and density should be as low as possible. It is not unexpected, therefore, that hot pressed Si_3N_4, with $K_{Ic} \approx 8$ mNm$^{-3/2}$, H $\approx$ 2,000 Kg mm^{-2}, E $\approx$ 310 GPa and $\sigma_f \approx$ 700 MPa fails by localized delamination mode in bearing applications, while SiC, with $K_{Ic} \approx 4$ mNm$^{-3/2}$, H $\approx$ 2,000 Kg mm^{-2}, E $\approx$ 410 GPa and $\sigma_f \approx$ 450 GPa, exhibits catastrophic failures.

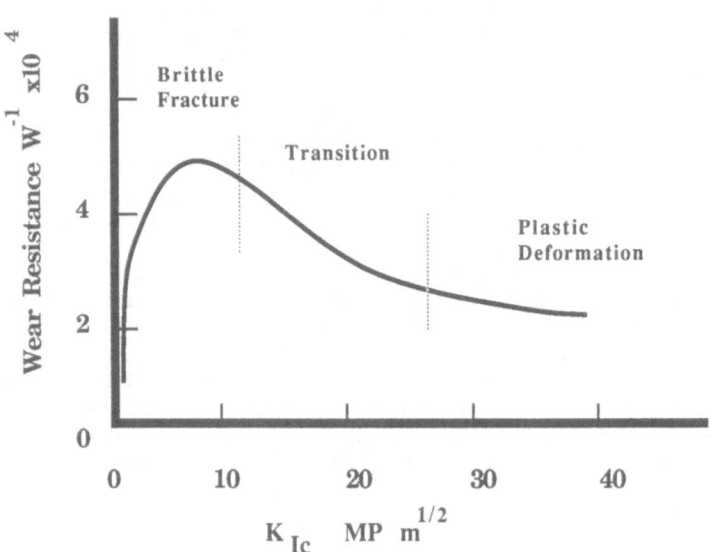

Figure 7. Generalized presentation of the dependence of wear resistance on fracture toughness (after Ref. 26).

2.4 Fracture Dynamics Approach

In discussing the popular Archard formulation, we have introduced the concept of fatigue as a possible explanation for the experimentally observed wide range in the parameter k. In the previous section we have discussed, in some detail, the fracture mechanics approach to wear of ceramics. This is certainly appropriate since these materials are brittle in nature and are quite limited in their ability to deform plastically. Combining the fracture mechanics concept with the fatigue analog brings us naturally to examine the potential of treating friction and wear of ceramics in terms of fracture dynamics.

The fatigue process in brittle materials is defined in terms of slow-fatigue-crack-growth process by the well known Paris relation,

$$da/dN = C(\Delta K)^n \quad . \tag{11}$$

The fatigue process may also be described by the general Wohler relation,

$$N_f = (\sigma_f/\sigma_a)^n \tag{12}$$

which has the same general form as the integral of Eq. (11), where N_f is the total number of cycles to failure, σ_f is the fracture strength of the material and σ_a is the average operating applied stress under which the fatigue process is progressing.

Following the general reasoning that led to the Archard equation, we note that the wear of a ceramic material resulting from a cyclic fatigue process will be derived from the cracking of asperities for which the fatigue limit of N_f cycles has been achieved under the operating stress. We thus write for the wear rate

$$W_N = kf(c)A_0/N_f \tag{13}$$

where $f(c)$ is a function describing the distribution of asperities and A_0 is the area of contact, or the total area of wearing asperities. Since $A_0 = \alpha E^{-2/3}L^{1/3}$ and $\sigma_a = \beta E^{2/3}L^{1/3}$, combining (12) and (13) we obtain

$$W = kf(c)\sigma_f^{-n} E^{2(n-1)/3} L^{(n+2)/3} \quad . \tag{14}$$

Expression (14) has the general form of the Archard relation but incorporates the specific properties of ceramics for which $n \geq 1$. The wear rate scales inversely with the strength of the material but increases with the elastic modulus [see also Eq. (10)]. The dependence of the wear rate on load is modified by the crack growth rate exponent n; wear would usually increase with load at a higher than linear rate.

3. HIGH TEMPERATURE FRICTION AND WEAR

3.1 Introduction

At elevated temperature, where chemical reactivities are likely to dominate, the chemistries of the surfaces in relations to the environment become important. Many of these environments are likely to be rather chemically complex. The obvious conclusion is that the chemistries of the contact surfaces play a major role in determining the behavior of a wearing couple [19,31-33]. Chemical reactions with the environment have been shown to alter the friction and wear of a number of ceramic-ceramic moving contact systems. Local contact temperature and changes in materials chemistry will also influence, or change, the mechanism of wear or the frictional forces operation in a given system. It is intuitively expected that oxygen will have a significant effect on the wear behavior of a metallic system. It is interesting to note that the presence of oxygen has been correlated with reductions in friction and wear in a number of ceramic systems, as well [19,32]. The presence of water vapor was shown to reduce the friction and wear rates in Al_2O_3, CrC, TiC, WC and Si_3N_4 [33]. On the other hand, when high temperatures prevail, such as measured on the surface of Si_3N_4 cutting tools during machining, sharp increases in wear rates of the ceramic were noted [34]. Machining in nitrogen environment resulted in significant improvements in the life expectancy of the ceramic tools. These changes were related to oxidation of the cut metal during operation and the absence of the mitigating effects of condensed water vapor. Segregation of free C to the surface of wearing SiC at elevated temperatures resulted in significant reductions in friction and wear [19,35]. Continuous improvements in wear were noted as temperature was increased to 600°C. The observations were related to increases in the volume fraction of the graphite form of carbon with increases in temperature. The formed layer of graphite provided an efficient lubricating medium.

3.2 Surface Modification by Ion Beams

Ion beam modification of ceramics has evolved into a major means of obtaining improved friction and wear behavior mostly, though, for room temperature applications. The development of a soft and/or amorphous structure is considered to be a means of improving the fracture toughness of the surface in many ceramics, which may lead to a reduction in cracking and delamination under moving contacts. Implantation of nitrogen into tungsten carbide-cobalt composites has been shown to induce a large defect structure thought to impart a degree of ductility to a surface [36]. Implantation of Cr or Ni into SiC causes amorphization, reducing the hardness, but increasing the fracture toughness [37-39].

Roberts and Page [40] studied indentation induced plasticities in silicon carbide implanted with nitrogen. Suppressions in the lateral mode of cracking under the microhardness indentation were related to implantation induced amorphization. The improved "plasticity" was then related to observed reductions in friction and wear. Similar changes in micro-toughness of the surface were reported for a sapphire implanted with Y [41]. McHargue and his colleagues [37,39,40] reported similar observations for alumina implanted with Cr, Ti or Zr. The studies of wear patterns in a number of ion implanted ceramic systems led McHargue to conclude that for those systems where ion implantation produced amorphous surface phases, hardness decreased, but the apparent fracture toughness and strength of the surfaces increased relative to unimplanted materials. The onset of surface amorphization was thus related to the observed reductions in friction.

Nastasi et al. [44] studied the wear behaviors of Al_2O_3, SiC, TiB_2 and B_4C following high dose implantation of N^+. These studies showed that while nitrogen implantation resulted in reduced coefficients of friction for the latter three materials, the Al_2O_3 materials showed significant increases in friction under the same implantation conditions. The authors attributed their results to implantation induced chemical and microstructural changes at the surfaces, although they did not rule out additional contributions from radiation damage, i.e., amorphization. Nastasi et al. [44] concluded that strong correlations were found between post implantation friction and the thermodynamic driving force to form "nitride-like" bonds within the implanted volume. For alumina, where the solid state incorporation of nitrogen was not thermodynamically favorable, implantation induced structural defects, i.e., voids, cracks, etc., were responsible for the increases in friction and wear. Although not explicitly stated, one may infer that the chemical and structural changes observed in these ceramics affected the strength and toughness of the implanted surface volumes.

The reader should now turn back to Eqs. (8), (9) and (10) and Fig. 7. These relations show that the tribological behavior of ceramics is indeed controlled by the fracture toughness and strength of the materials. Although the relations in Eqs. (8), (9) and (10) were developed with bulk properties in mind, it is reasonable to argue that it is the state of stress and the toughness prevailing in the volume of materials affected by the wear process that are important. We conclude that the tribological characteristics of ceramic surfaces should scale with their toughness and strength following surface modification treatment by energetic ion beams. These statements should hold true for a broad range of temperatures with the stipulation that elevated temperatures do not affect appreciably the thermodynamic equilibrium within the implanted volume.

Several studies [48-50] related to the tribological behaviors of structural ceramics at temperature about 350°C, reported the onset of high friction coefficients and severe wear for dry mating surfaces. Elimination of lower temperature environmental lubricating effects, i.e., water vapor, concomitant with enhanced chemical reaction across the bare interfaces, were a few of the explanation advanced to account for the observations. Wedeven et al. [49] showed that lubrication by graphite containing certain additives, could be remarkably effective for sliding silicon nitride components at elevated temperature, due to the formation of a "glassy-like" interfacial layer. These investigators also noted that the lubricating effectiveness of graphite was due to a large extent to the release of phosphorous pentoxide. This oxide was responsible for the strong adherence of the glassy transfer film to the silicon nitride surfaces.

Studt [51] compared the lubrication effectiveness of fatty acids with extreme-pressure additives applied to sintered α alumina and to silicon carbide. There was practically no lubrication in the case of SiC, while the oils were very effective in reducing friction and wear in the case of Al_2O_3. The marked difference in the behaviors of the two ceramic materials were attributed to the effective formation of lubricating adsorption layers. While the diamond bond structure of SiC prevented oil-ceramic interfacial reactions, long chains of zinc di-n-alkyl dithiophosphates (ZnDTP), formed strong bonds with the ionic bonded surface of alumina [51].

The above observations [45-51] are consistent with the well established principles of boundary film lubrication which state that effective film lubrication should (a) support the mating surfaces at the optimal separation to prevent interaction at asperities, (b) adhere well to the mating surfaces so that interfacial shear is controlled by the liquid film, and (c) be of optimal viscosity to provide a balance between the tendency for the film to be squeezed out and the dynamic resistance to shear.

These principles were behind the works of Lankford, Wei and Kossowsky [46,48]. They directed their work at developing ion beam mixed lubricating thin metallic film layers on structural ceramics operating at about 500°C in the partially oxidizing diesel atmosphere.

Kossowsky [52] proposed a model to describe the lubricating action of the oxide films, shown schematically in Fig. 8.

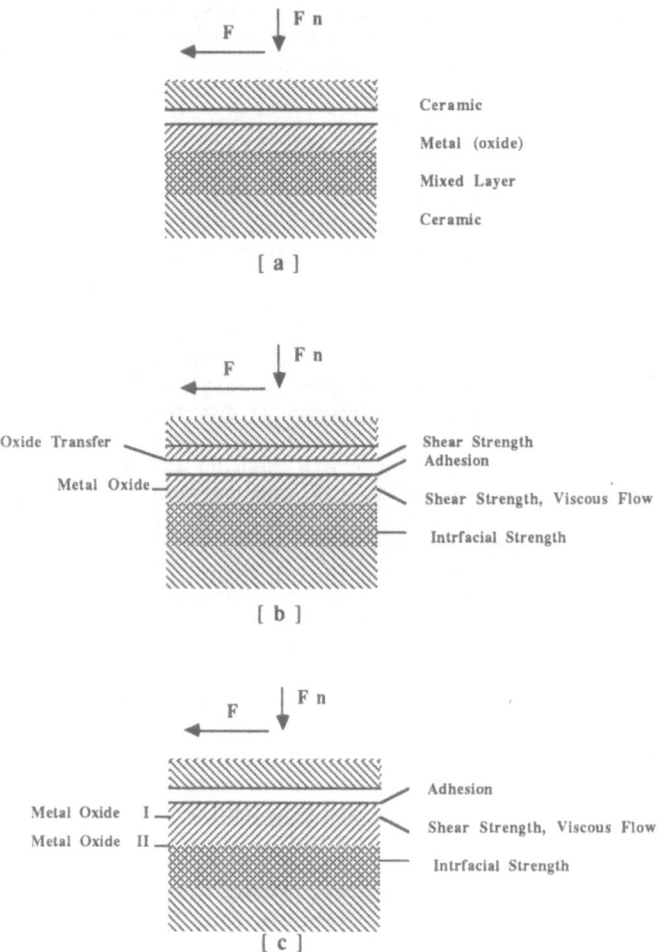

Figure 8. Schematic diagram of the lubrication process of ceramic/modified ceramic couples.

The initial state of the ceramic/ceramic couple is shown in Fig. 8(a). The modified ceramic may consist of up to three "layers", the metal film, the mixed metal in the ceramic substrate and the ceramic substrate. The actual thickness of each layer, and thus the specific contribution it makes during the wear process, depends essentially on the mixing efficiency of the specific metallic species in relation to the ceramic substrate [53]. The schematic in Fig. 8(a) refers, essentially, to adhesion controlled wear mechanism. The adhesive model is shown in Fig. 9 where the coefficient of friction would scale with temperature but should be independent of pressure and velocity within the normal ranges for these variables. Experimental evidence [46,48] points at two possible lubricating mechanisms, as shown in Figs. 8(b) and 8(c). In the first configuration, the oxide layer transfer to the unmodified rider. Lubrication is then provided, either by the reduced adhesion between the two oxide layers, or by the melting or flowing of the oxide layers. The model in Fig. 8(c) applies to conditions of high temperatures and pressures where the oxide layers adhere well to the ceramic surfaces due to enhanced diffusional reactivity, and the volume of the oxide flows and provides lubrication not unlike the action of a viscous oil film at low temperatures.

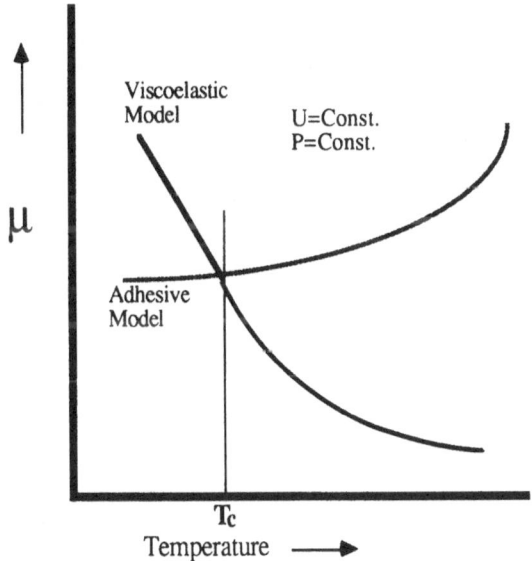

Figure 9.   Temperature dependencies of coefficient of friction for two wear mechanisms.

In rheological terms, Kossowsky wrote that

$$\dot{\gamma} = f(\tau,\eta) \tag{15}$$

where the exact functional relationship depends on the rheological characteristics of the fluid, i.e., Newtonian, non-Newtonian, etc. [54]. The coefficient of friction, μ, is related to the shear stress, τ, and the relative velocity of the wearing surfaces defines the shear strain rate or

$$u = \dot{\gamma} \quad, \quad \tau = \mu P \quad, \quad \text{and} \quad \tau = \alpha\dot{\gamma} \tag{16}$$

where P is the pressure and is given by the normal force divided by an effective contact area. The velocity and pressure dependencies of the coefficient of friction will thus be expressed as

$$\mu = f(u \ , \ \eta/P) \quad, \tag{17}$$

which is shown schematically in Fig. 10. For a Newtonian-like lubricating oxide, the dependence of μ on velocity will be linear. It is conceivable, though, that there would be a limiting value for μ, $\mu_a(T)$, above which the system will revert to the adhesive model behavior.

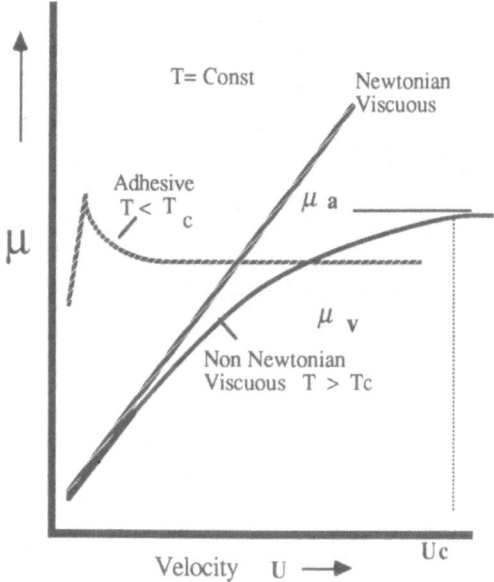

Figure 10.   Velocity dependencies of coefficients of friction.

4. SUMMARY
    The problems of friction and wear at mating ceramic surfaces have
been receiving increased attention, in step with the expanding field of
applications for structural ceramic components. One avenue for the
optimization and possible tailoring of ceramic surfaces wearing at
elevated temperatures is through the utilization of high energy
deposition schemes. Friction and wear are secondary to multitude of
parameters, intrinsic and extrinsic, which come together to dictate the
dynamic behavior of a particular system. These parameters were
reviewed against the background of prevailing friction and wear
theories. The evolution of friction and wear with temperature was
discussed, with emphasis on adhesive wear. We concluded with
a review of surface modification schemes by energetic ion beams and how
they affect the tribological behavior of ceramic materials. Special
attention was paid to the potential of the utilization of ion beam
mixing to introduce metal/oxide high temperature lubricating films.

5. REFERENCES
(1) Firestone, F. A. and E. J. Abott. Test for Smoothness of Machined
     Surfaces. Metal Progr. 21:57 (1932).
(2) Edmonds, M. J. Surfaces in Tribological Technology, Vol. I,
     P. B. Senholzi (ed), Martinus Nijhoff, Dordrecht, p. 290 (1982).
(3) Kossowsky, R. and W. Wei. Friction and Wear in Surface
     Modification Engineering, R. Kossowsky (ed), CRC Press,
     Boca Raton (1988).
(4) Tabor, D. Friction -- The Present State of our Understanding.
     J. Lub. Tech. 103:169 (1981).
(5) Archard, J. A. Surface Topography and Tribology. Tribology
     10:213 (1974).
(6) Greenwood, J. A. and J. B. P. Williamson. Contact of Normally
     Flat Surfaces. Proc. Roy. Soc. (London), p. 300 (1966).
(7) Wilks, E. M. and J. Wilks. The Resistance of Diamond to Abrasion.
     J. Phys. D. Appl. Phys. 5:1902 (1972).
(8) Thomas, T. R. Rough Surfaces, Longman, London (1982).
(9) Bowden, F. B. and D. Tabor. Friction, An Introduction to
     Tribology, Anchor Press/Doubleday, New York, p. 16 (1973).
(10) Holm, R. Electric Contacts, Springer, New York, p. 34 (1967).
(11) Archard, J. F. Contact and Rubbing of Flat Surfaces. J. Appl.
     Phys. 24:981 (1953).
(12) Samuels, L. E., E. D. Doyle and D. M. Turley. Sliding Wear
     Mechanisms in Fundamentals of Friction and Wear of Materials,
     D. A. Rigney (ed), Am. Soc. Met., Metals Park, p. 13 (1981).
(13) Tabor, D. Status and Directions of Tribology as a Science in the
     80's in New Directions in Lubrication, Materials, Wear and Surface
     Iterations, W. R. Lomis (ed), Noyes, Park Ridge, p. 1 (1985).
(14) Czichos, H. Importance of Properties of Solids to Friction and
     Wear Behavior in New Directions in Lubrication, Materials, Wear
     and Surface Iterations, W. R. Lomis (ed), Noyes, Park Ridge, p. 68
     (1985).

292

(15) Sutor, P. Tribology of Silicon Nitride-Steel Sliding Pairs in Proc. 9th Annual Conference on Composites and Advanced Ceramic Materials, F. E. Gac (ed), Amer. Cer. Soc., p. 460 (1985).

(16) Ishigaki, H., I. Kawaguchi and H. Iwasa. Friction and Wear of Hot Pressed Silicon Nitride and Other Ceramics in Wear of Materials 1985, K. C. Ludema (ed), Amer. Soc. Mech. Eng., New York, p. 13 (1985).

(17) Timoshenko, S. P. and J. N. Goodier. Theory of Elasticity, McGraw-Hill, New York, Chapter 13 (1970).

(18) Bowden, F. P. and D. Tabor. The Friction and Lubrication of Solids, Clarendon, Oxford, Vol. I (1950) and Vol. II (1966).

(19) Buckley, D. H. and K. Miyoshi. Friction and Wear of Ceramics. Wear 100:333 (1984).

(20) Adewye, O. O. and T. F. Page. Frictional Deformation and Fracture in Polycrystalline SiC and Si_3N_4. Wear 70:37 (1981).

(21) Finger, D. G. Contact Stress Analysis of Ceramic to Metal Interfaces. Final Report, ONR Contrat N00014-78-C-0547.

(22) Chiang, S. S. and A. G. Evans. Influence of a Tangential Force on the Fracture of Two Contacting Elastic Bodies. J. Amer. Cer. Soc. 66:4 (1983).

(23) Macmillan, N. H., A. R. Huntingdon and A. R. C. Westwood. Chemomechanical Control of Sliding Friction Behavior in Non-Metals. J. Mat. Sci. 9:697 (1974).

(24) Lawn, B. R. and D. B. Marshall. Indentation Fracture and Strength Degradation in Ceramics in Fracture Mechanics of Ceramics, Vol. 3, Flaws and Testing, R. C. Brandt, D. P. H. Hasselman and F. F. Lange (eds), Plenum Press, New York, p. 205 (1978).

(25) Frank, F. C. and B. R. Lawn. On the Theory of Hertzian Fracture. Proc. Roy. Soc., London, p. 291 (1967).

(26) Zum Gahr, K. H. and D. V. Doane. Wear Resistance and Fracture Toughness in Brittle Materials. Metall. Trans. 11A:141 (1980).

(27) Suh, N. P. Principles of the Delamination Mechanisms of Wear. Wear 44:1 (1977).

(28) Suh, N. P. Update on the Delamination Theory of Wear in Fundamentals of Friction and Wear of Materials, D. A. Rigney (ed), Am. Soc. Met., Metals Park, OH, p. 235 (1981).

(29) Hannink, R. H. J., M. J. Murray and H. G. Scott. Friction and Wear of Partially Stabilized Zirconia: Basic Science and Practical Applications. Wear 100:355 (1984).

(30) Yust, C. S. Tribology and Wear. International Metals Review 30:141 (1985).

(31) Buckley, D. H. Surface Effects in Adhesion, Friction, Wear and Lubrication, Elsevier Scientific Publishing Company, Amsterdam (1981).

(32) Buckley, D. H. and K. Miyoshi. Fundamental Tribological Properties of Ceramics in Ceramic Science and Engineering Proceedings, Vol. 6, pp. 7-8 (1985).

(33) Shimura, H. and Y. Tsuya. Effects of Atmosphere on the Wear Rate of Some Ceramics and Cermets in Proc. Int. Conference Wear of Materials, St. Louis, MO, p. 452 (1977).

(34) Tennenhouse, G. J. and R. B. Runkle. The Effect of Oxygen on the Wear of Si_3N_4 against Cast Iron and Steel. Wear 110:75 (1986).

(35) Miyoshi, K. and D. H. Buckley. XPS, AES and Friction Studies of Single-Crystal Silicon Carbide. Applications of Surface Science 10:357 (1982).

(36) Greggi, J., Jr. and R. Kossowsky. STEM Studies of Ion Implanted Cemented Co WC in Science and Hard Materials, D. J. Rowcliff and D. J. Gurland (eds), Plenum Press, New York, p. 375 (1983).

(37) Yust, C. S. and C. J. McHargue. Microstructure and Mechanical Properties of Ion-Implanted Ceramics in Materials Science Research, Vol. 17: Emergent Process Methods for High-Technology Ceramics, R. F. Davis, H. P. Plamour, and R. L. Porter (eds), Plenum Press, New York, p. 533 (1984).

(38) Rai, A. K., R. S. Bhattacharya, P. P. Pronko and T.-i. Mah. Application of Cross-Sectional Transmission Electron Microscopy in the Characterization of Ion Beam Processed Materials Surface. Surface and Interface Analysis 10:142 (1987).

(39) Bohn, H. G., J. M. Williams, C. J. McHargue and G. M. Begun. Recrystallation of Ion-Implanted α-SiC. J. Mater. Res. 2:107 (1987).

(40) Robert, S. G. and T. F. Page. The Effect of N^+ Ion Implantation on the Hardness and Behavior of Brittle Materials in Ion Implantation into Metals, W. Ashworth, W. A. Grant and R. P. M. Procter (eds), Pergamon Press, p. 135 (1982).

(41) Burnett, P. J. and T. F. Page. Changing the Surface Mechanical Properties of Silicon and α Al_2O_3 by Ion Implantation. J. Mat. Sci. 19:3524 (1984).

(42) Oliver, W. C., C. J. McHargue, G. C. Farlow and C. W. White. The Hardness of Ion Implanted Ceramics in Defect Properties and Processing of High-Technology Non-Metallic Materials, Materials Research Society Symposia Proceedings, p. 515 (1986).

(43) McHargue, C. J. Structural and Mechanical Properties of Ion Implanted Ceramics. Nucl. Instrum. Method B19/20:797 (1987).

(44) Nastasi, M., R. Kossowsky, J. P. Hirvonen and N. Elliot. Friction and Wear Studies in N Implanted Al_2O_3, SiC, TiB_2 and B_4C Ceramics. J. Mat. Res. [to be published in 1989].

(45) Sutor, P., C. Windisch and M. Starcher. Ceramic Friction Studies. MRI Final Report, Contract No. F-33615-80C-5190, Midwest Research Institute, Kansas City (1983).

(46) Wei, W., J. Lankford and R. Kossowsky. Friction and Wear of Ion-Beam Modified Ceramics for Use in High Temperature Adiabatic Engines. Mat. Sci. Eng. 90:307 (1987).

(47) Fischer, T. E. and H. Tomizawa. Interaction of Tribochemistry and Microfracture in the Friction and Wear of Silicon Nitride. Wear 105:29 (1985).

(48) Lankford, J., W. Wei and R. Kossowsky. Friction and Wear Behavior of Ion Beam Modified Ceramics. J. Mat. Sci. 22:2064 (1987).

(49) Wedeven, L. D., R. A. Pallini and N. C. Miller. Tribological Examination of Unlubricated and Graphite-Lubricated Silicon Nitride under Traction Stress. Wear 122:183 (1988).

(50) Sugita, T., K. Veda and Y. Kanemura. Material Removal Mechanism of Silicon Nitride during Rubbing in Water. Wear **97**:1 (1984).

(51) Studt, P. Influence of Lubricating Oil Additives on Friction of Ceramics under Conditions of Boundary Lubrication. Wear **115**:185 (1987).

(52) Kossowsky, R. Unpublished Work. The Pennsylvania State University, University Park, PA (August 1986).

(53) Appleton, B. R. Proceedings of the Second Workshop on Ion Mixing and Surface Layer Alloying, D. M. Fullstaedt, R. S. Averback and M. A. Nickolet, (eds), Sandia Report No. SAND85-2464, p. 99 (1986).

(54) Wilson, W. R. D. and S. Shea. Rheological Behavior of Lubricating Oil Films. J. Lub. Tech. **105**:187 (1983).

THIN FILM AND NEAR SURFACE CHARACTERIZATION USING INDENTATION SYSTEMS

D. L. JOSLIN, C. J. McHARGUE,* AND W. C. OLIVER*

University of Tennessee, Knoxville, TN 37996-2200

1. INTRODUCTION

In recent years, thin films and ion implanted layers have been the focus of much research. Such surface layers are used for a variety of applications in industry. For example, thin metallic films are used for electrical contacts and for diffusion barriers in the semiconductor industry. Thin ceramic films are used as optical coatings for light emitting diodes and for optical windows. Thin films are also used to enhance wear resistance, and to provide corrosion protection and lubrication. Ion implanted layers are created in both metals and ceramics to enhance the toughness and wear resistance of such things as medical implants and optical materials. Such layers can also provide corrosion resistance and lubrication.

As the number of applications for thin films and ion implanted layers grows, so does the need to know the mechanical and elastic properties of such surface layers. This need has rejuvenated interest in microindentation techniques for the analysis of these properties. Traditional microindentation tests, such as Knoop and Vickers, have been used for this purpose. Such tests do indicate trends in the hardness of these materials. However, due to the effect of the substrate on the measurements, they do not provide substrate-independent values for the hardness of the layers. For example, a Knoop test done on sapphire using a 0.147 N (15-g) load produces an impression with a long diagonal of about 12 μm and a depth of about 0.4 μm. A 0.147 N Vickers indent in sapphire will produce an impression 0.8 μ deep. Since the thickness of many technologically interesting thin films are on the order of 0.1 to 0.3 μm, the hardness usually measured by either the Knoop or the Vickers method is actually a composite of the hardness of the layer and its substrate. In order to avoid the influence of the substrate, the indent depth must be substantially less than the film thickness.

Because of the difficulties in imaging and measuring small indents, depth-sensing microindentation systems have been developed by a number of researchers (1-4). However, elimination of the indent imaging step is not the only advantage of depth-sensing microindentation systems. Continuous data acquisition during the loading portion of the test allows measurement of the properties of interest as a function of depth in a single indentation experiment. In addition, using data taken during the unloading portion of the indentation experiment, these systems can also provide information on elastic properties of materials (5).

*Oak Ridge National Laboratory, Oak Ridge, TN 37831-6118. This research is sponsored in part by the U.S. Department of Energy, Assistant Secretary for Conservation and Renewable Energy, Office of Transportation Systems as part of the Ceramic Technology for Advanced Heat Engines Project of the Advanced Materials Development Program, under contract DE-AC05-840R21400 with the Martin Marietta Energy Systems, Inc.

C.J. McHargue et al. (eds.), Structure-Property Relationships in Surface-Modified Ceramics, 295–302.
© 1989 by Kluwer Academic Publishers.

In order to obtain the shallow indentations necessary to test thin films and ion implanted layers, loads much smaller than those used for conventional microhardness testing are used. For example, the minimum practical load to give indents large enough to be imaged optically with reasonable accuracy in Knoop testing of ion implanted ceramics is 0.147 N. The standard ultra-low load microindentation system used in this study operates with loads between 3 μN and 120 mN.

The purpose of this research is to demonstrate the capabilities of ultra-low load microindentation testing for determination of the hardness and elastic modulus of thin films (both metallic and non-metallic) and ion implanted layers on ceramic substrates.

2. EXPERIMENTAL

2.1. Sample materials

Two substrates were used to support the metal and hard carbon films used in this study: sapphire and ALON ($Al_{23}O_{27}N_5$). The sapphire substrates were 25 mm diam, 1-mm thick, optically polished (0001) single crystal disks (Crystal Systems, Inc., Bedford, MA) which were annealed in oxygen for 120 h at 1400°C. The polycrystalline ALON substrates were also optically polished but larger (30 mm diam, 2 mm thick) disks (Raytheon, Inc., Lexington, MA) which were used as received.

For the thin film experiments, both metallic and non-metallic films were used. Diamond-like carbon coatings (DLC), also known as hard carbon coatings (HCC), were deposited onto each type of substrate by rf glow discharge of methane in argon. The coatings were approximately 100 nm thick. A chromium film approximately 50 nm thick was evaporated onto a sapphire substrate by a standard evaporation technique. All films were tested for adherence to their substrates using a standard pull test performed with a Sebastian I adherence tester (Quad Group, Santa Barbara, CA). For this test, a 3-mm-diam pull pin was epoxied to the film, and the tensile force necessary to remove the film was measured.

Two ion implanted samples were tested. Both substrates were sapphire. One sample was implanted at room temperature with a fluence of 4 x 10^{15} Cr ions/cm^2 at an energy of 150 keV, thus producing a damaged but still crystalline surface layer (6). A second sample was implanted at liquid nitrogen temperature with a stoichiometric mixture of aluminum and oxygen (4 x 10^{16} Al ions/cm^2 at 90 keV and 6 x 10^{16} O ions/cm^2 at 55 keV), thus producing an amorphous surface layer. An area on each of the specimens was masked off during implantation to provide a virgin substrate area (6).

2.2. Ultra-low load (ULL) indentation testing

The experiments were carried out on a commercially available computer-controlled ULL indentation system, the Nanoindenter (Nano Instruments, Inc., Knoxville, TN). A schematic of the instrument is shown in Fig. 1. The system uses a triangular pyramid-shaped diamond indenter, which is driven into the test material by a coil and magnet assembly. The indenter's position is determined by a capacitance displacement gauge with a resolution of 0.16 nm. The loading column is supported by springs, and the force on the column is controlled by varying the current in the coil. The load resolution is 0.3 μN. The motion of the indenter is damped by air flow around the center plate of the capacitor, which is attached to the loading column.

Although the system can be operated in other test modes (i. e., constant loading rate), these experiments were performed in a constant displacement rate mode. The test procedure involves driving the indenter toward the surface of the test specimen at a constant rate, detecting the surface contact as a change in velocity. After the surface has been detected, the indenter is

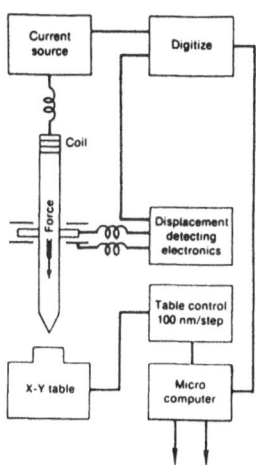

FIGURE 1. Schematic diagram of the Nanoindenter.

forced into the specimen at a different but still constant displacement
rate. The system continuously records the displacements and forces associated with the indentation process for a pre-set sequence (for example,
load to a certain depth, unload a specified amount, reload to a deeper
depth, hold for a period to detect thermal drift, unload totally). A typical load-displacement curve is shown in Fig. 2. With knowledge of the exact
geometry of the diamond indenter and the system compliance, the data can be
processed to give both the hardness and the Young's modulus of the test
material as a function of depth from the free surface (5).

3. RESULTS AND DISCUSSION
   Plots of the hardness and elastic modulus as a function of indenter
displacement for the HCC coatings are shown in Figs. 3 and 4. Samples 2, 4,
5, and 10 were DLC on sapphire substrates; sample 11 was HCC on an ALON
substrate. The films on samples 2, 4, and 5 were found to be poorly
adherent to their substrates by a standard pull test. In contrast, the
films on 10 and 11 were very adherent and this is reflected in the hardness
data. Samples 10 and 11 show substrate effects - the measured hardness
approaches the hardness of the substrate alone (about 23 GPa for sapphire
and 24 GPa for ALON) at increasing indentation depths. At depths shallower
than some shallow critical depth, the measured hardness is the hardness of
the film, and at some depth deeper than the film thickness, the hardness
measured is that of the substrate. All of the hardnesses measured at intermediate depths are some composite of the two. The hardness data from
samples 2, 4, and 5 show no substrate effects. The hardness values are virtually the same at each of the depths sampled. Since these films are not
well-adhering, the stress produced during indentation is not effectively
transferred across the film/substrate interface.
   Since the film/substrate interface is perpendicular to the direction of
indentation, it is subjected to substantial shear stresses during the indentation process. A weak interface cannot support shear stresses; thus, the
shear stresses produced during indentation are not effectively transferred
across the interface. Therefore, the substrate may not plastically deform
to the same extent that it would if the film were adherent.

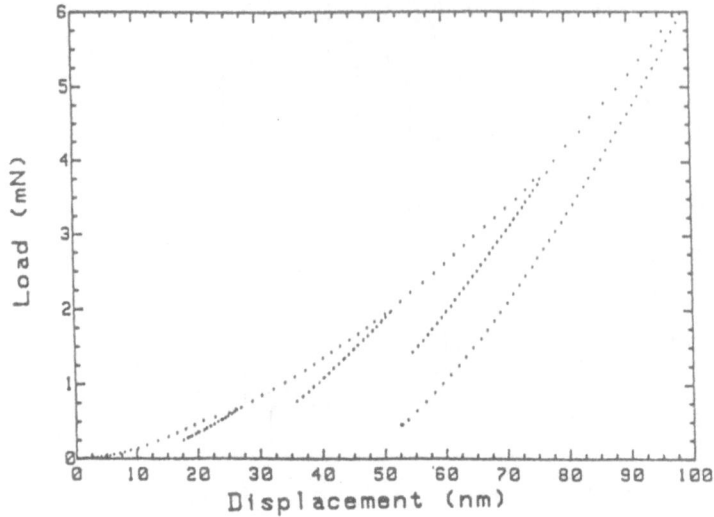

FIGURE 2. Typical load-displacement curve for sapphire.

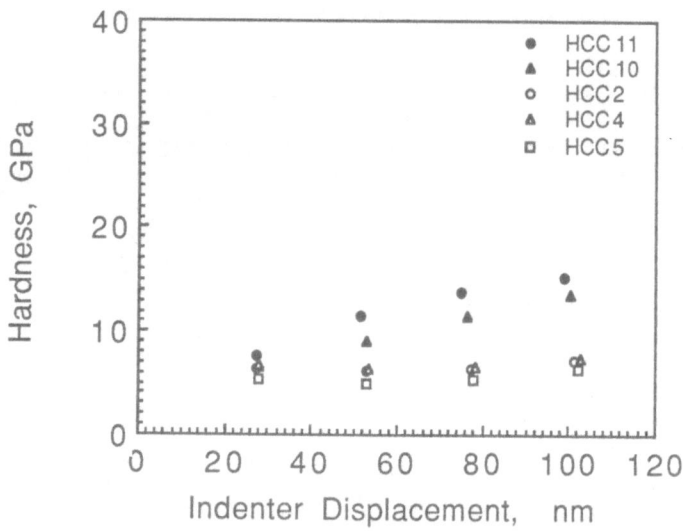

FIGURE 3. Hardness versus indenter displacement for five different diamond-like carbon films. Sample 11 was on an ALON substrate. Samples 2, 4, 5, and 10 were on sapphire. The films on samples 10 and 11 were adherent to the substrates. The films on samples 2, 4, and 5 were nonadherent. Adherence was determined by the pull test.

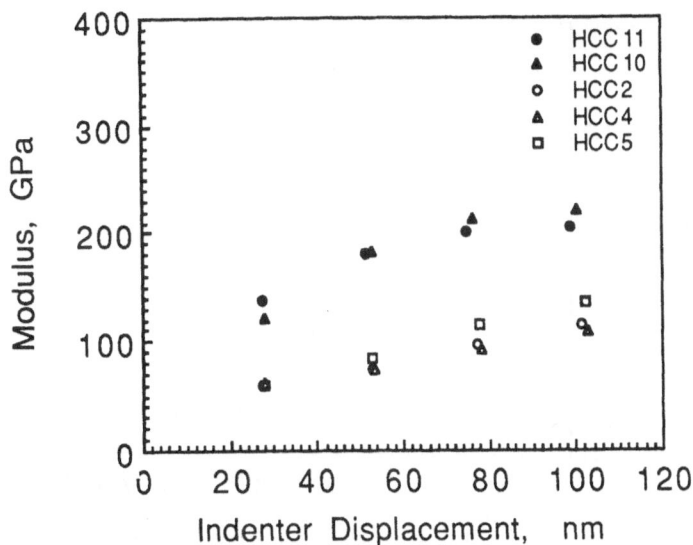

FIGURE 4. Modulus versus indenter displacement for the same diamond-like carbon films shown in Fig. 5.

Substrate-independent modulus values are not present in the data for either the adherent or non-adherent HCC films. This is because the measured modulus values depend more on the normal stresses exerted on the sample. Although the film/substrate interface is weak in shear, it can support compressive normal stresses. Thus, the substrate affects the modulus measurements for both adherent and non-adherent films.

The measured hardness of the HCC films is approximately 7 GPa, which is within the range of values quoted in the literature (7,8), although it is at the low end of the range. The spread of the values quoted in the literature may be due to actual differences in films. Alternatively, the substrate may have affected some of these data, since most were obtained from Knoop and Vickers systems. If one calculates the indentation depths used from the loads and measured hardnesses, it can be seen that the depths of most of the Knoop and Vickers hardness impressions were significant fractions of the film thicknesses. Data taken on the Nanoindenter for this experiment is in good agreement with other ULL data (7).

Figures 5 and 6 show data taken on an adherent 50 nm thick chromium film on a sapphire substrate. Both the hardness and modulus data show the effects of the substrate on the measured values. The measured hardness of the film decreases with decreasing depth, but never reaches a constant value. The hardness of the film alone was not measured. The data extrapolates to a value of approximately 5 GPa, which is within the range of values cited for chromium, 1 to 10 GPa (8). The measured modulus of the film also decreases, extrapolating to a value of approximately 225 GPa. This value is in good agreement with the bulk value, which is around 248 GPa (9).

Data taken from the ion implanted samples are shown in Figs. 7 and 8. The amorphous layer (stoichiometric Al and O implantation at liquid nitrogen temperature) is approximately half as hard as the sapphire substrate. However, the apparent hardness of the damaged but crystalline layer (Cr implantation at room temperature) is slightly higher than the substrate hardness. Thus, ion implantation can raise the apparent surface hardness of ceramics, as previously found by other workers using the Knoop method (10-13).

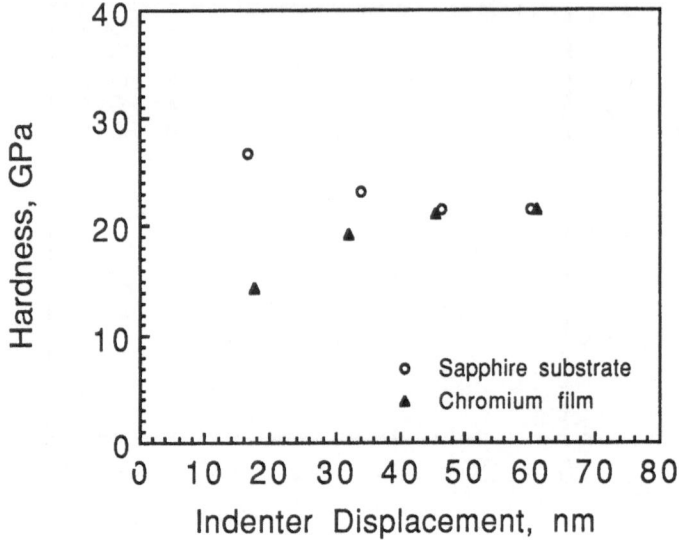

FIGURE 5. Hardness versus indenter displacement for a 50-nm-thick chromium film and its sapphire substrate.

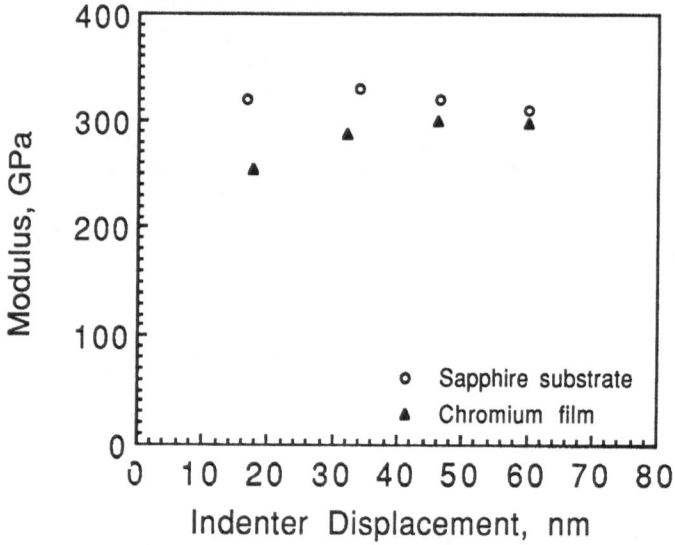

FIGURE 6. Elastic modulus versus indenter displacement for a 50-nm thick chromium film and its sapphire substrate.

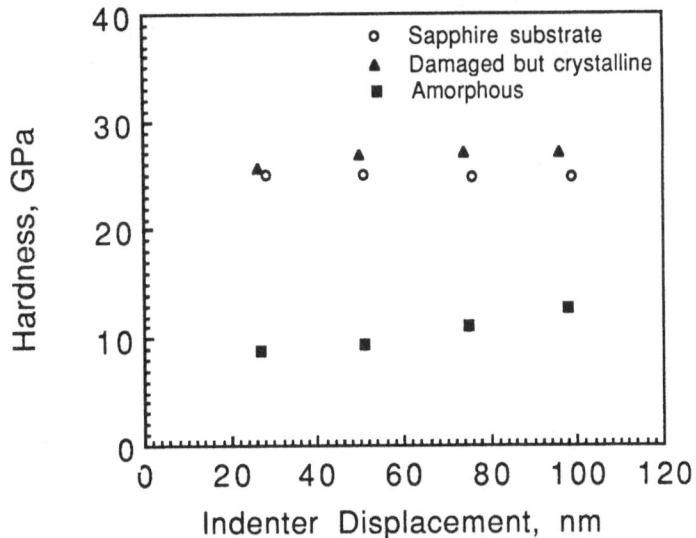

FIGURE 7. Hardness versus indenter displacement for two different implanted layers on the surface of a sapphire substrate. The damaged but crystalline layer was formed using a fluence of 4 x 10^{15} Cr ions/cm^2 at an energy of 150 keV at room temperature. The amorphous layer was produced by dual implantations of 1 x 10^{16} Al ions/cm^2 at 90 keV and 6 x 10^{16} O ions/cm^2 at 55 keV at liquid nitrogen temperature.

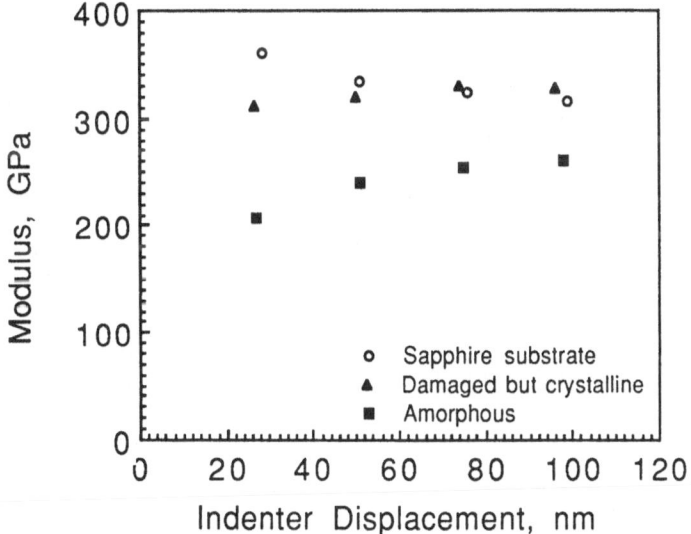

FIGURE 8. Modulus versus indenter displacement for the same samples as Fig. 7.

This increase in hardness can be attributed to a combination of solid solution and defect strengthening (14). The modulus of the amorphous layer is lower than that of the sapphire substrate, while the modulus of the damaged but crystalline layer is slightly higher.

4. CONCLUSIONS

As evidenced by these experiments, the hardness and elastic modulus of thin films and ion implanted layers on ceramic substrates may be determined using ultra-low load microindentation testing. Substrate effects are easily distinguished, and poor film adhesion can be detected by this method.

REFERENCES

1. Nishibori M, and K Kinosita: Thin Solid Films 48 (1978) 325.
2. Pethica JB, R Hutchings, and WC Oliver: Phil. Mag. A 48(4) (1983) 593.
3. Loubet JL, JM Georges, O Marchesini, and G Meille: J. Tribology 106 (1984) 43.
4. Newey D, HM Pollock, and MA Wilkins: Ion Implantation into Metals. New York: Pergamon Press, 1982, 157.
5. Doerner MF, and WD Nix: J. Mater. Res. 1(4) (1986) 601.
6. Oliver, WC, CJ McHargue, GC Farlow, and CW White: Defect Properties and Processing of High-Technology Non-Metallic Materials. Y Chen, WD Kingery, and RJ Stokes (eds). Pittsburgh: Materials Research Society, 1986, 515.
7. Tsai H-c, and DB Bogy: J. Vac. Sci. Technol. A 5(6) (1987) 3287.
8. Pethica, JB, P Koidl, J Gobrecht, and C. Schüler: J. Vac. Sci. Technol. A 3(6) (1985) 2391.
9. Metals Handbook, 8th ed.(1) Cleveland: American Society for Metals, 1961, 1201.
10. McHargue CJ, MB Lewis, BR Appleton, H Naramoto, CW White, and JM Williams: Science of Hard Materials. RK Viswanadham, DJ Rowcliffe, and J Gurland (eds). New York: Plenum Publishing Co., 1983, 451.
11. McHargue CJ, CW White, BR Appleton, GC Farlow, and JM Williams: Ion Implantation and Ion Beam Processing of Materials. GK Hubler, OW Holland, CR Clayton, and CW White (eds). Amsterdam: North-Holland, 1984, 385.
12. Farlow GC, CW White, CJ McHargue, PS Sklad, and BR Appleton: Nucl. Inst. Meth. Phys. Res. B 7/8 (1985) 541.
13. Burnett PJ, and TF Page: Plasticity of Ceramic Materials. RC Bradt and RE Tressler (eds). New York: Plenum Publishing Co., 1984, 669.
14. McHargue CJ: Ion Beam Modification of Insulators. P Mazzoldi and G Arnold (eds). Amsterdam: Elsevier Science Publishers, 1987, 223.

TRIBOLOGICAL PROPERTIES OF CERAMICS MODIFIED BY ION IMPLANTATION AND BY
ION BEAM-ASSISTED DEPOSITION OF ORGANIC MATERIAL

T. HIOKI, A. ITOH, S. HIBI, AND J. KAWAMOTO

Toyota Central Research & Development Laboratories, Inc.
Nagakute, Aichi-gun, Aichi, 480-11 Japan

1. INTRODUCTION

Ceramic materials are increasingly being used as rolling or sliding parts
of machines, for example, as bearings or seals, and surface treatments to
enhance the friction and wear performance of ceramics have become more and
more important.

Surface modifications of structural ceramics by ion implantation or by ion
beam-assisted techniques have recently been studied extensively (1–3]. Ion
implantation in ceramics has been demonstrated to significantly affect the
mechanical properties, such as surface hardness (4), indentation fracture
behavior (5), flexural strength (6), friction and wear characteristics (7),
and so on. It has also been found that coefficients of friction as low as
about 0.05 are obtained for ceramic surfaces modified by ion beam mixing (8)
or by metal film deposition and subsequent ion irradiation (9).

In this paper, we report the influences of energetic Ar^+ ion implantation
on the friction and wear properties of polycrystalline SiC and Si_3N_4 in
sliding contact with SUJ2 steel or Si_3N_4 ceramics. We also report the formation of an adhesive solid carbonaceous film onto ceramic surfaces by vapor
deposition of a silicone oil and energetic Ar^+ ion irradiation. The tribological properties of Si_3N_4 and sapphire coated with this carbonaceous film
are presented.

2. EXPERIMENTAL PROCEDURE

2.1. Materials

The ceramic materials studied were polycrystalline α-SiC (Kyocera, SC201),
polycrystalline α-Si_3N_4 (Kyocera, SN220) and single crystalline α-Al_2O_3
(sapphire). Disk samples (3 mm thick, 30 mm in diameter) of these ceramics
were surface-modified and were subjected to pin-on-disk type tribology
tests. The surface roughnesses of the disks were 0.1 μmRa (an average
deviation from a mean line of the surface profile) for SiC, 0.25 μmRa for
Si_3N_4, and less than 0.02 μmRa for Al_2O_3. The pin materials used were
SUJ2(US code; 52100) steel (0.1 μmRa) and Si_3N_4 ceramics (0.1 μmRa).

The organic material used for the vapor deposition was a silicone oil
[pentaphenyle-trimetyl-trisiloxane, $(Si_3O_2C_{33}H_{32})_n$] with specific gravity
1.09.

2.2. Ion implantation

Ions of Ar^+ at an acceleration energy of 800 keV were implanted to the
disks of SiC or Si_3N_4 at room temperature to an ion dose of 1 x 10^{16}
ions/cm^2.

C.J. McHargue et al. (eds.), Structure-Property Relationships in Surface-Modified Ceramics, 303–311.
© 1989 by Kluwer Academic Publishers.

2.3. Ion beam-assisted deposition of silicone oil

The vacuum chamber used for forming the carbonaceous films on the ceramic disks is illustrated schematically in Fig. 1. The base pressure of the system was 1×10^{-6} Torr. The disk samples were mounted on a sample holder which was cooled with liquid nitrogen. A reservoir for the silicone oil was heated in the chamber, and a shutter in front of the oil reservoir was opened when the reservoir temperature reached a given value ranging from 60 to 120°C. Simultaneously, Ar^+ ions at an acceleration energy of 1.5 MeV irradiated the samples to a fixed dose of 5×10^{16} ions/cm². During the ion irradiation, the reservoir temperature was kept constant. The coated film thickness was varied mostly by changing the oil reservoir temperature.

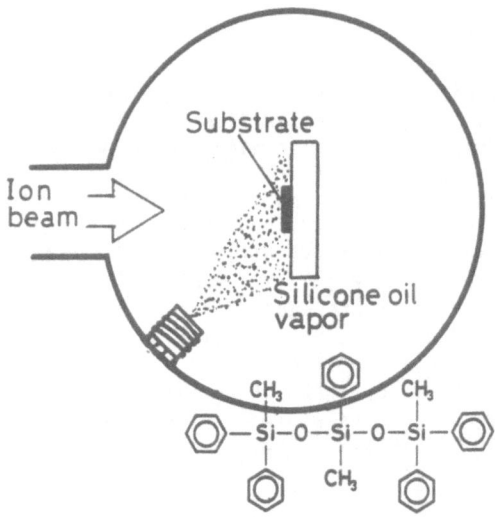

FIGURE 1. The system for silicone oil vapor deposition and Ar^+ ion irradiation.

2.4. Measurements

2.4.1. Friction and wear. A pin-on-disk type apparatus was used to eva-luate the tribological property of the surface-modified ceramics. The disk samples were slided against loaded, unimplanted, hemispherical-ended pins of SUJ2 or Si_3N_4 of 5 mm in diameter. Testing conditions included no lubri-cant, room temperature, loads of 2.2 N (otherwise specified), sliding velo-city of 0.05 m/s, disk rotation speed of 80 rpm and an ambient atmosphere of relative humidity of 50 to 70%.

2.4.2. Characterization. Surface layer analyses were performed by using Rutherford backscattering spectroscopy (RBS) with 2 MeV He^+ ion beam and by using Raman spectroscopy.

3. RESULTS AND DISCUSSION

3.1. Ion implantation

Figure 2(A) shows the coefficient of friction, μ, as a function of sliding time for Si_3N_4 disks unimplanted and implanted with 800 keV Ar^+ ions. The pin material is SUJ2. The coefficient of friction for the unimplanted disk is almost constant with sliding time and is 0.85. A similar pattern of μ versus sliding time is also seen for the Ar^+ ion-implanted disk. For this disk, μ is 0.80, which is slightly smaller than the unimplanted value. Similar tests were also performed for a variety of disk (Si_3N_4, SiC)-pin (SUJ2, Si_3N_4) combinations. The results are summarized in Table 1. Smaller values of μ are always obtained by the Ar^+ implantation into the disks.

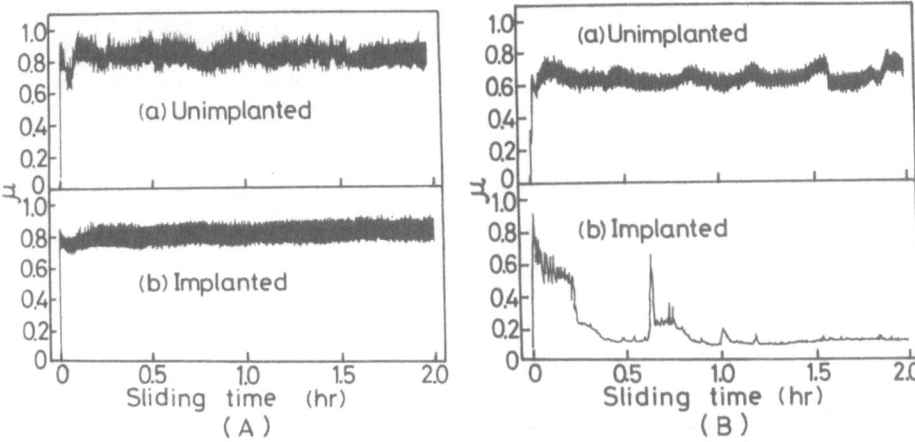

FIGURE 2. (A) μ versus sliding time for Si₃N₄ disks unimplanted (a) and implanted with 800 keV Ar⁺ ions to 1 x 10¹⁶ ions/cm² (b). The pin material is SUJ2. (B) μ versus sliding time for SiC disks unimplanted (a) and implanted with 800 keV Ar⁺ ions to 1 x 10¹⁶ ions/cm² (b). The pin material is SUJ2.

Table 1. μ and amplitude of fluctuation of μ (the figure in the parenthesis)

| | SiC disk | | Si₃N₄ disk | |
	unimplanted	implanted	unimplanted	implanted
SUJ2 pin	0.60(0.08)	0.15(<0.01)	0.85(0.16)	0.80(0.14)
Si₃N₄ pin	0.75(0.20)	0.50(0.20)	0.80(0.22)	0.75(0.20)

As shown in Fig. 2(B), a remarkable effect of the Ar⁺ ion implantation has been found for a SiC disk in sliding contact with a SUJ2 pin (10). For the unimplanted disk, both μ and the fluctuation of μ show large values of 0.6 and 0.08, respectively. On the other hand, for the ion-implanted disk, μ shows a high value at an initial stage of the test, but it decreases gradually with time and then reaches a value as low as about 0.15. The fluctuation of μ is also reduced as μ is decreased. These behaviors were observed for a broad range of the normal load changed (1 to 30 N). As shown in Fig. 3, the smooth sliding state with μ as low as 0.15 lasts for more than 1 x 10⁵ cycles. In Figs. 2(B) and 3, temporary, abrupt increases of μ are seen in the low friction state. The wear debris piled on both sides of the wear track is considered to sometimes drop in the track and disturb the smooth sliding state, resulting in the abrupt increases of μ. But, in the use of hemispherical pins, the wear debris tends to be removed from the track during the test, and the smooth sliding state recovers soon. Corresponding to the reduction in μ, the wear rate of the SUJ2 pin reduced

306

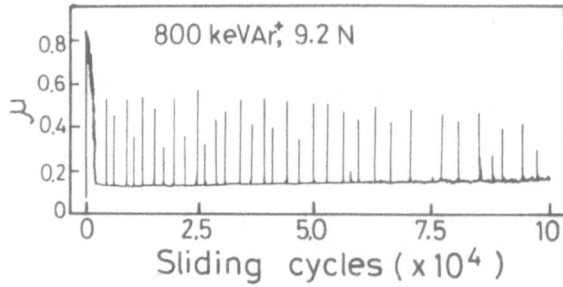

FIGURE 3. μ versus sliding cycle for a SiC disk implanted with 800 keV Ar⁺ ions to 1 x 10¹⁸ ions/cm². The pin material is SUJ2. The normal load applied to the pin is 9.2 N.

to about one-fifth of the value obtained for sliding against the unimplanted SiC disk (10). Raman spectroscopy measurements indicated amorphous phases of C, SiC, and Si on the wear scar surface at $\mu = 0.15$ and the crystalline phase of SiC on the surface at $\mu = 0.5$ (10). The low friction state maintained until the surface amorphous layer was worn away.

It has been known that covalently-bonded ceramic materials such as SiC and Si_3N_4 are easily amorphized by ion implantation, and that the amorphization is accompanied by a significant reduction of hardness (1–3). Microhardness measurements with a Knoop profile indenter at a load of 0.25 N showed that the hardnesses of the implanted surface layers are about 65 and 70% of the unimplanted values, for the SiC and Si_3N_4 disks, respectively. The reductions in μ by implanting Ar⁺ ions to SiC and Si_3N_4 disks are partly attributable to the formation of a soft amorphous surface layer on the hard substrate, i.e. the formation of a thin layer of a reduced shear modulus on the substrate of a high elastic modulus. It has been reported by Burnett and Page (7) that for sapphire, implantations of Ti⁺ and Zr⁺ ions to doses less than those required for amorphization results in increases in both the coefficient of friction and the near-surface microhardness.

As seen in Table 1, the observed reductions in μ are much larger for SiC than for Si_3N_4. For SiC, the Ar⁺ ion-irradiation produces an amorphous carbon phase in the implanted surface layer, as indicated from the Raman spectroscopy. This phase may work as a solid lubricant, resulting in the larger reductions of μ.

For the ion-implanted SiC disk, μ is lower in sliding contact with the SUJ2 pin than with the Si_3N_4 pin. This may be due to the easy formation of a well run-in surface for a metal pin. As seen in Figs. 2(B) and 3, the gradual decreases of both μ and the fluctuation of μ are preceded by an initial stage of the tests of about 15 min (10³ cycles), which is regarded as the running-in stage.

From the ion implantation experiments, it is suggested that an adhesive carbon film on ceramics may work as a good solid lubricant.

3.2. Ion beam-assisted deposition of silicone oil
3.2.1. Film characterization. Figure 4 shows the RBS spectrum for a Si_3N_4 disk coated with a carbonaceous film prepared by the silicone oil vapor deposition and 1.5 MeV Ar⁺ ion-irradiation. The film was black and showed a low electrical resistivity of about 10^{-1} Ω cm, indicating that the vapor-deposited silicone oil was carbonized by the ion-irradiation. As seen in Fig. 4, the film contains Si and O, as well as C. The RBS yields from Y

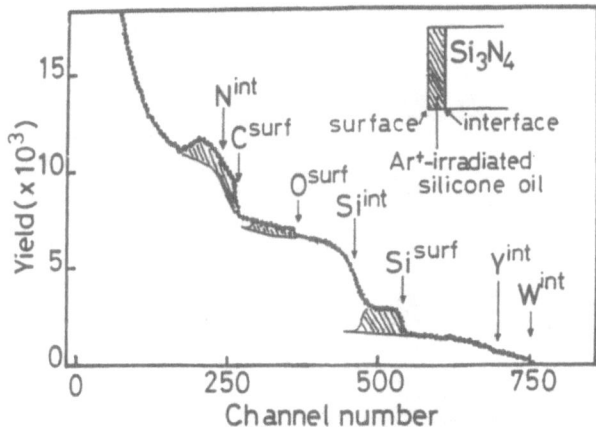

FIGURE 4. RBS spectrum for a Si₃N₄ disk coated with a carbonaceous film
prepared by silicone oil vapor deposition and 1.5 MeV Ar⁺ ion irradiation.

and W atoms are due to the sintering additives and the contaminants in the
Si₃N₄ substrate used. From the RBS spectrum, the film thickness is determined to be 0.27 μm by assuming that the film is made of C alone and that
the density of the film is equal to that of graphite, i.e. 2.3 g/cm³. It
has been known that ion irradiation of an organic material causes the
emission of volatile molecules made up of the composing elements of the
material (11). Consequently, gaseous elements such as H, N, O are preferentially removed from the material, resulting in a carbonization. A computer
simulation (12) of the RBS spectrum gave the atomic ratio for the film
$Si:O:C = 3:2.0:32$, which agrees well with that of the original silicone oil
(3:2:33). This suggests that volatile molecules containing Si, O, and C are
hardly generated in the Ar⁺ ion-irradiated silicone oil. It is probable
that the ion-irradiation of the silicone oil mainly causes the emission of
hydrogen molecules.
    In Fig. 5, the Raman spectrum for a Si₃N₄ disk coated with a 0.11 μm-thick
carbonaceous film is compared to that for a Si₃N₄ disk coated with a 0.15
μm-thick pure carbon film vapor-deposited by electron beam heating a
graphite rod. The asymmetric broad peak around 1500 cm⁻¹ in the spectrum for
the pure carbon film shows that the film is amorphous (13). A similar broad
peak is seen in the spectrum for the ion-irradiated silicone oil, indicating
that the film contains amorphous carbon phase.
    From these results, it is suggested that the film prepared by the silicone
oil vapor deposition and Ar⁺ ion-irradiation is similar to amorphous carbon
and that it contains considerable amounts of Si, O, and probably a small
amount of H.
    3.2.2. Tribological property. Figure 6 shows μ as a function of sliding
time for Si₃N₄ disks uncoated and coated with a carbonaceous film 0.27 μm
thick. The pin material used is SUJ2. As seen from the figure, for the
uncoated disk, μ is 0.85 and the fluctuation of μ is quite large. On the
other hand, for the coated disk, a coefficient of friction as low as 0.04 is
obtained and the fluctuation of μ is very small.
    In order to examine the durability of the film, long-term tests were con-
ducted. Further, in order to examine the effect of the film thickness, the
tests were repeated on a Si₃N₄ disk coated with a 0.11 μm-film. The results
are summarized in Fig. 7(A). For the disk coated with the 0.27 μm-film, the

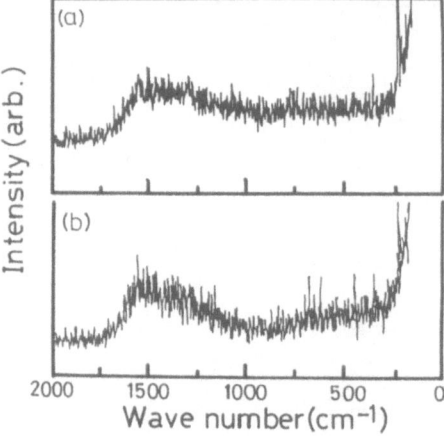

FIGURE 5. Raman spectra for Si_3N_4 disks coated with a 0.11-μm-thick carbonaceous film (a) prepared by silicone oil vapor deposition and 1.5 MeV Ar^+ ion-irradiation, and with a 0.5-μm-thick pure carbon film (b) prepared by electron beam-heating a graphite rod.

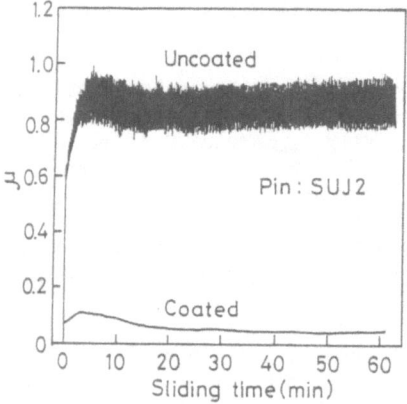

FIGURE 6. μ versus sliding time for Si_3N_4 disks uncoated and coated with a 0.27-μm-thick carbonaceous film prepared by silicone oil vapor deposition and 1.5 MeV Ar^+ ion-irradiation. The pin material is SUJ2.

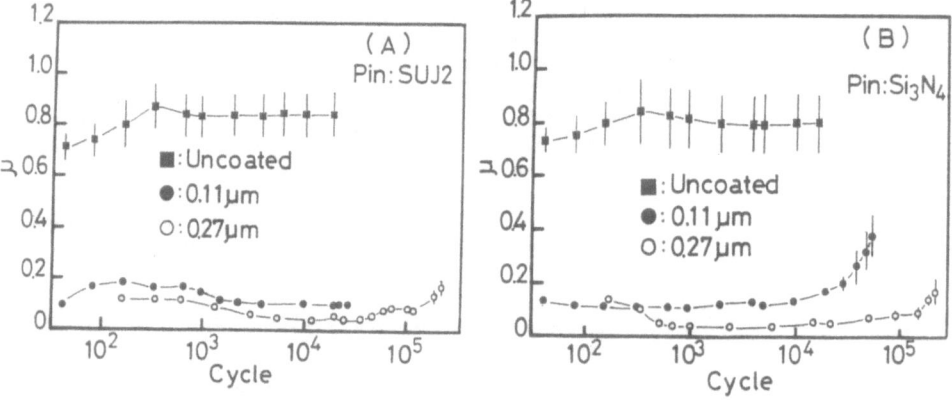

FIGURE 7. μ versus sliding cycle for Si_3N_4 disks uncoated (□) and coated with a 0.11-μm-thick (●) or a 0.27-μm-thick (○) film prepared by silicone oil vapor deposition and 1.5 MeV Ar^+ ion-irradiation. The pin materials are SUJ2 (A) and Si_3N_4 (B). The vertical bars denote the amplitude of fluctuation of μ.

sliding state with μ as low as 0.04 to 0.1 lasts about 1 x 10⁵ cycles, followed by gradual increases of both μ and the fluctuation of μ. For the disk coated with the 0.11-μm-film, the duration of the low friction state with μ of about 0.1 is more than 2 x 10⁴ cycles. The gradual increase of μ at larger sliding cycles corresponded to gradual removal of the coated film from the slided disk surface. The comparatively high durability and the gradual removal of the film indicate that the film is strongly adhered to the ceramic disk.

In order to evaluate the solid lubricant performance of the carbonaceous film for ceramics-ceramics couples, the coated Si_3N_4 disks were slided by Si_3N_4 pins. The results are shown in Fig. 7(B). It is seen that for the Si_3N_4-Si_3N_4 couple also, the film works as a good solid lubricant. For the uncoated Si_3N_4 disk, μ is about 0.8, whereas μ reduces to 0.1 for the disk coated with the 0.11 μm-film and to 0.04 for the disk with the 0.27 μm-film. The duration of the low friction state is about 10⁴ cycles for the 0.11 μm-film and about 10⁵ cycles for the 0.27 μm-film. The results are quite similar to those obtained by using SUJ2 steel pins. It has been demonstrated in Fig. 7 that the solid carbonaceous film coated on ceramic surface by the silicone oil vapor deposition and Ar⁺ ion-irradiation works as a good solid lubricant for both ceramics-ceramics and ceramics-metal couples.

In Fig. 7, it is also seen that the lowest value of μ obtained is dependent on the film thickness. The lowest value of μ is 0.1 for the 0.11 μm-film and 0.04 for the 0.27 μm-film.

In the former case, the film thickness is only about one-half of the surface roughness of the Si_3N_4 disk used, and there may be direct contact area of the pin and disk materials, resulting in an increased frictional force. It is considered that to obtain a good solid lubricant performance of the film, the thickness should be larger than or comparable to the surface roughness of the substrate material. A sapphire disk with surface roughness less than 0.01 μmRa was coated with a carbonaceous film 1.2 μm thick and was slided by a SUJ2 pin. In Fig. 8, μ versus sliding cycle for this disk is shown, together with the result for the uncoated sapphire. It is seen that μ reduces to a value as low as 0.05 and that the low friction state lasts about 2 x 10⁵ cycles.

3.2.3. <u>Comparison to pure carbon film</u>. In order to investigate the origin of the low μ value of the carbonaceous film, a Si_3N_4 disk was coated with a 0.15-μm-thick pure carbon film by electron beam heating a graphite rod and was slided against a SUJ2 pin. The coefficient of friction was around 0.2 at an initial stage of the test (1 min or 10² cycles), and then it abruptly increased to values of 0.4 to 0.7, indicating a breakage or a removal of the carbon film. To increase the adhesivity of the film, the Si_3N_4 disk coated with the 0.15 μm-thick carbon film was irradiated with 1.5 MeV Ar⁺ ions to a dose of 5 x 10¹⁶ ions/cm². The irradiation condition is the same with that used in producing the solid carbonaceous film. In Fig. 9, μ as a function of time is shown for this disk in sliding contact with a SUJ2 pin. A coefficient of friction as low as 0.2 was obtained, and the duration of this low friction state is about 3 x 10³ cycles. A replacement of the pin material by Si_3N_4 also resulted in a μ value of 0.2, but the duration was less than 4 min (3 x 10² cycles).

These results show that carbon plays an important role in reducing μ of ceramic surface. It is noted, however, that the μ value for pure carbon film is higher by a factor of more than two than that for the carbonaceous film produced by the Ar⁺ ion irradiation of the silicone oil. This suggests that the value of μ as low as 0.04 for the carbonaceous film cannot be

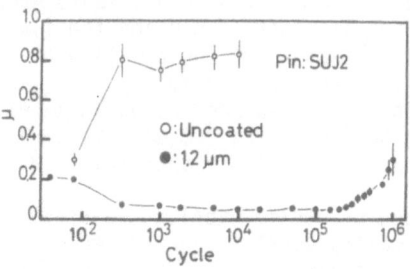

FIGURE 8. μ versus sliding cycle for sapphire disks uncoated (O) and coated with a 1.2-μm-thick carbonaceous (●) film prepared by silicone vapor deposition and 1.5 MeV Ar+ ion irradiation. The pin material is SUJ2. The vertical bars denote the amplitude of fluctuation of μ.

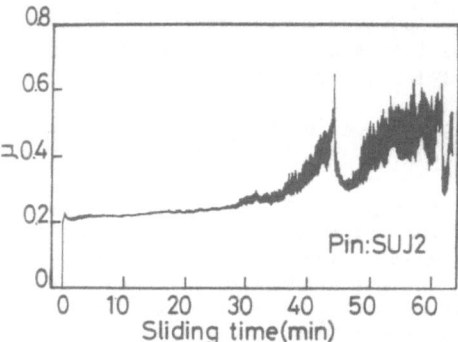

FIGURE 9. μ versus sliding time for a Si_3N_4 disk coated with a 0.15-μm-thick carbon film prepared by electron beam-heating a graphite rod. The pin material is SUJ2.

explained entirely by the amorphous carbon phase contained in the film. It seems that in addition to C, Si, O and/or H in the film work to reduce μ.

4. CONCLUSIONS

Two techniques were adopted as means for modifying ceramic surfaces in an attempt to improve the friction and wear property: (i) implanting energetic Ar+ ions and (ii) forming solid carbonaceous films onto ceramic surfaces by vapor deposition of a silicone oil (pentaphenyle-trimetyl-trisiloxane) and energetic Ar+ ion-irradiation. The friction and wear properties of the modified ceramics were tested by using the pin-on-disk method. The disk materials studies were SiC, Si_3N_4, and Al_2O_3 and the pin materials used were SUJ2 steel and Si_3N_4 ceramics. The results obtained are as follows:

1. Implantation of 800 keV Ar+ ions to disks of SiC and Si_3N_4 results in reductions of μ for sliding contact against pins of both SUJ2 metal and Si_3N_4 ceramics. This effect is partly attributed to the formation of a soft amorphous surface layer on the hard substrate by ion implantation.
2. A remarkable reduction of μ from 0.6 to 0.15 is observed for a SiC disk in sliding contact with a SUJ2 pin by implanting Ar+ ions to the disk. The durability of the low μ state is more than 1×10^5 cycles for a disk implanted with 800 keV Ar+ ions to a dose of 1×10^{18} ions/cm². The amorphous carbon phase generated in the surface layer of SiC as a result of Ar+ ion implantation may work as a solid lubricant, resulting in the large reduction in μ.

3. An adhesive, solid, electrically conductive, carbonaceous film can be formed on the surface of Si_3N_4 or Al_2O_3 by vapor deposition of the silicone oil and 1.5 MeV Ar^+ ion-irradiation. The film obtained is similar to amorphous carbon but contains considerable amounts of Si and O and probably a small amount of H.
4. Si_3N_4 disks coated with these carbonaceous films show coefficients of friction as low as 0.04 for sliding against pins of both Si_3N_4 and SUJ2. The durability of the low μ state is more than 1×10^5 cycles for a Si_3N_4 disk coated with a 0.27-μm film. The coefficient of friction for the carbonaceous film is lower than that for a vapor-deposited and subsequently Ar^+ ion-irradiated pure carbon film by a factor of more than two.

5. ACKNOWLEDGEMENTS

The authors would like to thank S. Noda, H. Doi, Y. Kido, Y. Shimura, and Y. Mizutani for discussions, and J. Mizuno for Raman spectroscopy.

REFERENCES

1. McHargue CJ, BR Appleton, and CW White: Surface Engineering. R Kossowsky and SC Singhal (eds). Dordrecht: Martinus Nijhoff Publishers, 19894, p. 228.
2. Burnett BR and TF Page: Proc. Br. Ceram. Soc. 34 (1984) 65.
3. McHargue CJ: Ion Beam Modification of Insulators. P Mazzoldi and GW Arnold (eds). Amsterdam: Elsevier, 1987, p. 223.
4. Roberts SG and TF Page: Ion Implantation into Metals. V Ashworth, WS Grant, and R Procter (eds). London: Pergamon Press, 1982, p. 135.
5. Burnett PJ and TF Page: J. Mater. Sci. 19 (1984) 3524.
6. Hioki T, A Itoh, S Noda, H Doi, J Kawamoto, and O Kamigaito: J. Mater. Sci. Lett. 3 (1984) 1099.
7. Burnett PJ and TF Page: Wear 114 (1987) 85.
8. Lankford J, W Wei, and R Kossowsky: J. Mater. Sci. 22 (1987) 2069.
9. Kohzaki M, S Noda, H Doi, and O Kamigaito: Mater. Lett. 6 (1987) 64.
10. Itoh A, T Hioki, and J. Kawamoto: Proc. 7th Int. Conf. on Ion Implantation Technology, Kyoto, 1988, to be published.
11. Venkatesan T, L Calcagno, BS Elman and G Foti: Ion Beam Modification of Insulators. P Mazzoldi and GW Arnold (eds). Amsterdam: Elsevier 1987, p. 301.
12. Kido Y and A Kawano: J. Appl. Phys. 61 (1987) 956.
13. Elman BS, MS Dresselhaus, G Dresselhaus, EW Maby, and H Mazurek: Phys. Rev. B24 (1981) 1027.

MODIFIED PULL TEST FOR THE TESTING OF VERY ADHERENT FILMS

J. E. PAWEL AND C. J. McHARGUE*

Vanderbilt University, Nashville, TN 37212

1. INTRODUCTION

In many applications, the effectiveness of a thin film on a substrate depends largely on the degree of adhesion of the film to its substrate. Because of this, much effort has been put into developing reproducible, quantitative adhesion tests (for reviews see references 1-6). The pull test is a direct measure of the tensile force required to remove the test film normal to the substrate surface. Pins or studs with flat ends are attached to the film surface and then pulled off by a calibrated instrument that measures the force of removal.

In principle, the pull test is the most straightforward of the many quantitative adhesion measurement techniques. In practice, however, it is not so easily used. One of the major difficulties of this method is to apply a truly normal force to the film rather than a combined force and moment due to misalignment of the tensile load. The attachment of the pin to the film surface must be such that it does not grossly affect the film under investigation or the film/substrate interface (3). The stress limit of the test is the strength of the adhesive used to secure the pin to the film surface (4,6-8); however, in many instances the de-adhesion stress of the film/substrate interface is greater than the strength of the epoxy that bonds the pins to the film surface. Another problem arises if the fracture occurs in a complex manner involving several of the interfaces and layers between the pin, epoxy, film, and substrate (9,10). If partial failure occurs in some layer or at some interface, the actual stress on the remaining intact portion of the system changes. Since it is not always known where failure first occurred, it may be difficult to determine the meaning of the maximum stress recorded by the test instrument.

Because of these problems, a technique was developed to introduce a reproducible flaw at the film/substrate interface in an attempt to lower the apparent adhesion of the film to within the range of the pull tester. This technique also promotes film failure at the artificial flaw and subsequent crack propagation along the film/substrate interface.

2. EXPERIMENTAL PROCEDURE AND METHOD OF ANALYSIS

Zirconium films approximately 500 nm thick were electron beam evaporated onto the basal plane of single crystal sapphire substrates. The pull test was performed using a Sebastian I adherence tester manufactured by the Quad Group of Santa Barbara, CA. The pins were 3 mm in diameter and coated with

*Oak Ridge National Laboratory, Oak Ridge, TN 37831-6118. This research is sponsored in part by the Division of Materials Sciences, U.S. Department of Energy, under contract DE-AC05-84OR21400 with the Martin Marietta Energy Systems, Inc., and under a National Science Foundation Graduate Fellowship.

313

epoxy. The adhesion of these zirconium films to their sapphire substrates was greater than the strength of the epoxy used to bond the test pin to the specimen. In each case, the largest tensile stress applied was equal to or slightly greater than the manufacturers' specified epoxy strength of 62 to 69 MPa but none of the test film was removed. The modified method introduced above was therefore employed in order to obtain a quantitative measure of the film adhesion.

A thin layer of gold approximately 20 nm thick was sputter deposited onto a small portion of the sapphire substrate after the substrate had been cleaned. The test film (500 nm of zirconium) was then deposited through a mask such that a portion of this new film overlapped the gold layer (see Fig. 1). Since the gold layer was non-adherent (experimentally) to the sapphire substrates used in this experiment, this served as a release layer for the test films. The mild shadowing effect of the mask used during gold deposition caused the edge of the gold layer under the test film to be tapered and thus this was treated as a reproducible crack and the analysis made using the concept of fracture toughness as indicated by the stress intensity factor, K. For Mode I loading, K_I is given (11) as

$$K_I = A\sigma \sqrt{\pi a} \ ,$$

where "a" is the crack length, σ is the normal stress, and "A" is a constant related to the geometry of configuration (see Fig. 2). The specimen geometry used in the modified pull test is an analogous situation: because the gold layer is non-adherent to the sapphire substrate, it does not transmit any stress across the interface and thus serves as a "crack" between the film and substrate (see Fig. 3). The stress intensity necessary for de-adhesion, rather than normal stress, was taken as the measure of film adhesion.

There may be other "cracks" at the film/substrate interface also, due to voids and other defects created during film growth, but the one created by the release layer is by far the largest and is, therefore, the failure initiation site. Because cracks tend to extend normal to the direction of maximum tensile stress (12) the crack will propagate along the film/substrate interface.

By using K_I, which incorporates the geometry of the crack, minor differences in pin placement over the gold layer were not significant since they could be accounted for by careful measurement of the crack length.

Because the goal of the analysis was not to determine a material property but to find some means of comparison of the fracture strength for similar but flawed systems, no attempt was made to ensure that the specimens used in this research met the specimen size requirements for fracture toughness testing (11). All specimens used for comparison were of almost identical shape and size and therefore the geometric factor "A" remained the same for each of the test specimens.

3. RESULTS

For comparison, some pull tests were performed on areas of the test film that did not overlap the release layer. None of the tests performed without a release layer resulted in any film loss. The tests performed utilizing the release layer resulted in varying amounts of film loss over a wide range of applied stresses (see Table 1).

The gold layer does have an effect on the stress intensity. Table 1 reveals that the adhesive strength of the film to the substrate was greater than the strength of the epoxy if no release layer was present and thus no zirconium film was removed. However, the use of the release layer resulted

ORNL–DWG 88–12659

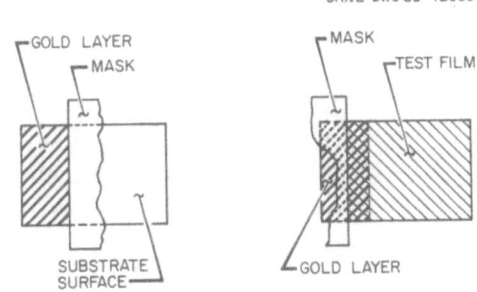

FIGURE 1. The gold layer is deposited on a portion of the specimen surface. The test film is then deposited such that it overlaps the gold layer. (View is normal to deposition surface.)

ORNL–DWG 88–12660

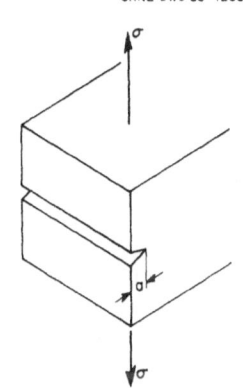

FIGURE 2. Schematic representation of Mode I crack opening for an edge crack of length "a."

ORNL–DWG 88 6556

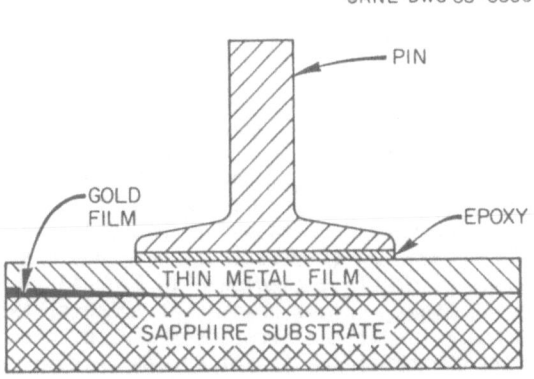

FIGURE 3. Schematic representation of the modified pull test. Because the gold film is non-adherent to the substrate, it can be modeled as a crack under the film.

TABLE 1. Pull test results. Different identification letters refer to
separate tests on the same specimen. "Test Film Removed" refers to the test
area not including that over the release layer.

Specimen Number	Gold Release Layer	Stress at Failure (MPa)	Maximum Crack Length (mm)	Stress Intensity Factor (MPa $\sqrt{m}$)	Test Film Removed (%)
477 c	no	64.74	---	---	0
477 d	no	59.57	---	---	0
478 b	no	69.98	---	---	0
480 d	no	56.75	---	---	0
478 a	yes	5.45	0.40	0.21	26
479 c	yes	35.10	0.77	1.90	20
480 a	yes	38.61	1.39	2.81	13
480 b	yes	25.65	0.47	1.09	65

in some test film removal. If the effect of the gold layer were solely to
reduce the effective area supporting the applied load, thus increasing the
stress on the system, recalculation of the stresses based on the area under
the pin but not over the gold layer should reveal maximum stresses of about
the same order of magnitude as when no release layer was utilized and,
again, the epoxy should fail before the test film. Table 2 shows this is
not the case: the stresses recalculated for the specimens omitting the pin
area over the gold material were in general much less than the strength of
the epoxy. Failure of the test film would not have occurred at these low
stresses if the release layer did not have an effect on the stress intensity. Even specimen 480a, which had a recalculated stress of 71 MPa, showed
film removal, not just epoxy failure.

4. DISCUSSION

In each case, the crack did not propagate along a single plane: there
were regions of film/substrate failure and of epoxy/film failure (see
Fig. 4). If the epoxy failed first, the stresses supported by the remaining
bonded area (Region B in Fig. 4) are much higher than the tensile strength
of the epoxy. This was deemed unreasonable because some epoxy is still
intact to transmit these high stresses. It was thus assumed that part of
the zirconium/sapphire interface (Region B in Fig. 4) failed first at some
stress less than the final maximum reading at which stress the remaining
epoxy failed. Calculation of the stresses based on this assumption gave
maximum stresses on the remaining intact system that were approximately
equal to the tensile strength of the epoxy. This latter, then, was a reasonable assumption and supported the conclusion that the crack propagated
along the film/substrate interface for some distance and then stopped.

TABLE 2. Stresses recalculated assuming reduced area under the
pin due to the gold release layer.

Specimen Number	Stress at Failure Recalculated Omitting Pin Area Over Gold Layer (MPa)
478 a	5.79
479 c	38.68
480 a	71.29
480 b	38.47

ORNL–DWG 88–12661

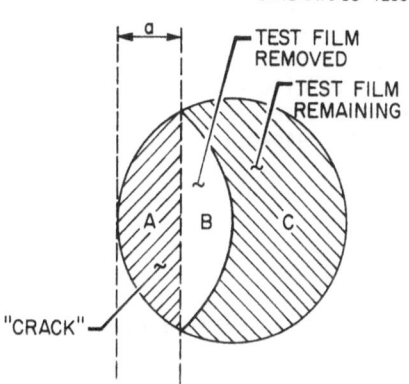

FIGURE 4. Schematic representation of the pull-pin site on the specimen
surface after pin removal. Region A is the area overlapping the gold layer
and represents the "crack." Region B is that of film/substrate failure and
Region C is the region of epoxy/film failure.

The load continued to increase until the stress on the remaining intact portion exceeded the strength of the epoxy. Since the stress to cause failure
of at least part of the test film occurred at an unknown smaller stress than
the maximum shown in Table 1, it can be concluded that the stress intensity
factors in Table 1 represent an upper limit of those required to cause film
de-adhesion.
    Some reasons that the crack resulting from the gold layer did not propa-
gate further between the test film and substrate include: (a) the gold
layer may not have been completely continuous at a thickness of 20 nm;
(b) the crack tip caused by the edge of the gold layer may not have been
sharp enough; and (c) the zirconium film may not have been continuous over
the edge of the gold. Also, gold on sapphire tends to agglomerate during

heating and this may have occurred during the test film deposition or during the epoxy cure cycle. It is hoped that more extensive crack propagation will be obtained after the investigation and optimization of these factors.

5. TENSILE ADHESION OF ION BEAM MIXED FILMS

The modified pull test was used to investigate the effect of ion beam mixing on the adhesion of zirconium films to sapphire substrates. The experimental procedure described above was followed. The specimens were then masked in such a way that the incoming ion beam (Fe^{++} ions at 1 MeV to a fluence of 1×10^{17} ions/cm^2) bombarded a portion of the test film but none of the film that overlapped the gold release layer. This last precaution was taken because ion beam mixing might enhance the adhesion of gold to sapphire.

Gold was chosen for the release layer material because it does not "wet" sapphire. Because of its non-wetting characteristics, and thus non-adherence, the gold film and the superimposed test film were often removed during sample preparation and handling, particularly during the masking procedure. As a result, data for the ion beam mixed case were gathered for only one ion beam condition. In each of the pull tests in the ion beam mixed region, the only part of the zirconium test film removed was that directly over the gold layer: the crack caused by the presence of the gold layer did not propagate at all between the mixed film and the substrate. For the unmixed case, there was some crack propagation along the film/substrate interface with the removal of 20% of the zirconium test film. The crack length for the test performed on the same specimen in the unmixed region was intermediate to the crack lengths utilized in the mixed region. The average maximum stress at failure and the average stress intensity factor for the mixed region were greater than the upper limits of these for the unmixed region by 40 and 30%, respectively (see Table 3). This indicated significant adhesion enhancement by ion beam mixing since both regions had similar flaws but the mixed region withstood a greater stress and a greater stress intensity factor with no zirconium film removal.

TABLE 3. Pull test results for ion beam mixed specimens

Specimen Number	Stress at Failure (MPa)	Crack Length Factor (mm)	Stress Intensity Factor (MPa $\sqrt{m}$)	Film Removed (%)
479 a (mixed)	37.03	0.94	2.02	0
479 b (mixed)	60.68	0.55	2.76	0
479 c (unmixed)	35.10	0.77	1.90	20

6. CONCLUSIONS

The gold release layer results in a reproducible crack that creates a stress intensity at the interface under investigation. This method was used successfully to cause failure of the film/substrate interface in systems for which this adhesion is normally greater than the tensile strength of the epoxy, thus allowing a reproducible, quantitative measurement of the stress intensity necessary for de-adhesion. This technique was used to show a 30% improvement in the tensile adhesion strength of zirconium films to sapphire as the result of ion beam mixing.

REFERENCES

1. Pawel JE and CJ McHargue: J. Adhesion Sci. Tech., to be published.
2. Campbell DS: Handbook of Thin Film Technology. Leon I. Maissel and Reinhard Glang (eds). New York: McGraw-Hill Book Company, 1970, Chap. 12.
3. Chapman BN: J. Vac. Sci. Technol. 11 (1974) 106.
4. Perry AJ, P Laeng, and HE Hintermann: Proc. 8th Intern. Conf. on Chemical Vapor Deposition. John M. Blocher, Jr., Guy E. Vuillard, and Georg Wahl (eds). Pennington, NJ: The Electrochemical Society, Inc., 1981, 475.
5. Valli J, U. Makela, and A. Matthews: Surface Eng. 2(1) (1986) 49.
6. Valli J: J. Vac. Sci. Technol. A 4(6) (1986) 3007.
7. Jacobsson R: Thin Solid Films 34 (1976) 191.
8. Collins LE, JG Perkins, and PT Stroud: Thin Solid Films 4 (1969) 41.
9. Bunnell LR, JC Crowe, and PE Hart: The Science of Ceramic Machining and Surface Finishing. Washington, DC: National Bureau of Standards Special Publication 348, 1972, p. 341.
10. Noda S, H Doi, N Yamomoto, T Hioki, J Kawamoto, and O. Kamigaito: J. Mater. Sci. Letters 5 (1986) 381.
11. Brown WF Jr. and JE Srawley: Plane Strain Crack Toughness Testing of High Strength Metallic Materials. Philadelphia, PA: American Society for Testing Materials, ASTM Special Technical Publication No. 410, 1967, p. 12.
12. Cannon RM, RM Fisher, and AG Evans: Thin Films-Interfaces and Phenomena. R. J. Nemanich, P. S. Ho, and S. S. Lau (eds), Pittsburgh, PA: Materials Research Society, 1985, p. 799.

ION IRRADIATION OF POLYMER-DERIVED GRAPHITIC CARBONS

J.T.A. POLLOCK, M. J. KENNY, L. S. WIELUNSKI, AND M. D. SCOTT

CSIRO, Division of Applied Physics, Private Mail Bag 7,
Menai, New South Wales 2234, Australia

1. INTRODUCTION

Carbon atoms exist in the natural state as graphite and diamond, two of
the most interesting allotropes found among the elements. Diamond, the
metastable, highly crystalline structure, is the hardest known substance and
prized for its beauty as well as its utility. Graphite has a layer structure and although the bonds within the layers are probably stronger than
those of diamond, weak layer-linking bonds produce a material which slips
easily and exhibits high lubricating properties. Each of these carbon
structures may be formed by industrial processing.

Polymeric carbons are formed by heating high molecular weight polymers to
temperatures where the non-carbon content is lost as a gas and the resulting
carbon mass has a "tangled" largely graphitic microstructure which relates
to the polymer chain configuration. These graphitic carbons include fibers,
glassy carbon and many of the chars (1). In these materials, the graphite
layers are short in length and stacking sequence (< 10 nm) preventing the
weak inter-layer bonding from dominating and producing uniquely strong, yet
highly elastic materials. Thin carbon coatings may be produced by a number
of vapor deposition techniques and exhibit properties in the range diamond
to diamond-like depending upon processing parameters (2). However, our
interest lies with the bulk (> 1 mm thick) polymer-derived carbons and
includes carbon/carbon composites processed from the curing of carbon fiber
mats interspersed with a polymer glue to produce a unique graphite-based
material. When combined with the renowned bio-compatibility of carbon, the
mechanical characteristics of these polymer derived graphites lead to a
powerful material which can be used to fabricate biomedical devices
including heart valves and mechanical components such as pins for broken
bones (3).

Work in our laboratory has concentrated on glassy carbon, a low density
(1.5 g cm^{-3}) material formed by the controlled heat-induced degradation of
resinous hydrocarbons (1). Although it is less strong and wear resistant
than pyrolytic (CVD) carbons (4), it has thickness, shaping and forming
advantages, making enhanced wear resistance an attractive goal. This goal
has been substantially achieved by our group using ion irradiation.

In this paper, we present a brief summary of our earlier reported conclusions followed by recent work on the effect of ion species and dose, the
irradiation induced structural changes observed by ion channelling and
scanning electron microscopy (SEM) of highly oriented pyrolytic graphite
(HOPG), comparative scratch testing results and some ideas on the reasons
for the observed increased hardness leading to enhanced wear resistance.

C.J. McHargue et al. (eds.), Structure-Property Relationships in Surface-Modified Ceramics, 321–330.
© *1989 by Kluwer Academic Publishers.*

2. SUMMARY OF CONCLUSIONS DRAWN FROM PREVIOUS WORK

Wear resistance enhancement is associated with the end-of-range zone where dense atom displacement occurs. For example, using 2 MeV He[+], enhanced wear resistance is found on diamond polishing to be almost 8 μm below the surface (5). However, the same He[+] dose at 12 keV produces wear resistance at the surface (6).

Improved wear resistance is observed with N, He, and C (5) reinforcing the hypothesis that the enhanced resistance is related to displacement damage rather than a chemical effect.

The enhanced wear resistance is dose dependent. For 50 keV N[+], the improved wear resistance is first observed at 10^{15} ions cm^{-2} increasing rapidly to a factor of 400 improvement by about 5 x 10^{15} ions cm^{-2} (7).

Other carbon materials having a graphite-based structure such as electrode graphite, HOPG and carbon fibers also show enhanced wear resistance followoing irradiation (6), the extent of which depends upon the initial hardness and macrostructure of the material used.

In comparative wear testing, using both diamond abrasive and ruby ball-on-disc, modified glassy carbon is substantially more wear resistant than pyrolytic carbon which is normally preferred in biomaterial applications such as artificial heart valves (4).

Upon irradiation of low density glassy carbon, a small surface compaction is usually observed indicating densification. (5). A probably related observation is the substantially increased visual reflectivity.

Direct hardness measurements are difficult due to both the thinness of the modified layer and the elastic recovery after indentation. The latter is very important and has been thoroughly reported by Hawthorne (9). We have, however, reported qualified hardness data for modified glassy carbon indicating increased hardness (10). Indirectly, we have measured the higher hardness of modified glassy carbon compared with pyrolytic carbon in ruby ball-on-disc wear testing (8). The ruby ball remained unmarked while cutting a groove in a pyrolytic carbon disc, but had a flat worn onto it under the same conditions while failing to cut a groove in an ion irradiated glassy carbon disc.

3. EXPERIMENTAL TECHNIQUE

All irradiations were carried out in vacuum with a beam area of a few cm^2 in most cases. With glassy carbon samples, masks were used to define irradiated and unmodified areas. Masks containing 2 mm diam apertures were employed to irradiate samples with different species at various doses (Fig. 1). The energy of the ion species was selected so that the modified layer depth for each was the same (Table 1), about 100 nm. As has been previously described (5), on polishing these zone irradiated samples with 1 μm diamond, the modified area resists abrasion and a plateau develops relative to the surrounding unmodified region. The step heights of these capped regions were measured at various stages of wear using optical interferometry and provided an assessment of wear resistance as a function of species and dose. Other samples were masked so that half the area was irradiated with 50 keV N[+] to a dose of 10^{16} ions cm^{-2}. These samples were used to determine the relative resistance to fine abrasives such as SiC, Al_2O_3, B_4C, and diamond. These powders were applied to a small plush pad and drawn across the boundary. The load required to give a dense pattern of visible scratches on the non-irradiated area was assessed.

Highly oriented pyrolytic graphite (HOPG), a very soft material, has already been reported as having very substantially increased wear resistance following modification (6). Samples of similar material were irradiated with 50 keV N[+] at doses in the range 2 x 10^{16} to 2 x 10^{16} ions cm^{-2} and examined by Rutherford backscattering and channelling using 1 MeV He[+]. These samples were also examined by optical microscopy and SEM.

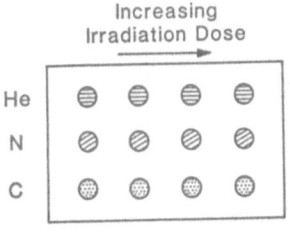

≥ 3 μm High Plateaus

FIGURE 1. Schematic of zone irraidated and polished glassy carbon used to determine the critical irradiation dose, D*.

TABLE 1. Critical dose for a 3-μm high wear resistance plateau, D*, density of energy per incident ion available for nuclear collision, Es, and the density of deposited energy, Es·D*, at the critical dose.

Ion species	Ion energy (keV)	Critical dose D* (ions cm^{-2})	Displacement energy density per ion Es (eV 10^{-15} cm^{-2})	Total displacement energy density Es · D*
Helium	10	2.4 x 10^{16}	1.3	31
Carbon	33	4.9 x 10^{15}	9.0	36
Nitrogen	39	2.4 x 10^{15}	11.3	27
Oxygen	44	2.5 x 10^{15}	14.0	35
Fluorine	50	1.9 x 10^{15}	16.0	31
Neon	50	1.2 x 10^{15}	21.3	27

4. EXPERIMENTAL DATA

Figure 2 shows the resistance to diamond wear of glassy carbon samples zone-irradiated with various ions as a function of dose. All observed cases of substantial wear resistance were at the surface. Small surface compactions of about 30 nm, were observed at the higher doses for all species in the as-modified condition. An unworn irradiated zone 3 μm high was used to determine a critical wear resistance dose for each ion species. Since the modified depth was held at ~100 nm for each species, by varying the irradiation energy, this plateau height represents an enhanced wear resistance factor of ≥30. These data confirm that enhanced wear resistance is not element specific and that for a given element a relatively sharp dose threshold exists. However, the data clearly reveal the important role of atom mass in determining the critical dose. For example, a dose of 1.1 x 10^{15} Ne$^+$ cm^{-2} is effective compared with a required critical dose of 2.6 x 10^{16} ions for helium.

The ion doses required for each species to resist diamond abrasion until a plateau ≥3 μm developed are listed in Table 1, together with calculated surface deposited energy densities per incident ion. The latter have been determined according to Ziegler et al. (11). The product of these parameters provides a measure of the total energy density deposited in nuclear collision cascades for each ion species and reveals that, within the accuracy of the dose measurement and the polishing technique employed, this nuclear collision displacement energy density is a constant. These calculations make it very clear that the enhanced wear resistance is related to the

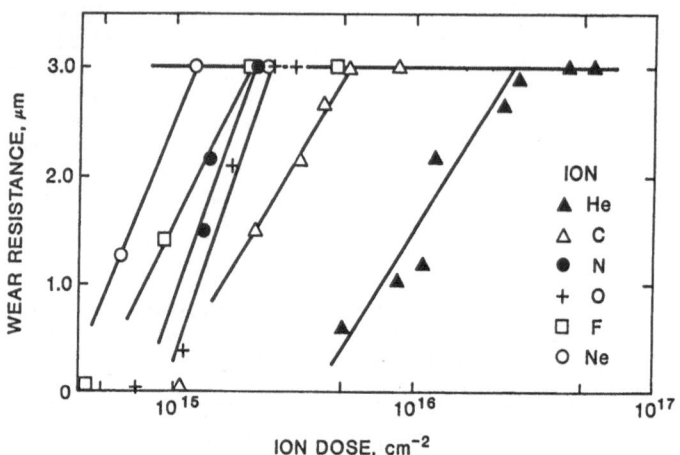

FIGURE 2. Wear resistance of zone irradiated glassy carbon for various ions as a function of ion dose. A 3-μm high wear resistant platuau defines D*.

total value of the density of deposited energy and is independent of the ion type. It is very likely that similar increases in wear resistance are available with neutron irradiation but at much higher doses owing to the low mass and greater penetration depth.

Scratch tests carried out on the half modified glassy carbon samples did not define a Moh hardness value since the irradiated areas, 10^{16} N cm^{-2} at 50 keV, resisted all grits including B$_4$C (4–10 μm) and diamond (1–6 μm). Although the thin ion modified area could be gouged through by an occasional large grain, numerous scratches were produced on the untreated area which stopped completely at the boundary with the modified zone as shown in Fig. 3 for B$_4$C and diamond.

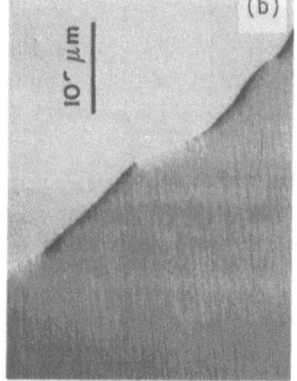

FIGURE 3. Photomicrographs showing the resistance of modified glassy carbon to B$_4$C (a) and diamond (b) abrasion.

The ion channelling data measured with irradiated and unirradiated HOPG samples are shown in Fig. 4. Good channelling is obtained with the unirradiated sample normal to the basal plane. However, even at the lowest dose of 50 keV N^+, 2 x 10^{14} ions cm^{-2}, loss of channelling is observed indicating that atomic rearrangement has taken place and highlighting the sensitivity of the technique for graphitic materials. With increasing dose the depth of channelling loss reaches a saturation value of ~200 nm at a dose of ~2 x 10^{15} N^+ cm^{-2}. SEM micrographs of the surfaces of irradiated samples are presented in Fig. 5 which reveal the topographical outcome of the displacement damage created at the surface. Fine cracks are present at the lowest doses, which widen and develop into delaminating layers for doses >5 x 10^{15} N^+ cm^{-2}. Optical interferometry of the medium dose sample indicates that the step height across the crack boundary is about 200 ± 50 nm and is the likely thickness of the delaminating layer. Examination of the high dose samples at higher magnification reveals a banded pattern in the still laminated surface areas representative of puckering. It is likely that the cracking and delamination observed occur after implantation or a progressive stepped topography would be observed.

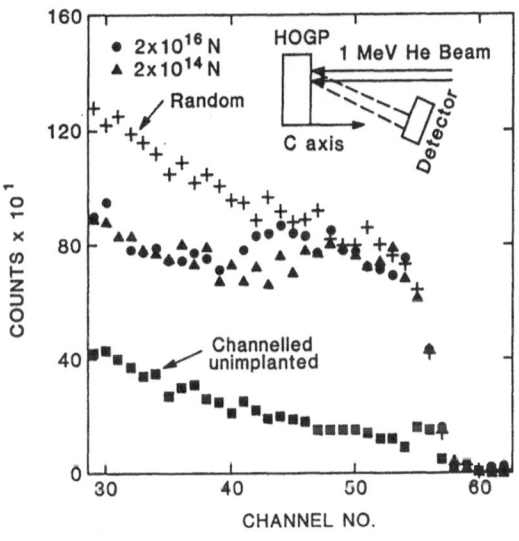

FIGURE 4. RBS/channelling spectra for HOPG, unimplanted and implanted with 50 keV N^+ to doses of 2 x 10^{14} N^+ and 2 x 10^{16} N^+ cm^{-2}. Loss of channelling and increased damage depth is observed with increasing dose.

5. DISCUSSION

Wear resistance generally derives from hardness for a range of materials. Toughness, a measure of the energy required for crack propagation, can also be altered by structural modification and lead to enhanced wear resistance. Owing to the thinness of the modified layers and the extraordinary elastic recovery [>90%, (ref. 9)] of the substrate it is difficult to assess the hardness of the irradiated carbons. Nevertheless, the observation that scratches produced using diamond and B_4C powders stop at the boundary between the unirradiated and irradiated zones [Fig. 3(a) and (b)] is a clear indication that the ion irradiated materials approach or exceed the hardness of diamond.

326

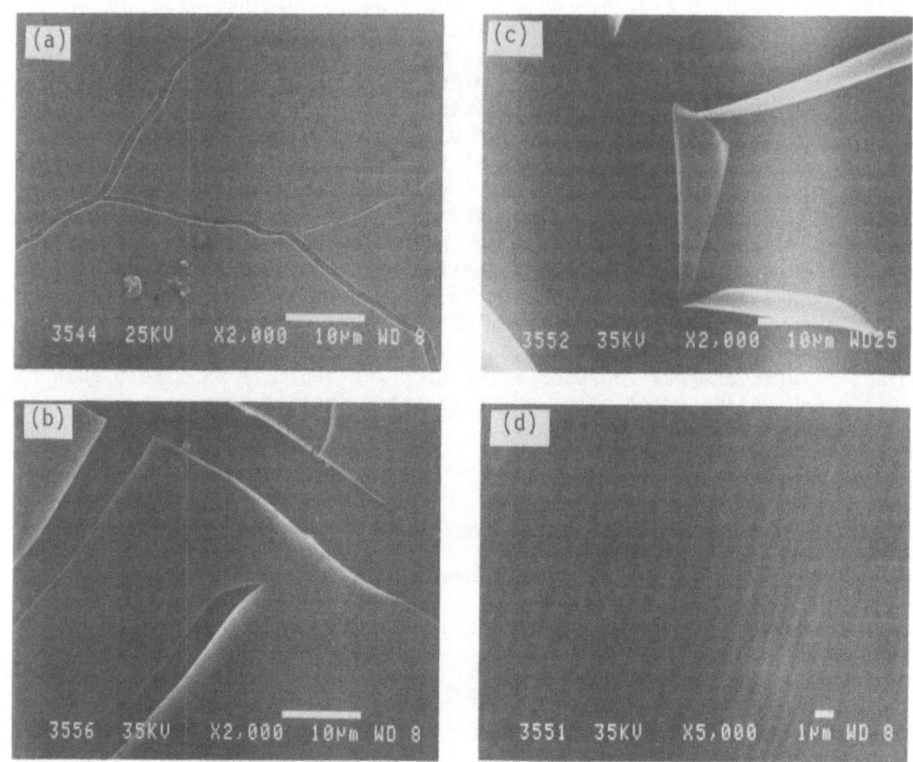

FIGURE 5. SEMs of HOPG with 50 keV N$^+$ to 2 x 10^{14} (a), 10^{15} (b), and 2 x 10^{16} (c) ions cm^{-2}. Banding is observed within the still laminated regions of the high dose sample (d).

We have observed changes in reflectivity, compaction and wear resistance as a function of ion dose with many graphite based materials and conclude that a common structural modification is responsible. We, therefore, feel that it is reasonable to apply data obtained with, for example, HOPG in the interpretation of glassy carbon results.

Together with the earlier reported observation that the enhanced wear resistance is encountered at the end of the ion track for 2 MeV He$^+$ implants (5), our current data point clearly to an atom displacement mechanism. This conclusion is significant since each incident ion can cause many (> 100) atomic, i.e., carbon displacements, and explains the much lower dose requirements for property changes in these materials compared with the alloying mechanisms associated with ion implantation-induced wear resistance in metals. The significance of the atom displacement mechanism has also been recognized in other covalently bonded ceramic materials (12). To reflect the distinction between low dose displacement mechanisms and high dose alloying, it is appropriate to use the term "ion irradiation" for the former, reserving "ion implantation" for the latter.

The sensitivity of the graphite structure to irradiation is well described by the effect of ion damage upon HOPG samples where atom disorder, leading

to a loss of ion channelling, and bulk disorder, leading to surface
upheaval, are clearly described in Figs. 4 and 5, respectively. This diso-
dering is also supported by changes in Raman spectra following irradition of
many graphite-based carbons (13) including glassy carbon samples prepared in
our laboratory (14). The loss of both ion channelling and the characteristic grapite peak in the Raman spectra, is usually interpreted as
amorphization of the crystal structure within the modified region (15,16).
High resolution transmission electron microscopy (HRTEM) examination of
modified layers in other materials, e.g., Si (ref. 17) generally indicates
the loss of crystal structure. In the case of irradiated glassy carbon this
does not appear to be the situation. The tangled, c-plane microstructure of
the unirradiated material, consistent with the polymer-derived turbostratic
arrangement proposed for glassy carbon (13), is observed even after irradiation with 2×10^{18} He$^+$ cm^{-2} (ref. 14). Venkatesan et al. (15) used ion
channelling to study annealing regrowth kinetics in damaged HOPG and
concluded that two damage regrowth regimes existed; (i) a low activation
energy regime (0.15 eV) where, although the layer structure was preserved,
atom disorder (~ 0.05 dpa) had destroyed the channelling characteristics and
(ii) a higher activation energy regime (1.2 eV) following the substantial
disordering of the graphite layer structure at higher ion doses. These
energies correspond closely with those reported by Reynolds for annealing of
neutron damage in graphite (18). Reynolds attributed the higher damage
annealing energy to the mobilization of interstitial clusters in a damaged
but largely intact graphite structure. This type of damage would not be
easily observable by HRTEM due to the already tangled c-plane structure of
glassy carbon.

It is clear that irradiation displacement damage is the controlling mechanism in producing the improved mechanical properties. Figure 6 presents a
schematic summary of the outcome of the structural changes which we propose
to be responsible for the increased hardness and wear resistance. This
model is based on both reported work with polymeric carbons (1,18) and our
own results. It is not ion specific, except in the sense that the damage
produced scales with incident species mass. We might therefore expect
neutron irradiation to provide a similar effect but are not aware of any
reports indicating that enhanced wear resistance follows neutron irradiation
of graphite. However, as summarized by Reynolds (18), there is a great deal
of information available in the literature dealing with damage effects and
physical property changes following neutron irradiation of graphite (single
and polycrystalline) in nuclear reactors. Much of this data is pertinent to
the present study. Interstitials formed as a result of atom displacements
are mobile, even at room temperature, and either recombine with immobile
vacancies or nucleate to form stable clusters which eventually become
visible as dislocation loops in TEM. The size and concentration of these
loops is strongly dependent upon irradiation temperature. Irradiation at
room temperature or below produces a dense mass of tiny clusters repre-
senting a significant retention of damage whereas higher temperatures
produce fewer and larger loops.

The accumulation of these dislocation loops causes dramatic changes in
many physical properties of graphite. Thermal expansion coefficients,
internal friction and elastic moduli change significantly, especially the
basal plane shear modulus. At the 0.5 dpa displacement level, corresponding
to the critical wear resistance dose of the current work, and equivalent
to a neutron dose of 1.3×10^{21} cm^{-2}, crystalline graphite swells by up to
30% in the c-axis direction while shrinking by up to 5% in the a-axis direction (18). The weak basal plane shear modulus increases by a factor of 20
(ref. 1). In polycrystalline graphite the response is more variable,

328

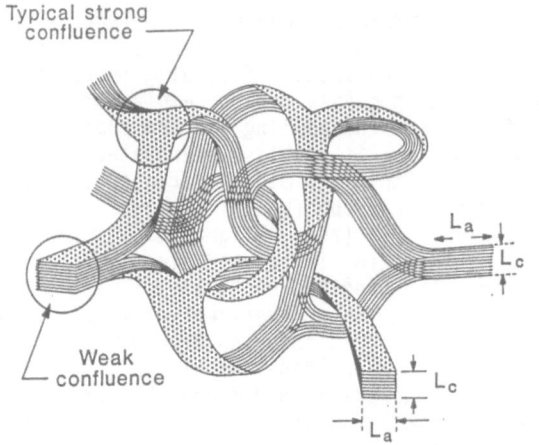

(a) Tangled structure proposed for many (e.g. glassy carbon) polymer-derived graphitic carbons. Note presence of pores (hence low density) and likely sources of mechanical weakness (1).

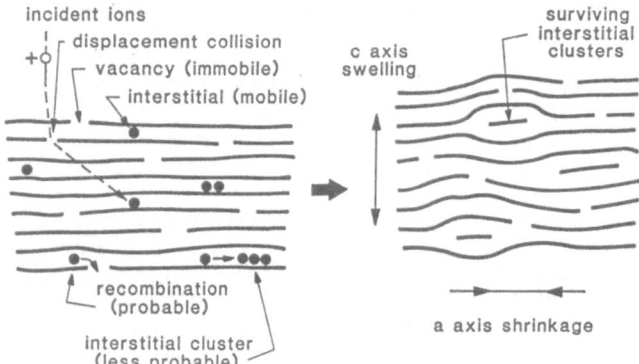

(b) Atomic displacement damage in graphite leading to swelling, shrinkage, and long-range wrinkling.

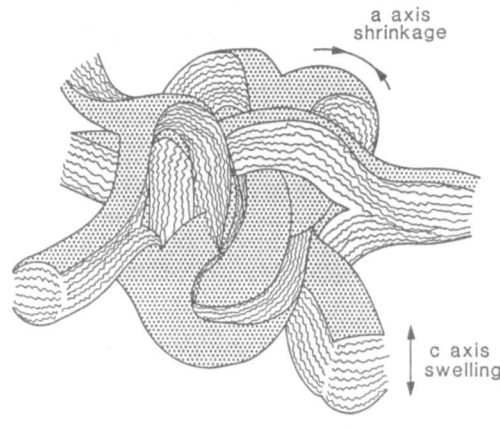

(c) Net result of damage in tangled structure following irradiation. Note, wrinkling of planes and swelling to close off voids.

FIGURE 6. Schematic representation of the effect of ion irradiation on the tangled graphitic structure proposed for polymer-derived carbons.

however, shrinkage (compaction) is common and a peak incrase in Young's modulus of 2.5X has been reported (1).

These property changes are consistent with our studies in both HOPG and glassy carbon. At high (wear-resistant) dose levels a-axis shrinkage of ≤20% is observed, much higher than that reported for neutron irradiated crystals. A correspondingly larger change in elastic moduli could be expected and would result in the increased hardness and abrasive wear resistance observed with obht HOPG and glassy carbon. It is also likely that the generation of a complex set of expansions and contractions in the polycrystalline glassy carbon, leading to the net compaction observed, suppresses the crack generating potential of the 0.1 nm diam pores (19) and layer mismatch regions in this material, leading to a harder and more wear resistant material.

Loss of ion channelling in graphite following irradiation with 2×10^{14} N^+ cm^{-2} (0.05 dpa) is in excess of that expected and may be the result of a "puckering" effect associated with defects. These defects, interstitials or clusters, could generate lateral distortions extending many atom distances from the defect source because of the weak Van der Waals forces linking the graphite planes. This multiplying effect would explain the extra sensitivity to ion channelling and the atomic scale wrinkling or puckering would make sliding between planes less easy. The banding structure observed int he SEM of the heavily irradiated HOPG [Fig. 5(d)] is the final outcome of this lateral distortion.

This model, the outcome of which is described schematically in Fig. 6(c), is consistent with the reported HRTEM, Raman and ion channelling data. The swelling and tightening suggested accounts for the hardening (enhanced wear resistance) and compaction observed with glassy carbon. However, this model cannot entirely explain the extraordinary increase in resistance to abrasion by B_4C and diamond which implies a hardness enhancement to values approaching that of diamond. It must, therefore, be the result of some atomic rearrangement leading to the formation of strong bonds between the graphite planes. These bonds could be SP2, SP3, or the other strong bonds proposed for carbon materials formed by ion beam deposition methods (2) and are likely to form between neighboring out-of-plane atoms at ribbon edges and/or interstitials. Direct evidence for this bond formation is not yet available.

In summary, the increased hardness and wear resistance of graphitic-based carbons following irradiation is most likely the result of three interacting processes; crack suppression and hardening through compaction, puckering of the c-plane making sliding less easy and the formation of new strong bonds between the planes and at plane edges.

6. ACKNOWLEDGEMENTS

It is a pleasure to acknowledge the assistance of D. Stevenson for irradiation and R. A. Clissold for sample preparation and wear studies. We thank A. W. Moore of the Union Carbide Corporation for supplying the HOPG samples and M. V. Swain for discussions on hardness.

REFERENCES

1. Jenkins GM and K Kawamura: Polymeric Carbons-Carbon Fibre, Glass and Char. Cambridge University Press, 1976.
2. Devries RC: GE Report 86CRO247, Schenectady, New York, March 1987.
3. Haubold AE, SH Shim, and JC Bokros: Biocompatibility of Clinical Implant Materials. DF Williams (ed). Florida: CRC Press, 1981, Vol. II, p. 3.

4. Shim HS and FJ Sheen: Biomat. Med. Dev. Artif. Organs 2 (1974) 103.
5. Pollock JTA, M Farrelly, and LS Wielunski: Materials Modification and Growth Using Ion Beams. U Gibson, A White, and P Pronko (eds). Pittsburgh: Materials Research Society, 1987, Vol. 93, p. 317.
6. Kenny MJ, JTA Pollock, and LS Wielunski: Nucl. Instr. Meth. B. (1988) in press.
7. Wielunski LS, JTA Pollock, and M Farrelly: Beam-Solid Interacitons and Transient Thermal Processing. M Aziz, L Rehn, and B Strizker (eds). Pittsburgh: Materials Research Society, 1988, Vol. 100, p. 213.
8. Pollock JTA and LS Wielunski: Biomedical Materials and Devices. JS Hanker and BL Giammara (eds). Pittsburgh, 1988, in press.
9. Hawthorne HM: Carbon 13 (1975) 215.
10. Farrelly M and JTA Pollock: Mater. Forum 10 (1987) 198.
11. Ziegler JF, JP Biersack, and U Littmark: The Stopping and Ranges of Ions in Solids. New York: Pergamon Press, 1985, p. 53.
12. Burnett PJ and TF Page: Radiat. Eff. 97 (1986) 283.
13. Tunistra F and JL Koenig: J. Chem. Phys. 53 (1970) 1126.
14. Prauer S and Roussow CJ: J. Appl. Phys. 63 (9) 4435.
15. Elman BS, MS Hayesan, MS Dresselhans, H Maznrek, and G Dresselhans: Phys. Rev. B 25 (1982) 4142.
16. Venkatesan T, BS Elman, G Braustein, MS Dresselhaus, and G Dresselhaus: J. Appl. Phys. 56 (1984) 3232.
17. Mayer JW, LE Rickson, and JA Davies: Ion Implantation in Semiconductors. New York: Academic, 1970.
18. Reynolds, WS: Physical Properties of Graphite. New York: Elsevier, 1968.
19. Bose S and RH Brass: Carbon 19 (1981) 289.

CHEMICAL PROPERTIES OF ION IMPLANTED CERAMICS

G. K. WOLF

Physikalisch-Chemisches Institut, Universitat Heidelberg,
Im Neuenheimer Feld 500, 6900 Heidelberg, FRG

1. INTRODUCTION

It is a difficult task to define exactly what "chemical properties of ion implanted ceramics" means. Firstly, it may mean: the chemical composition of compound changes as a consequence of ion bombardment. Or it may mean: the surface chemistry of implanted or ion beam modified compounds is different from the non-bombarded case. This change may be wanted or unwanted.

It is also a difficult task to define what the expression "ceramics" means. Firstly, it may mean: ceramics are structural materials with high temperature resistance defined by their function. Or secondly it may mean: ceramics are compounds somewhere in the middle between metals, pure ionic compounds and pure covalent compounds, mostly insulators (but not always) and mostly thermally and chemically very stable (but not always). This second definition includes many nitrides, carbides, borides, silicides of transition metals, and even amorphous carbon and amorphous silicon.

If the author of this contribution would prefer the first definition of "chemical properties" he would have to realize that the majority of all other contributions already deal at least partly with chemical composition changes, and he would come to the conclusion that there is not much left to discuss in this paper. If he prefers the second definition, the situation is a little better, but still he has to select special aspects of surface chemistry not covered in other contributions. The same argument forces him to use the second broader definition of the word "ceramics." The consequence of these considerations is to split the paper into two parts:

• A "mechanistic" part on particle/solid interactions where some older ideas coming from "hot atom chemistry" are discussed with respect to their potential for explaining chemical bombardment effects. This part is supplemented by examples for compound formation of bombarding particles with solid targets.

• The main part on the use of ion beam techniques for the following applications: catalytic activity and corrosion and oxidation behavior of ion beam processed ceramics (according to definition 2), and applications and performance of ion beam produced ceramic coatings on metals.

2. CHEMICAL ASPECTS OF THE INTERACTION OF ENERGETIC PARTICLES WITH INORGANIC SOLIDS

During the last years, most people working in the field of particle/solid interactions realized that in addition to diffusion, thermal spikes, ballistic processes, etc. chemical driving forces were necessary to explain the final state of an ion bombarded compound. Typical examples for chemical reactions are as far as the solid target is concerned:

        --breaking and reforming of bonds;
        --change of oxidation state; and
        --loss of loosely bound elements from the surface.

C.J. McHargue et al. (eds.), Structure-Property Relationships in Surface-Modified Ceramics, 331–353.
© 1989 by Kluwer Academic Publishers.

As far as the bombarded particle is affected:
        --stabilization in an unusual oxidation state;
        --segregation and precipitation of new phases;
        --synthesis of new compounds with the target.
While the chemical changes in the target are mostly taken account of, the chemical states of the implant or of secondary knock-on atoms are still very often ignored. To give a very simple example: the precipitation of gas bubbles in compounds bombarded with rare gases, or of metal clusters in the case of bombardment with noble metals can be easily understood on the basis of physical models as well as chemical ones.

Not a very complicated example: energetic cobalt ions interacting with a complex molecular solid like $[Co(en)_2Cl_2]NO_3$ which exists in two isomeric forms "cis" and "trans" may be inserted as new central atom into the complex (1,2). In case of little radiation damage they always prefer the original configuration of the target, in case of more damage the chemically more stable trans-form is preferentially formed independent on the target configuration. No purely physical theory could ever explain this result (Fig. 1).

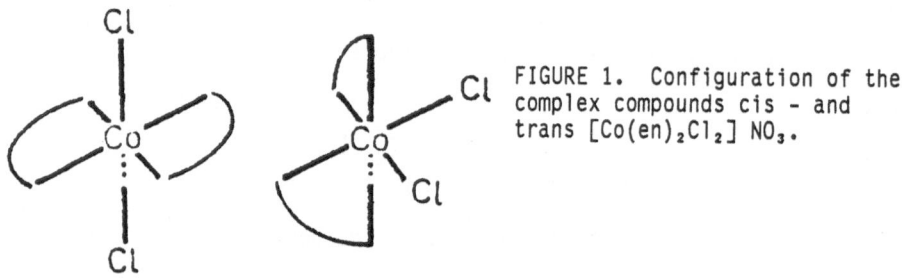

FIGURE 1. Configuration of the complex compounds cis - and trans $[Co(en)_2Cl_2]$ NO_3.

In the following, a few ideas are presented arising from the research field of "hot atom chemistry". There the chemical state of energetic recoil atoms from nuclear reactions is analyzed, using the fact that these recoils can be distinguished from the target atoms because of their radioactivity. Surface and concentration effects can be ignored because the recoils are formed inside the target in rather low concentration. Apart from this there should be no major difference between "hot atoms" and accelerated "hot ions." The energy regime that the energetic particles are subjected to is shown in Fig. 2. There are many different types of chemical reactions taking place during the slowing down process of an atom or ion.

1. In the region $E_{kin} > 100$ eV mainly energy transfer processes to the target atoms occur by electron excitation or elastic collisions. Consequently, target atoms are ionized or excited and displaced, respectively. This causes often braking of chemical bonds (coulomb explosion, collision effects) followed by either restoration or formation of new compounds.

2. Atoms or ions with kinetic energy of 1 to 100 eV may interact themselves with target atoms: the energetic particles can be inserted in target molecules either under destruction of bonds or undestructive. In addition exchange reaction between target atoms and the slowed down atom may proceed via excited states. This is the classical energy regime of "hot atom chemistry."

KINETIC ENERGY, eV

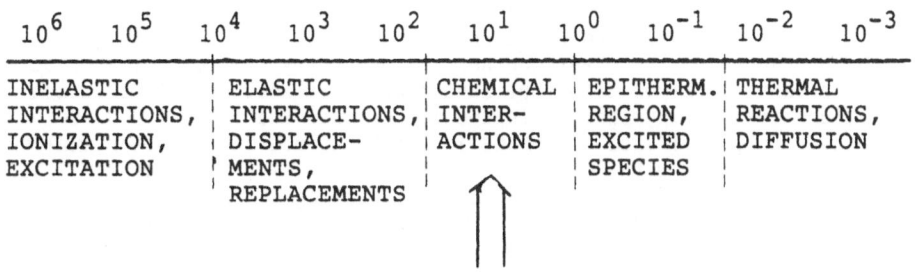

$$10^6 \quad 10^5 \quad 10^4 \quad 10^3 \quad 10^2 \quad 10^1 \quad 10^0 \quad 10^{-1} \quad 10^{-2} \quad 10^{-3}$$

INELASTIC INTERACTIONS, IONIZATION, EXCITATION	ELASTIC INTERACTIONS, DISPLACE- MENTS, REPLACEMENTS	CHEMICAL INTER- ACTIONS	EPITHERM. REGION, EXCITED SPECIES	THERMAL REACTIONS, DIFFUSION

HOT ATOM CHEMISTRY

FORMATION OF:

IONS,FRAG- MENTS RADICALS (COULOMB EX- PLOSION)	FRAGMENTS RADICALS (COLLISIONS)	KNOCK ON ATOMS AND PRI- MARY ATOM MAY REACT (COLLI- SION)	FAST RE- COMBINA- TIONS. EXCHANGE REACTION (EXCITED)	SLOW RECOM- BINATION. ANNEALING

FIGURE 2. Energy regions passed during slowing down of an energetic ion.

3. Below 1 eV thermally activated secondary reactions set in. Fast thermal rearrangement of metastable reaction products may occur as well as slow rearrangement by diffusion of defects.

Looking at the many existing possibilities for reactions which exist in a target consisting of complex molecules, one may realize the theoretical predictions of product yields are very difficult. Even applying pure statistical considerations in a collision cascade results in manifold products. Luckily enough the number of these products is considerably reduced taking into account chemical driving forces like concentration effects and especially bond stability. On the basis of the latter consideration some predictions are rather simple:

--Bombarding a rare earth fluoride with another rare earth ion will result in the formation of REF_3, perhaps also some REF_4.
--Bombarding a rare earth fluoride with gold or platinum will result in no AuF_x, but gold or platinum lusters.

Others are very difficult:

--Bombarding a rare earth fluoride with Ta (Nb,W) may result in the formation of TaF_2 ... TaF_5. However, the yield of the different species cannot be predicted.

In order to illustrate the situation further a few examples from "hot atom chemistry results" will follow:

1. The introduction of energetic Co in a cobalt complex has already been mentioned above (1,2).

2. Transition metal recoil atoms which may form complex compounds of the type A_2MX_6, where A are alkali metals, M = Sn, Re, Os, and X halogens like chlorine and bromine, are easily inserted in neighboring complexes (3,4). When mixed crystals are used where the ratio of chlorine/bromine is gradually changed one can test whether the product yield follows statistically the varying composition (hot zone model), or whether ballistic considerations are of importance. In the latter case the mass of the atoms must play a role (billiard ball model). It turns out that neither explanation is in agreement with the results. The probability for the recoil atoms of finding a suitable lattice side, and of breaking and forming bonds seems to be the best way to understand the measurement. Figure 3 illustrates how the reformation of the complex may take place.

3. Metal atoms or ions recoiling in compounds like $K_2Cr_2O_7$ or $KMnO_4$ where the metal has a high oxidation state may practically adopt any oxidation state after slowing down. However, the experimental results show that primarily low oxidation states are favored, Mn^{2+}, Cr^{2+}, Cr^{3+}, MnO_2, etc. The fraction of the recoils adopting the higher +6 and +7 states increases only during annealing with time (5,6).

The following simple conclusions can be drawn now:

--The interaction of a recoil or implanted atom or ion after slowing down with a compound target is not a pure statistical one.

--The interaction can also not be understood only on ballistic considerations.

--Chemical considerations or driving forces like local concentration gradients, heat of solution, strength or stability of bonds play a major role.

--The more complex the target molecules, the more important are chemical considerations.

--The slowed down recoil or implanted ion ends up initially as neutral or singly charged ion. If chemically stable, it tends to adopt primarily a low oxidation state. Thermal secondary reactions are responsible for the final oxidation state.

--Predictions for product formation and yields can only be made for rather simple systems, and only for those systems a precise control of the reaction parameters is possible in order to obtain a distinct product.

In the next section a few examples for simple cases of ion beam syntheses will be presented.

2.1. Compound formation

There are two major possibilities to synthesize compounds by ion beam techniques. The first one is the implantation of energetic particles of one component into a thin film of another one. The second one is the simultaneous deposition of one component in atomic or molecular form, and the other one in form of accelerated ions. In the first case one is very flexible in the choice of the target compound, however, the concentration of the implanted second component is limited in general by the sputter effect to 50 at. %. By using ion beam assisted deposition techniques this limitation is overcome and 1:1 compounds can be obtained in the surface region. Therefore, this technique is very well suited for synthesizing compound coatings as described later on. Figure 4 shows the vacuum chamber of an apparatus for producing compound coatings by ion beam assisted evaporation.

The following examples are grouped in three categories; the first one describes the interaction of metal atom with non-metal ions under compound formation, the second one the interaction of semimetal atoms with non-metal ions and the last one of metal oxide with non-metal ions.

2.1.1. Non-metal ions interacting with metals. Implantation of boron in iron causes at lower concentrations (up to 20 at. % B) the formation of

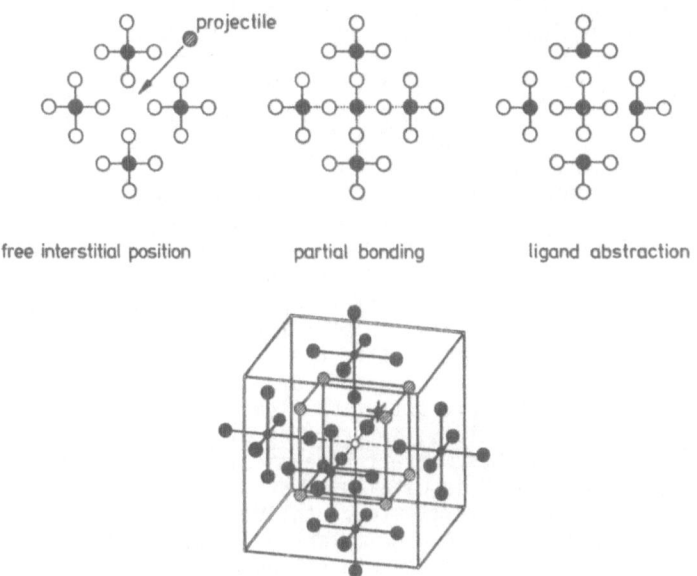

FIGURE 3. Incorporation of a recoil atom into A_2MX_6 compounds.

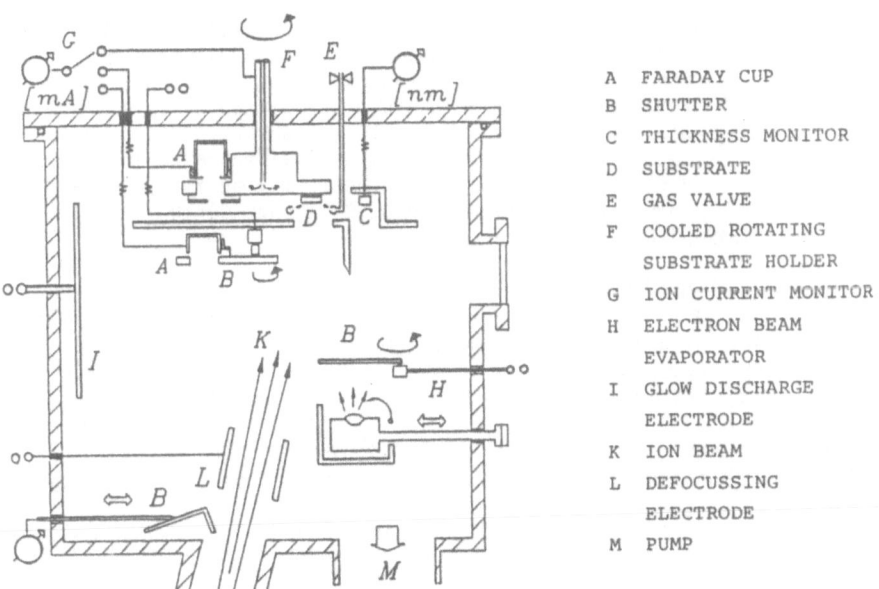

A   FARADAY CUP
B   SHUTTER
C   THICKNESS MONITOR
D   SUBSTRATE
E   GAS VALVE
F   COOLED ROTATING
    SUBSTRATE HOLDER
G   ION CURRENT MONITOR
H   ELECTRON BEAM
    EVAPORATOR
I   GLOW DISCHARGE
    ELECTRODE
K   ION BEAM
L   DEFOCUSSING
    ELECTRODE
M   PUMP

FIGURE 4. Vacuum chamber used for ion beam assisted evaporation (IBAD).

amorphous iron alloys (7). We investigated with Mössbauer spectroscopy the results of higher implantation doses up to 6×10^{17} B^+/cm^2 (ref. 8). Thin iron films of 80 nm thickness were bombarded at room temperature with 30 keV B^+-ions. The target chamber could be transferred directly to the spectrometer without changing vacuum conditions or temperature. Figure 5 shows the spectrum of the samples bombarded with 8×10^{16} to 6×10^{17} B^+/cm^2. In addition to the iron signal the lines of two amorphous $Fe_{1-x}B_x$ and 30% of the non-magnetic compound. Upon annealing they are transformed into the crystalline form above 450 K. Figure 6 shows the change of the line width and the average hyperfine field as function of temperature. Decreasing width indicates high energy ions have to be used to produce burried layers. The synthesis of SiO_2 is possible, for example, by bombarding silicon wafers with 1.8×10^{18} O^+/cm^2 of 200 keV at 500°C (10). The RBS spectra in Fig. 8 show for condition (C) the formation of a 400-nm-thick layer with uniform composition below a 300-nm-thick silicon surface film. Similar experiments have been performed for the production of Si_3N_4.

2.1.3. <u>Non-metals (semi-metals) interacting with metal oxides (ceramics)</u>. There are only very few cases where the chemical nature of implanted species in ceramics has been studied in detail. Mostly only structural changes of the implanted layers were investigated. An interesting example is the recent work of Noda et al. (11) who implanted Si^+ and N_2^+ ions into sapphire. The implantation was done in sequence at 400 keV, the dose was 2×10^{17} Si^+/cm^2 and $1,4 \times 10^{17}$ N_2^+/cm^2. Upon annealing to 1673 K compound formation could be detected by x-ray diffraction. When the bombardment was done at 100 K the resulting reaction product was aluminum oxynitride. However, room temperature implantation lead to the formation of a solid solution of β-Si_3N_4 and alumina, called β-sialon. Figure 9 shows a comparison of a part of the x-ray diffraction spectra taken at room temperature, 1473 and 1673 K. The fact that the implantation temperature controls the compound formation even though the same additional heat treatment was given to the samples, is consistent with the experience one gained from the old hot atom chemistry experiments. At low temperature the percentage of radiation enhanced spontaneous reactions is very small, and the implant remains a metastable (amorphous) solution. At room temperature spontaneous reactions are possible and Si-N-bonds may be already preformed, a situation well known from the formation of C-N-bonds during interaction of energetic C-atoms with solid NH_3 or NH_4X (ref. 12).

3. CHEMICAL APPLICATIONS OF ION BEAM TREATED CERAMICS

Reactions taking place on the surface of a material involving the material and the environment are usually chemical reactions. This may mean:
--two or more reaction partners react on the surface, and the surface remains macroscopically unchanged. Such reactions one would call <u>catalytic reactions</u>;
--the reaction takes place in a solid/liquid interface and parts of the materials are consumed or transformed in another compound. This one would call <u>corrosion</u>;
--the reaction proceeds in a solid/gas interface and parts of the materials are consumed or transformed. In this case <u>oxidation</u> is the proper expression.

Following this distinction the section is subdivided in (1) catalytic reactions, (2) corrosion, and (3) oxidation. In all three cases different types of ion beam modified surfaces or thin films will be discussed:
--ceramic substrates modified by ion implantation and ion beam mixing.
--metal substrates transformed in the near surface region by ion beam
   techniques into ceramics or comparable compounds,

Boron Ion Implantation at RT
with different Fluences

Subspectra : ················· $Fe_{(1-x)}B_x$(magnetic)

– – – – – – $Fe_{(1-x)}B_x$(non-magnetic) + $Fe_{(1-x)}Be_x$

Difference is α–Fe

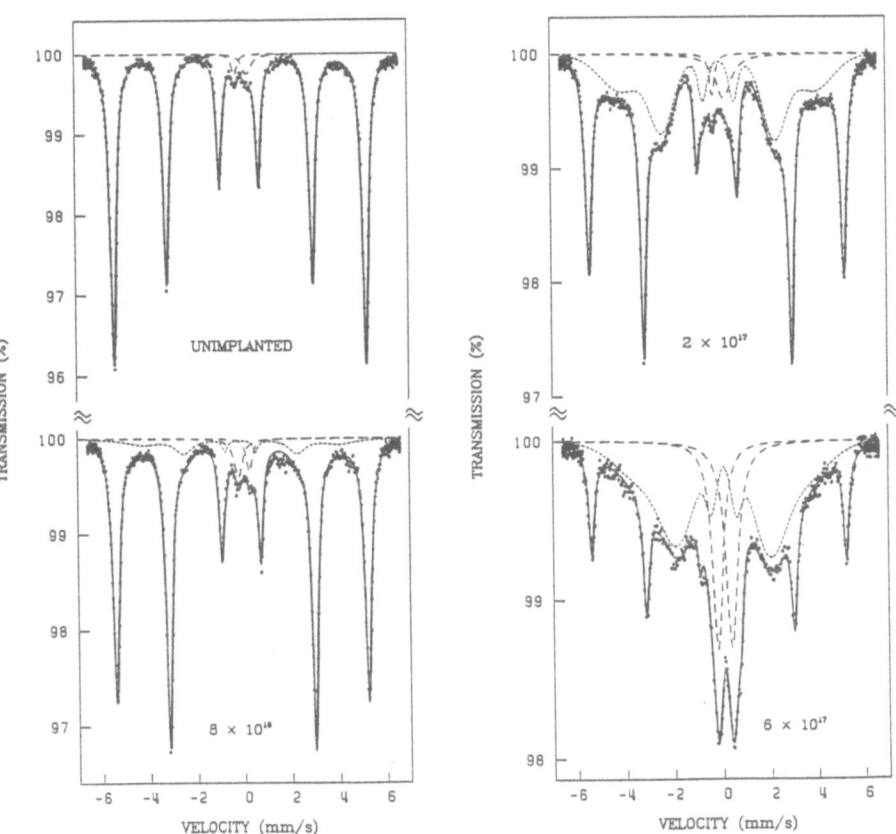

FIGURE 5. Mössbauer spectra of Fe-films bombarded at RT with boron ions (30 keV).

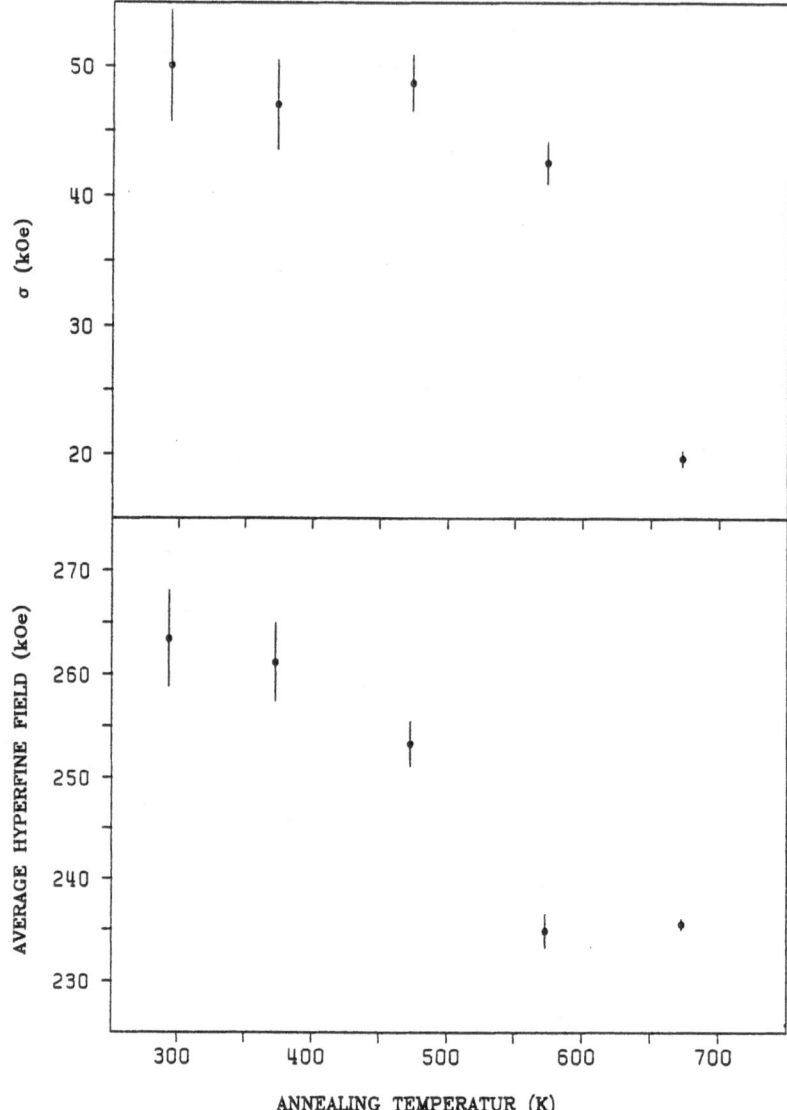

FIGURE 6. Line width and average hyperfine field of boron-bombarded Fe as function of annealing T.

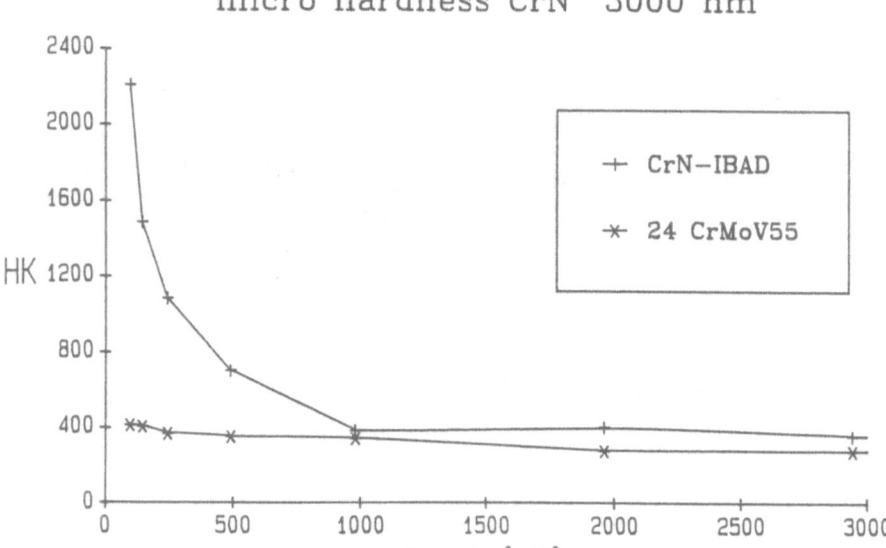

FIGURE 7. Microhardness of a 3-µm-thick CrN layer on steel (IBAD).

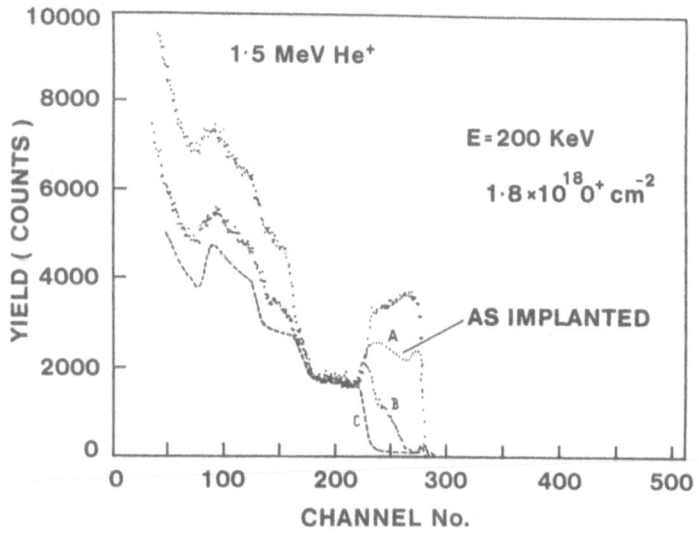

FIGURE 8. RBS channeled spectra of Si bombarded with 1.8 x 10¹⁸ O⁺/cm²
(200 keV, 500°C). A: as implanted, B: annealed 1200°C for 2 h, and
C: annealed 1300°C for 6 h.

340

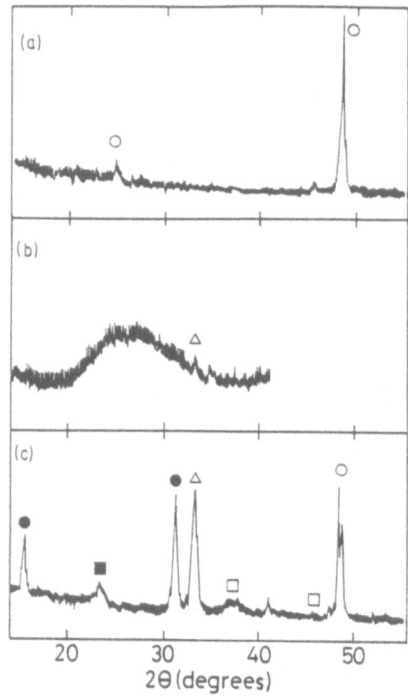

FIGURE 9. X-ray diffraction spectra of Si^+ - and N_2^+ - bombarded Al_2O_3 taken at room temperature, 1473 K and 1673 K O alumina, ● β´-sialon, Δ silicon, □ 15R sialon, ▩ unidentified.

--metal substrates with ceramics or comparable coatings.
3.1. Catalytic reactions on ion beam modified surfaces.
    Why nearly nobody works in the field of catalysis is one of the great mysteries of the ion beam community. Beyond the question of practical applicability, ion beams are ideally suited for investigating catalytic reactions. Any element can be introduced in any substrate surface. Hence, chemical metal/substrate interactions may be studied in whatever extent. In addition any combination of two or three different elements can be implanted simultaneously in a great variety of substrates. This type of ternary compound is nearly not accessible by conventional doping methods. Also the distribution of the implant in the substrate may be quite different. If one starts with a low to medium dose bombardment at or below room temperature one will usually obtain single atoms distributed in the substrate, a system with strong catalyst/substrate interaction. High dose implantation or subsequent heat treatment induces precipitation of clusters or phases the size of which can be controlled to a certain extent. In some cases strong heating forces the implant to diffuse to the surface. The result is a situation comparable to a substrate with an evaporated surface layer. One may also start with the latter situation and distribute the surface layer by ion beam mixing in the substrate. In cases where the active element is not soluble the chance to end up with single dispersed atoms or small clusters is very low, however medium sized clusters may be obtained in that way. Unfortunately, there are no recently published results on ion implantation in ceramics for catalytic purposes. The old data of Matzke et al. (13,14) - platinum implantation in MgO and Al_2O_3 - illustrate the above mentioned considerations on the reaction sequence: implantation, reordering, clustering

and precipitation upon heating, diffusion to the surface at high temperature. The measured catalytic activity exactly reflected the structural changes obtained.

A more recent example is not concerned with a pure ceramic substrate, but with the metal carbide WC. Platinum was implanted with different dose values and energy of 100 keV in thin films of WC. The activity of the resulting catalyst was monitored by the electrochemical hydrogen production reaction from acidic solutions. Figure 10 shows the results. The activity is dependent on the electrochemical potential applied and on the dose which is proportional to the surface concentration. However, beyond this simple correlation one realizes that the implanted catalyst is considerably more active than smooth platinum with 100% Pt on the surface. In Fig. 10 the data point for smooth Pt is marked with an asterisk. By means of Auger surface analysis it could be shown that in no implanted case the Pt surface concentration was above 30 at. %. The only way to understand the higher activity as against 100% smooth platinum is to postulate small Pt-clusters because of electronic interaction with the substrate to be more active than a bare metal surface. Such behavior, namely a dependency of the activity on the size of metal particles, is well known from many studies on heterogeneous gas phase catalysis. It is interesting to note (Fig. 11) that samples made by ion beam mixing of an evaporated Pt-layer on WC with rare gas atoms did not show this extraordinary high activity. With increasing dose the activity decreased considerably.

3.2. Corrosion of ion beam modified surfaces

In this subsection, I would like to start with the aqueous corrosion behavior of modified metals and move slowly towards the leaching of modified ceramics.

3.2.1. Corrosion inhibition by ion beam tailored interface compounds. It is well known that a number of "refractory" iron compound as borides, silicides or phosphides and phosphates have protective power against corrosion in aqueous solutions. Several workers have studied in the past the behavior of iron and steel after boron or phosphorous implantation (17,18). they found positive effects, but the long-term protection was not satisfactory because the amorphous alloy formed was very shallow and the boron or phosphorous concentration not very high. Therefore, we tried ion beam mixing and ion beam assisted deposition in order to obtain high boron concentrations in the interface region (19). The iron and steel samples were coated with electron beam evaporated boron under continuous bombardment with 6 keV argon ions. The result is an 80 and 100 nm, respectively, thick boron coating with a slightly intermixed iron/boron interface (Fig. 12).

Because the boron concentration changes gradually it is not possible to exactly predict the composition of the interface. However from the Mössbauer measurements mentioned already in section two, one may guess that in the iron-rich part of the interface an amorphous solid solution and/or an amorphous Fe_2B phase have been formed. In the boron-rich part in addition FeB could be present. The corrosion behavior in buffered acetic acid is shown in Fig. 13, and is consistent with this interpretation. The dissolution rate of iron determined by polarographic analysis is higher in the beginning, slows down afterwards and increases again after 30 days. In the beginning iron is corroding because of pin holes and defects in the boron surface layer. Afterwards the hydrogen forming in the pin holes removes by its pressure parts of the boron layer, and the pure boron-rich interface alloy remains. The loss of boron from the surface either as element or more probable as oxide can easily be seen with the naked eye. The interface alloy is now very corrosion resistant, and only when it is consumed the corrosion rate increases again. The ion energy and the ion/atom arrival ratio have direct influence on the interface composition.

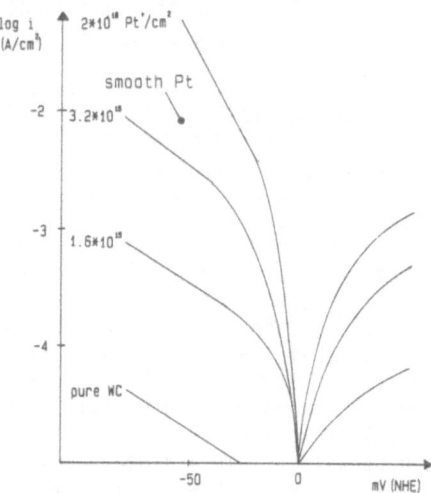

FIGURE 10. Current-density potential plots for the H+-reduction in HClO₄ at Pt-implanted WC. The value for smooth Pt at −50 MV is given for comparison.

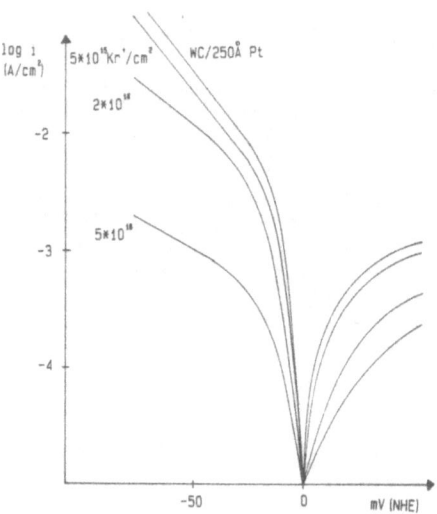

FIGURE 11. Current-density potential plots for the H+-reduction in HClO₄ at WC avered with 25 nm evaporated Pt and bombarded subsequently with Kr+-ions.

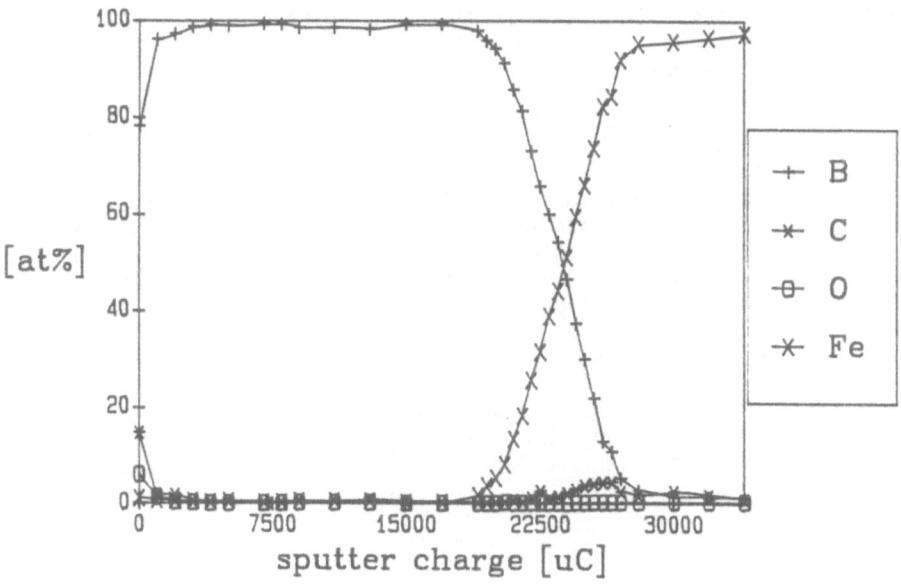

FIGURE 12. Auger depth profile of 80 nm boron evaporated under 6 keV Ar$^+$-bombardment on iron.

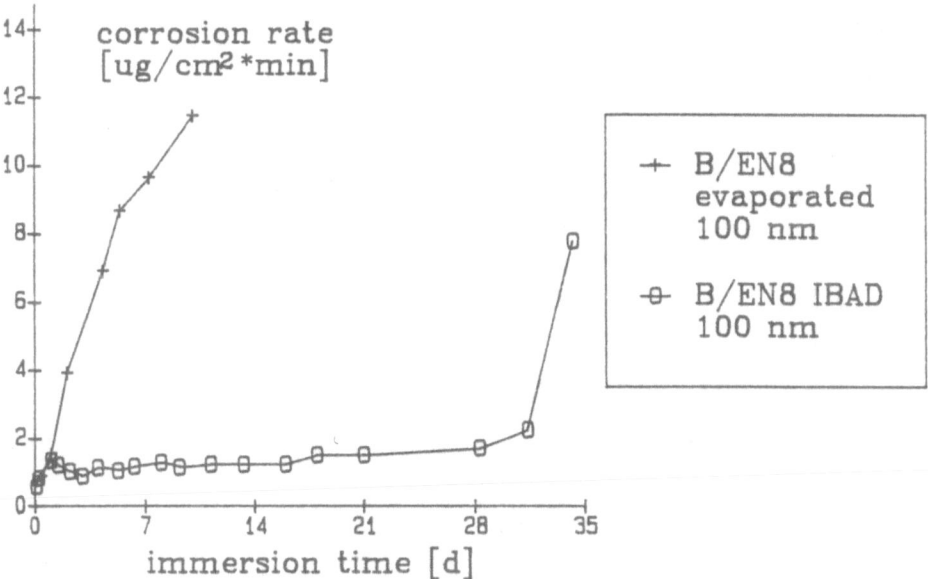

FIGURE 13. Long-term dissolution of boron-modified steel in buffered acetic acid (pH = 5,6).

344

Because of different coating parameters for the Fe/Si system a very broad
interface was obtained. Figure 14 shows the Auger depth profile of a
100-nm-thick Si coating evaporated under strong bombardment with 6 keV
Ar+-ions (20). Here the pure silicon layer disappeared completely and an
intermixed broad interface with gradually changing composition was
established. It seems again reasonable to expect regions containing an
amorphous solid solution and/or amorphous iron silicide phases. The corrosion behavior is comparable to boron (Fig. 15). In the beginning the
remaining surface iron corrodes fast, afterwards slow corrosion of the
"silicide" region, and after 20 days loss of the protective layer. The
figure contains also a comparison with the behavior of a pure evaporated Si
layer and of a Si layer bombarded subsequently with 5 x 10¹⁶ Ar+/cm². Both
films do not present a good protection, because the iron substrate had a
high surface roughness. Under these conditions the Si-layer is removed
quickly by means of undercreeping of the corrosive solution. Only the uniform "silicide" layer obtained by the IBAD process is resistant enough.
Consequently, for a good long-term protection a very broad interface region
covered by a dense pin hole free pure boron or silicon overlayer would be
the best solution.

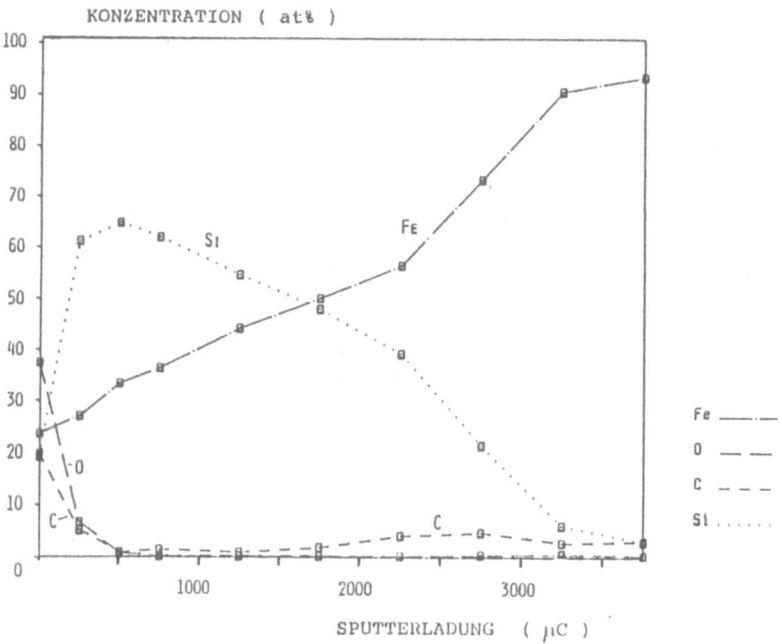

FIGURE 14. Auger depth profile of 100 nm silicon evaporated under 6 keV
Ar+-bombardment on iron.

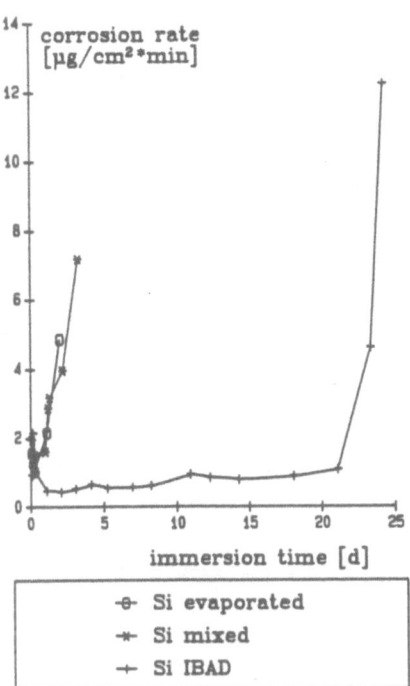

FIGURE 15. Long-term dissolution of
silicon-modified steel in buffered
acetic acid (pH = 5,6).

3.2.2. Corrosion inhibition by "ceramic" coatings

Instead of forming a protective compound by implantation or ion beam
mixing in the subsurface region of a metal, one may synthesize a compound
coating on top of the metal. In addition to well known techniques like PVD,
sputtering or CVD reactive ion beam assisted deposition (RIBAD) is a useful
method. Even though coatings like CrN or TiN are not specially meant for
corrosion protection, but rather for wear protection, their corrosion behavior gives some indication for the quality and porosity of the coating. In
our laboratory, a 300-nm-thick TiN layer on iron was made by titanium evaporation under 10 keV nitrogen bombardment using an ion/atom arrival ratio
of 1:1. Figure 16 shows the evaluation of the reduction of the RBS titanium
signal because of nitrogen (21). This is a strong indication for the formation of nearly stoichiometric TiN. The adhesion of these layers was
excellent as illustrated by the results of the pull-off test represented in
Fig. 17. Under similar conditions CrN-films of up to 3 μm thickness were
grown on steel substrates. Their high microhardness was already shown in
Fig. 7. These coatings have also very reasonable corrosion behavior.
Figure 18 summarizes the results of a long-term immersion test in artificial
sea water. Over a period of 50 days the percentage of the surface showing
visible corrosion products from the steel substrate is negligible. This
means that the 2 μm thick CrN-layer had very good adhesion and acted as protective cathode.

Figure 18 also contains results from carbon coatings on steel. These
coatings are not particularly suited to be treated under the heading corrosion, but because the nitrides are hard compounds it is interesting to
compare them with "hard carbon." The carbon coatings were obtained by

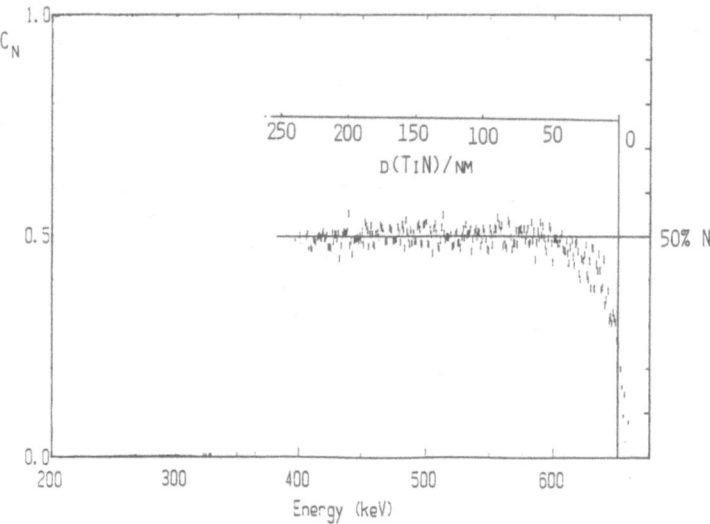

FIGURE 16. Evaluation of a RBS specturm of TiN on iron, produced by reactive ion beam assisted evaporation (RIBAD).

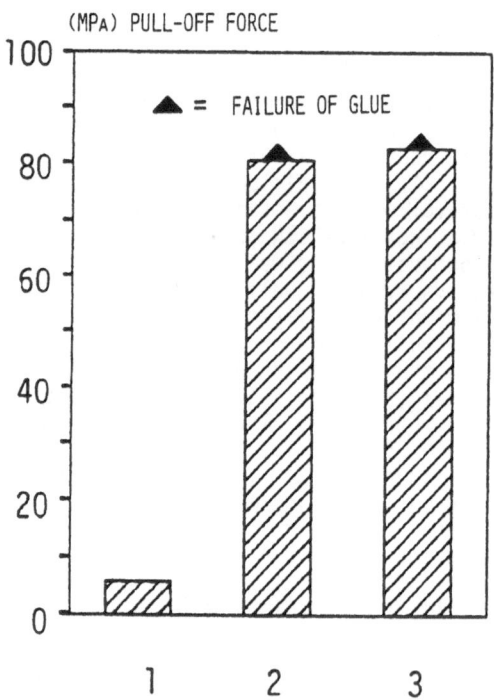

FIGURE 17. Results of pull-off adhesion tests for 300 nm Ti on Fe, Ti layers obtained by ion beam assisted CVD and TiN layers obtained by RIBAD.

Immersion test (art. seawater)

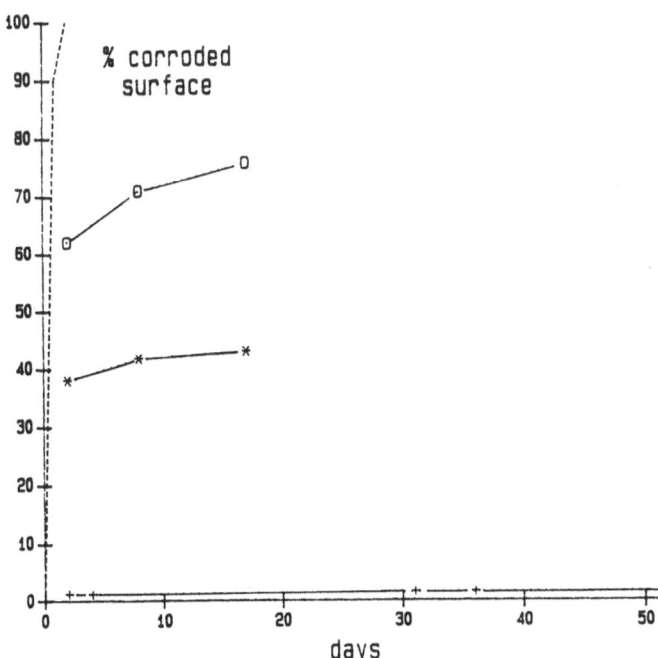

FIGURE 18. Long-term immersion test in artificial sea water. +2 μm thick CrN on low alloy steel; ☐ * 200 nm hard carbon on steel (two different samples); – – – unimplanted steel.

evaporation of graphite under continuous argon ion bombardment. Under these conditions one does not obtain graphite films, but amorphous hard carbon, usually called a-C:H. The mixed microhardness of the system 360 nm carbon on steel (EN 8) is shown in Fig. 19 (ref. 20). With decreasing load, one obtains a steep increase for the coated sample as opposed to the untreated one. Other arguments for the formation of a-C:H were the optical band gap of 0.3 eV and the refractive index of 2.9, values typical of hard carbon with low hydrogen content (24). Actually the H-concentration in the film was measured by the ^1H $(^{15}N,\alpha\gamma)$ ^{12}C-reaction to be about 2 at. %. On the other hand, the corrosion measurement in Fig. 18 indicates that the 200 nm film still contains a fair number of defects or pin holes. Nearly from the beginning of immersion a rather large part of the surface is covered with corrosion products from the substrate dissolution. In order to taylor universal coatings being hard, corrosion resistant and adherent one usually has to have film thicknesses of at least 1 μm. Of all ion beam techniques only ion beam assisted deposition (IBAD) meets these requirements.

   3.2.3. Corrosion (leaching) of ion beam modified ceramics. The influence ion implantation or ion beam mixing processes may have on the leaching of the surfaces of ceramics and comparable compounds are probably manifold. The microstructure may be changed, the composition, the stress state and the defect content. All these modifications should have some implications on the interaction of the surface with solutions. Unfortunately, the knowledge of this subject is very scarce, and only a few examples for measurements can be presented in the following.

   The corrosion of WC was measured in our laboratory as function of the electrochemical potential (25) at the surface of a thin WC-film in

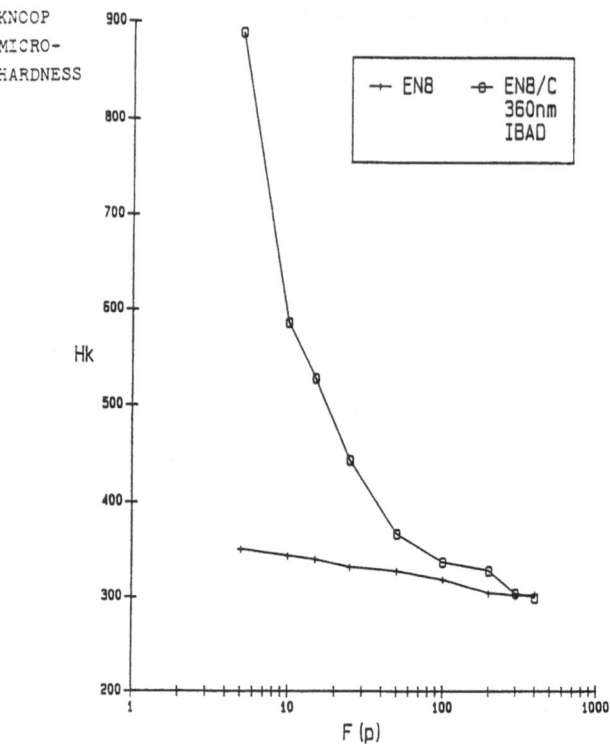

FIGURE 19. Microhardness of 360 nm-C:H on top of a low alloys steel as a function of applied load.

perchloric acid. The corrosion current density is plotted as function of this potential in Fig. 20. The virgin WC-film does not corrode signifi-cantly up to a potential of 600 mV/vs NHE (normal hydrogen electrode). If the sample, however, is bombarded with 10^{16} Xe^+-ions of 60 keV, one obtains an oxidation peak between 400 and 500 mV. There are strong indications of decomposition of the WC by ion bombardment into $W_2C + W$. Both compounds are less corrosion resistant than WC, and therefore the oxidation under for-mation of WO_x and CO_2 occurs at lower potentials. Similar observations of increased corrosion have been made after nitrogen bombardment of Co-cemented WC under open circuit conditions (26).

McHargue et al. (27) studied the etching rate of SiC with and without implantation of 2 to 3 x 10^{16} Cr^+-ions/cm² (200 keV). For polycrystalline material as well as for single crystals the etching ratio implanted/unimplanted in 50% KOH containing 50% $K_3Fe(CN)_6$ was between 2.4 and 4.1. The authors concluded that the main reason for the enhanced corro-sion was the amorphization of the implanted SiC surface layer. Whether the composition changes, the etched layer contained 1 to 2 mol % chromium, contributed to the enhancement was not studied in detail. Even from these few examples it is clear that one has to take a different leaching behavior after ion bombardment into account. Ignoring this possibility could mean that a positive effect of bombardment like increased fracture toughness or reduced coefficient of friction might go at the expense of chemical stabi-lity against aqueous solutions.

current density

$(\mu A/cm^2)$

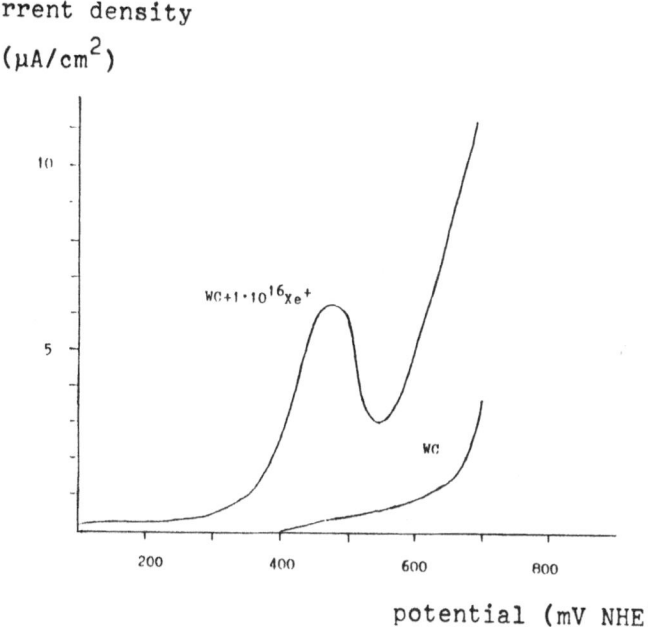

potential (mV NHE)

FIGURE 20. Corrosion current density as a function of the electrochemical potential for untreated and and Xe+-bombarded WC.

4. OXIDATION OF ION BEAM MODIFIED SURFACES

The same that is true for the leaching behavior of ion beam modified ceramics and comparable compounds is also true for high temperature oxidation: there exist nearly no published results on this subject. This is even more sorry because one can forecast many interesting applications for ceramics the oxidation resistance of which is improved. Therefore in the first part of this section results on high temperature oxidation of ion beam modified metals will be discussed, because the surface compounds resulting from this treatment have some similarities with ceramics as we have already seen.

4.1. Oxidation of ion beam modified metals

A specially interesting case is the high temperature oxidation of titanium, because it is a construction material for air and space applications. Galerie et al. (28,29) were able to show that a number of implanted elements are able to reduce the high temperature oxidation in oxygen containing gases. Figure 21 shows the gravimetric mass gain during oxidation as a function of time. The strongest reduction of oxidation rate is caused by Al-, B-, and P-implantation. These elements can form upon oxidation ceramic-like or refractory compounds on the titanium surface. Another interesting case is silicon or the resulting silicides. It turned out that an ion beam mixed Si-layer is much more efficient than Si-implantation because of the rather low concentration of the latter in the surface film (30). In Fig. 22 the mass gain of Si-implanted Ti, ion beam mixed Si/Ti and untreated Ti is compared in the temperature region 600 to 800°C. Si-implantation has only a small effect, while ion beam mixing causes a considerable reduction of the oxidation rate. However, this is only true for the temperature region up to 800°C. Above 800°C the effect is reduced again and around 850°C the reduction drops below a factor of 2.

350

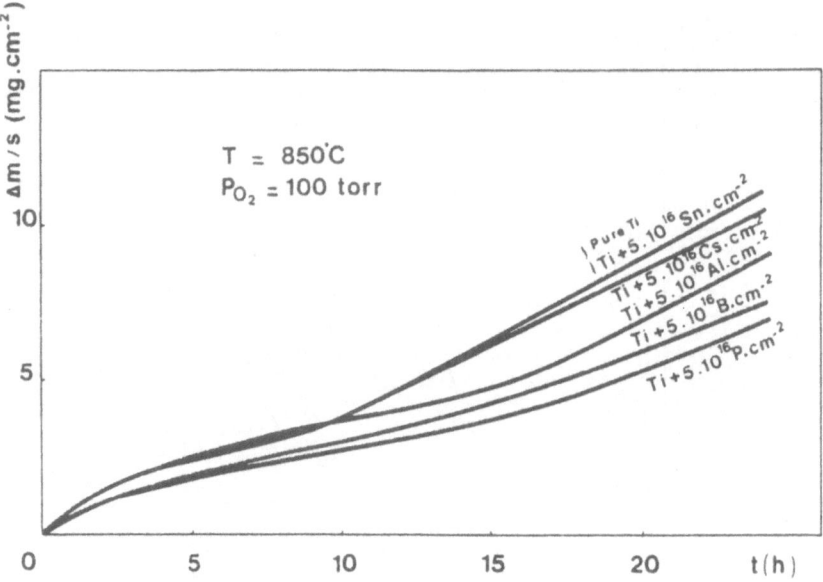

FIGURE 21. Oxidation of titanium implanted with different species: gravimetric curves.

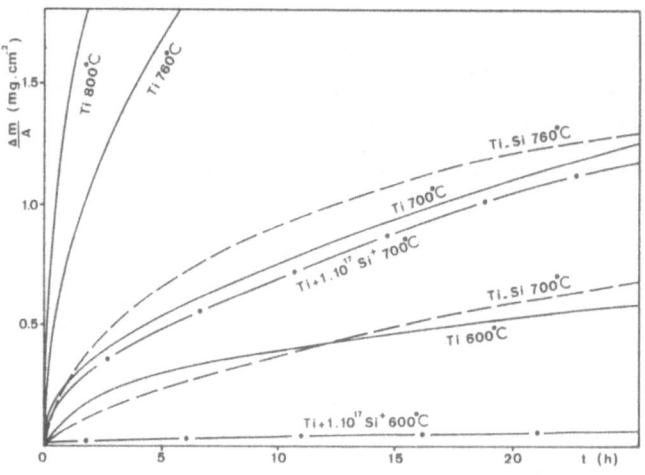

FIGURE 22. Mass gain vs time curves recorded during oxidation in pure oxygen of pure, implanted (1 x 10¹⁷ Si⁺/cm², 150 keV) or ion beam mixed titanium (2 x 10¹⁷ Ar⁺/cm², 200 keV).

From these and many other studies, one can draw the following conclusions which should be generally valid:
--Surface compounds generated by ion implantation or ion beam mixing may reduce the oxidation rate and influence the oxidation mechanism.
--The effectiveness of a specific "compound" depends on the temperature regime.
--Surface compounds can impede oxidation because they block the outward diffusion paths for the metal atoms. In this case the protective layer slowly migrates to the surface of the oxide scale and is then lost.
--Surface compounds can block also the inward diffusion for oxygen. In this case the protective layer stays for longer time near the metal/scale interface.

4.2. Oxidation of ion beam modified ceramics

The most interesting materials to be protected against high temperature oxidation are the ones without oxygen like SiC, Si_3N_4 or other carbides and nitrides because their oxidation resistance in the high temperature region is usually poor. The first example is Cr-implanted α-SiC (27). This work was already mentioned in connection with the leaching of implanted ceramics. The authors also studied the high temperature oxidation behavior at 1300°C. The samples implanted with 2×10^{16} Cr^+/cm^2 showed in the initial state of oxidation a higher oxidation rate compared to the unimplanted case. These data were interpreted primarily as consequence of the microstructure changes. The unimplanted sample is crystalline, the implantation turns it amorphous.

Si_3N_4 was investigated very carefully by Noda et al. (31). they attempted to implant Si_3N_4 containing up to 5% sinter additives like $MgAl_2O_4$ and Y_2O_3 with high doses of Cr^+-ions of 200 keV. The samples, bombarded with 1×10^{17} ions/cm² - 5 10^{17} ions/cm² were oxidized in air at 1173 to 1473 K. At 1273 K the surface region of the unimplanted sample is oxidized to SiO_2, and cations like Y, Fe, Ca diffuse into the surface oxide. In the implanted samples (5×10^{17} Cr^+/cm^2 a part of the chromium atoms diffuses to the surface and is oxidized. The oxidation of the other compounds and the cation diffusion is suppressed. In Fig. 23 the Cr/O, Si/O, and Y/O ratios in the surface region are plotted as function of the implantation dose.

AT 1473 K the situation changes. In the unimplanted sample one finds again SiO_2 and the other cations. The surface is strongly enriched in Mg, and XRD measurements proved the formation of enstatite ($MgSiO_3$) an yttrium silicate (Y_2O_3 2 SiO_2). In the implanted case the surface concentration of Cr_2O_3 decreased considerably, probably because of oxidation and vaporization of CrO_3. In the implanted specimen no compound formation could be detected below 1373 K because the implanted area was amorphous. AT 1473 the same compounds were identified as in the unimplanted case in addition some indications for the presence of -$MgCr_2O_4$ were detected. The presence of chromium seems to suppress the formation of enstatite. This study shows that Cr^+-implantation protects Si_3N_4 against oxidation below 1300 K. Above that temperature because of a complicated chemistry Cr is in principle still protective, but gradually lost from the surface by evaporation.

Other than in the SiC case, it is not the microstructure but the chemistry of the system that is responsible for the oxidation behavior.

In summary one may say that there are several possibilities for oxidation protection:
--The most simple one is protection by the self-oxide formed at the beginning of the oxidation process. It is an open question whether one can influence the density and structure of this oxide by ion bombardment.
--A little more complicated is the formation of surface oxides with higher oxidation resistance than the base metal. Here Al_2O_3 or ZrO_2 are of interest.

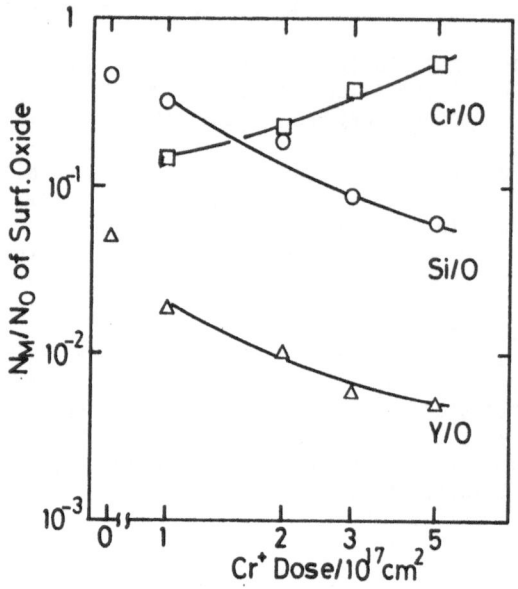

FIGURE 23. The ratios of the surface concentrations of Cr, Si, and Y atoms to that of oxygen as a function of the Cr$^+$-dose.

--The most complex method is doping or coating with species influencing the inward diffusion of oxygen or outward diffusion of metal ions, and forming in a complicated way protective compounds with the substrate, respectively.

5. FUTURE TRENDS

The chemical interactions of implanted or ion beam mixed atoms with solid compounds are usually complex and still not understood. Only in rather simple cases a basic knowledge exists. Therefore, "ion bombardment chemistry" and "interface chemistry" are scientifically underdeveloped regions needing more attention in the future.

Chemical applications of ion beam processed ceramics and comparable compounds having to the authors knowledge high potential for the future are:
--Ceramics and other oxidation resistant compounds as <u>carriers for catalytically active, dispersed metals</u>. Especially the processing of powders is of great interest.
--<u>Ceramic coatings on metals</u>. Under ambient conditions they should serve as universal coatings with good adhesion, wear and corrosion resistance and eventually low friction coefficient. Under high temperature conditions they should be oxidation resistant.
--<u>Sandwich coatings on metals</u>. A combination of metal layers with ceramic layers may provide the desired protection more easily than a one-phase layer. Also combinations of conducting (metal) and isolating (ceramics) films are of interest.
--<u>Surface chemistry of ion beam modified ceramics</u>. This today nearly not existing field is the key to tailor-made ceramic surfaces. It involves oxidation and corrosion as well as friction and oxidative wear. Besides ion implantation ion beam mixing seems to be the most promising technique.
--For the processing of ceramic-fiber reinforced alloys ion beam assisted coatings might be of future importance.

REFERENCES
1. Wolf GK: Radiochim. Acta 6 (1966) 39.
2. Wolf GK and T Fritsch: Radiochim. Acta 11 (1969) 194.
3. Rössler K and L Pross: Radiat. Eff. 48 (1980) 207.
4. Müller H: Angew. Chem. 79 (1967) 128.
5. Andersen T and K Olesen: Trans. Faraday Soc. 61 (1965) 781.
6. Rieder W, E Broda, and J Erber: Mech. Chem. 81 (1950) 657.
7. Rauschenbach B and K Hohmuth: Phys. stat. sol. (a)72 (1982) 667.
8. Hans Diplomarbeit M: Universität Heidelberg, 1986.
9. Wolf GK: Contribution to PSE Conference, Garmisch, September 19-23, 1988, to be published.
10. Reeson KJ, PL Hemment, RF Peart, CD Meekison, C Marsh, GR Booker, RJ Chater, JA Kilner, and J Davis: Radiation Effects in Insulators 3. I Wilson and R Webb (eds). London: Gordon and Breach, 1986, p. 555.
11. Noda S, H Doi, T Hioko, J-I Kawamoto, and O Kamigaito: J. Mater. Sci. 22 (1987) 4267.
12. Rössler K: Radiation Effects in Insulators 3. I Wilson and R Webb (eds). London: Gordon and Breach, 1986, p. 505.
13. Matzke HJ, A Turos, and P Rabette: Radiat. Eff. 65 (1982) 1.
14. Matzke HJ, A turos, P Rabette, and O Meyer: J. Phys. C 14 (1981) 3333.
15. Wolf GK, R Spiegel, and K Zucholl: Nucl. Instr. Meth. B 19/20 (1987) 1030.
16. Zucholl K and GK Wolf: Dechema-Monographien 102, VCH Verlag (1986) 413.
17. Giacomozzi F, L Guzman, F Marchetti, A Molinari, M Sarkar, and A Tomasi: Mater. Sci. Engr. 90 (1987) 197.
18. Hubler GK, JK Hirvonen, CR Gosset, I Singer, CR Clayton, YF Wang, HE Munson, and G Kuhlmann: Naval Research Laboratory Report NRL-Report 4481, Washington, DC, 1981.
19. Schröer A: Diplomarbeit, Heidelberg, 1988.
20. Wolf GK, M Barth, and W Ensinger: Contribution IIT'88, Kioto, Japan, June 7-10, 1988. To be published in Nucl. Instr. Meth. B.
21. Bolse W, W Ensinger, KP Lieb, and GK Wolf: unpublished results, 1988.
22. Ensinger W: Ph.D. Thesis, Heidelberg, 1988.
23. Fischer W: Diplomarbeit, Heidelberg, 1988.
24. Savvides N: Proc. E-MRS Meeting, June 1987, Vol. XVII, p. 275.
25. Spiegel R: Diplomarbeit, Heidelberg, 1986.
26. Ballhause P and GK Wolf: unpublished work, 1987.
27. McHargue CJ, MB Lewis, JM Williams, and BR Appleton: Mater. Sci. Engr. 69 (1905) 991.
28. Pons M, M Caillet, and A Galerie: J. Less Common Metals 109 (1985) 45.
29. Galerie A and G Dearnaley: Mater. Sci. Engr. 69 (1985) 381.
30. Galerie A, M Caillet, M Pons, and G Dearnaley: Nucl. Instr. Meth. B19/20 (1987) 708.
31. Noda S, H Doi, T Hioki, J-I Kawamoto, and O Kamigaito: J. Japan Soc. Powder and Powder Met. 35 (1988) 12.
32. Habig KH: Reibung und Verschleiß bei metallischen und nichtmetallischen Werkstoffen. KH Zum Gahr (ed). Oberursel: DGM Verlag, 1986, p. 203.

EFFECTS OF ION BEAM PROCESSING ON OPTICAL PROPERTIES

P. D. TOWNSEND

Mathematical and Physical Sciences, University of Sussex,
Brighton, BN1 9QH, United Kingdom

1. INTRODUCTION
   A major impetus for the development of optical and electro-optic signal
processing has come from the use of low loss optical fibers for signal communication. Fiber fabrication is now sufficiently routine that in order to
make full use of the potential signal capacity there has been a major worldwide effort to devise optical components which will generate light, modulate
it with electrical signals, couple it to the fibers and subsequently detect
and decode the signals. By analogy with electrical signal processing and
miniaturization techniques this field has variously been termed integrated
optics or photonics (e.g. 1,2,3). The consequent research and commercial
pressures have raised interest in many new optical materials, new routes to
the formation of modulators, solid state lasers, optical detectors and optoelectronic, opto-acoustic and piezo-optic devices. For electronic cir-
cuitry, the fabrication methods used for component integration, such as the
formation of high purity planar devices in which impurities are introduced
in combination with lithographic methods, are well developed. Hence, these
same procedures have been incorporated into the optical component production. Optical structures, which are written into a surface plane, are typically a few wavelengths in cross section. From the viewpoint of ion
implantation this is interesting (4-7) as, at least for light ions, there
are many accelerators which can implant ions to this micron depth scale.
Commercial interest in ion beam technology can be further supported as many
of the new and novel materials are not readily modified by other routes.
Ion implantation initially gained acceptance in semiconductor technology
because the intrinsic value, per unit area, of the product is high and hence
there was less inhibition in considering a high cost technology. In addition there are benefits which accrue from the use of ion beams. If ion
implantation is to be equally successful with optical applications, then a
similar supportive argument will be helpful. By comparison with semiconductor applications, it will appear that for optical applications in insulators
the ion doses may be large. However, the processing with ion beams may
still be simpler, faster and economic, compared with alternative treatments.
It is appropriate to note that, at least at the research stage, novel
crystals are expensive, for example beta barium borate or KTP are as much as
40,000 lb/cc (in 1988), hence processing costs are not a primary concern
initially. Therefore it is advantageous to promote the case for ion implantation at the present time, before there is any entrenched opposition from
proponents of other techniques, which have had a costly development phase.
   Only a relatively small number of references are given here, but many more
are discussed in detail in references (4-7).

C.J. McHargue et al. (eds.), Structure-Property Relationships in Surface-Modified Ceramics, 355–370.
© *1989 by Kluwer Academic Publishers.*

2. ADVANTAGES AND DISADVANTAGES OF ION IMPLANTATION

The advantages of ion implantation in insulators and semiconductors are persuasive, and a brief list of them is repeated here. Namely, the ion beam can be accurately defined by masks, dose and depth profiles controlled by selection of energy and deposited charge, there is relatively little lateral beam spreading (e.g. compared with thermal diffusion), there are only minor limits to the choice of dopant, more than one dopant profile can be introduced into the target, chemical solubility is not initially important, implants may be made into a substrate which is at a chosen temperature. In practice this means that there is a high degree of reproducibility. The process is very clean, even to the point of only introducing a single isotope. For the optical applications of waveguide formation it has been effective in nearly every system that has so far been attempted. Such wide ranging applicability is particularly interesting for the rapid introduction of new materials, and in many cases the implantation route may be the only one which has been developed. The flexibility of the ion beam approach for control of the depth profile of the property changes are familiar from the semiconductor examples, and they are equally useful in optical technology. For example, it is quite feasible to tailor the refractive index profile. An extreme case is the formation of a double buried waveguide layer (8,9). Finally, ion implantation across an interface can produce enhanced bonding and is valuable for both metal/insulator and insulator/insulator interfaces.

Limitations are set for the maximum dopant concentration by sputtering of the target. In practice this limiting level is typically measured in the order of 10^{17} ions/cm^2, and is higher than normally required for optical applications. Alternative deposition methods must be employed if higher concentrations or thick optical layers are required (10). Newer methods have included a combination of implantation with conventional deposition techniques.

An unavoidable consequence of all ion beam implantation is the displacement of host atoms from lattice sites, and ionization of the lattice. When such features are unwanted they are referred to as radiation damage, alternatively they may themselves provide the material modification that is needed. For semiconductor applications the ionization and defect problems are not trivial (e.g. 11). For example, if one implanted a 1-MeV boron ion into GaAs then, not only is a single dopant ion introduced into the lattice, but also there are some 200 carriers removed per implanted ion. Similarly, during the implantation, there may be a comparable number of lattice atoms displaced, but many of these will only be in metastable configurations and so will promptly return from interstitial sites to the vacancies created by their displacement. Controlled removal of the unwanted ionization features, and lattice defects, is achieved by annealing. Variations include furnace annealing or laser and electron beam pulses (12). Each route has advantages for specific devices.

3. OPTICAL PROPERTIES

The other chapters in this volume will detail many of the damage and sputtering problems that are briefly sketched above, and so the remainder of this chapter will be devoted to the specifically optical changes produced by ion beam implantation. To offer a perspective of the changes which can be induced, we may first catalogue optical properties which are currently being exploited. The passive features include changes in refractive index, birefringence, reflectivity, optical absorption, photoconductivity and luminescence. Dynamic properties include the use of the Pockels effect to electrically modulate the refractive index, interactions of light with surface acoustic waves, four wave mixing, hologram formation, and other non-linear effects such as frequency doubling (1,2,3,13). Light sources, in the form of LEDs or lasers, must be fabricated in transparent material and their confinement to selected regions can be conveniently made by the use of ion beams.

4. EXAMPLES OF PASSIVE STRUCTURES

A primary step for many devices is the definition of an optical waveguide. In principle all that is required is a region of high refractive index, surrounded by material of lower index. If signals are to travel unattenuated then both regions should be transparent to the wavelength used. Figure 1 shows cross sections for a variety of structures in which this can be achieved. The sketches of Fig. 1(a) and (b) simply rely on topographic structures for the lateral confinement of the light. The ridge structure is inherently more lossy than the trapezoidal one as there will be power coupled from the base of the guide into the support material. In the sketch of Fig. 1(a) only a single material is used, but for Fig. 1(b) the guiding layer is formed by structuring an overlayer with a higher refractive index than the substrate. Implantation, in conjunction with mask patterns, can be used in the construction of both types. Using low energy heavy ions, ion beam milling can sputter the patterns. Alternatively, radiation damage in the layers can change the chemical stability of the insulators and hence subsequent chemical etching will preferentially remove the bombarded material. The scale of the enhancement of chemical etching varies from a factor of 3 for silica glass to up to 1000 times for garnets and $LiNbO_3$, (e.g. 2,7).

In Figs. 1(c) and (d), the vertical confinement of the light is again made by the use of an overlayer of higher refractive index than the substrate. lateral confinement is then attempted by ion implantation of the layer. Two possibilities exist. Either the implant increases, or reduces, the index. In either case implantation can be used to define a waveguide. In Fig. 1(c) the index is assumed to be enhanced and so the implant is directly into the region of the guide. By contrast, in Fig. 1(d) the material is reduced in index, and so the implants are used to define the boundary regions. Examples of materials which behave in the Fig. 1(c) type manner are silica, and some silica-rich glasses. In these materials, damage causes compaction of the glass and an increase in index of 1 or 2%. Index enhancement is not found in all glass material, particularly if there is any associated movement of the component ions, for example in work with soda lime glasses index decreases of several percent are quoted which correlate with loss of alkali ions.

While the formation of optical coating layers is a well established art (10) the layers so deposited are generally amorphous and, even if crystal growth is possible, the surface films may lack the perfection and low loss found in bulk single crystals. A further problem for electro-optic applications is that, even for single crystals, the material may not be poled into a single domain. Consequently, the development of waveguides in electro-optic crystals has commenced by starting with single crystals, and introducing impurities into the surface to enhance the index. A particularly important material at this point is $LiNbO_3$. Here the doping is made by in-diffusion of Ti or ion exchange with protons for Li. Historically, the Ti technology has become well established, despite a number of serious deficiencies in terms of scattering and optical damage. However in diffusion technology, to form waveguide patterns, there are some inherent limitations in that diffusion through a mask will not only proceed into the sample, but also laterally beneath the mask. There is also the fact that peak Ti concentrations will occur near the surfaces, and this

358

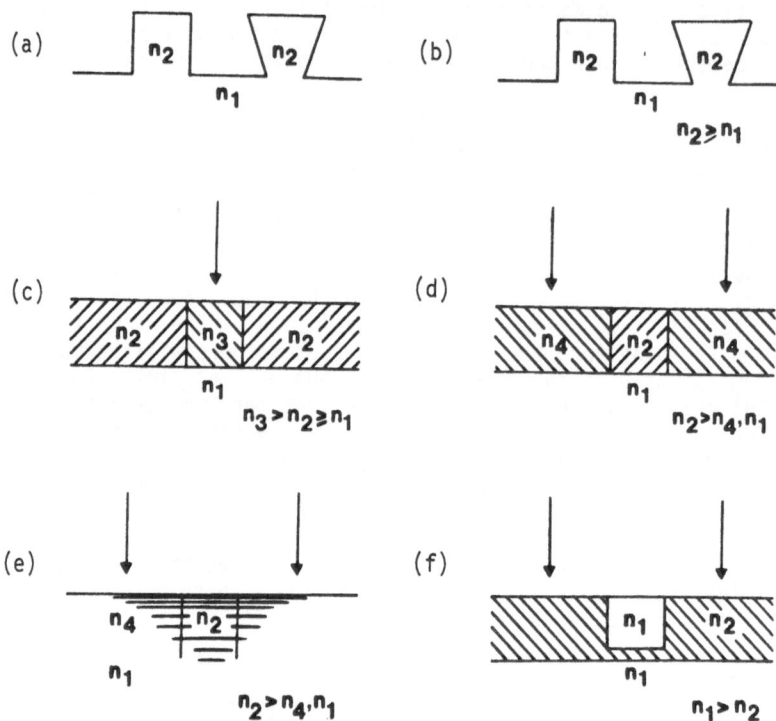

FIGURE 1. Waveguide structures: (a,b) geometric with $n_2 = n_1$ or $n_2 > n_1$, (c,d,e) overlayer or diffusion with changes formed by implantation to form index differences, and (f) an all-implant boundary layer of low index.

contributes to the scatter losses from the guide (14). However, an advantage of Ti diffusion is that it can extend several microns into the surface. Ion implantation in $LiNbO_3$ reduces the refractive index by as much as 7%, in the region of nuclear collision damage. Therefore, one may combine advantages from both technologies by first forming a Ti diffused planar layer of enhanced index, and then using a mask pattern, to ion implant the boundary regions which define a waveguide pattern, Fig. 1(e) (15,16). A less obvious benefit of this route is that the depth of the guide is determined by the planar guide formation, but for lateral confinement the decrease in index does not need to extend throughout the entire depth. In practice for a monomode guide, of say 1 µ scale, a low index notch extending only 10% into the layer will force optical confinement. In ion implantation terms this is valuable as it means that the guide can be patterned using an ion beam of very modest energy (e.g. 200 keV He^+).

A further example of combining the two technologies is provided by a study of loss in Ti diffused $LiNbO_3$ strip waveguides (17). In the diffused guide the electric field of the light is concentrated near the surface, where there is the greatest concentration of Ti. Unfortunately, Ti does not simply enter into the $LiNbO_3$ lattice. Instead, at high concentrations it forms a range of Ti-Li-Nb-O compounds which can cluster, or precipitate, and hence produce surface scatter and power loss in the guide. In an attempt to overcome this, ion implantation has been used into the upper layer of a Ti diffused strip guide surface. By selecting the ion energy and dose, the index profile was smoothly lowered from the body of the guide to the surface (i.e. the opposite of a normal Ti guide profile). The resulting guide was more closely symmetric in index profile and had losses which were only half those seen prior to implantation. Overall, the measured loss was as low as 0.07 dB/cm.

In the final section of Fig. 1, diagram (f) shows how one could form the entire waveguide pattern by ion implantation into a crystalline target. As a first step, low mass high energy ions are used to form a buried damage layer, of low refractive index. Lateral confinement is then made by selecting lower energy ions in combination with a mask pattern. The underlying principle for this will now be discussed.

5. DAMAGE FORMATION

Energy transfer between the incident ion and the target can conveniently be described in terms of electronic excitation and nuclear collisions. Figure 2 shows the standard idealized picture in which the nuclear collision events only occur at the end of the ion range when the ion energy falls below about 100 keV for lighter ions. This situation is ideal for the purposes of waveguide formation in crystals. For the more strongly bonded systems of oxides, such as $LiNbO_3$, electronic excitation does not induce defects. Hence, the only major damage ensues at the end of the ion range in the region of nuclear collisions. There is a fall in refractive index because the destruction of the lattice, towards an amorphous structure, reduces the density of the material, and hence the index. Bond breakage will also change the bond polarizability which in turn may further lower the refractive index. Consequently, the conversion of a well-ordered crystalline structure into a highly disordered, or amorphous, damage layer can cause a fall in index of several percent. Minor color center problems occur within the region of electronic excitation but, as these are primarily the result of charge redistribution, the general result is that such color centers may be annealed at a much lower temperature than is needed to cause recovery of the structural damage (7). This is the ideal situation for waveguide construction as the guide is formed in a region of "perfect" crystal and surrounded by a low index amorphous optical barrier. Because crystallinity is preserved, all the desirable electro-optic features of the guide material are retained.

6. WAVEGUIDE FORMATION IN QUARTZ

The ideal situation, in which nuclear damage produces amorphization, and electronic damage only produces color centers, is realized in the bombardment of quartz. Here the crystalline material is destroyed and it converts into amorphous silica. Figure 3 shows the refractive index profile of quartz bombarded to different dose levels with MeV He^+ ions. One must first note that, in our method of displaying the refractive index data, it is plotted as a decrease on the ordinate axis. This has the advantage that the depth profile for the trapped light displays an optical well, and hence intuition gained from experience with quantum mechanical wells for electrons, is readily applied. At low ion dose levels, the refractive index change occurs at the projected ion range and the width of the barrier is given by the range straggling. The magnitude of the change is proportional to the ion dose for low doses but saturates when the individual damage cascade regions overlap. The rate of approach to saturation is simply determined by the displacement energy for the system. In the case of quartz, the refractive index falls from 1.54 to 1.48. This latter value is not the index of commercial silica (1.46), but corresponds to the value found in ion or electron bombarded silica. (Damaged silica is more compact

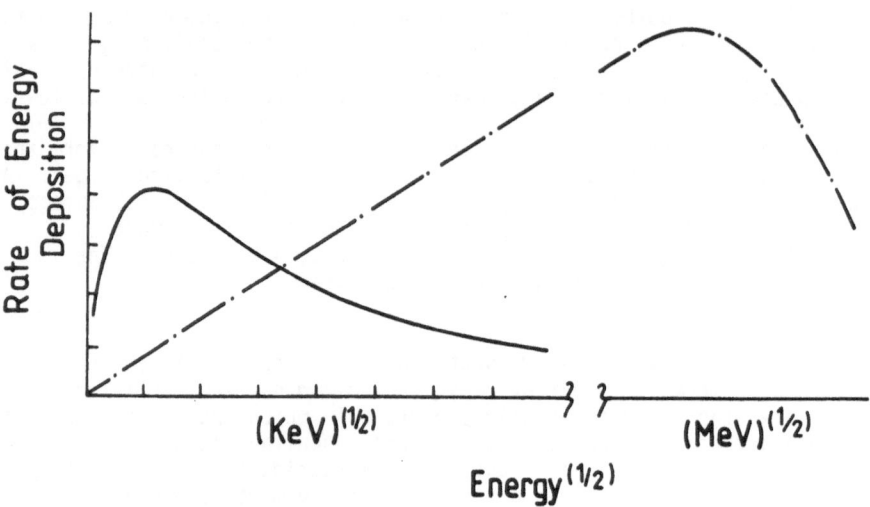

FIGURE 2. A separation of the rate of energy loss into electronic processes
(−·−·) and nuclear collisions (——).

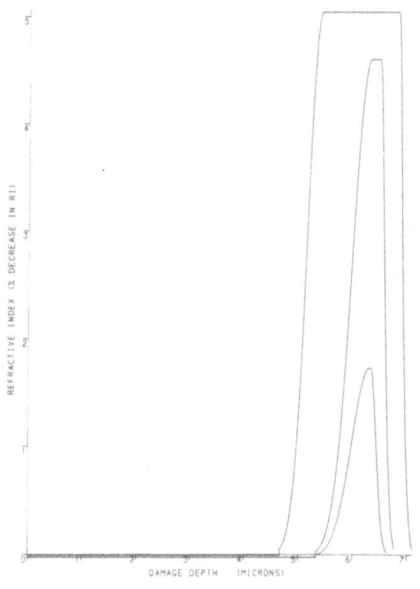

FIGURE 3. An example of the
reduction in refractive index for
quartz as a function of ion dose,
using He ions. Note that on
increasing the dose the Gaussian
profile saturates, and then
broadens towards the surface. The
index falls towards that of
amorphous silica.

than the original material and therefore shows a higher refractive index).
Continued bombardment of the quartz increases the width of the damage layer
but does not change the saturation value. Once formed, this buried
amorphous layer is highly stable and in annealing studies of such guides
there is no obvious recovery to crystalline quartz below 1200°C. Such stability is interesting as it means that any color centers formed in the
quartz are removed. Note also that although the quartz passes through two

phase transitions, at 573 and 850°C, this does not induce a regrowth of the amorphized layer. Overall, these properties allow one to form low loss quartz waveguides (18,19). Variations on this basic approach are to use a range of ion energies, and dose, to control the width, and height of the barrier. An excellent demonstration of how one may tailor the index profile in quartz waveguides is given by Fig. 4 in which a pair of guides have been constructed at different depths from the surface (8,9). In this particular example, the central barrier was deliberately chosen to allow coupling between the two guides. To emphasize the coupling between the wells, the measured mode spacings of the composite well are shown as solid lines, and for comparison the dotted lines are used to indicate where one would expect modes in the individual wells.

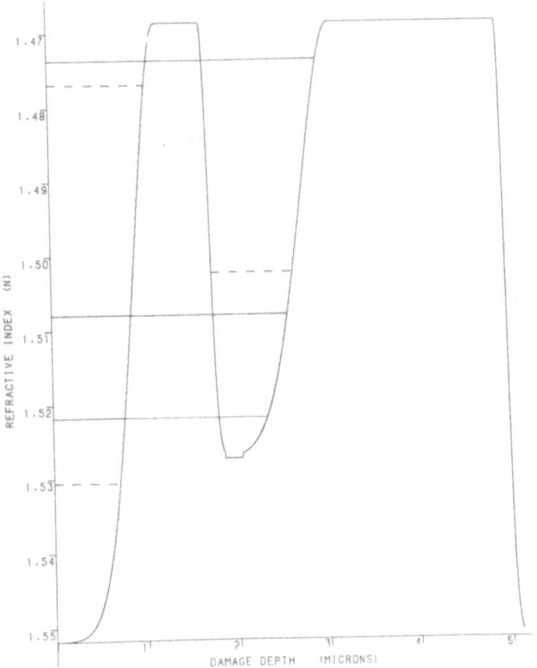

FIGURE 4. By control of ion energy and dose, a double-well structure may be formed in quartz. The solid lines show the measured waveguide modes for waveguide modes for the double guide and the dashed lines those measured for separate guides.

7. WAVEGUIDES IN SILICA

By contrast with the crystalline version of SiO_2, implants in amorphous silica cause compaction and index enhancement. Subtle electronic routes to defect formation exist in many insulators and silica shows such effects, so that both the nuclear collision damage and electronic excitation, lead to atomic displacements. In silica glass this results in a relaxation of the glass network, and compaction. However, the efficiency of electronic energy transfer from the ion beam, to create defects, is lower by a factor of about 10^3, than for nuclear collision processes. Because both types of excitation are effective in displacing lattice atoms, the combined effects of electronic and nuclear collision damage are not simply additive. Indeed, if electronic excitation occurs in previously collision damaged material, there can be some defect annealing or dissociation of larger defect complexes.

Similar effects occur in many other oxides, such as MgO, Al_2O_3, or $LiNbO_3$, but are rarely as easily resolved.

In the case of silica, there has been some interest in changing the refractive index by impurity doping. To some extent this is less interesting from an economic viewpoint as for each incident ion one forms some 200 displacements, so the total implant dose needed to form a waveguide can be quite modest. For silica ion doses of only 5×10^{15}, with He, achieve a 1 or 2% index change. However, this is an upper limit and although relatively stable, to say 300°C, the guides may show some loss or scatter from the defect structures. Changing the index by polarizability modifications caused by impurities could, in principle, allow larger index changes and greater thermal stability. A successful example of this approach was to use nitrogen implantation, as at high concentration, this forms a silicon oxynitride. This has a higher refractive index than silica. Guides so formed, and annealed at 450°C, cease to show losses from intrinsic silica defects but retain the index enhancement of the nitrogen dopant. Consequently, such guides are of very low loss (20,21).

The ultimate limit of impurity doping is to form new compounds and this possibility has been exploited to vary the refractive index throughout the range from Si to SiO_2. The intermediate cases have been reached from either end of the range by Si or O implantation (4,22).

8. ELECTRONIC EFFECTS IN OTHER MATERIALS

Electronic excitation is the dominant process for energy deposition when bombarding with high energy light ions. Therefore, even if electronically induced effects are of low efficiency, they are likely to be apparent, either by the production of damage, or by synergistic effects in which they modify defect structures formed by nuclear collision damage. Quite spectacular effects are reported for $LiNbO_3$, though they are absent in the isomorphous $LiTaO_3$. Figure 5 contrasts the refractive index profiles obtained by He^+ implants into these two materials (23). The materials have an approximately rhombohedral unit cell and therefore show rather different indices for n_0 and n_e, the ordinary and extraordinary indices. In each case, the maximum damage occurs at the projected ion range and the destruction of lattice perfection causes a decrease in index, by up to some 5%. However, for $LiNbO_3$ the n_0 damage extends up to the surface of the crystal and, in the electronic stopping region, n_e is enhanced. Simple electron irradiation of $LiNbO_3$ does not produce these index changes. Consequently, one must be viewing changes which are only triggered by the production of the defects by nuclear collisions. Electronically driven migration, or lattice relaxation, then occurs while these defects exist in the lattice. In the niobate, it is not a trivial problem to separate structural relaxations from effects of radiation enhanced diffusion. Indeed in attempts to do so, using Raman spectroscopy or x-ray analysis, no differences have yet been detected between the original bulk material and the electronic stopping region, even though there are major changes in index. The contrast between the niobate and tantalate is interesting as this behaves very much like quartz, in which electronic effects were absent. A possible source of the difficulty may be that the niobate is not grown as a perfect stoichiometric material, which implies that an intrinsic defect concentration may facilitate radiation-enhanced diffusion of atoms which are displaced by the nuclear collisions. Attempts to form guides in the so-called stoichiometric niobate, as well as the normal congruent melt material, gave similar index profiles. However, since defect concentrations may be important on the scale of parts per million, this is not a conclusive experiment. The view that electronically induced effects become of greater importance in

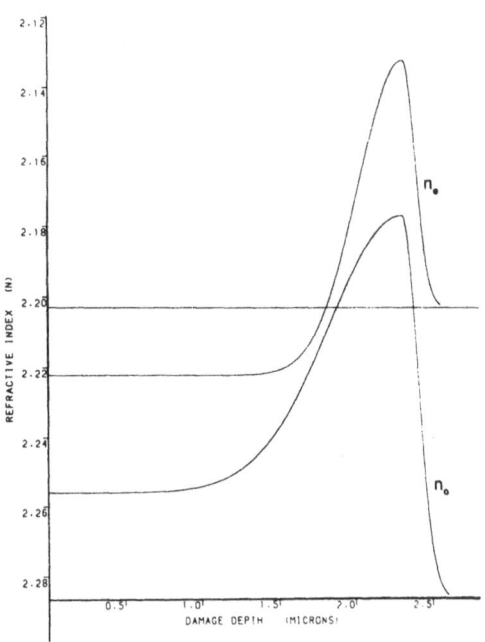

FIGURE 5. Refractive index profiles in LiNbO$_3$. Note the n_e index is enhanced in the electronic stopping region. The nuclear collision damage region is broader for n_0.

tate radiation-enhanced diffusion of atoms which are displaced by the nuclear collisions. Attempts to form guides in the so-called stoichiometric niobate, as well as the normal congruent melt material, gave similar index profiles. However, since defect concentrations may be important on the scale of parts per million, this is not a conclusive experiment. The view that electronically induced effects become of greater importance in imperfect materials is in accord with the observations of complex damage generation in the silica glass, in which electronic and nuclear processes strongly interact.

An even more pronounced example of index enhancement in the region of electronic excitation is provided in Bi$_4$Ge$_3$O$_{12}$, Fig. 6 (24). In this example, the material forms excellent waveguides with light tightly confined in the optical well. However, the shape of the well does not reflect the rate of electronic energy deposition, but instead shows a pronounced maximum index increase near the computed projected range of the ions. The enhancement extends several microns, up to the surface, but no significantly to depths beyond the ion range. Annealing of the material reveals that the stability of the changes is different for the electronically generated and collision generated damage. (A similar effect is reported in silica.) In the case of BGO there is a gradual loss of the well on heating towards 400°C. Between 400 and 450°C the well suddenly disappears, and instead of a well, a very clear damage peak emerges. This index minimum is positioned closer to the surface than the R_p of the ions. An initial interpretation is that ions displaced by the nuclear collisions are moved by electronically driven radiation-enhanced diffusion. These rearrangements alter the index most noticeably in the region adjacent to the damage, but some transport is possible even up to the surface. In index terms, these ionic shifts totally swamp the index decrease, caused by atomic defects. On annealing, ionic movement towards the original defect forming region, leaves only the damage

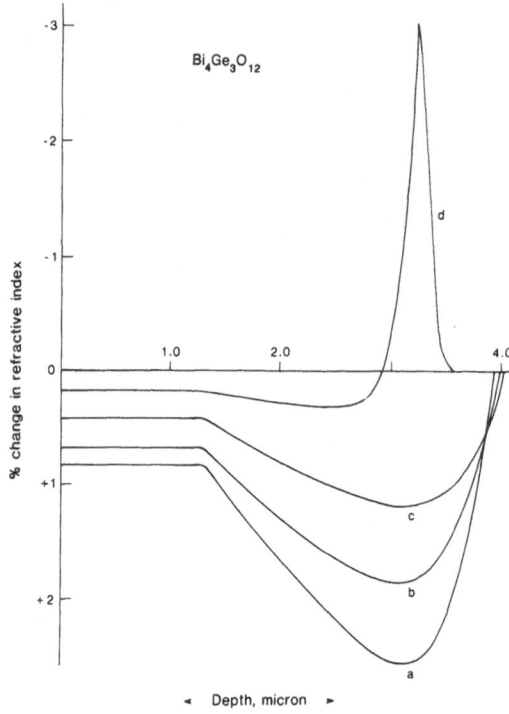

FIGURE 6. In BGO there is a major enhancement of the index but on annealing the profile changes. Anneal temperatures were (a) 20, (b) 325, and (c) 400.

contribution. It should be noted that at present the quality of BGO crystals is less than perfect. This was apparent by the presence of minor variations in the waveguide profiles formed in different crystals.

Although the preceding examples have concentrated on the narrow problem of optical waveguide formation in insulators by ion beam bombardment, several more general problems of ion beam processing have emerged. These include several trends, such as: (a) energy deposited by nuclear collisions during bombardment of crystalline insulators results in a lower density, lower index, disordered (or amorphous) material; (b) electronic energy deposition generally forms some color centers, which are removable by heat treatment, without leading to structural recovery of the disordered zone; and (c) additions of impurity ions produce changes in index which, at sufficiently high dose levels, may exceed the damage effects.

9. LOSS AND OPTICAL DAMAGE IN OPTICAL WAVEGUIDES

Loss in optical waveguides may occur from several processes such as optical absorption, scattering from inclusions or refractive index inhomogeneities, surface scattering or evanescent wave tunnelling through boundary layers. The latter effect may take on extra importance if there is mode conversion or polarization conversion. Simple thermal annealing will remove many of the color centers but will not reduce other losses. Optical absorption may of course be an objective of the implant and it is used to produce selective photoconductive or luminescence centers.

In some instances, heat treatments may exacerbate the losses, for example in materials with excess metal ions the thermal energy will lead to growth of colloidal precipitates. Recent examples of absorption band features noted for silver implants in Al_2O_3 and $LiNbO_3$ are particularly interesting in that they show that colloidal band growth is highly dependent on the crystal plane which is being implanted, Fig. 7 (25). A simple explanation may be that during implantation there is radiation-enhanced diffusion which transports the silver preferentially in one plane. If the implant direction is parallel to this plane, then silver diffuses away from the implant region. Conversely, if the diffusion is in the same plane as the implant then there is no net loss of silver, hence the colloidal features appear. In the examples quoted the absorption band strength can vary by a factor of two for identical implant conditions in perpendicular crystal cuts.

Optical waveguide loss processes can be minimized by increasing the barrier widths, but surface and inhomogeneity scattering are more difficult to combat, particularly when using the diffusion or layer growth routes to guide formation. Ion implantation can be used to bury the guides, and as mentioned earlier for a combined Ti diffused and implanted guide, this can produce exceedingly low losses. Losses less than 1 dB/cm are normally acceptable for optoelectronic devices as they are relatively small (i.e. one does not need the low losses that are seen for communication type optical fibers).

At high optical power levels, many new features of optical damage and non-linearity can occur. High power in this context may be a milliwatt. It must be noted that a 1-mW laser beam, when confined within a waveguide of 1 μ dimensions, corresponds to power levels of 10^5 w/cm^2. "Optical damage" appears as a change in refractive index, additional light scattering, and loss of transmitted power. Such features are undesirable unless they are used in a more controlled version to record image information. In this case, they are termed photorefractive effects and have the virtue of providing holographic stores.

In non-linear applications, high power densities are required and hence the optical damage properties of the guides must be avoided. Once again using LiNbO$_3$ as a reference, one can contrast guides formed in Ti diffused niobate, which show a very low damage threshold, with ion implanted guides formed in pure material. In the latter case the damage threshold is some 100 times greater (26,27). Proton exchange guides show comparable high power handling capability. Therefore in the case of Ti doped material, the problem may arise from the precipitation of impurity phases, which have been identified by compositional analysis.

Solid state lasers in insulators generally operate by stimulating optical transitions at defect sites (28). If the defects are intrinsic in character then they may be introduced by ion beams with quite modest ion doses. However if the transition occurs at an impurity site, the the implantation methods have some drawbacks. For example, there will be a need for high concentrations of dopant, low loss, and controlled depth profiles. If the impurity is of high mass (e.g. Nd) then commercially available implant energies will be insufficient to introduce the impurities to waveguide depths. Post implant heat treatments are needed to diffuse the impurities, restore lattice order and avoid clustering or defect scattering. Nevertheless, the implantation route may still be desirable, particularly if more than one impurity ion is required.

10. FURTHER OPTICAL CHANGES

Control of the refractive index allows one to define optical interactions into waveguide, rather than bulk, features. Inevitably, the ionization and displacement damage associated with the implantation modified nearly all the other optical properties as well. The loss of crystallinity will in general degrade many of the interesting electro-optic and non-linear features. The extent to which this occurs depends on both the residual damage in the guide, and the extent to which the optical field extends into the boundary

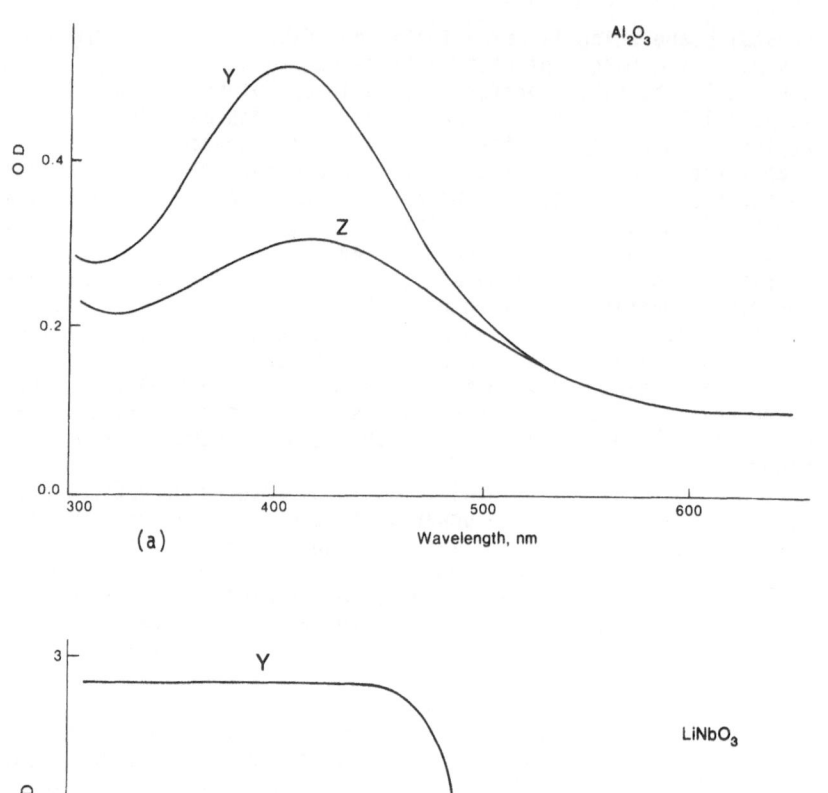

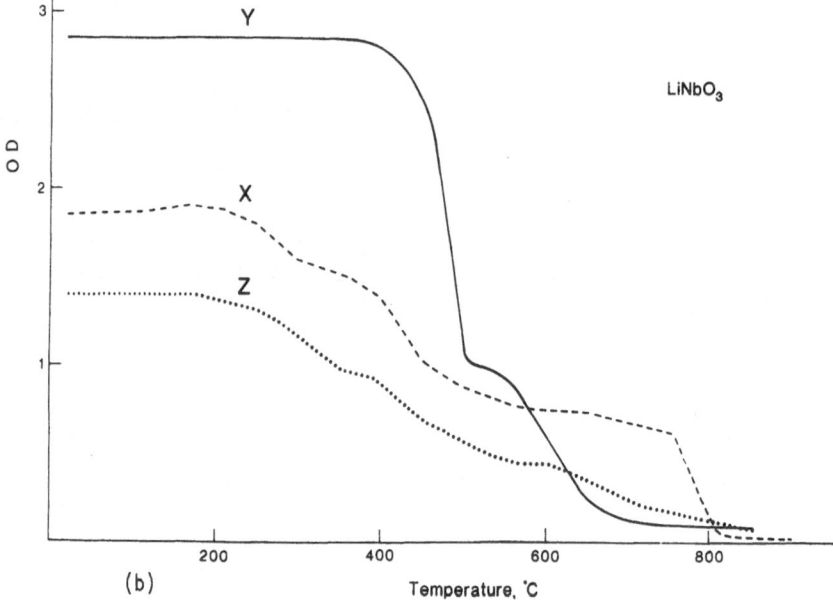

FIGURE 7. (a) Examples of colloid optical absorption bands formed in sapphire by Ag implants. Note that the same dose produces different strength bands in the Y and Z cut crystals. (b) Similar behavior exists for X,Y and Z cut LiNbO$_3$. The crystal cuts show differences in annealing behavior.

layers. From the example of LiNbO$_3$ the measured decrease in the electro-optic coefficient is some 30 to 40%. At first sight this may seem quite serious, however, the range of coefficients used in the material ranges covers a factor of four. Doping methods frequently lead to the use of the lower value but, by suitable selection of crystal axes, ion implanted guides can be designed to utilize the highest values. Loss of bulk crystal performance is not a relevant consideration if there are no alternative methods of forming guide regions. Birefringence changes occur in anisotropic material because, as noted above, n_0 and n_e indices are often modified differently. For high beam doses one expects the material to approach an amorphous state, in which case the birefringence will be removed. A measure of control in the degree of birefringence is useful for applications of second harmonic generation, or frequency mixing of two lasers. The index matching conditions for the various frequencies are often difficult to control, either by the choice of propagation axes or crystal temperature. For example in some cases the crystal must be temperature stabilized to within hundredths of a degree. A relaxation of this stringent control is therefore advantageous. Other possibilities exist if one has mode conversion between the pumping and second harmonic beams, or, as is conceivable by control of the refractive index profile, the two guides effectively propagate in different, but interacting, paths.

Diffraction of light by surface acoustic waves can be the basis of spectrum analysis for radio frequency waves, for example in radar detection systems. In such systems the light must be confined to a surface waveguide which is crossed by the SAW. The diffraction efficiency of the structure will depend on the material and the detailed optical and SAW design. A demonstration of such a structure, using a quartz crystal and an ion implanted optical waveguide, as shown in Fig. 8, produced a diffraction efficiency of nearly 30% (29). This matched the theoretical estimate for the geometry used. A more interesting extension of implantation technology for components of the spectrum analyzer is to consider how to generate planar optical lenses for beam expansion and focussing. Early attempts, using conventional technology of layer deposition or grinding of geodesic lenses, were extremely difficult and costly (30). Geodesic lens production in a brittle material such as LiNbO$_3$ may cost some 6,000 lb/lens. Even more problematic, is that the diamond tip grinding has a low success rate, and two lenses are required per device. A more direct route in which the surface remains planar is therefore preferable. Ion implants, to change the index, or provide a damaged pattern which can be easily etched into the geodesic shape, offers an easier low-cost option.

11. LASER SOURCES

There is currently a major effort worldwide to produce visible lasers, and tunable lasers, which are entirely made from solid state components. In the ideal situation, the primary laser source would be a high power semiconductor laser; this in turn would be frequency doubled to give visible laser light. For some applications, such as addressing compact discs, the development of a modest power (mW) blue laser is the final objective. However, there are clearly advantages to obtaining high power visible laser sources as these in turn can be used to drive further wavelength conversion stages. Particularly tempting possibilities occur when the final stage relies on the pumping of a broad band fluorescence system. In this case the presence of a tunable cavity will allow pumping and laser emission at any frequency within the envelope of the luminescence band. The current developments of infrared laser diodes which produce a watt or more of power means that such speculations can now be turned into reality. In the first stage the second

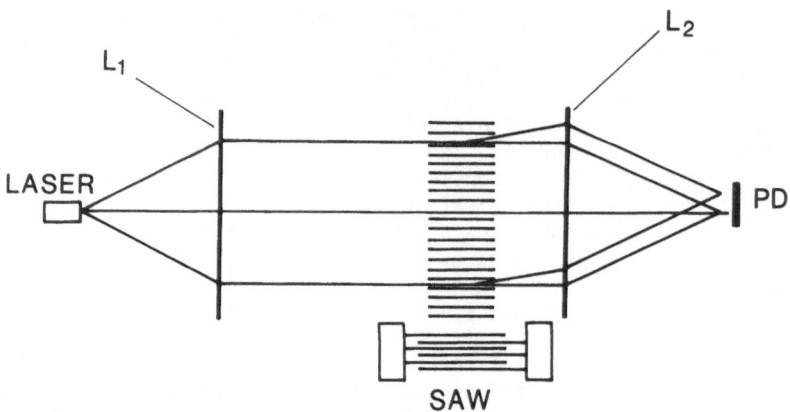

FIGURE 8. An rf spectrum analyzer using laser diffraction with a SAW.
L_1 and L_2 are lenses, PD is a photodetector array.

harmonic components were bulk crystals but there are obvious power density
advantages for the non-linear process of second harmonic generation if one
uses crystal waveguides. Indeed, the geometry of the laser diode source
favors coupling to a crystal waveguide laser. Successful accounts of diode
pumping to produce visible laser light from optical fibers and $LiNbO_3$ waveguides have recently appeared. At this stage in the development of the components of ion beams are already used in fabrication of the initial laser
diodes, as discussed below, and have been used to define low loss waveguide
pathways, in the key materials, which are the probable contenders for the
other elements of the total system.

12. SOLID STATE LASERS

Semiconductor laser diodes are variously made with a complex multilayer
heterostructure arrangement. The sketch of Fig. 9 indicates one such structure fabricated with n-AlGaAs, p-GaAs, p-AlGaAs layers. The planar deposition of the layers is straightforward but in order to define a laser stripe
a high resistive boundary region is required. Proton irradiation generates
a lower carrier density than the original semiconductor material and hence
produces the high resistivity. Equally important, there is a reduction in
the refractive index, hence the laser light is guided. In principle the n-
and p-type doping can be introduced into the layers by implantation but for
very narrow layer structures the doping may also be performed during molecular beam epitaxy. Progress in semiconductor laser diode design has lead to
low power quantum well structures, which operate at the limit of the visible
spectrum and, more interestingly for the present discussion, development of
high power lasers with output powers measured in watts. At this stage they
are still expensive but clearly this situation will change.

There are a multitude of laser transitions in inorganic materials, both
crystalline and amorphous. As mentioned earlier, the defect or impurity
states may be introduced by ion beam implantation but little work has yet
been attempted for systems in which lasing had not already been
demonstrated. An obvious advantage of the implant route may be the ability
to write distinct pathways of specific impurities across the same surface.
Hence, by addressing differently doped regions, with a pump laser, it would

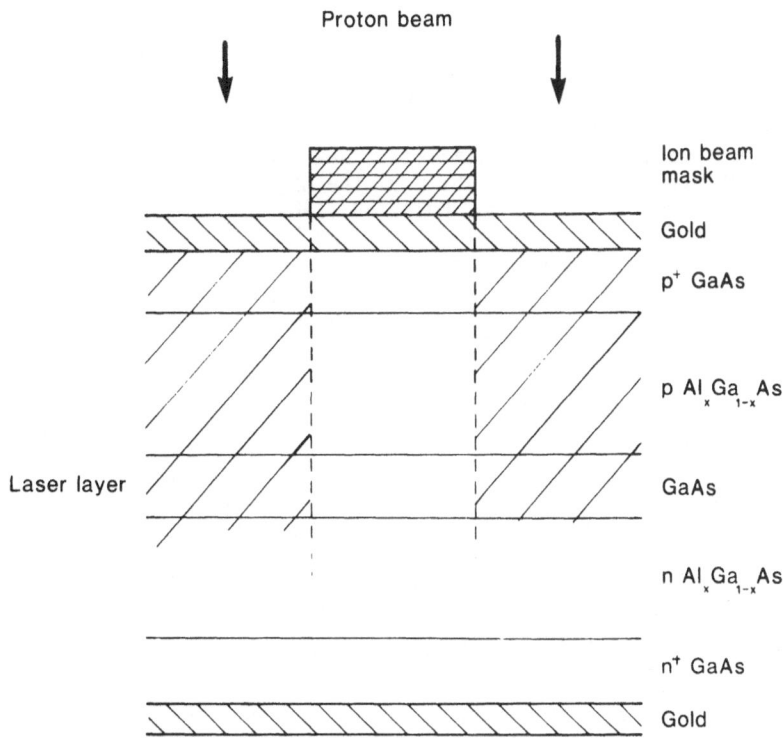

FIGURE 9. Proton isolation to define a laser stripe in a heterojunction.

be possible to generate a range of specific laser emissions from the same
sample. The more flexible route, using a cavity resonance for a wavelength
within a broad luminescence band, is equally possible. Waveguide lasers in
which there is second harmonic generation have already been demonstrated.
Although none of these systems has yet been sufficiently optimized for commercial applications it is clear that these products will appear, and there
is no reason why ion beam techniques should not be used in their manufacture.

13. SUMMARY
In summary, we have shown that ion implantation is an acceptable way to
form many of the optical device components that are currently being considered. We have only discussed waveguide and laser-type applications but
it should be apparent that there are very many more possibilities of property change resulting from ion implantation of insulators. From the literature of the optical waveguides it has already emerged that there are some
very positive advantages in using implantation, for example, high power capacity waveguide structures can be defined in a very wide range of materials.
The large changes in refractive index are ideal for the design of modulators
and optical circuits, losses and so electro-optic, acousto-optic and piezo-
optic features may be exploited. Finally, and certainly not least, the
developing area of all-solid-state visible, or tunable, laser devices
appears to be ideally poised for exploitation by ion beam technology.

REFERENCES

1. Yariv A: Optical Electronics. New York: CBS Publishing, 1985.
2. Martellucci S and AN Chester (eds): Integrated Optics, NATO-ASI 91. New York: Plenum, 1981.
3. Ostrowsky DB and E Spitz (eds): New Directions in Guided Wave and Coherent Optics. NATO-ASI E79. The Hague: Martinus Nijhoff, 1984.
4. Townsend PD: Rep. Prog. Phys. 50 (1987) 501.
5. Mazzoldi P and GW Arnold (eds): Ion Beam Modification of Insulators. Amsterdam: Elsevier, 1987.
6. Brown WL: Nucl. Instr. Meth. B32 (1988) 1.
7. Agullo Lopez F, CRA Catlow, and PD Townsend: Point Defects in Materials. London: Academic Press, 1988.
8. Chandler PJ and PD Townsend: Proc. IPAT 1987. Edinburgh. CEP Consultants, 1987, p. 476.
9. Chandler PJ, L Lama, PD Townsend, and L Zhang: Appl. Phys. Lett. 53 (1988) 89.
10. Pulker HK: Coatings on Glass. Amsterdam: Elsevier, 1984.
11. Morgan DV, FH Eisen, and A Ezis: Proc. IEEE 128 (1981) 109.
12. Cullis AG: Rept. Prog. Phys. 48 (1985) 1155.
13. Günter P: Phys. Repts. 93 (1982) 199.
14. de Sario M, MN Armenise, C Canali, A Carnera, P Mazzoldi, and G Celotti: J. Appl. Phys. 57 (1985) 1482.
15. Heibei J and E Voges: IEEE QE18 (1982) 820.
16. Burns WK, J Comas, and RP Moeller: Optics Lett. 5 (1980) 45.
17. Drummond E and PD Townsend: unpublished 1985.
18. Red'ko V, LM Shteingart, VI Soroka, MV Artsimovich, and AI Mal'ko: Sov. Tech. Phys. Lett. 17 (1981) 399.
19. Zhang, L, PJ Chandler, and PD Townsend: to be published.
20. Naik IK: Appl. Phys. Lett. 43 (1983) 519.
21. Faik AB, PJ Chandler, PD Townsend, and RP Webb: Radiat. Eff. 98 (1986) 233.
22. Heidemann KF: Radiat. Eff. 61 (1982) 235.
23. Glavas E, L Zhang, PJ Chandler, and PD Townsend: Nucl. Instr. Meth. B32 (1988) 45.
24. Mahdavi M, PJ Chandler, and PD Townsend: in preparation.
25. Rahmani M, LH Abu Hassan, PD Townsend, IH Wilson, and GL Destefanis: Nucl. Instr. Meth. B32 (1988) 56.
26. Glavas E, PD Townsend, G Droungas, M Dorey, KK Wong, and L Allen: Electron Lett. 23 (1987) 73.
27. Glavas E, JM Cabrera, and PD Townsend: in press.
28. Chen Y, VM Orera, R Gonzalez, and C Ballesteros: Cryst. Latt. Defects 14 (1987) 283.
29. Pitt CW, JD Skinner, and PD Townsend: Elect. Lett. 20 (1984) 4.
30. Doughty GF, RM delaRue, J. Singh, JF Smith, and S Wright: IEEE CHMT5 (1982) 205.

CHARACTERIZATION OF PLANAR OPTICAL WAVEGUIDES IN ION-IMPLANTED QUARTZ

L. ZHANG, P. J. CHANDLER, P. D. TOWNSEND, AND F. L. LAMA

Mathematical and Physical Sciences, University of Sussex,
Brighton, BN1 9QH, United Kingdom

1. INTRODUCTION

Optical waveguides may be fabricated in many transparent insulators by means of ion implantation with energetic light ions (1-3). The usual mechanism, with a few exceptions (4), is due to the radiation damage produced by the ions, and not the presence of the implanted ions themselves. The energy deposited at the end of the ion track where direct "nuclear" collisions occur, is able to produce atomic displacements. In a crystalline matrix this leads to partial amorphization resulting in a reduction in physical density and hence refractive index. Thus, a low-index buried layer is formed which acts as an optical isolation "barrier" between the surface guiding region and the substrate. Very little nuclear stopping energy is deposited in the guiding region, and the electronic stopping which dominates there is unable to produce displacement damage in most materials. Therefore after a low temperature anneal (~200°C) to remove ionized states (color centers), the result is usually a low-loss barrier-confined surface waveguide.

This description is ideally demonstrated in the case of crystalline quartz (2,5-7), where practically no index change occurs in the guiding region and the sharp damage peak reaches a saturation height of $\Delta n \sim -5\%$ due to amorphization, eventually producing a broad flat-topped barrier.

The purpose of the work reported here is a comprehensive characterization of planar waveguides produced in quartz by He$^+$ ion implantation. A detailed analysis has been made of the refractive index profiles as a function of the ion dose and energy, and of the implant temperature. Their thermal stability has been measured up to high temperatures, and the optimum implant conditions have been ascertained for low attenuation due to tunnelling and scattering losses.

2. REFRACTIVE INDEX PROFILES

Helium ions were implanted at 300 and 77 K into Y-cut quartz supplied by The Roditi Corporation, using energies from 0.7 to 2.2 MeV and doses from 5×10^{15} to 1.6×10^{17} ions·cm^{-2}. The beam had an area of ~1 cm^2 and a current of ~1 µA. In order to achieve a uniform dose the beam was scanned over the implant area (~0.5 cm^2) and this was surrounded by an aluminum mask to minimize surface charge and heating effects. The temperature rise was estimated at less than 10°C.

The dark mode positions were measured for these waveguides using X-propagation at 0.6328 and 0.488 µm for both TM and TE polarizations, giving the ordinary (n_0) and extraordinary (n_e) indices, respectively. From these experimental mode spacings, the complete refractive index profiles of the planar waveguides were determined, including the shape of the confining barrier. This was achieved by means of our recently developed technique (8)

371

C.J. McHargue et al. (eds.), Structure-Property Relationships in Surface-Modified Ceramics, 371–378.
© *1989 by Kluwer Academic Publishers.*

involving a calculation of the reflectivity at the coupling prism interface which thereby evaluates the positions of all the real and "substrate" dark modes.

This is demonstrated in Fig. 1(a) which is a mode index plot at 0.488 μm showing the experimental dark modes (triangles) for a He$^+$ implanted sample of crystalline quartz. This shape is typical of an ion-implanted waveguide with a flat-topped barrier. The first seven modes lie nearly on a straight line — these are the real modes contained within the almost square well. The higher (substrate) "modes" are produced by interference effects from multiple reflectivity at both barrier edges, and so they lie on an intricate curve. The computer reflectivity analysis is able to find, by means of a least-squares optimization routine, a barrier shape which has theoretical modes (dotted curve) corresponding to these experimental values (triangles). This selected profile together with mode levels, is shown in Fig. 1(b). The analytic shape used here is a flat-topped barrier with two half-Gaussian edges and a flat base level in the guiding region. Such a model is about to reproduce the experimental dark mode positions to considerable accuracy ($\Delta n \sim \pm 0.0002$). It must be realized that the shape of the barrier top and far edge is entirely determined by the unusual upper (substrate) mode spacing, which our method is able to faithfully model. This fit could obviously not be achieved by less rigorous approximations such as the WKB method.

The growth of damage in quartz was investigated by implanting samples to different doses, from 5×10^{15} to 1.2×10^{17} ions·cm^{-2}. Figure 2 shows a set of these waveguide profiles for n_e measured at 0.488 μm. For low doses the barrier is a narrow skewed Gaussian ($\sigma_1 \sim 0.2 : \sigma_2 \sim 0.1$ μm) roughly corresponding in position to the peak in nuclear stopping power for the ions, but somewhat narrower. The lack of damage in the earlier part of the ion track suggests that the point defects produced there are annealed out, possibly by the prevailing ionization energy. At doses above $\sim 1 \times 10^{16}$ ions·cm^{-2} the barrier reaches a saturation height ($\Delta n \sim -5$) and becomes flat-topped, once a stable quasi-amorphous region has been produced. For higher doses, the barrier width increases as might be expected from a saturating Gaussian, but at very high doses ($>10^{17}$ ions·cm^{-2}) the width is seen to increase faster than would be produced by linear damage rate, especially on the incident edge. This can be explained by a mechanism involving defect diffusion and association which is enhanced in the ionized pre-damaged regions, thereby building up a stable defect network on the leading edge of the barrier.

Once the barrier reaches a saturation height it becomes isotropic, as can be seen from a comparison of n_0 and n_e profiles for the same sample. Figure 3 shows practically identical profiles excepting that the saturated region has become isotropic.

The damage mechanism is temperature dependent, both in its rate and in its final degree. This is demonstrated in Fig. 4 where a comparison is made of two profiles for identical implants at 77 and 300 K. The 77 K implant produces a broader barrier because the increased defect lifetime against self-annealing encourages the formation of more stable complex defects. The difference in barrier height simply implies that the degree of amorphization attainable increases with temperature.

The depth of the refractive index barrier beneath the surface will depend on the mechanism involved. For a chemical effect it should coincide with the projected ion range and have a width determined by the range straggling. For a purely damage-dependent mechanism it should be at a slightly shallower position where the maximum in nuclear stopping power occurs. Any index

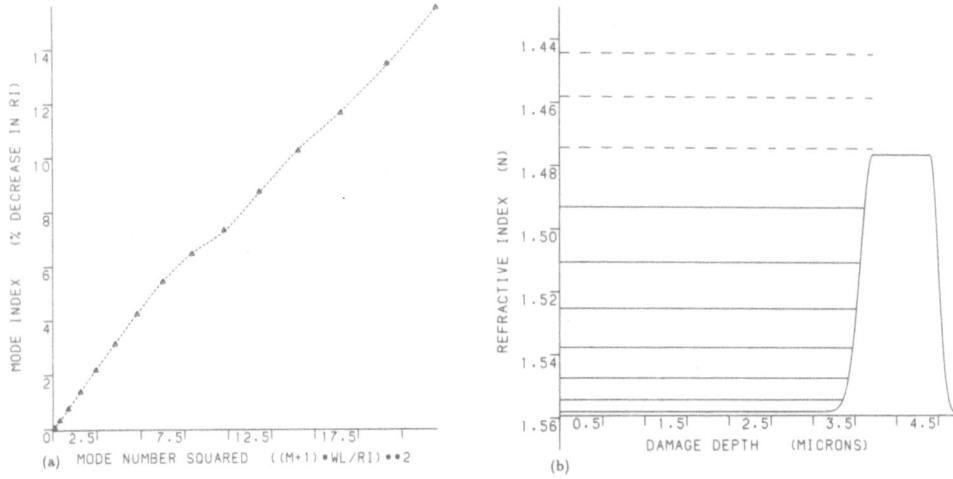

FIGURE 1. (a) Mode index plot for quartz implanted at 77 K with 1.5 MeV He+ to a dose of 3 x 10¹⁶ ions·cm⁻². The triangles show the experimental dark mode positions and the dashed curve represents the theoretical modes corresponding to the computer-fitted profile shown in (b).

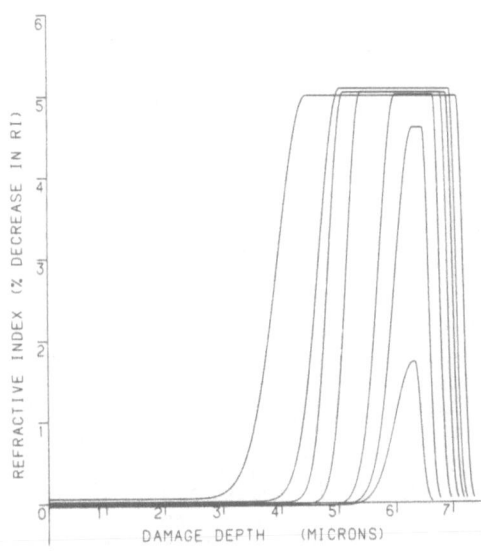

FIGURE 2. Computed waveguide profiles measured at 0.488 µm for quartz samples implanted at 77 K with He+ of energy 2.2 MeV and doses 0.5, 1, 2, 4, 6, 8, 12 x 10¹⁷ ions·cm⁻².

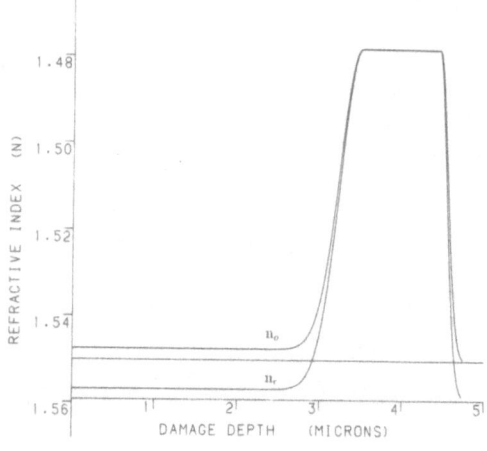

FIGURE 3. Comparison of n_0 and n_e profiles for a 77 K implant with 1.5 MeV He+ to a dose of 3×10^{16} ions·cm⁻².

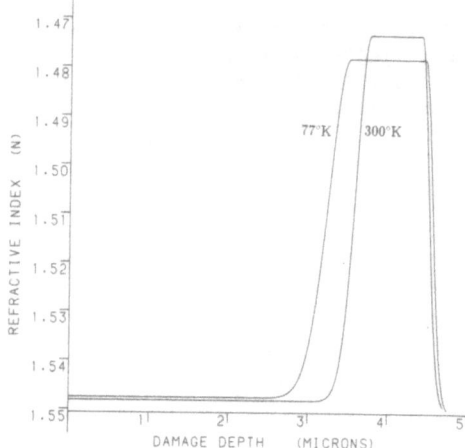

FIGURE 4. Comparison of n_0 profiles for 77 and 300 K implants with 1.5 MeV He+ to a dose of 3×10^{16} ions·cm⁻².

change due to stress will be superposed on this, and will effect the profile shape nearer to the surface. The profiles in Fig. 5(a) are for separate samples implanted with the same dose (2×10^{16} ions·cm⁻²) but energies from 0.7 to 2.2 MeV. The peak positions are shown in Fig. 5(b) where the points represent the experimental barrier peaks, and the dashed curve shows the He+ ion range as given in tables published by Arnold and Mazzoldi (3). The damage peak occurs at ~90% of the quoted ion range, and follows the non-linear range/energy function governed by the electronic stopping-power variation with energy. This suggests that the index change is predominantly a nuclear damage phenomenon.

3. THERMAL STABILITY

Useful waveguides need to have long-term stability at ambient temperatures. Furthermore, in the case of ion implanted guides the barrier must be able to withstand the annealing cycle required for the removal of color center absorption in the guiding region.

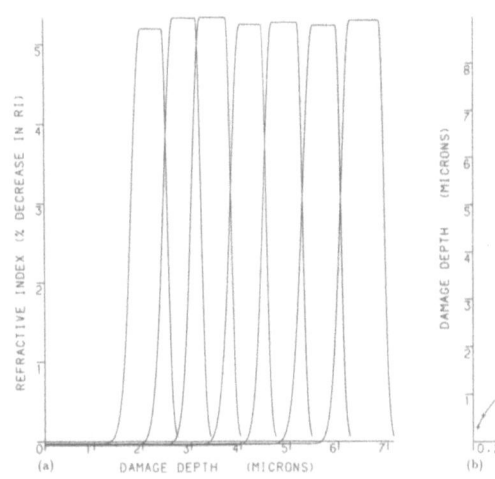
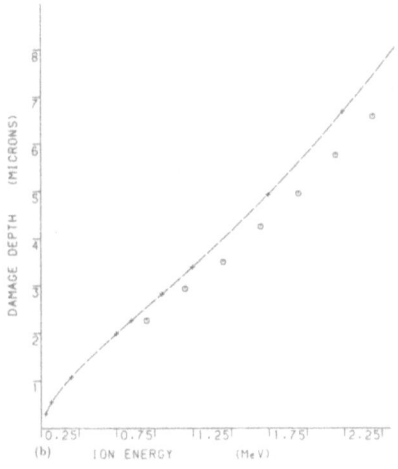

FIGURE 5. (a) Variation of barrier position for doses of 2×10^{16} ions·cm^{-2} and energies from 0.7 to 2.2 MeV in 0.25 MeV steps. (b) The circles give the experimental barrier centers and the curve shows the He$^+$ ion range as given in tables (3).

Tests were carried out on various implanted samples by isochronally annealing them in 100°C stages for 60 min up to 1100°C. The resulting profile changes are summarized in Fig. 6. Low dose barriers, e.g. 5×10^{15} ions·cm^{-2} [curve (a)] recover towards the substrate index of crystalline quartz. Higher barriers with doses above ~1×10^{16} ions·cm^{-2} [curve (b)] do not recover their crystallinity, but become more amorphous and tend towards fused silica (horizontal line at $n = 1.46$). Curve (c) shows the annealing of a sample with a high dose of 3×10^{16} ions·cm^{-2}. As expected, the top of the damage peak increases in height, but the bottom (low damage) recovers, thus producing a more vertical barrier. The annealing curves of the two regions are shown in Fig. 6(d) in terms of percentage index change. The bottom curve plots the low damaged region in the barrier tail recovering towards crystalline quartz, and the upper curve shows the quasi-amorphous region at the peak of the barrier moving towards fused silica.

4. ATTENUATION

The results presented above for the refractive index profiles suggest that quartz should be a good host for low-loss ion-implanted waveguides. The existence of a broad square barrier should reduce tunnelling losses, and the barrier's stability at high temperature means that it will survive during the thermal annealing of point defects and color centers in the guiding region, thus producing low scatter and absorption losses. The only remaining attenuation should be due to surface scattering.

In order to perform loss measurements, planar waveguides were fabricated with 1.5 MeV He$^+$ using 30 mm long Y-cut crystals at doses from 1 to 6×10^{16} ions·cm^{-2}, giving barrier widths from 0.4 to 1.5 μm. These guides exhibited from 2 to 4 guiding modes, and their losses were determined at 0.6328 μm with X-propagation for TE (n_e) and TM (n_0) polarizations. A method similar to the three prism technique was used, but replacing the third prism by a direct end-spot monitor.

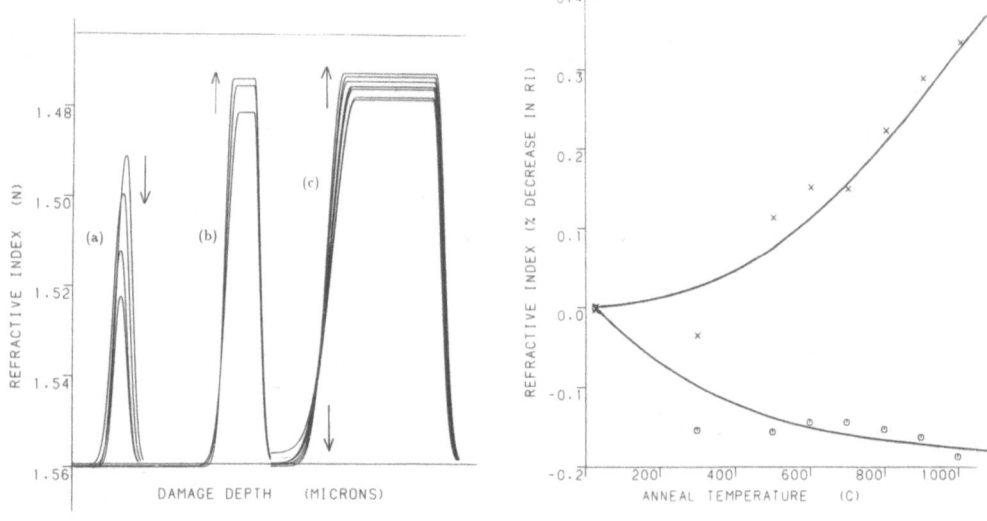

FIGURE 6. Thermal annealing of barriers with (a) low dose (5 x 10¹⁵ ions·cm⁻²), (b) medium dose (1 x 10¹⁶ ions·cm⁻², (c) high dose (3 x 10¹⁶ ions·cm⁻², and (d) annealing curves for regions with low damage (bottom) and high damage (top).

   The as-implanted guides showed losses of ~3 dB cm⁻¹ for the broad barriers (doses 3 and 6 x 10¹⁶ ions·cm⁻²) but as high as 24 dB cm⁻¹ for the low dose narrow barriers (see Table 1). This is attributed to significant tunnelling for barriers with doses below ~3 x 10¹⁶ ions·cm⁻².
   After tunnelling out the color centers in the guiding region, the loss values were considerably improved, eventually reaching ~1.5 dB cm⁻¹ for the broad barrier (dose 6 x 10¹⁶ ions·cm⁻²) after a 600°C anneal (see Table 1 and Fig. 7 plots 1-3). The sample with dose 3 x 10¹⁶ ions·cm⁻² however achieved an even better loss (1.2 dB cm⁻¹) after a 600°C anneal (plot 4). This is because at the higher dose, losses due to scattering and absorption are beginning to dominate those due to tunnelling.
   An optimum ion-implanted waveguide thus requires a broad barrier (≥1 μm) to minimize tunnelling, but a low ion dose in order to minimize scattering and absorption. These conditions are not ideally achieved with a single energy implant, because a high proportion of the ion energy is wasted in post-amorphization damage. The broad barrier is more efficiently produced by a carefully selected multiple-energy implant. The top curve (5) in Fig. 7 (loss 0.2 dB cm⁻¹) is for the waveguide of Table 2. This uses four energies (1.51 − 1.72 MeV) with a total dose of only 2 x 10¹⁶ ions·cm⁻² to produce a 1.45 μm barrier, equivalent to a single energy barrier needing a dose of 6 x 10¹⁶ ions·cm⁻². Barriers produced by single and multiple energy implants are compared in Fig. 8.
   Tests are being performed at present to reduce this loss even further by post-implant polishing to cut down surface scatter, and also by the use of low energy implants (~0.1 MeV O⁺) to confine the guide away from the surface. In the latter case an end-coupling insertion-loss method must be applied to estimate the attenuation.

TABLE 1. Loss in single-energy implanted samples
(He; 1.5 MeV; 300 K; 0.6328 μm)

Dose ($\times 10^{16}$ ions/cm^2)	Barrier Width (μm)	Before Anneal (dB/cm)					Anneal Temp (°C)	After Anneal (dB/cm)				
		n_e (TE)			n_o (TM)			n_e (TE)			n_o (TM)	
		0	1	2	0	1		0	1	2	0	1
0.2	0.30	14.3										
1.0	0.55	4.9	9.9		3.9							
3.0	1.05	3.6	3.9		3.2	3.9	600	1.2			1.7	
6.0	1.45	3.2	4.7	6.3	3.2	4.9	300	2.6	2.7	2.6	3.2	3.4
							400	2.1	3.9	3.3	2.8.	2.8
							600	1.5	2.7	3.5	2.1	2.0

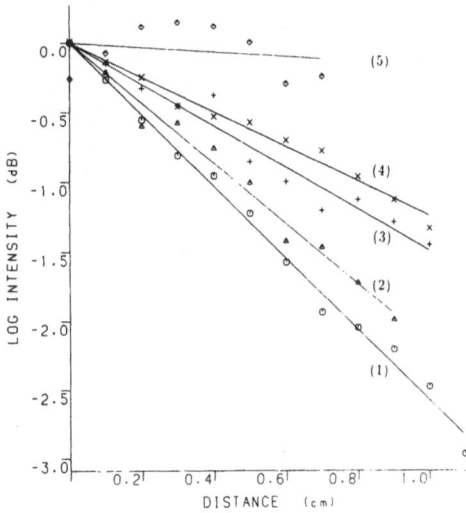

FIGURE 7. Loss plots for annealed
waveguides. Samples (1)-(3) with
dose 6 x 10^{16} ions·cm^{-2} annealed for
30 min at 300°C, 400°C, and 600°C.
Sample (4) with dose 3 x 10^{16}
ions·cm^{-2} after a 600°C anneal.
Sample (5) is a multi-energy implant
using four energies (1.51 - 1.72
MeV) with a total dose of only
2 x 10^{16} ions·cm^{-2}.

TABLE 2. Loss in multi-energy implanted sample
(He; 1.51, 1.58, 1.65, 1.72 MeV; 300 K; 0.6328 μm)

Dose ($\times 10^{16}$ ions/cm^2)	Barrier Width (μm)	Before Anneal (dB/cm)		Anneal Temperature (°C)	After Anneal (dB/cm)	
		n_e (TE)	n_o (TM)		n_e (TE)	n_o (TM)
0.5×4	1.45	1.5	2.0	800	0.2	0.3

378

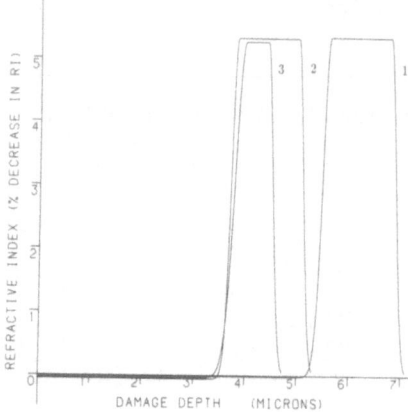

FIGURE 8. Index barrier profiles for
(1) single energy giving barrier width
1.45 μm needing a dose 6 x 10¹⁸
ions·cm⁻², (2) multiple energy giving
the same barrier width needing a total
dose of only 2 x 10¹⁸ ions·cm⁻²,
(3) single energy dose 2 x 10¹⁸
ions·cm⁻² producing a barrier with a
width of only 0.75 μm.

5. CONCLUSION

Waveguides produced by ion implantation in crystalline quartz have been
shown to possess almost ideal properties. Index profiles may be produced
with broad square barriers having a confinement of $\Delta n \sim$ -5% and with practically no damage in the guiding region, thereby retaining the properties of
the original crystal. The barriers are stable up to at least 1100°C and in
fact after heat treatment they increase in height and become more square due
to the divergent annealing mechanisms for low and high damage regions, recovering towards crystalline and fused quartz, respectively. Broad barriers
may be produced with low total dose by using multiple-energy implants, and
after suitable annealing these guides give losses of ~0.2 dB cm⁻¹.

One advantage of these highly stable square barriers is that two or more
of them may be superposed at different depths to produce overlaid waveguides
(7). These may be used as passive interconnects, or as precision coupling
devices. Work is at present under way to study the feasibility of this
technique in device applications.

6. ACKNOWLEDGEMENTS

We are grateful to the Science and Engineering Research Council, the Great
Britain-China Educational Trust, the Universities' China Committee in
London, the Henry Lester Trust, and the Committee of Vice Chancellors and
Principals for financial support.

REFERENCES
1. Chandler PJ and PD Townsend: Proc. Conf. Ion Plating and Allied
   Technology. Edinburgh: CEP Consultants Ltd., 1987, pp. 476-481.
2. Townsend PD: Rep. Prog. Phys. 50 (1987) 501-558.
3. Arnold GW and P Mazzoldi (eds): Ion Beam Modification of Insulators.
   Amsterdam: Elsevier, 1987.
4. Faik AB, PJ Chandler, PD Townsend, and RP Webb: Radiat. Eff. 46 (1986)
   235-240.
5. Red'ko VP, LM Shteingart, VI Soroka, MV Artsimovich, and AI Mal'ko:
   Sov. Tech. Phys. Lett. 7 (1981) 399-400.
6. Lax SE: D Phil Thesis, Sussex University, 1987.
7. Chandler PJ, FL Lama, PD Townsend, and L Zhang: Appl. Phys. Lett. 53
   (1988) 89-91.
8. Chandler PJ and FL Lama: Optica Acta 33 (1986) 127-143.

FABRICATION OF ELECTROOPTICAL WAVEGUIDES BY MEANS OF ION IMPLANTATION

T. BREMER

Universität Osnabrück, Fachbereich Physik,
Postfach 4469, D-4500 Osnabrück, FRG

1. INTRODUCTION

Due to the drastically increasing transmission rates in telecom-
munications, the capacities of electrical circuits and cable technology have
reached their limits. Modern long distance communication is performed via
monomode fibers which enable transmission rates of several GBits/s. The
creation, modulation and switching of light has to be performed in local
area networks. In the last twenty years an increasing effort has been spent
in research and development of optical data circuits, or integrated optical
circuits. Materials are required, which offer the possibility of external
manipulation of light via the electrooptic, acoustooptic or magnetooptic
effect.

Besides the GaAlAs systems, being especially important due to the possibility of stimulated light emission in semiconductor lasers, electrooptic
crystals exhibit a wide range of suitable characteristics. Among several
electrooptic materials, $LiNbO_3$ is the preferred because it offers suitable
electrooptic and acoustooptic constants, high chemical and mechanical
resistance, a high Curie temperature of 1210°C, and large single domain
crystals are available at low cost (1). The basic element of integrated
optics is the optical waveguide, i.e. a higher index layer embedded into an
environment of lower refractive index which allows the propagation of optical modes (2). Since the first demonstration of waveguiding in $LiNbO_3$ in
1972, various alternative methods for waveguide fabrication have been
published. The most widespread is Ti indiffusion at typically 1000°C
leading to an enhancement of both, the ordinary and the extraordinary
refractive indices (3). Excellent low-loss waveguides, as well as a manifold of passive and active devices, have been fabricated and are now
becoming commercially available (4). Another method for waveguide fabrication is proton exchange in benzoic acid. Although this production method is
cheap and easy, it is not yet in widespread use because of long time insta-
bilities and because it only effects the extraordinary index.

This paper deals with the third important method, ion implantation.
Because of several reasons to be discussed later, there is a growing
interest in implantation during the last years. The state of the art is
summarized and a special focus is posed on recent results concerning implantation in $LiNbO_3$ and $KNbO_3$.

2. REFRACTIVE INDEX PROFILES OF IMPLANTED WAVEGUIDES

The first results of ion implantation in $LiNbO_3$ were published in 1974
(5). With a 60 keV Ar^+ beam and an ion dose of $10^{16}/cm^2$ an index decrease
of approximately 5% was observed in the damaged layer. By using higher ion
energies thus burying the damaged layer deeper in the substrate, waveguiding
could be achieved in the nearly unaffected layer between the surface and

C.J. McHargue et al. (eds.), Structure-Property Relationships in Surface-Modified Ceramics, 379–387.
© 1989 by Kluwer Academic Publishers.

the ion deposition region a few years later (6). Detailed investigations
followed to evaluate the energy and dose dependence of ion implantation in
LiNbO$_3$. Here we summarize the most important facts, for a more detailed
evaluation see the recent review by Townsend (7).
(1) The induced refractive index alterations do not depend on the ion
    species.
(2) The change of physical properties depends on the ion dose and can be
    divided into three regions:
    (a) The predamage state, characterized by the creation of point defects
        for a dose of D $\leq$ 10^{15}/cm^2.
    (b) The heavy damage state, with overlapping defect clusters up to a
        dose of D $\leq$ 5 x 10^{15}/cm^2.
    (c) The saturation stage for doses D $\geq$ 5 x 10^{15}/cm^2 with a large volume
        expansion of up to 11%.
    The ion induced volume expansion can be measured directly with a surface
    profilometer, or the resulting refractive index profiles can be eva-
    luated. For this purpose, two alternative methods have been published:
    measurements of the angular dependence of the reflectivity of the sample
    surface (9), as well as mode spectroscopy in combination with a profile
    reconstruction algorithm (10).
(3) The ion energy determines the penetration depth and thus the width of
    the refractive index profile. This effect can either be measured by
    channeling experiments (11) or by optical investigations (12).
    In contract with the other methods, ion implantation offers the possibi-
lity of producing nearly any desired refractive index profile, because ion
dose and energy can be varied independently. Another important advantage is
the reproducibility, since the ion beam parameters can be determined with
high accuracy. Furthermore, ion implanted waveguides show no optical damage
effects (13), which degrade the quality of Ti diffused waveguides (14).
Recently, investigations have been published which demonstrate the production of channel waveguides by means of He$^+$ implantation with propagation
losses comparable to diffused guides (15). It must, however, be stated that
the possibilities of device fabrication have not yet been exploited in a
wider range.

3. NEW RESULTS OF OPTICAL INVESTIGATIONS

In order to increase the reliability of optical investigations, we developed a new algorithm for the reconstruction of monotonically decreasing
refractive index profiles (16). It is based on the inverse WKB procedure,
but several improvements allow the reconstruction of profiles with an error
less than $\delta n = 1$ x 10^{-4} for at least two modes as input. for the
reconstruction, the behavior of the profile at the surface and in greater
depth is exactly taken into account. The waveguides are measured with dark
or bright line spectroscopy, providing the mode spectra as input for the
refractive index profile reconstruction. All investigations were performed
with y-cut LiNbO$_3$ waveguides, in which the waves propagated perpendicular to
the crystal axis. The polarization of the laser beam enabled a clear
distinction between TE and TM modes. In this geometry, TE modes are guided
by the extraordinary (n_e), TM modes by the ordinary (n_0) refractive index
profile of the birefringent LiNbO$_3$ crystal. A detailed description of the
experimental procedure is given in ref. (17). The surface side flank of the
refractive index profiles is reconstructed with high accuracy. It is,
however, impossible to reconstruct the part beyond the minimum. The total
decrease of the refractive index depends on the highest order mode used as
input parameter for the reconstruction. As we shall see, the distinction
between guided and radiation modes number is not unambiguous for ion
implanted waveguides.

We have fabricated several waveguides by He⁺ implantation at room temperature. The ion dose varied between $10^{15}/cm^2$ and $10^{16}/cm^2$, the energy between 1.2 and 3.17 MeV. In contrast with other methods, our measurements
give absolute information about the refractive index profiles, not only
changes. We have quantitatively evaluated the influence of ion energy and
dose on the surface refractive indices as well as on the profile widths.
Two sets of experiments have been performed with different doses and
energies. The dose dependence of the ordinary and extraordinary surface
refractive indices is shown in Fig. 1 for an energy of 3.17 MeV.

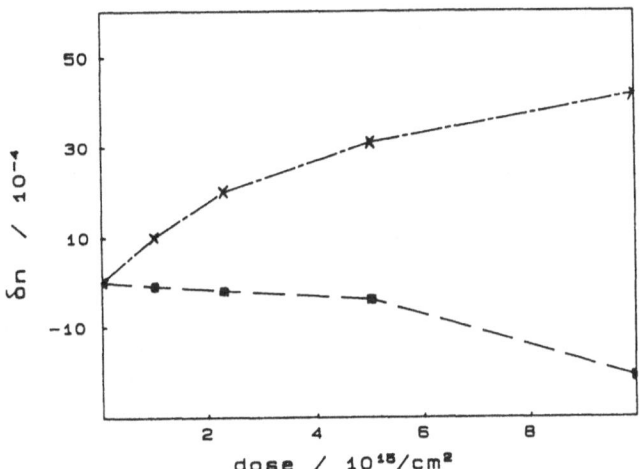

FIGURE 1. Dose dependence of the surface refractive index alterations $\delta n(D)$
of He⁺ implanted $LiNbO_3$ waveguides for an energy of 3.17 MeV. The ordinary
index, represented by boxes, is lowered while the extraordinary index
(crosses) is enhanced at the surface.

While the ordinary index is lowered at the surface, the extraordinary
index increases. The different sign of the extraordinary index alteration
at the surface is caused by the superposition of several effects: The
nuclear damage decreases both refractive indices in the region of lower ion
velocity near the end of the ion track. However, in the region of high ion
velocity near the surface, electronic collision effects and implantation
induced Li loss lead to an enhancement of the extraordinary refractive
index. The sum of all effects is measured by mode spectroscopy. In the
range between 10^{15} and $10^{16}/cm^2$ the extraordinary surface refractive index
increases logarithmically with the ion dose. The ordinary index decrease in
linear up to 5 x $10^{15}/cm^2$, but the slope of $\delta n(D)$ increases faster for
higher doses.
The energy dependence of the ordinary refractive index profiles for a dose
of 5 x $10^{15}/cm^2$ is shown in Fig. 2. Higher energies result in a reduced
damage near the surface. Higher ion velocities mean shorter coulomb
interaction times per path length, and thus the electronic damage decreases
for higher energies. Furthermore, the flank of the nuclear damage distribution gives only small contributions to the total index change at the surface
when it is buried deeper in the sample. Quantitatively, the damage is lower
at for example 3.17 MeV than expected from the Bethe-Bloch expression for

382

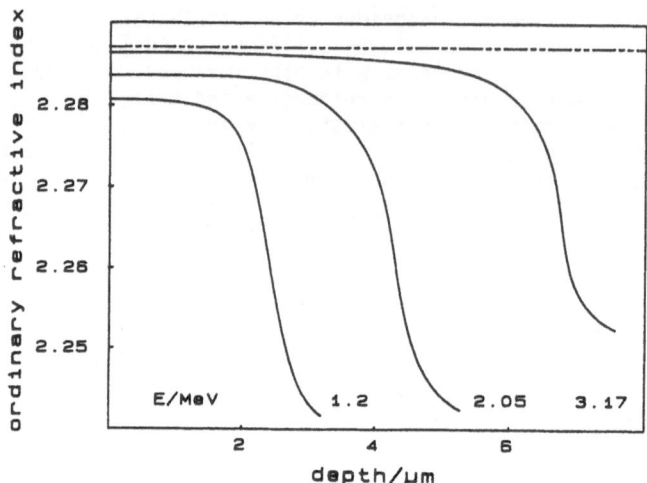

FIGURE 2. Ordinary refractive index profiles over ion energy (17). The dose is 5 x 10¹⁵/cm².

the electronic stopping power extrapolated from 1.2 MeV. Also the nuclear stopping varies slowly in this range. We suggest a radiation induced annealing (18-20).

Figure 3 demonstrates that the refractive index profile width depends linearly on the ion energy.

Our data agree with the results of Faik et al. (12) but extend to higher energies. Compared to the mean ion ranges calculated by Monte Carlo simulations with the TRIM code (21), also shown in Fig. 3, the optically measured widths are always about 15% lower. The explanation is rather simple: The results of TRIM calculations correspond to the mean ion range, while the refractive index profile widths are given by the surface edge of the nuclear damage distribution.

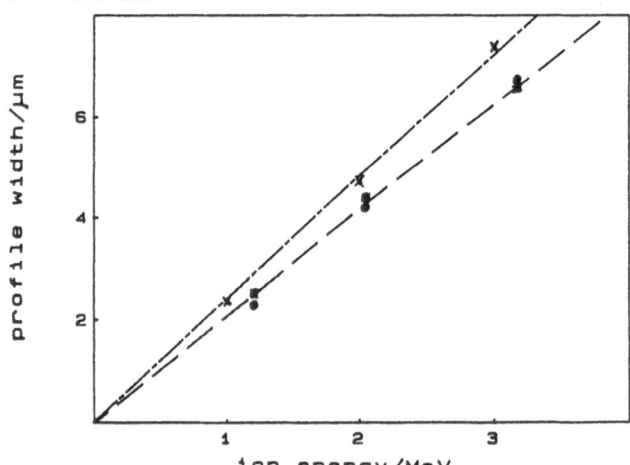

FIGURE 3. Refractive index profile width over ion energy for a dose of 5 x 10¹⁵/cm². Boxes refer to the ordinary, circles to the extraordinary profiles. Crosses represent the mean path lengths resulting from TRIM simulations.

Figure 4 shows the ordinary refractive profiles induced by different ion doses. Up to 5 x 10^{15}/cm² the profile widths are independent from the ion dose, but when the dose is increased further, the flank is considerably shifted to the surface. This saturation effect, also visible for the extraordinary profiles, fits very well the results in ref. (8). However, the influence on the refractive index distribution has been measured for the first time. The saturation value also corresponds with the change in the slope of the ordinary surface refractive index change given in Fig. 1. The investigations of the energy and dose dependence allow the choice of implantation parameters for the production of electrooptical waveguides with moderate damage at the surface, i.e. low losses. We propose high energies of at least 2 MeV and doses in the lower saturation stage, i.e. about 10^{16}/cm².

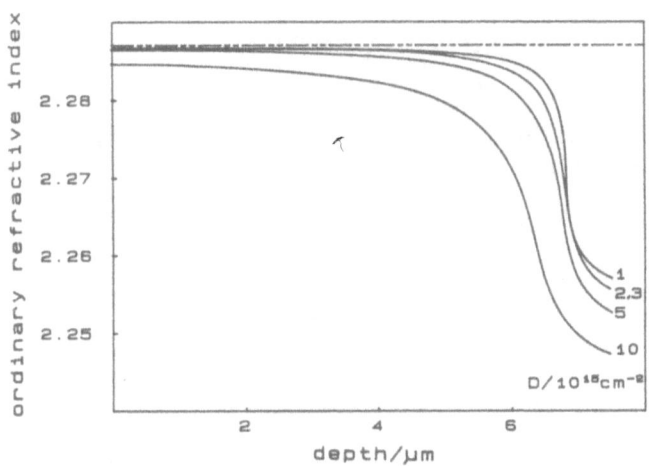

FIGURE 4. Ordinary refractive index profiles for an energy of 3.17 MeV and doses of 1, 2.3, 5, and 10 x 10^{15}/cm² (ref. 17).

4. REDUCTION OF LEAKAGE

A severe problem in ion implanted electrooptical waveguides is the tunneling of electromagnetic energy through the damage barrier, an effect which is caused by the finite thickness of the barrier itself. With increasing mode number, i.e. decreasing effective refractive index, there is a smooth transition from guided via leaky to resonant modes. A large amount of the electromagnetic energy of higher order modes can radiate into the substrate, this fact becomes directly visible in the mode spectrum in Fig. 5. The waveguide was implanted with an energy of 3.15 MeV and a total dose of 10^{16}/cm². The implantation was performed at an impact angle of 10°. The lower order modes are well guides, i.e. the dark lines are sharp and the reflected intensity decreases drastically when the resonance condition is fulfilled. For higher order modes, the dark lines are broadened and the amount of incoupled energy decreases considerably. Seven well guided modes can be excited, followed by a transition range of about three radiation modes until only weak resonances are stimulated.

In a recent paper, Weiss et al. proposed multiple energy implantations to broaden the barrier (22). With this method the problem of leakage could be

384

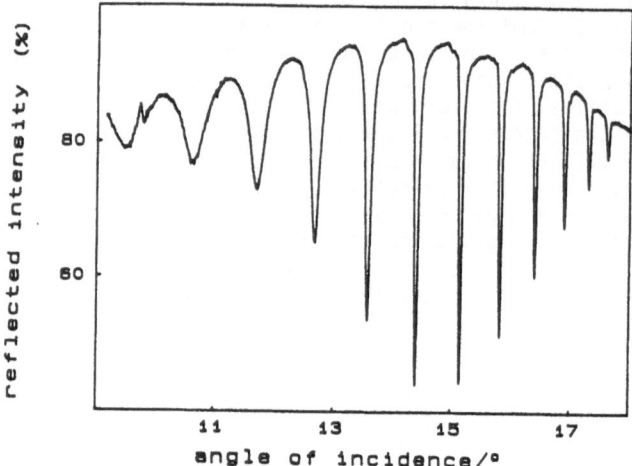

FIGURE 5. Dark line TM spectrum of a conventionally implanted He:LiNbO₃ waveguide (23). The reflected intensity is plotted as a function of the angle of incidence. Modes are characterized by the minima of the intensity. The effective refractive index increases with increasing angle.

considerably reduced, but the alteration of the beam parameters is unavoidable during the implantation. We followed a different strategy (23): While all the beam parameters are held constant, the samples are tilted stepwise, thus dividing the total ion dose into several peaks which are shifted towards the surface with increasing angle of incidence. The penetration depth perpendicular to the surface depends on the cosine of the angle of incidence with respect to the surface normal (24). A correction for the ion dose has to be taken into account due to the fact that the implanted area increases with the sine of the cited angle. Therefore, the dose per time unit decreases. Taking these very simple corrections into account, any desired profile shape can be made only by combining the angles of incidence and the irradiation time per angle. We have implanted several LiNbO₃ samaples with varying angle implantations of He⁺. A typical result is shown in Fig. 6. The sample was irradiated with an energy of 3.15 MeV and a total dose of 5×10^{16}/cm² increasing the impact angle from 7 to 37° in steps of 1°. Thus, the penetration depth is decreased stepwise down to 80% of the starting value. Although these parameters are not optimized for a flat bottom of the refractive index profile, the dark line spectrum exhibits a drastically reduced leakage for higher order modes. The guide supports six guided modes, and the transition area is formed by only one radiation made. Note that the lowest order mode occurs at a much lower angle, i.e. at lower effective refractive index. This is due to the higher dose which increases the damage (see Fig. 4).

5. WAVEGUIDES IN KNbO₃

KNbO₃ offers very large Pockels coefficients and a pronounced nonlinear electrooptic effect. Due to fast recording times and large photorefractive effects, KNbO₃ is the most promising material for storage applications in electrooptics (25). However, so far several obstacles have prevented the large scale application of KNbO₃. One of the main difficulties (compared with LiNbO₃) is the growth and poling procedure. Furthermore, the phase

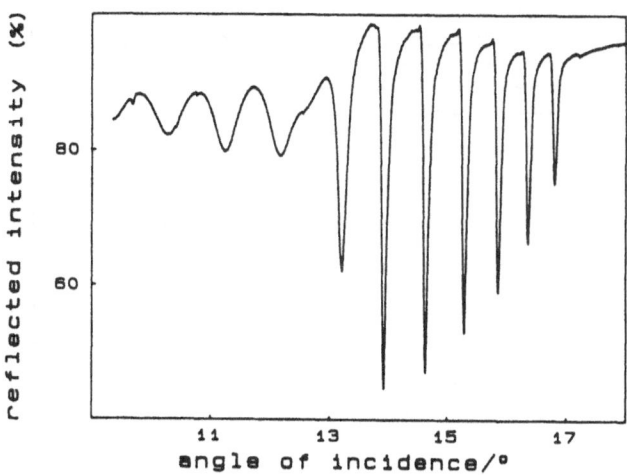

FIGURE 6. Dark line spectrum of a He:LiNbO₃ waveguide produced by varying angle implantation (23).

transition from the orthorhombic to the tetragonal phase, occurring at 490 K, does not allow high temperature processes for waveguide production. The only published result of waveguiding in KNbO₃ has been the application of an external electric field via outer electrodes (26), while diffusion and ion exchange did not work at all.

By implanting He⁺ ions, we succeeded in producing the first permanent multimode waveguide in KNbO₃ (27). The crystals were supplied by the crystal growth laboratory of the University of Osnabrück. They were grown from melt solutions containing 53 mol % K₂CO₃ and 47 mol % Nb₂O₅, cut into rectangular plates and polished. After poling with a dc-field the sample edges coincided with the b, a, and c axes.

Guided by the results of He⁺ implantation in LiNbO₃, we chose an ion energy of 2 MeV, a total dose of 10¹⁶/cm² and an angle of 10° with respect to the surface normal to avoid ion channeling. The rather high energy should result in a moderate damage plateau near the surface, making annealing almost unnecessary. This fact is especially important in the case of KNbO₃, because this crystal should not be heated due to the phase transition mentioned above. The chosen dose is a good compromise between high refractive index alteration and low scattering losses. As the KNbO₃ crystal structure is by far not as stable as LiNbO₃, heating during implantation has to be avoided. We monitored the sample temperature, and reduced the ion flux whenever 310 K were exceeded. For a flux of 10¹³/cm²s, no depoling occurred. However, implantations with higher ion fluxes demonstrated that KNbO₃ is much more sensitive than LiNbO₃. Increasing the flux by an order of magnitude leads to depoling and cracking of the substrates.

The optical investigations were performed "as implanted" using dark line spectroscopy. The light propagated along the c-axis and was polarized perpendicular to the plane of incidence (TE modes). Figure 7 shows the result of the profile nₐ(x), reconstructed with our IWKB algorithm. The measured effective indices are indicated by the horizontal lines. The profile is very flat for the first 3 μm below the surface and decreases rapidly in the region of ion deposition. The profile width is in good agreement with the results in LiNbO₃ for the same energy. However, at the same dose, the guide

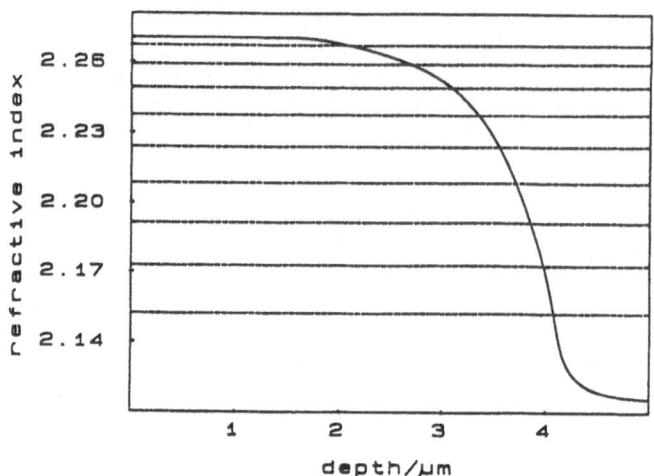

FIGURE 7. Reconstructed refractive index profile $n_a(x)$ for an He+ implanted KNbO$_3$ waveguide (27). Implantation conditions see text.

supports a larger number of modes indicating a stronger decrease of the refractive index compared with LiNbO$_3$.

6. CONCLUSIONS

We have presented recent results of He+ implanted LiNbO$_3$ waveguides and discussed the influence of the beam parameters on the surface refractive indices and the profile widths. The refractive index profiles were evaluated by means of dark line spectroscopy and an improved inverse WKB method, leading to an accuracy of $\delta n = 1 \times 10^{-4}$. At doses above $5 \times 10^{15}/cm^2$, a saturation effect has been observed for both, the ordinary and extraordinary index profiles.

A severe problem for He+ implanted waveguides is the tunneling of electromagnetic energy through the low index barrier. This effect can be reduced by broadening the barrier. Two strategies have been applied by now: multiple energy implantations and varying angle implantation. Results of the latter method have been discussed in this paper. They demonstrate that the leakage can be reduced without any change of the beam parameters only by varying the angle of incidence of the ion beam. By properly choosing the implant parameters energy, dose and angle, nearly any desired profile should be achievable. No other fabrication method offers the versatility of ion implantation.

Implantation is not only useful for waveguide fabrication in materials where other methods have given acceptable results, it furthermore offers the possibility of waveguide fabrication in more complex crystals. An important example is KNbO$_3$, where all other methods failed so far. With He+ implantation, the first permanent multimode waveguide has been formed in this excellent electrooptic material. We demonstrated that KNbO$_3$ is very sensitive to ion implantation, so that only low doses are required for sufficient index alterations.

7. ACKNOWLEDGEMENT

The author wishes to thank D. Kollewe from Universität Stuttgart for his assistance with the ion implantations, B. Hellermann and H. Hesse for

KNbO$_3$ crystal growth, and W. Heiland and P. Hertel for helpful discussions. The work was performed within the program of SFB 225. The financial support from Deutsche Forschungsgemeinschaft is gratefully acknowledged.

REFERENCES
1. Weiss RS and TK Gaylord: Appl. Phys. A37 (1985) 191-203.
2. Tien PK: Rev. Mod. Phys. 49(2) (1977) 361-410.
3. Schmidt RV and IP Kaminov: Appl. Phys. Lett. 25(8) (1974) 458-460.
4. Becker RA: Opt. News Feb.88 (1988) 27-28.
5. Wei DTY, WW Lee, and LR Blom: Appl. Phys. Lett. 25(6) (1974) 329-331.
6. Destefanis GL, PD Townsend, and JP Gailliard: Appl. Phys. Lett. 32(5) (1978) 293-294.
7. Townsend PD: Rep. Progr. Phys. 50 (1987) 501-558.
8. Götz G and H Karge: Nucl. Instrum. Meth. Phys. Res. 209/210 (1983) 1079-1088.
9. Wenzlik K, J Heibei, and E Voges: Phys. Status Solidi A 61 (1980) K207-K211.
10. Chandler PJ and PD Townsend: Nucl. Instrum. Meth. Phys. Res. B19/20 (1987) 921-926.
11. Barfoot KM and BL Weiss: J. Phys. D:Appl. Phys. 17 (1984) L47-L52.
12. Faik A, PG Dawber, DJ O'Connor, and PD Townsend: Radiat. Eff. 64 (1982) 235-240.
13. Glavas E, PD Townsend, G Droungas, M Dorey, KK Wong, and L Allen: Electron. Lett. 23(2) (1987) 73-74.
14. Bremer T, W Heiland, A Klekamp, and E Krätzig: Phys. Status Solidi A105 (1988) K17-K20.
15. Reed GT and BL Weiss: Electron. Lett. 23(15) (1987) 792-794.
16. Hertel P and HP Menzler: Appl. Phys. B44 (1987) 75-80.
17. Bremer T, P Hertel, and D Kollewe: Nucl. Instrum. Meth. Phys. Res. B34 (1988) 62-67.
18. Linnros J, G Holmen, and B Svensson: Phys. Rev. B32 (1985) 2770-2777.
19. Kool WH, HE Roosendaal, LW Wiggers, and FW Saris: Radiat. Eff. 36 (1978) 41-48.
20. Elliman RG, JS Williams, WL Brown, A Leiberich, DM Maher, and RV Knoell: Nucl. Instrum. Meth. Phys. Res. B19/20 (1987) 435-442.
21. Biersack JP and LG Haggmark: Nucl. Instrum. Meth. Phys. Res. 174 (1980) 257-269.
22. Weiss BL and CN Ahmad: Nucl. Instrum. Meth. Phys. Res. B30 (1988) 51-55.
23. Bremer T, W Heiland, and D Kollewe: Ferroelectr. Lett. Sect., in press.
24. Fink D, JP Biersack, H Kranz, J DeSouza, M Behar, and FC Zawislak: Radiat. Eff. 106 (1988) 165-181.
25. Baumert JC, C Walther, P Buchmann, H Kaufmann, H Melchior, and P Günter: Appl. Phys. Lett. 46(11) (1985) 1018-1020.
26. Günter P and JP Huignard: Photorefractive Materials and Their Applications: Berlin: Springer, 1988.
27. Bremer T, W Heiland, B Hellermann, P Hertel, E Krätzig, and D Kollewe: Ferroelectr. Lett. Sect. 9(1) (1988) 11-14.

ION IMPLANTATION OF ELECTRO-OPTICAL CERAMICS

CH. BUCHAL

KFA-ISI, D-5170, Jülich, FRG

1. INTRODUCTION
 Single-crystalline oxide ceramics are very useful for electro-optical
applications. Good crystals are transparent for a wide range of wavelengths, display useful non-linear properties and show a noticeable Pickels
effect, i.e. the refractive index changes if an electrical field is applied
to the crystal. This effect is exploited for the fabrication of modulators,
switches, etc.
 A general prerequisite for device fabrication is the formation of an optical waveguide within the material. This is a region of enhanced refractive
index which acts as a light pipe. In the main part of this contribution we
will discuss waveguide fabrication by ion implantation into $LiNbO_3$.
However, as the desirable properties of electro-optical ceramics are a consequence of their complex and interesting structure, we will start with a
discussion of their crystal lattice.

2. THE STRUCTURE OF PEROVSKITE-RELATED OPTICAL CERAMICS
 Perovskite-related ceramics have the chemical composition ABO_3 with A and
B being metal ions. As there are many metals in the periodic table and as
the metal ion stoichiometry may be divided up into $AC_xD_{1-x}O_3$ or similar,
there is an infinite wealth of compounds imaginable and Galasso discussed
more than 500 different perovskites already in 1969 (1). Today this
situation has become amazingly more complex with the discovery of the superconducting ceramics. Therefore, it is useful to go back to basics (2).
Figure 1 shows a comparison of ionic radii. For our discussion we need the
following values:

$$O^{--} = 1.4 \text{ Å} \qquad Nb^{5+} = 0.70 \text{ Å} \qquad Li^+ = 0.60 \text{ Å}$$
$$K^+ = 1.33 \text{ Å} \qquad Ti^{4+} = 0.68 \text{ Å}$$

The true cubic Perovskites (e.g. $BaTiO_3$, $SrTiO_3$, $KNbO_3$) form a close-packed
lattice of oxygen anions together with the big metal cations (Ba^{2+}, Sr^{2+},
K^+). Figure 2 shows a (111) plane of $KNbO_3$ as an example. In this arrangement the Coulomb energy of the big cations is minimized by maximizing their
separation. The positions of the K^+-ions in the next layer are marked by K
and again every K^+ is surrounded by O^{2-}, giving a total of 12 nearest oxygen
neighbors (3+6+3). The small Nb^{5+}-ions are added half way between adjacent
close-packed layers in such a way that their repulsive Coulomb energy is
minimal. Due to their small size they fit into a pocket between six
surrounding oxygens forming a metal-oxygen octahedron. While the preceding
discussion has assumed the crystal to be truly ionic, now some small covalent contribution has to be added. The metal ion in the octahedral position
has s, p, and d orbitals available for covalent bonding and an spd-hybride
wavefunction forms six equivalent bonds with the surrounding oxygens. This

389

C.J. McHargue et al. (eds.), Structure-Property Relationships in Surface-Modified Ceramics, 389–397.
© *1989 by Kluwer Academic Publishers.*

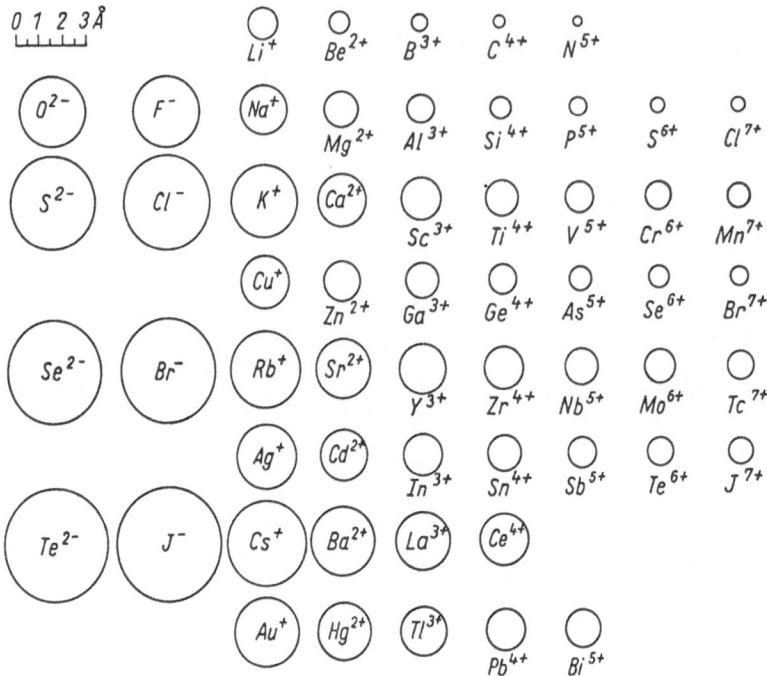

FIGURE 1. Compilation of ionic radii (from ref. 2).

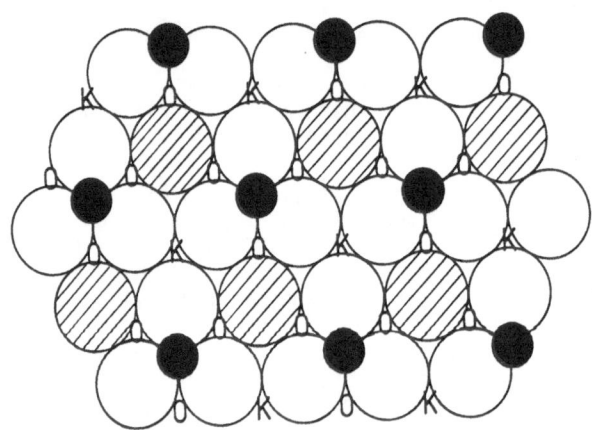

FIGURE 2. Lattice of $KNbO_3$, looking at (111)-planes. First hcp-layer of O^{2-} (○) and K^+ (⊘) is fully drawn. Next hcp-layer is denoted by letters O and K for O^{2-} and K^+. In between these two hcp layers Nb^{5+} (●) is located in octahedral places.

situation is shown in Fig. 3. The close-packed big ions form the fcc-structure of Fig. 3. The electrical effects (piezoelectric, ferroelectric, electro-optic, ...) of these crystals come about by small distortions of the unit cell and by shifting the middle ion somewhat out of the center of its oxygen octahedron, lifting its crystal field degeneracy and generating a permanent dipolar moment. This permanent polarization is achieved by cooling a crystal in an electrical field through its Curie temperature. In this case all distorted octahedra are frozen in uniformly. If a lightwave passes through these structures, its electrical field generates ionic displacements and displacements of the electrons, especially in the octahedral bonds. A quantitative calculation of the observed optical indices due to atomic polarization, ionic displacement and bond polarizability is not yet available. Nevertheless there is evidence that especially the non-linear bond polarizability increases in the order $XTaO_3 \rightarrow XNbO_3 \rightarrow XTiO_3$ (3,4). Therefore, a local replacement of Nb in $LiNbO_3$ or $KNbO_3$ by Ti is a valid concept for waveguide formation. As of now, this works very well for $LiNbO_3$ but not at all for $KNbO_3$.

3. THE STRUCTURE OF LiNbO₃

$LiNbO_3$ is somewhat unique and it is related to the Perovskites only in a wider scope (5). Both metal ions are very small and the huge oxygens dominate the structure by forming hcp layers just as shown in Fig. 2, if the K^+ ions were substituted by oxygens. Now both Li^+ and Nb^{5+} are regularly distributed into the oxygen pockets, resulting in a more dense ionic arrangement ($LiNbO_3$: 31.8 cm^3/mole; $KNbO_3$: 39 cm^3/mole). Niobium moves into an octahedral position exactly as discussed before. There is also an octahedral position for the Li-ion available, but it has no covalent d-orbitals to stabilize the center position and shifts very closely to three of its six oxygen neighbors. The resulting crystal arrangement is shown in Fig. 4. $LiNbO_3$ has no cubic phase, it is always hexagonal. One third of the octahedral interstices remains empty, being arranged in the c-direction as Li-Nb-vacant-Li-Nb-. The generated dipolar asymmetry is very stable, resulting in a Curie temperature of 1050 to 1200°C, depending on Li/Nb-ratio (5). This permits very high processing temperatures of T < 1050°C for device fabrication, which comes close to the melting temperature of 1240°C.

Surprisingly, this stability is contrasted by an onset of Li diffusion at 200°C, which permits the Li ↔ H exchange at these low temperatures. But as these ion-exchanged crystals are not long-term stable, the procedure has not gained much application.

At temperatures above 500°C, oxygen vacancies become mobile in $LiNbO_3$ as well as in the true Perovskites. In all these structures vacancies in the close-packed oxygen sublattice are needed to open a path for oxygen and cation diffusion (with the exception of Li^+ and H^+ which seem to move freely via interstitials).

At temperatures between 800 and 1000°C direct diffusion may be used to introduce Ti into $LiNbO_3$. As mentioned before, the introduction of Ti-O-octahedra increases the optical indices and the formation of low-loss stable waveguides becomes possible. This process is of great technological importance and has been studied by numerous groups very carefully (9,10,11). The formation of a mixed Ti-Nb-oxide and its incorporation into the $LiNbO_3$ crystal is described by reference (12). It is generally assumed that the balance in stoichiometry ($Nb^{5+} \leftrightarrow Ti^{4+}$) is established by introducing vacancies into the oxygen sublattice. The equivalent octahedral lattice locations of Nb and Ti have been proved by EXAFS (13) and channeling/PIXE (14), as shown in Fig. 5.

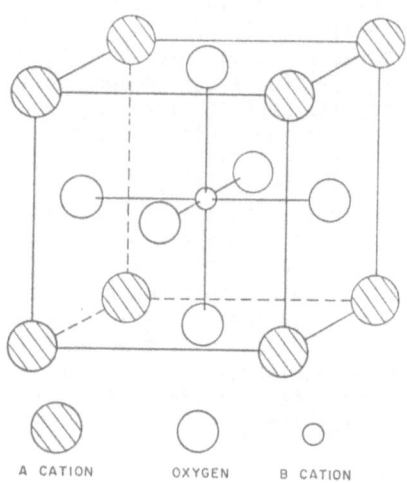

FIGURE 3.   Unit Cell of Perovskite ABO₃ (from ref. 1).

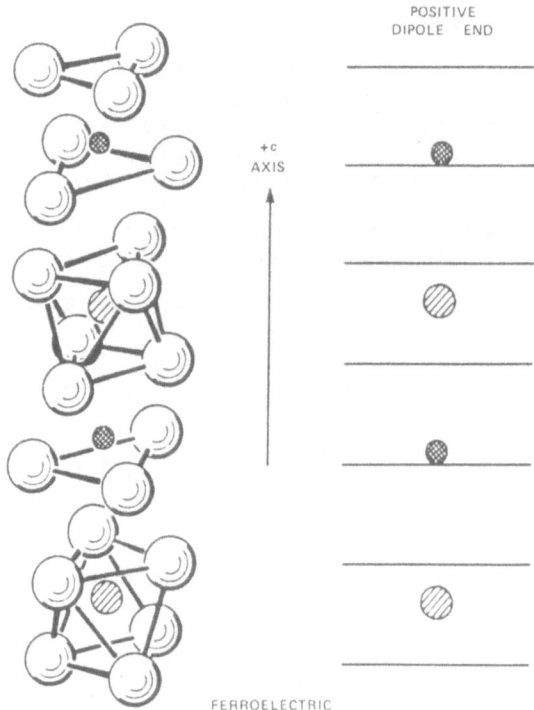

FIGURE 4.   Positions of the lithium atoms and niobium atoms with respect to the oxygen octahedra in the ferroelectric phase (T < T_c) of lithium niobate (from ref. 6,7,8).

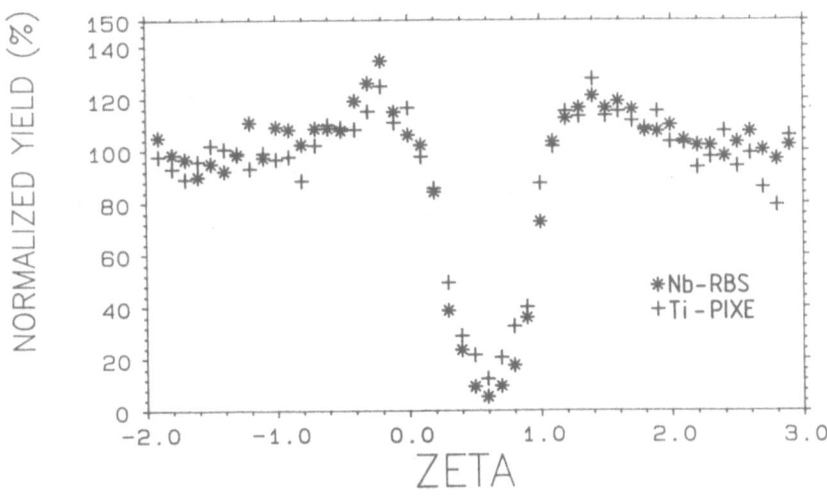

FIGURE 5. Angular channeling scans across sample normal (z-axis) of implanted waveguide. This can shows that Ti ions display the same channeling behavior as Nb, demonstrating an equivalent lattice location (ref. 14).

4. DILUTION OF THE LATTICE BY ION IMPLANTATION

Townsend and coworkers have pioneered this type of ion implantation for waveguide fabrication (15,16). The formation of a strip waveguide by MeV helium implantation is demonstrated in Fig. 6. It is the virtue of this method that the guiding volume is only mildly damaged by the first MeV helium implant, forming a deep layer of reduced optical index. A low-temperature anneal of 200°C is sufficient to remove the lattice damage in the guiding volume. Therefore this method may be applied to various substrates and Bremer very recently fabricated the first waveguide in KNbO$_3$ by MeV helium implantation (17).

Unfortunately some wave propagation losses due to absorption in the damaged material of the walls are observed.

5. DIRECT TITANIUM IMPLANTATION AND SOLID-PHASE EPITAXY

Stimulated by the desire to raise the local Ti concentration above the limits imposed by diffusion kinetics, the Oak Ridge group has developed the concept of direct ^{48}Ti implantation into LiNbO$_3$ (18,19). The necessary masking technique may be readily converted from the standard device fabrication line for diffused waveguides. Normally the lithographical technique will provide a Ti pattern on top of the substrate and a subsequent heat treatment is employed to diffuse this pattern into LiNbO$_3$ and generate the guiding structures. Now the mask has been "inverted" in such a way that the channels will remain open for future implantation and the deposited Ti is employed solely as the masking metal for the implantation process. The choice of Ti metal as the masking material is very important. It is known that different ions may generate active scattering center and corresponding optical losses. Obviously a mask of Ti metal is a very safe choice, since some sputtering and subsequent ion beam mixing during implantation will always be present. This procedure is shown in Fig. 7.

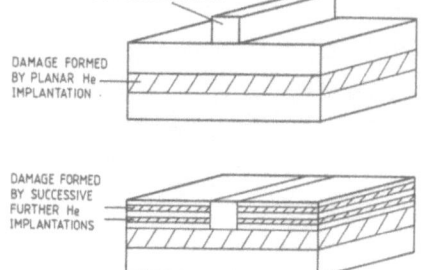

FIGURE 6. Ion implantation of a strip waveguide. The material <u>surrounding</u> the guide is implanted with He and its index of refraction is lowered, while the guiding volume stays virtually unchanged.

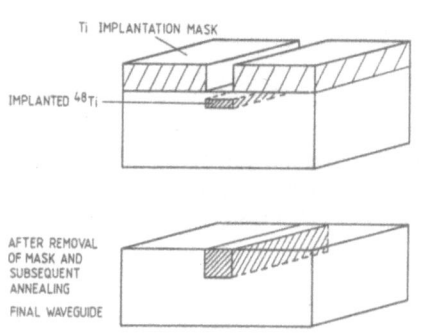

FIGURE 7. Direct implantation of a strip waveguide. We use the same lithographical masking techniques as successfully applied for standard diffused waveguide fabrication. The element Ti is chosen as the masking material, since it guarantees no foreign ion contamination.

In brief, our process employs the following steps:
1. Photolithographical masking of the optical structures.
2. Sputter deposition of Ti to generate the implantation mask. We typically use 1 μm of Ti as the masking metal.
3. Removal of photoresist, leaving the channels accessible to implantation.
4. Titanium implantation, while the substrate is cooled to 100 K to reduce radiation enhanced diffusion. Typical ion energies are 200 to 400 keV, resulting in an implant depth of 0.1 to 0.2 μm. Typical doses range from 1.21×10^{17} to 3×10^{17} Ti/cm². Note that 1.2×10^{17} Ti/cm² equals the number of atoms in a Ti film of 200 Å thickness.
5. Removal of the mask.
6. High temperature anneal to accomplish solid phase epitaxy of implanted regions.

After implantation the LiNbO$_3$ crystal is amorphized over the implanted depth and a high amount of Ti is introduced into the bulk. In this state the crystal is dark and opaque and of no use for any optical application. It is of critical importance to restore the oxygen crystallinity and to move the Ti ions into their regular lattice positions. This may be achieved by annealing, but of course long annealing schedules have to be avoided, as they will result in excessive diffusion and undue spread of the Ti distribution.

We found best results for short anneals at a temperature of 1000°C in water-saturated oxygen atmosphere. The Li deficiency in the implanted layer is replenished by diffusion from the bulk. At a dose of 2.5×10^{17} Ti/cm² it takes approximately t = 1 h at 1000°C to accomplish the necessary annealing and the HWHM of the Ti distribution is increased to 1 µm due to diffusion. The near-surface concentration of Ti gets as high as 12%, which should be compared to typical values of 3% for diffused guides. The measured effective index for the sample with 2.5×10^{17} Ti/cm² is 2.218 and the calculated change in material index is 0.030 for propagation in the y direction. For a dose of 2×10^{17} Ti/cm² these numbers are reduced to 2.213 and 0.025, respectively.

A detailed comparison of implanted and diffused guides, an analysis of the process parameters with respect to the int4ended wavelength of operation and a discussion of device performance will be found in references (20 and 21). By lowering the annealing temperature to 800°C we could demonstrate that solid-phase epitaxy starts at the interface to the undamaged bulk and proceeds towards the surface during annealing.

Poker has investigated the nature of this process by lowering Ti dose and regrowth temperature even further (22,23). Figure 8 shows channeling studies of the movement of the regrowth front after implanted 10^{16} Ti/cm² at 360 keV and annealing at 400°C only. It is surprising that the lattice may be restored at these low temperatures, even if the dose of 10^{16}/cm² does not introduce a huge change in stoichiometry. For higher doses higher annealing temperatures become necessary in order to achieve the same crystalline perfection. As can be seen from Fig. 9, the activation energy seems to be lowered at higher doses. This observation is not yet understood.

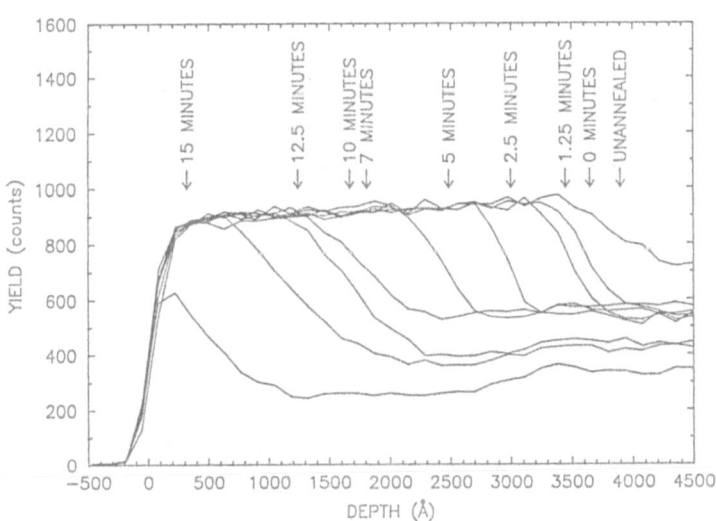

FIGURE 8. RBS spectra of SPE regrowth of implanted LiNbO₃ at 400°C. The arrows represent the approximate positions of the interface at various annealing times. The Ti dose was 10^{16} Ti/cm², which is a relatively small disturbance of the stoichiometry (from ref. 22).

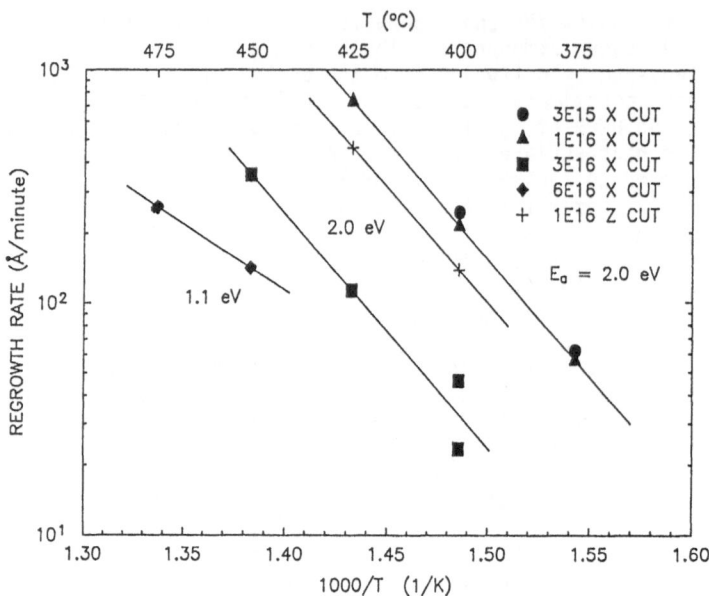

FIGURE 9. Arrhenius plot of the SPE regrowth of Ti-implanted LiNbO$_3$, indicating the trend toward slower, nonactivated regrowth at higher implant doses. Filled symbols represent X-cut crystals, while pluses represent Z-cut (from ref. 22).

6. CONCLUSION

The structure of Perovskite-type ceramics and their characteristic close-packed oxygen layers with embedded metal-oxygen octahedra has been discussed. LiNbO$_3$ seems to possess unique features: high Curie temperature, good lattice stability, tolerance for diffusion or even implantation of high doses of foreign ions, and the ability of epitaxial regrowth to beautiful crystalline perfection. This all is accompanied by very useful electro-optical properties, which allow the fabrication of devices for integrated optics.

ACKNOWLEDGEMENT

KNbO$_3$ samples have been supplied by P. Günter, ETH Zürich and some implants on those samples were conducted byR. Irmscher. This work will be published in a forthcoming paper.

REFERENCES

1. Galasso FS: Structure, Properties and Preparation of Perovskite-Type Compounds. Oxford: Pergamon, 1969.
2. Pauling L: The Nature of the Chemical Bond. Ithaca: Cornell University Press, 1969.
3. Günter P (ed): Electrooptic and Photorefractive Materials. Berlin: Springer Proc. in Phys. 18, 1987.
4. Bergman JG and GR Crane: J. Sol. State Chem. 12 (1975) 172.
5. Räuber A: Current Topics in Materials Science. K Keldis (ed). Amsterdam: North Holland, 1978, vol. 1, p. 481.
6. Weis RS and Gaylor TK: Appl. Phys. A37 (1985) 191.
7. Abrahams SC et al: J. Chem. Phys. Sol. 27 (1966) 1019.

8. Abrahams SC et al: J. Chem. Phys. Sol. 34 (1973) 521.
9. Hunsperger RB: Integrated Optics: Theory and Technology. Berlin/New York: Springer, 1985.
10. Mentzer MA (ed): Integrated Optical Circuit Engineering V. Proc. SPIE. Bellingham, WA, 1988, p. 835.
11. Canali C, A Carnera, G Celotti, G Della Mea, and P Mazzoldi: Mat. Res. Soc. Symp. Proc. 24 (1983) 459.
12. Mazzoldi P and A Carnera: Cryst. Latt. Def. Amorph. Mat. 14 (1987) 319.
13. Skeath P, W Elam, W Burns, F Stevie, and T Briggs: Phys. Rev. Lett. 59 (1987) 1950.
14. Buchal Ch, S Mantl, and DK Thomas: Mat. Res. Soc. Symp. Proc. 100 (1988) 317.
15. Townsend PD: Optical Effects of Ion Implantation. Rep. Prog. Phys. 50 (1987) 501.
16. Destefanis GL et al: J. Appl. Phys. 50 (1979) 7898.
17. Bremer T et al: Ferroelectr. Lett. 9 (1988) 11.
18. Buchal Ch, PR Ashley, and BR Appleton: J. Mat. Res. 2 (1987) 222.
19. Buchal Ch, PR Ashley, DK Thomas, and BR Appleton: Mat. Res. Soc. Symp. Proc. 88 (1987) 93.
20. Ashley PR and Ch Buchal: Integrated Optical Circuit Engineering V. Proc. SPIE. Bellingham, WA, 1988, p. 113.
21. Ashley PR, W Chang, Ch Buchal, and DK Thomas: to be published.
22. Poker DB: Mat. Res. Soc. Symp. Proc. 93 (1987) 293.
23. Poker DB and DK Thomas: Mat. Res. Soc. Symp. Proc. 100 (1988) 311.

PLASMA- AND ION-BEAM ASSISTED PHYSICAL VAPOR DEPOSITION: PROCESSES
AND MATERIALS

J. M. RIGSBEE

Department of Materials Science and Engineering, University of Illinois at
Urbana-Champaign, 1304 West Green Street, Urbana, IL 61801

1. INTRODUCTION

Physical vapor deposition (PVD) includes any thin film process involving
the deposition of physically generated atoms or molecules onto a substrate
in a vacuum environment. Evaporation, sputtering, and ion plating, the fundamental PVD processes, are characterized by the physical mechanism by which
the vapor flux is generated. Modification of these basic processes (e.g.,
by addition of an ion beam) accounts for the apparent numerous PVD processes
in the literature. In spite of minor processing variations, PVD processes
have many common features including: (1) a high vacuum system with low
impurity gas levels and with the ability to input controlled flow rates or
partial pressures of one or more working gases; (2) a coating material
source(s) with a well-controlled and often monitored vapor flux; and (3) a
substrate mounting assembly which controls substrate temperature and
distance/orientation to the coating source. Plasma and ion-beam assisted
PVD processes are derived from the fundamental evaporation, sputtering, and
ion plating processes through, for example, incorporation of a bias voltage
to create a glow discharge.

PVD processes are used to deposit thin films for purposes such as
microelectronics circuit fabrication, control of light reflectance and
absorption, corrosion and oxidation mitigation, decorative coatings, and
tribology. PVD fabricated materials include: metals, ceramic compounds,
including superconductor thin films; composites; and semiconductors,
including layered superlattice structures. Suitable substrates range from
metals to ceramics to polymers. Nearly every imaginable industry from automotive to textiles to microelectronics actively utilizes PVD technologies in
their products and manufacturing processes. Recently developed ion assisted
PVD technologies which are rapidly finding applications include ion plating
(a plasma-assisted process), activated reactive evaporation, and ion-beam
assisted molecular beam epitaxy. These processes offer possible advantages
in film adhesion, microstructure, chemistry, and mechanical/electrical
properties.

Bombardment of a PVD film with energetic particles (ions and/or neutrals)
prior to and during film deposition can dramatically and beneficially alter
the structure, chemistry, and physical properties of the film and the
film/substrate interface. Applications of plasma-assisted and ion beam-
assisted thin film PVD processes are increasing in areas such as
microelectronics and aerospace where improved performance requirements are
continually being placed upon materials. Energetic particle bombardment
during film growth has also been found to enhance formation of new materials
and alloys, such as compound semiconductors, whose metastable microstructures result in unique physical properties.

399

C.J. McHargue et al. (eds.), Structure-Property Relationships in Surface-Modified Ceramics, 399–416.
© 1989 by Kluwer Academic Publishers.

The objectives of this paper are to review (1) the processes (equipment and operation) for plasma and ion-beam assisted PVD and (2) the structure, chemistry and physical properties of films deposited with these processes. Discussion will include ceramic (nitride and oxide), metallic (elemental and layered), and compound semiconductor films deposited onto metallic and ceramic substrates. Topics to be discussed include the effects of energetic particle bombardment on film characteristics such as: porosity; residual stresses; film nucleation, growth, and preferred orientation; and, adhesion, including chemical mixing and compound formation at the film/substrate interface. Emphasis will be placed on the microstructural (analysis with SEM and cross-section TEM) and microchemical (analysis with EDX, Auger, and SIMS) effects of energetic particle bombardment. For additional in-depth discussions on PVD and related topics, the reader is directed to references (1-6) [note especially the extensive and detailed discussions of PVD deposition processes and materials by J. A. Thornton in references (4 and 5)].

2. EVAPORATION

Evaporation is a process involving the heating of a source material to a sufficiently high temperature, usually in a high vacuum, so that atoms or molecules are liberated from the source, move through the vacuum with little scattering and low average kinetic energies, and deposit onto a substrate (7,8). A basic evaporation system consists of a vacuum chamber, a vacuum pumping system capable of achieving high vacuum, a gas inlet system for chamber venting, an evaporation source(s), and a specimen holder. Vacuum technology is a common and essential feature for all PVD processes and the reader is directed to Glang, Holmwood, and Kurtz (9) for a thorough review.

2.1. Equipment and operation

Processing is done inside a vacuum chamber which may range in size and shape from a small glass bell jar to a room-sized cube. Fixturing is required inside the vacuum chamber for supporting and positioning the substrates during film deposition. This fixturing may be a simple stationary rod or plate or it may utilize motion along one or more axes to enhance coating thickness uniformity over complex-shaped substrate surfaces. Substrate fixturing systems often incorporate heating and cooling capabilities because temperature is a key parameter in controlling substrate cleanliness and film adhesion, residual stress, and microstructure. If a bias voltage is to be applied to the substrate for plasma assisted deposition, electrical connections and shielding requirements can make the fixturing quite complex.

2.2.1. Vacuum system.

Straight evaporation is typically done in a high vacuum environment so the need for gas introduction into the chamber is limited mainly to initial pump-down cycles where introduction of an inert gas serves to purge the chamber of unwanted gases. Depending on purity requirements the inert gas may need to be further purified of reactive gases such as oxygen, nitrogen, or water vapor prior to introduction into the chamber by passing it through a heated titanium sponge system. To understand how residual gases affect the film microstructure and properties, it is important to know the composition of the residual gases in the vacuum chamber during deposition. This is done with a residual gas analyzer (RGA) which is simply a mass spectrometer designed to measure ion current as a function of atomic mass (9). Because RGA's require vacuums of 10^{-2} Pa or less to operate, auxiliary pumping and valving may be required when analyzing the gas compositions of poor vacuums often found in plasma-assisted deposition systems.

The vacuum pumping system must be capable of producing a high vacuum of approximately 10^{-3} Pa (7.5 x 10^{-6} Torr or 7.5 x 10^{-3} μm) or better and this system usually consists of a roughing pump/high vacuum pump combination. For systems using oil-filled rotary mechanical pumps and diffusion pumps, an optically opaque, liquid nitrogen-cooled baffle should be used between the diffusion pump and the deposition chamber to eliminate substrate or coating contamination by backstreaming of pump oils. A properly designed and maintained diffusion pumped vacuum system is reliable, cost effective, and will attain ultimate pressures in the 10^{-4} Pa range. For critical PVD applications such as semiconductor thin films where parts-per-million impurity levels may be required, serious contamination of the depositing film may occur by backstreaming of pump oils or by incorporation of residual gases produced by such mechanisms as desorption of physisorbed molecules from surfaces within the chamber. At a pressure of 10^{-4} Pa (about 10^{-6} Torr) the flux of residual gas atoms incident on the substrate surface will equal one monolayer of coverage in roughly 2 sec. The potential for unacceptable contamination levels for film growth in a high vacuum environment is clear since at a chamber pressure of 10^{-4} Pa the flux of unwanted gas atoms will be roughly equivalent to the flux needed to grow a 5×10^{-4} mm (0.5 μm) thick film in 1 h. For these critical PVD applications it is necessary to eliminate oil-based pumps and use, singly or in combination, ion pumps, turbopumps, sublimation pumps or cryopumps which are capable of producing clean ultrahigh vacuums (UHV) of 10^{-7} Pa or better.

2.1.2. <u>Evaporation sources</u>. An evaporation source produces a controlled flux of vapor from a liquid by evaporation or from a solid by sublimation, the choice of which for a given material depends on the relationship between vapor pressure versus temperature and melting point. One Pa (approximately 10^{-2} Torr) is a typical vapor pressure required for reasonable film deposition rates. Metals such as Cr, Sb, As, Be, Si, Mo, W, and Zn which have ratios of $T(1 \text{ Pa})/T_{mp}$ of less than unity (this ratio for Cr is about 0.77) will deposit by sublimation. This is also true for ceramic and semiconductor compounds such as alumina, silicon dioxide, magnesium fluoride, zinc selenide, and cadmium sulfide.

Types of evaporation sources are numerous with the most common being: resistance heating (including crucible sources), electron beam (thermionic and plasma beam) heating, cathodic arc, induction heating, laser heating, and exploding wire (6,7). Resistance-heated wire or metal foil sources are normally laboratory-scale and are made from refractory metals, such as tungsten, which have high melting points and low vapor pressures. Commercial-scale evaporation-based processes have been developed to ion plate aluminum (or other low melting point materials) onto large aircraft sections for corrosion mitigation (10,11). Deposition of controlled chemistry alloy films by evaporation from multiple sources requires use of piezoelectric quartz crystal oscillators or ionization/mass analysis rate monitors for each source. This is essential because small temperature changes (100°C) typically produce order of magnitude or larger changes in the vapor flux.

Electron beam evaporation sources allow the high power densities and extremely high temperatures needed to evaporate refractory metals and oxides. These systems, which are well suited for commercial production, are compatible with ion beam and plasma-assisted deposition processes. The major difficulties encountered with electron beam evaporation are: maintaining a constant evaporation rate; dealing with differences in vapor pressure between different elements in an alloy evaporant source material; preventing coating and eventual electrical shorting of high voltage insulating electron gun components; and preventing expulsion of macroscopic droplets from the melt pool because of transient excessive heating or entrapped gas.

The arc evaporation source also is a form of ion assisted deposition process. Arc evaporation sources, which operate at low voltages and high currents (around 20 to 30 V and 200 to 500 Å), have been shown to efficiently "evaporate" material through numerous and very transient arc discharges over flat metal substrates. The metal substrates are cathodes in this system and the resulting locally intense, arc-induced plasma causes a high degree of ionization (estimated as high as 90%) of the metal and non-metal flux. These arc sources are principally used for ion plating applications (discussed later in this review) and have the following advantages: (1) their simple design and operation allows virtually any size source to be created and (2) because the localized melting event is very brief and for practical considerations no liquid exists on the source surface, the arc sources can be freely placed in any orientation. This process allows uniform coverage of even very complex-shaped surfaces through the use of properly placed and oriented multiple arc sources. The primary disadvantage of this process is its tendency to generate micron and larger sized metallic droplets, called macroparticles, which deposit unreacted on the substrate surface producing a rough and dull-appearing coating. Although these particles have little apparent effect on the coating's tribological properties, control of macroparticles is a major challenge for this process for decorative and corrosion protective coatings.

2.2. The evaporation process

The Hertz-Knudsen equation (12), derived using concepts of reaction kinetics, thermodynamics, and solid state theory, provides the best model of the evaporation mechanism. The Hertz-Knudsen equation is:

$$W = \alpha_V \ (2\pi m k T)^{-0.5} \ P* \ ,$$

where W is the evaporation rate [atoms/(cm$^2\cdot$sec)]; α_V is the evaporation coefficient (varies from 0 to 1); m is the molecular weight (grams); P* is the vapor pressure (Torr); and k and T have the usual meanings. Substituting appropriate values for these terms, this equation becomes:

$$W = 3.5 \times 10^{22} \ \alpha_V \ (mT)^{-0.5} \ P* \ .$$

Knudsen's work in the early 1900s on mercury evaporation showed that the evaporation coefficient is very sensitive to the evaporant source surface cleanliness. The evaporation coefficient varies from near unity for very clean surfaces (such as the continuously produced mercury droplets studied by Knudsen) to as small as 10^{-3} for dirty surfaces (5,12).

Evaporated atoms and molecules have only the kinetic energy corresponding to when they left the evaporant source and their average energies will be 3/2 kT, where T is the source temperature. Typical energies for evaporated atoms and molecules are below 0.3 eV. As examples, the energies of elemental Cd, Cu, and W at source temperatures corresponding to partial pressures of 1.33 Pa are: 0.047 eV (538 k), 0.133 eV (1533 K), and 0.303 eV (3503 K), respectively. Energies of the evaporant particles increase continuously with increasing vapor pressure (equivalent to increasing source temperature). It is clear that addition of a plasma or ion beam can result in a major increase (several orders of magnitude) in the average energy of the depositing atoms.

The geometrical factors of distance from the source to the substrate and the location/orientation of the substrate relative to the direction normal to the source also contribute significantly to film deposition rate. Assuming a small evaporant source located a distance r from the substrate, the deposition rate per unit area of substrate is given by:

$$W_d = (W\ A_e\ \cos\phi\ \cos\theta)/(\pi\ r^2)$$

where W_d is the deposition rate per unit area of substrate or the flux of evaporant atoms onto the substrate [atoms/(cm^2·sec)]; W is the flux of atoms from the evaporation source [atoms/(cm^2·sec)]; A_e is the area of the evaporation source (cm^2); ϕ is the angle between the evaporation source normal and the vector between the source and the substrate surface; and, θ is the angle between the substrate surface normal and the vector between the source and the substrate surface. It is clear from this equation that the deposition rate decreases rapidly with increasing source to substrate distance and decreases slowly (for the first 25 degrees or so) as the substrate surface rotates away from directly facing the evaporation source.

The previous paragraphs have discussed how the deposition rate is controlled by changes in the specimen position and orientation and for changes in the material being evaporated and operation of the evaporation source. The deposition rate or film thickness growth rate may be calculated using the following equation:

$$D = (M\ W_d)/(\rho\ N_a)\ ,$$

where D is the thickness growth rate (cm/sec); M is the molecular weight (g/mole); W_d is evaporant atom flux onto the substrate [atoms/(cm^2·sec)]; ρ is the mass density (g/cm^3); and N_a is Avogadros number (atoms/mole).

As an example of film deposition by evaporation, consider evaporation from a 1-cm^2 Au source with an evaporation coefficient of unity and heated to 1400°C (this corresponds to a vapor pressure of 1.33 Pa or 10^{-2} Torr). The 0.5 m diam flat substrate surface is centered 0.5 m directly above the source with the surface normal parallel to the source normal. The effects of position and orientation on film growth rates are shown by comparing the film growth rates at the sheet center (r = 0.5 m, ϕ = θ = 0 degrees) versus the sheet edge (r = 0.559 m, ϕ = θ = 26.57 degrees). Calculating the source evaporation rate, W, of Au at 1400°C gives 6.1 x 10^{17} atoms/(cm^2·sec), which for a 1-cm^2 source yields 6.1 x 10^{17} atoms/sec. The evaporant atom flux onto the substrate will be 7.7 x 10^{13} atoms/(cm^2·sec) at the sheet center and 5.0 x 10^{13} atoms/(cm^2·sec) at the sheet edge. Assuming the evaporant atoms all stick, a sticking coefficient of unity, the atom deposition rate at the sheet edge is about 65% that at the sheet center. Calculation of the film thickness growth rate at the sheet center gives a value of 0.13 Å/sec which is approximately equivalent to a monolayer of film growth every 15 sec. This film growth rate would be increased an order of magnitude by either heating the Au source to 1550°C or reducing the substrate to source distance from 0.5 m to about 0.16 m. Reduction in substrate to source distance markedly increases the center to edge nonuniformity in film growth rate.

Relatively low deposition rates of less than several angstroms per second are generally unacceptable for high vacuum systems due to the potential for extensive contamination of the film by residual gas incorporation. A high vacuum deposition chamber pressure of 10^{-4} Pa corresponds to a residual gas flux on the order of 10^{14} molecules per cm^2·sec. Deposition of high-purity films requires high-purity evaporation source materials, careful substrate cleaning procedures, and either deposition rates much greater than 10^{14} atoms/(cm^2·sec) or, equivalently, very low residual gas pressures. For processes incorporating plasmas and ion beams, it is essential to consider the potential for impurity contamination from the working gas.

404

2.3. Coating nucleation and growth

Nucleation and growth of an evaporated coating occurs by individual atoms or molecules striking the substrate surface, condensing onto the surface by equilibrating energetically with the surface atoms, diffusing about the surface, and (for the case where a nucleation barrier exists) clustering to form nuclei which grow and evolve into the final coating. PVD thin film nucleation and growth literature is extensive and the reader should see reference (13) for a thorough review of this topic.

Evaporated atoms or molecules which strike the substrate surface generally have low enough energies (<0.3 eV) to allow them to adsorb onto the substrate surface, if only briefly, and become adatoms. These adatoms will quickly (on the order of 10^{-12} sec) equilibrate energetically with the surface atoms and will diffuse over the surface until they desorb, are trapped energetically at a surface adsorption site (usually a defect such as a vacancy or a ledge), or join an existing coating nucleus. The probability that an evaporated (or sputter deposited) atom will re-evaporate before it equilibrates energetically with the substrate surface atoms is small since these atoms generally have quite low energies and the accommodation process is very fast. Thus with the normally low substrate temperatures used for evaporation it is reasonable to assume that a very high fraction (approaching unity) of atoms striking the substrate stick for some time much greater than the lattice vibration time of 10^{-13} sec. As an aside, PVD processes provide the ultimate form of rapid quenching since equilibration in 10^{-12} sec for impinging atoms with energies equivalent to more than 1000°C yields quench rates on the order of 10^{15} C/sec.

The probability that a surface atom will have sufficient energy to desorb is given by:

$$\exp\ (-Q_a/kT)\ ,$$

where Q_a is the adsorbed adatom atom binding energy to the substrate, and k and T have the usual meanings. The mean residence time for desorption of a substrate atom, τ_r, is simply the lattice vibration frequency, τ_0, times the probability that the atom will have sufficient energy for desorption during any one of its vibrations. This relationship shows that the mean residence time of a deposited atom is a strong function of binding energy and substrate temperature. The residence time at a substrate temperature of 200°C is 5×10^{-3} sec for a binding energy of 1 eV and 5×10^{-12} sec for a binding energy of 0.15 eV.

Assuming a crystallographically perfect substrate surface with no defects to serve as trapping sites, it is possible to calculate the equilibrium adatom density, n_a, as the product of the incident evaporant flux, R, times the mean residence time. For the 1 eV adatom binding energy and 200°C substrate temperature case, a flux of 2×10^{17} atoms/cm²·sec is required to obtain an equilibrium adatom density of one monolayer. At 0.15 eV adatom binding energy the flux required to obtain an equilibrium monolayer of coverage is too large to be physically reasonable. A flux of 2×10^{17} atoms/cm²·sec for the 0.15 eV adatom binding energy case would produce an equilibrium coverage of 10^6 atoms/cm²·sec, equivalent to about one adatom per every 10^9 substrate surface lattice sites. A recent examination of thin film nucleation theory by Lee and Rigsbee (14,15) concluded that thin film nucleation kinetics are almost always controlled by the substrate surface defect density and this effect increases with increasing temperature or decreasing adatom flux.

2.3.1. Nucleation and growth models. Nucleation and growth of a PVD thin film has been observed to occur by the three distinctly different processes shown in Fig. 1 (ref. 13). The first process, called island growth or

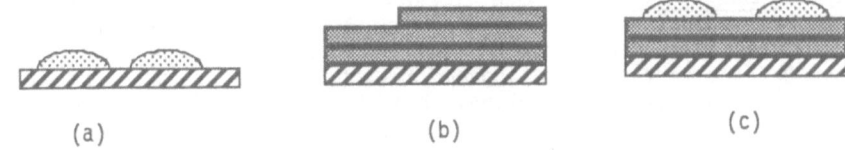

(a)   (b)   (c)

FIGURE 1. Film growth models: (a) Volmer-Weber three-dimensional island
growth; (b) Frank-van der Merwe two-dimensional layer-by-layer growth; and
(c) Stranski-Krastanov mixed mode growth.

Volmer-Weber growth, involves nucleation of distinct and separate three-
dimensional clusters of the condensed phase on the substrate. With time and
continued adatom flux the clusters proceed to grow into larger clusters or
islands, which upon continued growth coalesce into a continuous film. This
growth model is consistent with existence of a nucleation barrier as defined
by traditional nucleation theory. Cluster formation from a flux of adatoms
proceeds along the traditional nucleation theory path where only those
clusters which reach a critical size are stable and grow.
  When the adatom flux onto the substrate, R, exceeds the flux of evaporating adatoms from the substrate, R_e, then film nucleation and growth can
occur. The flux of evaporating adatoms, R_e, from the coating surface is
given by:

$$R_e = (constant) \exp (-Q_{aa}/kT) ,$$

where Q_{aa} is the adatom-adatom binding energy for the bulk coating material.
The adatom jump frequency, v, is proportional to [exp {(-surface diffusion
energy)/kT}] and when the substrate is at an elevated temperature high adatom surface mobility exists. Island growth will occur when the binding
energy between coating atoms, Q_b, is greater than the binding energy between
coating atom and the substrate, Q_a. This is the common growth mode for metal
films grown on insulators.
  The next coating growth model involves two-dimensional layer-by-layer formation of the coating in the apparent absence of a nucleation barrier. This
model, called two-dimensional growth or Frank-van der Merwe growth, is for
the case where coating atoms bind more strongly to the substrate than to
each other, Q_a is greater than Q_b, and adatom surface mobility is high. It
is also important that atomic matching across the coating/substrate interface be good so the interfacial defect density and, correspondingly, the
interphase boundary energy will be low. Such a growth mode is necessary for
heteroepitaxial thin film formation in semiconductor-semiconductor systems
such as $(Ga_{1-x}Al_x)As$ on GaAs and in metal-metal systems such as Cd on W 16).
  The third coating growth mode, called Stranski-Krastanov, is essentially a
combination of the layer-by-layer and island growth modes. In this process
the film begins with layer-by-layer growth and after several layers have
formed epitaxially then growth continues by nucleation of discrete islands.
The basis for this transition is not well understood but it is thought that
lattice mismatch between the coating and substrate could generate stored
elastic energy which would increase with increasing film thickness and would
eventually generate sufficient stress to nucleate defects at the interphase
boundary. These defects could serve as heterogeneous nucleation sites and
thereby inhibit epitaxial film growth after formation of some system-
dependent critical thickness.

2.3.2. <u>Impurity effects</u>. Impurities include defects such as dislocations, ledges, and vacancies on the substrate surface as well as contamination of the substrate surface by processes such as adsorption of unwanted residual gases. Defects in the crystallographic structure of the substrate surface are the rule and not the exception. Even for very carefully prepared surfaces it is likely that the coating nucleation kinetics will be controlled by defect density because these are preferential trapping sites and therefore preferential cluster nucleation sites. The dominance of cluster formation at surface defects will increase with increasing substrate temperature or low incident adatom flux. Controlling the distribution and density of the surface defect sites is possible in plasma and ion beam assisted processes through control of energies and fluxes. This may allow control of coating characteristics such as grain size.

Contamination of the substrate surface by impurities such as hydrocarbons from improper solvent cleaning procedures or a monolayer of water molecules from the residual gas environment in the deposition chamber can significantly affect the film nucleation and growth and other parameters such as film adhesion. An example of the effect of contamination is the case of Cd deposited onto W where contamination of the W surface with less than one monolayer of oxygen reduces the adatom to substrate binding energy relative to the adatom to adatom binding energy and changes the film growth mode from layer-by-layer to island (17). Regarding the effects of contamination on film adhesion, an extensive investigation (18) is in progress to study the effects of substrate surface preparation, including cleanliness, on film adhesion for Al films deposited on a wide range of metallic and non-metallic substrates. The results show improvements of more than 100 times for chemically polished and sputter cleaned surfaces relative to oxidized or otherwise contaminated surfaces. Because adsorbed impurities have relatively weak binding energies (about 1 eV for a water molecule), it is possible to clean impurities and even oxides off surfaces through application of a medium energy plasma or ion beam.

2.4. <u>Coating microstructures</u>

After formation of a continuous surface film and with continued adatom deposition, the microstructure which evolves during evaporation is mainly based on substrate temperature (since this parameter controls the key growth mechanisms of adatom surface diffusion and bulk diffusion) and on surface roughness generated shadowing of surface regions from the line-of-sight evaporated atoms. Increasing substrate temperature increases both surface and bulk diffusion kinetics since the activation energies for these processes are both related to the melting point of the bulk coating material. At high temperatures where surface diffusion is rapid, the effects of topographic shadowing are decreased and a smoother film results. High temperatures also promote recovery, recrystallization, and grain growth processes within the growing film thus producing a more dense and defect-free coating.

2.4.1. <u>Zone models</u>. Several useful studies over the past two decades have examined and developed models to explain PVD coating microstructure evolution. The first detailed microstructural zone model was devised by Movchan and Demchishin (19) for evaporated Ti, Ni, W, ZrO_2, and Al_2O_3 coatings deposited for a series of substrate temperature, T, to melting point temperature, T_{mp}, ratios where the temperatures are in degrees kelvin. Thornton later extended this zone model (20) and made it applicable to sputtering by adding an additional axis to account for gas pressure. In each of these models, multiple zones with distinct microstructures were found to be roughly defined by T/T_{mp} ratios and gas pressures, and the microstructures for the various zones ranged from isolated, columnar crystallites in a low-density film to large, equiaxed grains in a fully dense film.

Zone 1 microstructures, which are porous and consist of tapered crystal-lites with domed tops and separated by voids, occur for low substrate temperatures corresponding to T/T_{mp} values below about 0.15. This open, columnar microstructure occurs because the low substrate temperature does not allow sufficient adatom surface diffusion to compensate for shadowing of the intercrystallite valleys by the crystallite peaks. Transmission electron microscopy studies of numerous metallic coating cross-sections reveal that these micron-sized crystallite columns typically evolve from a very thin (<50 nm) coating/substrate region whose microstructure consists of small (<10 nm), equiaxed grains. Also, the crystallite columns are often polycrystalline with the individual grains being small and equiaxed.

The next microstructural zone, called zone T, occurs for temperatures in the 0.15 to 0.45 T/T_{mp} range and because surface diffusion is easier at these temperatures the films are more dense and smoother than for zone 1. The film still is composed of fibrous, columnar grains which have high defect densities, mainly dislocations and point defects.

At substrate to melting point temperature ratios in the 0.45 to 0.75 range, adatom surface diffusion is very rapid and the films now are fully dense and have a coarse columnar microstructure. The larger columns form because of coarsening from bulk diffusion. The grain size is larger and the intercolumnar regions are essentially grain boundaries.

Above 0.75 T/T_{mp} a zone 3 microstructure exists where coating growth is dominated by bulk diffusion. The final microstructure may consist of large, equiaxed or columnar grains which have low defect densities typical of well-annealed bulk materials. The films will have low internal residual stresses because new stress-free and strain-free grains form easily by recrystallization. Grain growth may produce a large final grain size for long deposition times.

2.4.2. Ion bombardment effects. Experimental and theoretical studies on a variety of material systems over the past several years have shown that bombardment with relatively low energy ions or energetic neutrals during deposition of an evaporated film can significantly increase film density and alter residual stresses. In effect, this bombardment contributes energy to the surface and near-surface atoms of the growing film enhancing atom mobility and producing a microstructure more typical of a much higher deposition temperature.

Using a two-dimensional molecular dynamics model to study the atomic arrangements of atoms deposited onto a crystalline substrate, Muller (22) has clearly demonstrated that epitaxy and density both increase with increasing impinging atom energy. The energies examined ranged from 0.05 of the Lennard-Jones potential (typical of the energy for an evaporated atom) to 1.5 times the L-J potential (roughly equivalent to several times the value for a sputter atom before any collisions occur with gas atoms). At the higher energies density was enhanced because the impinging atoms eroded the tips of evolving columns and thereby prevented development of columns by self-shadowing.

Another modeling study, also by Muller (22), involved evaporation of Ni in combination with an impinging Ar^+ ion beam of specific energy and flux. Again the addition of energy to the growing surface enhanced film density and epitaxy with both increasing as the ion energy increased from 0 to 100 eV and as the Ar/Ni ratio increased from 0 to 0.16. These results are qualitatively consistent with experimental results on a variety of systems. Another interesting effect of low energy ion bombardment is its ability to enhance incorporation of low solubility constituents in a growing film. Greene and coworkers (16) have used this approach to produce epitaxial films at low deposition temperatures and various semiconductor compounds and metallic alloys which cannot be grown by conventional evaporation or sputter deposition techniques.

3. SPUTTER DEPOSITION

Sputtering occurs when highly energetic ions or neutral atoms strike the surface of a solid substrate causing ejection of one or more atoms or molecules by momentum transfer. The sputter deposition process, which has been studied for more than three decades, is widely used because of its simplicity and because it offers capabilities sometimes superior to evaporation for applications such as alloy film deposition. Sputter deposition process has an advantage over evaporation because evaporation is a thermally activated process and sputtering is a mechanical process. The mechanical, momentum transfer nature of sputtering alloys greater control of the resulting atom flux with much less variation in flux between different materials identically sputtered. Many aspects of the sputter deposition process and the resulting films are essentially identical to the just completed evaporation discussion and will not be repeated. The discussion in this section will concentrate only on those unique aspects of sputter deposition. For more thorough reviews of the process of sputter deposition and sputter deposited thin films, the reader is directed to references (3,4,20).

3.1. Equipment and operation

A conventional diode sputter deposition system consists of a vacuum chamber, a target (cathode) which is the coating material source, a substrate holder (anode), a working gas supply, and a high voltage rf or dc power supply. In sputtering the source (target) material is biased to a high rf or dc negative potential, typically around −1000 v, and if the deposition chamber is at a pressure on the order of 1 Pa the target voltage will initiate a plasma. The positive ions within the plasma will be accelerated across this voltage potential striking the target with large enough kinetic energies to knock atoms off the target surface. In addition to the sputter deposition process described above, modifications of this basic process may include: substrate heating or application of a substrate bias voltage to enhance adatom surface diffusion; addition of a reactive gas(s) into the plasma to cause deposition of a compound such as titanium nitride; and incorporation of multiple sputter targets to produce multilayer or chemically graded films. Biasing of the substrate causes low and medium energy ion bombardment, for cleaning, and bombardment of the growing film. The same result is obtained by addition of an independently controlled ion source, although this is geometrically difficult for sputtering.

Although there are several distinct sputter source designs (planar diode and magnetron are the two major types), nearly all sputter sources operate by highly negatively biasing the cathode target to produce a low pressure glow discharge plasma. This plasma is adjacent to the cathode and provides a uniform and copious supply of positive ions which are accelerated toward and strike the target causing the sputter event (3,20). A planar diode sputter source consists of two flat, parallel electrodes, one being the target electrode and the other being the substrate or substrate holder. The substrate electrode serves as the anode in this diode arrangement and will either be grounded or negatively biased to a low voltage (−100 v is a typical value) if low energy ion bombardment of the growing film is desired. Magnetron sputter sources provide major advantages over planar diode sources in the areas of greatly increased rates (values as high as 30 nm/sec have been reported) and reduced substrate heating. A magnetron source is essentially a diode source which has been modified by addition of a moderate strength magnetic field (few hundred gauss) to "trap" secondary electrons emitted by the cathode. The magnetic field causes the trapped, energetic electrons to spiral increasing their path length and thus increasing the probability that they will interact with and cause ionization of a neutral atom. This increased ionization probability allows the glow discharge plasma to be sustained at lower working gas pressures thereby increasing the mean free path and average kinetic energies of sputtered atoms.

3.2. The sputter process

Sputtering is a physical process where highly energetic ions and neutrals bombard a negatively biased target causing ejection of surface and near-surface target atoms through momentum transfer. For atom removal from the target surface to occur, collision of the impinging inert gas ion/neutral must transfer sufficient kinetic energy to a target atom to overcome the local bonding forces. When an energetic ion/neutral strikes the substrate surface, its momentum is distributed by a collision cascade process over primarily the atoms within a few atomic distances from the incident site. For a substrate atom to be ejected it is necessary for the atom to have acquired a sufficiently large momentum component oriented out of the substrate surface.

Assuming a hard sphere elastic collision model, the exchange of momentum among atoms in and around a collision cascade is a statistical process similar to that encountered in billiards. It is expected that creation of momentum vectors in a direction out of the surface will occur by reflection upwards of energetic target atoms, by a sequential series of collisions, and by the lateral momentum of surface atoms. This discussion illustrates how sputtering of a single atom must be a multiple collision process. Considerations of the transfer of momentum between incident and target particles reveals that this transfer is greatest when the masses are equal. Maximum momentum transfer from the incident ions will correlate to maximum substrate atom sputter rate.

Atoms sputtered from a surface will have average energies in the 10 to 40 eV range, with energy increasing almost linearly with increasing atomic number (20). Increasing the energy of the bombarding ions does not significantly increase the average sputtered atom energy. Sputter yield, the number of target atoms ejected per incident ion, is a function of the relative atomic masses, the energy and angle of incidence of the impacting ion, and the target surface topography. Experiments have shown that there is a threshold energy of about 25 eV below which no sputtering occurs and above this threshold energy sputter yield increases steadily with increasing ion energy. Typical sputter yields for metals are in the 0.5 to 2.0 atoms/ion range. Theoretical treatment of sputter yield by Sigmund (21) show good correlation with experimentally measured values. Parameters used in Sigmund's model are: atomic masses of target and incident ions, the energy of the incident ion, and the binding energy of the target atom (typically 5 to 10 eV).

When the incident particles are much lighter than the target atoms the incident particles will reflect from the target surface. The energetic neutrals (ions will generally be neutralized just prior to impact with the target) reflected from the target are very important since at low working pressures they are unlikely to be scattered and can impact the growing coating surface with energies large enough to alter the coating microstructure or to become entrapped within the growing coating. Experimental data (23) for the atomic fraction of Ar entrapped in films of several metal sputter deposited with a range of target atom mass, M_t, to gas atom mass, M_g, ratios show <0.1 at. % Ar for M_t/M_g <1.5 and ~2 at. % for a ratio of 5.

3.3. Coating nucleation and growth

Nucleation and growth of a sputter deposited coating, as discussed for evaporated films, occurs by individual atoms or molecules striking the substrate surface, condensing and equilibrating energetically onto the surface, diffusing about the surface, and clustering to form nuclei which grow and evolve into the final coating. Nucleation and growth examples, for different coating/substrate combinations, have been observed for each of the three basic mechanisms: Volmer-Weber (three-dimensional island growth), Frank-van der Merwe (two-dimensional layer-by-layer growth), and Stranski-Krastanov (first layer-by-layer and then island).

The major differences in nucleation and growth processes between evaporation and sputter deposition lie in the fact that the sputter deposited particles have one to two orders of magnitude greater average energies. This is particularly true for magnetron sputtering and ion beam assisted sputtering, where an additional ion beam source is used in conjunction with the sputter source. The effects of this low energy ion irradiation prior to and during film nucleation and growth include: removal of loosely adsorbed contamination from the substrate surface, a form of sputter cleaning; improved film adhesion due to enhancement of chemical mixing at the coating/substrate interface; increased density of active nucleation sites and decreased film grain size; and enhanced adatom surface mobility. This enhanced adatom mobility has been found to allow reduction of growth temperatures for growth of epitaxial films (16).

3.4. Coating microstructure

Sputter deposited films exhibit all of the microstructural features previously identified for evaporated films. That is, these films follow the zone growth model originally devised by Movchan and Demchishin (19) and expanded by Thornton (20). Thornton's contribution involved addition of a pressure axis to the Movchan and Demchishin microstructure versus substrate temperature (T) divided by melting point (T_{mp}) ratio. This modification was specifically directed to sputter-deposited films. Basically, Thornton found that increasing argon pressure during film deposition was equivalent to reducing the deposition temperature (i.e., reducing T/T_{mp}).

The effect is explained based on the concept of adatom mobility. Reduction of the T/T_{mp} ratio will reduce adatom mobility since this is a thermally activated process (i.e., an activation energy, Q, exists because of the binding forces between the adatom and its neighboring substrate atoms) which is exponentially related to temperature. As discussed previously, increasing pressure causes more collisions of the sputtered atoms to occur prior to impacting the substrate. The initial 10 eV or so energy of the sputtered atom will be reduced by these collisions to a value closer to the average energy of the residual gas (<1 eV). The result of the increased scattering with increasing pressure is to reduce the energy of the impacting particles on the substrate surface and thereby reduce the energy deposited in the surface. This effectively reduces the local substrate temperature and correspondingly reduces adatom mobility. Hence, increasing pressure for sputter deposition raises the T/T_{mp} values needed for transition from zone 1 to zone T to zone 2 to zone 3 microstructures.

Ion assisted sputter deposition involves modification of the microstructure and chemistry of a deposited film through continuous, low energy (~20 to 200 eV) ion bombardment before and during film growth. The continuous collisional mixing produced in the upper few atomic layers by low energy ion bombardment has been shown to be a way to effectively achieve high adatom mobilities at low substrate temperatures and to increase the incorporation probabilities of high vapor pressure or low solubility elements. Extensive studies by Greene and co-workers (16) have clearly shown that low energy ion bombardment greatly alters the growth and compositions of complex semiconductor alloy films such as $InSb_{1-x}Bi_x$ and $(GaAs)_{1-x}Ge_x$. In the later section on ion plating, examples of the microstructural (film density and grain size), microchemistry (chemically graded interface and intermetallic compound formation), and mechanical property (adhesion) effects of low energy ion bombardment will be presented.

4. ION PLATING (PLASMA-ASSISTED PVD)

The ion plating process was developed by Mattox in the early 1960s (24,25) as a means of producing adherent thin film coatings on materials such as uranium. More recently, Spalvin (26) and Mattox (27) have reviewed the ion plating process and its applications. For a thorough discussion of all aspects of ion plating, the reader is directed to the detailed review by Mattox which also includes a very useful bibliography on ion plating literature.

Ion plating is a plasma-aided physical vapor deposition process capable of depositing highly adherent metal, alloy and ceramic coatings onto virtually any substrate at relatively low temperatures. This process incorporates low energy ion and energetic neutral atom bombardment prior to and during coating growth. Sputtering caused by this energetic particle bombardment of the substrate surface prior to coating deposition allows removal of surface contaminants (oxides, hydrocarbons, etc.) and production of active nucleation sites (at defects). Energetic particle bombardment during coating nucleation and growth: (1) aids in formation of compounds such as titanium nitride; (2) produces a structurally and chemically graded coating/substrate interface, resulting in excellent film adhesion, even where relatively large coating/substrate thermal expansion coefficient differences exist; and (3) produces dense, equiaxed-grained coatings at substrate temperatures considerably lower than those required for evaporation or non-biased sputter deposition.

4.1. Equipment and operation

Figure 2 shows the basic components of an ion plating system: (1) a relatively large vacuum chamber with high vacuum capability; (2) a coating source(s), which in this system is a pair of thermionic electron beam evaporation sources but could be anything from a resistance-heated wire source to a hollow cathode electron beam source to a magnetron sputter source (30); (3) a controlled gas source(s), which may include both high-purity inert gas such as argon and reactive gases such as nitrogen or oxygen; (4) a heated/cooled, electrically biased substrate holder, which may have a dc or an rf applied bias; and (5) a ground shield around the substrate holder to control location of the plasma and optimize plasma density. Process control is greatly enhanced by use of such items as a film thickness deposition rate monitor, an optical emission spectrometer for plasma chemistry analysis, and a residual gas analysis system.

4.2. Ion plating process

During ion plating the substrate holder is negatively biased between 500 and 5000 v, and for gas pressures in the range of 1 Pa, a glow discharge plasma is established around the substrate holder. This situation is identical to the diode sputter deposition process except that for sputter deposition the target is the cathode and for ion plating the substrate holder (and similarly the substrate) is the cathode. Ion plated substrates will therefore be bombarded with a flux of positive ions and energetic neutrals. The origin of the energetic neutrals lies in the fact that the relatively high pressures used for ion plating result in short mean free paths and extensive ion/neutral charge exchange reactions. A recent experimental study by Machet and co-workers (28) into the ion energy distribution for ion

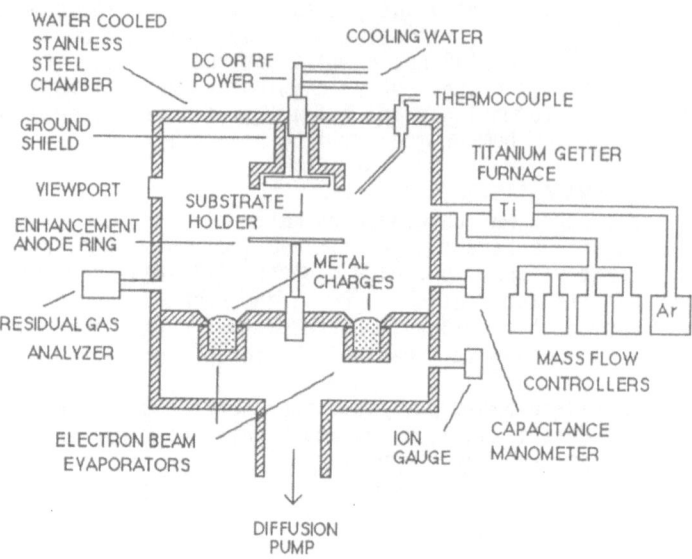

FIGURE 2.   Schematic showing components of a typical ion plating system.

plating shows that for typical chamber pressures and simple substrate
geometries the ions impacting the substrate have average energies of around
10 to 20% of the applied bias voltage. With a typical applied substrate
bias of 1500-v dc, this gives an average ion energy around 200 eV, which is
sufficient to sputter clean the substrate prior to coating deposition. It
is important to remember that the majority of the energy is carried by the
energetic neutrals since the fraction of particles which are actually
ionized at any given instant is estimated at no more than 1% (and usually
much less). Deposition of a coating commences when the flux of atoms depositing onto the substrate from the coating material source exceeds the rate
at which the substrate is being sputtered. Reactive ion plating allows
deposition of ceramic compounds such as TiN, ZrN, and Al_2O_3 through introduction of reactive gases such as nitrogen or oxygen into the deposition
chamber. Dissociation of the reactive gas molecules and subsequent compound
formation with the metal flux are both enhanced by existence of the plasma.

4.3. Coating and growth

Ion plated PVD coatings follow basically the previously described Thornton
and Movchan-Demchishin zone models. Deposition under conditions approaching
evaporation (low levels of energetic particle bombardment) results in
coatings which are porous and columnar when deposited at low T/T_{mp} temperatures and become more equiaxed and dense with increasing deposition temperature. Similarly, with a fixed T/T_{mp} ratio, increasing the degree of
energetic particle bombardment results in increased film density and a more
equiaxed grain structure.

Explanations of these microstructural changes are based upon the effects
of the energies of the depositing ions and energetic neutral atoms. As
discussed previously for ion assisted sputter deposition, accommodation of
this kinetic energy into the first few atomic layers of the surface would
increase the average energy of the surface atoms thereby enhancing adatom
mobility and causing desorption of weakly bonded impurity atoms/molecules.

The effects of this enhanced adatom mobility and improved surface cleanliness would be to improved film density and improved film adhesion. These impacting energetic particles would also sputter the tips of any rapidly growing columnar grains thereby causing the valleys between the grains to be filled and resulting in increased film density.

4.4. Coating microstructure

It is clear at this point of the review that evaporation and sputter deposition PVD processes are in many ways quite similar to ion plating, especially when a bias is applied to the substrate. It is therefore not unexpected that the microstructures of ion plated coatings are frequently similar to those of the other PVD coatings. It is also true that the very wide range of processing parameters possible with ion plating results in a correspondingly wide variation of possible microstructures. Ion plating processing parameters which have the most direct effect on microstructure are applied substrate bias, plasma density (or ion current density at the substrate), deposition rate, and substrate temperature. Critical microstructural factors include film density, columnar versus equiaxed grain structures, residual stresses within the film, interdiffusion/reaction at the coating/substrate interface, and stoichiometry (degree of reaction) of a reactively deposited compound. The following paragraphs will briefly discuss, aided with experimental results, how ion plating processing parameters affect several of the above mentioned microstructural features. Microstructural studies of the films and film/substrate interfaces have principally involved transmission electron microscopy (TEM) on film/substrate cross-sections. Microchemical studies have mainly concentrated on the film/substrate interface and the techniques used include energy dispersive x-ray analysis (in the TEM), Auger electron spectroscopy (AES), and secondary ion mass spectroscopy (SIMS).

Detailed microstructural studies of metallic films ion plated onto metallic and ceramic substrates show that increasing the applied substrate bias and ion current density increases the film density, alters the film grain structure from columnar to equiaxed, and increases film adhesion. As part of a general study of ion plated metal/ceramic interfaces, copper films have been deposited onto cordierite glass ceramic substrates under a range of ion plating conditions (29,30). Cross-section TEM analyses showed increasing film density as the bias voltage was increased from 1 to 3 to 5 kV. The columnar grains became coarser and more defect-free and at 5 kV some equiaxed grains were found. As an additional benefit of the energetic particle bombardment of the growing copper film, the adhesion was markedly improved. Measurements of film adhesion using an epoxy bonded stub pull test technique showed a more than 100 times increase comparing an evaporated copper film to a film deposited with a 3-kV or greater negative bias (30). Copper films deposited with applied substrate biases of negative 3 kV or greater had adhesion failure stresses in excess of 70 MPa (10 ksi).

Development of a chemically graded film/substrate interface was generally observed to occur for both metal/ceramic and metal/metal systems. For the case of copper on cordierite, this interface was typically 10 to 30 nm wide. For aluminum ion plated onto copper, cross-section TEM analysis showed the interface width to be more than 100 nm. Secondary ion mass spectroscopy (SIMS) analysis of copper films evaporated and ion plated onto fully recrystallized copper substrates also indicated that the evaporated interface was quite abrupt chemically while the ion plated interface was much wider. It was further found that an Al-Cu intermetallic compound was formed at the film/substrate interface. This is direct evidence of diffusion which is being enhanced by defects created by energetic particle bombardment during ion plating. The same pseudodiffusion effect promotes

development of chemically graded film/substrate interfaces even when compound formation does not occur.

As a final example of ion plating and the effects that the plasma has on
the deposition process, an ongoing study is examining the structure and properties of mixed Ti and Ru oxides deposited by simultaneous evaporation of
both metals in the presence of an oxygen-rich plasma (reactive ion plating).
Results have shown that a variety of crystal phases can be produced
depending on the operating parameters employed. Figure 3 is an SEM
micrograph of a fracture cross-section of an 85 at. % Ti, 15 at. % Ru oxide
film grown with an rf power density of 3.3 W/cm² on Corning 7059 glass. The
deposition temperature was approximately 400°C, the deposition rate was 0.6
nm/s, the oxygen-to-argon ratio was 3:1, and the chamber pressure was
0.5 Pa. This film, which is clearly columnar and very dense, was unusual in
that x-ray and electron diffraction showed it to be single-phase rutile
(TiO_2), with the ruthenium in solid solution. Films grown under conditions
where increased oxygen levels exist normally are mixtures of titanium oxide
and ruthenium oxide phases (31).

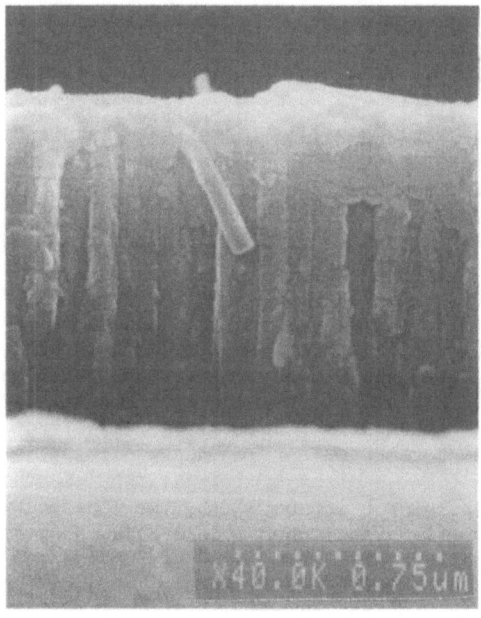

FIGURE 3. SEM micrograph showing fracture cross-section of a (Ti,Ru) oxide
film. The structure is single-phase rutile, with the 15 at. %Ru in solid
solution in the TiO_2 lattice.

5. SUMMARY

The objective of this review has been to summarize our present
understanding of basic and ion/plasma-assisted PVD processes and the thin
films produced by these processes. This review has discussed equipment
design and operation for each process and similarities between the processes
have been stressed. The basic atomic mechanisms of evaporation, sputtering,
and ion plating and film nucleation/growth processes and microstructures

have been presented and discussed, again with similarities being stressed. Finally, references have been suggested to the reader to allow a deeper understanding of specific aspects of the topics covered in this review.

6. ACKNOWLEDGEMENTS

The author gratefully acknowledges partial support for this review by the U.S. Army Research Office under contract number DAAL03-87-K-0006 through the University of Illinois Advanced Construction Technology Center and by the U.S. Army Corps of Engineers Construction Engineering Research Laboratory. This review has been made possible by discussions with and the research of my students and co-workers.

REFERENCES
1. Holland L: Vacuum Deposition of Thin Films. New York: Wiley, 1956.
2. Maissel LI and R Glang (eds): Handbook of Thin Film Technology. New York: McGraw-Hill, 1970.
3. Chapman B: Glow Discharge Processes: Sputtering and Plasma Etching. New York: Wiley, 1980.
4. Bunshah RF (ed): Deposition Technologies for Films and Coatings. Park Ridge, NJ: Noyes Publications, 1982.
5. McGuire GE (ed): Semiconductor Materials and Process Technologies. Park Ridge, NJ: Noyes Publications, 1984.
6. Hill RJ (ed): Physical Vapor Deposition. Berkeley, CA: Temescal, 1986.
7. Bunshah RF: Deposition Technologies for Films and Coatings. Chap. 4. Park Ridge, NJ: Noyes Publications, 1982, p. 83.
8. Maissel LI and R Gland (eds): Handbook of Thin Film Technology. Chap. 1. New York: McGraw-Hill, 1970.
9. Glang R, RA Holmwood, and F. Kurtz: Handbook of Thin Film Technology. Chap. 2. New York: McGraw-Hill, 1970.
10. Steube K and L McCrary: J. Vac. Sci. Technol. 11 (1974) 362.
11. Muehlberger DE: Ion Plating and Implantation. RF Hochman (ed). Metals Park, OH: American Society for Metals, 1986, p. 63.
12. Knudsen M: Ann. Physik 35 (1911) 389.
13. Lewis B and JC Anderson: Nucleation and Growth of Thin Films. New York: Academic Press, 1978.
14. Lee YW and JM Rigsbee: Surf. Sci. 173 (1986) 3.
15. Lee JW an JM Rigsbee: Surf. Sci. 173 (1986) 49.
16. Greene JE: Solid State Technol. 30(4) (1987) 115.
17. Venables JA and GL Price: Epitaxial Growth. Part B. JW Matthews (ed). New York: Academic Press, 1975, pp. 381-486.
18. Savage HS and JM Rigsbee: University of Illinois, to be published.
19. Movchan B and A Demchishin: Phys. Met. Metalog. 28 (1969) 83.
20. Thornton JA: Semiconductor Materials and Process Technologies. GE McGuire (ed). Park Ridge, NJ: Noyes Publications, 1984.
21. Sigmund P: J. Vac. Sci. Technol. 17 (1980) 396.
22. Muller K-H: Materials Modification and Growth Using Ion Beams. U Gibson, AE White, and PP Pronko (eds). Vol. 93. Pittsburgh: Materials Research Society, 1987, p. 275.
23. Thornton JA and DW Hoffman: J. Vac. Sci. Technol. (A)4(6) (1986) 2717.
24. Mattox DM: Electrochem. Tech. 2 (1964) 295.
25. Mattox DM: Design Considerations for Ion Plating. Sandia Corporation Report No. SC-R-65-997, 1966.
26. Spalvins T: J. Vac. Sci. Technol. 17 (1980) 315.
27. Mattox DM: Deposition Technologies for Films and Coatings. Chap. 6. Park Ridge, NJ: Noyes Publications, 1982, p. 244.
28. Machet J, P Saulnier, J Ezquerra, and J Guille: Vacuum 33 (1983) 279.

416

29. Rigsbee JM, PA Scott, RK Knipe, CP Ju, and VF Hock: Vacuum 36 (1986) 71.
30. Scott PA and JM Rigsbee: Ion Plating and Implantation Applications to
    Materials. RF Hochman (ed). Metals Park, OH: American Society for
    Metals, 1986.
31. Suarez JE and JM Rigsbee: University of Illinois, to be published.

ION BEAM ASSISTED DEPOSITION OF THIN FILMS AND COATINGS

F.A. Smidt

Naval Research Laboratory
Washington, DC   20375-5000

1.INTRODUCTION

The bombardment of a growing film with energetic particles has been observed to produce beneficial modifications in a number of characteristics and properties critical to the performance of thin films and coatings such as improved adhesion, densification of films grown at low substrate temperatures, modification of residual stresses, control of texture (orientation), modification of grain size and morphology, modification of optical properties, and modification of hardness and ductility. The earliest systematic studies of these effects are found in the work of Mattox (1,2). Today a wide variety of deposition techniques such as biased sputtering, activated reactive evaporation, ion plating, cathodic arc deposition, and plasma assisted chemical vapor deposition utilize energetic particle bombardment of the growing film to modify the properties of the films (3,4).

In spite of the widespread use of the plasma based techniques listed above, our understanding of the physical processes and mechanisms by which the beneficial changes in film properties are produced is rather limited due to the complex nature of the plasma processes and the inability to independently vary the major process variables (5). A number of reviews and topical articles have been written on ion beam based techniques and the property modifications produced by the techniques (6-21).

One of the first issues which needs to be addressed is that of terminology since a wide variety of terms is used in the literature to describe essentially the same technique. The process of simultaneous film deposition and bombardment with an independently controlled ion beam has been called ion beam assisted deposition, ion beam enhanced deposition, ion beam activated deposition and dynamic recoil mixing. All the above provide a technically accurate description of the process. However, ion assisted deposition and ion vapor deposition which have also been used to describe the process are not accurate terminology since they emphasize the effects of the ionized species rather than the energetic beam and are more appropriately applied to processes such as plasma assisted chemical vapor deposition (22) or ion plating. Ion beam assisted deposition is widely used in the optical thin film literature (14-16) and appears to have growing support among the other segments of the community so it will be adopted in this review. Ion beam deposition is a technique in which the ion beam is also the source of atoms deposited in the film (6,20). This requires a beam in which the charged particle energy is less than the energy at which the sputtering yield exceeds unity if any of the deposited film is to remain. Ionized cluster beam deposition is a technique (11,12) in which clusters of atoms are produced, partially ionized and accelerated to a biased substrate. Ion beam mixing is the process in which a film is first deposited on the substrate and then subsequently bombarded with an

C.J. McHargue et al. (eds.), Structure-Property Relationships in Surface-Modified Ceramics, 417–454.
© *1989 by Kluwer Academic Publishers.*

ion beam to "mix" the film with the substrate by elastic recoils.

The earliest description of an ion beam assisted deposition system appears to be in 1976. Weissmantel (21-23) described the use of a dual ion beam system to deposit films of Si, with one ion gun used for sputtering Si from a target and a second ion gun used to bombard the growing film with a beam of 680 eV N_2^+ ions. Dudonis and Pranevicius (24) presented a paper at the same conference which described a system using an electron beam to evaporate Si or Al films and an auxiliary ion gun to bombard the growing film with oxygen ions at 5 keV. The field of ion beam assisted deposition has thus developed over the past 12 years with earlier origins in ion plating, biased sputtering and other plasma based methods. The past four years in particular have seen a large increase in the number of papers published in the field.

2. ION BEAM ASSISTED DEPOSITION (IBAD) SYSTEMS

A variety of experimental configurations have been used in IBAD systems and the most suitable system depends strongly on the objectives of the investigator's research. All current systems have advantages and limitations which require trade-offs; there is no ideal all-purpose system. The basic motivation for development of the IBAD process is the need for independent control of the major process variables - vapor flux, ion flux, ion energy, system pressure and substrate temperature. The IBAD system combines some method of producing a vapor flux with some method of producing an ion beam with the option of introducing a reactive gas at partial pressures compatible with operation of the ion source.

Four configurations have been used extensively for ion beam assisted deposition experiments. These are (a) a low energy Kaufman source used with an electron-beam evaporator, (b) a high energy ion beam from an implanter used with an electron beam evaporator, (c) a dual ion gun system (where one Kaufman source produces an atom flux by sputtering and the other from a target Kaufman gun is directed at the surface) and (d) an ionized evaporant stream alone or with MBE sources.

An IBAD system of the Kaufman gun/e-beam evaporator type is illustrated in Fig. 1. This system described by Van Vechten et al. (25) consists of a 3 cm Kaufman gun, operated at 0.5-1.0 keV with a N_2 gas pressure of 2.7×10^{-2} Pa and a beam current of 35 mA, and an electron beam evaporator in a cryopumped Bell jar system with a base pressure of 2.7×10^{-5} Pa. The source produced a mixed beam of 89% N_2^+ and 11% N^+ ions at a discharge voltage of 50V. An area of 3 cm diam with a flux uniformity within $\pm$ 5% and a particle flux density up to $1\mu A/cm^2$ was obtained on a substrate holder 31 cm from the source. The deposition rate of Si was monitored with a quartz crystal thickness monitor masked from the ion beam and the ion current was measured with a Faraday cup. These sensors were interfaced with a computer to control the ion source power supply and the evaporator power supply so the ion/atom arrival rate ratio could be accurately controlled. The system has been used to deposit SiN_x films with various predetermined values of x up to stoichiometric Si_3N_4. Similar configurations have been used by Martin et al. (14) and McNeil (26). Advantages of this system are a high ion current, nearly normal incidence of the ion beam on the specimen, large area coverage by the evaporation source, high deposition rate, compatibility with computer control of the deposition system, and relatively low cost. Disadvantages of the system are sensitivity of the beam shape and uniformity to adjustment of the grid system and gun orientation, relatively high gas pressure (10^{-4} Torr), some contamination of the film due to sputtering of the grids, multiple charge states and molecular species and the presence of neutrals in the beam.

A Dual Ion Gun System was one of the first IBAD configurations developed by Weissmantel (23). A system of this type built by Harper, Cuomo and Hentzell (27) is illustrated in Fig. 2. The system uses two ion guns of the Kaufman type, one gun to sputter atoms from the target for deposition on the substrate, and the other to directly bombard the growing film. The sputtering gun usually uses Ar as the sputtering gas and is operated at a fixed voltage. The bombarding gun may use Ar or a reactive gas and typically operates between 0.5 and 1.5 kV. The system is pumped continuously during deposition and the gas pressure is low enough to avoid degradation of the beam but with two ion guns the gas load is higher than the electron beam evaporator system. A quartz crystal film growth rate monitor is used to measure the deposition flux and a Faraday cup is used to measure the ion flux. Kay and his co-workers have used a similar system (28). Advantages of this configuration are a high flux of ions for bombardment of the surface, and low cost of the system. The disadvantages are the same as those listed for the Kaufman/e-beam evaporator system plus cross-contamination of the sputtering and direct beam ions. Another possible complication is separating the effects due to bombardment by energetic ions from the effects due to energetic sputtered atoms.

The third configuration consists of a high energy ion beam, either from an implanter or a dedicated source and lens system, used with an electron beam evaporator. This configuration frequently uses analysis to provide a beam with well-defined mass and charge state (29-31), but a number of non-analyzed high energy systems have also been reported in the literature (32,33). A schematic of the system built by Kant et al. (29) is shown in Fig. 3 to illustrate the geometry of a typical system of this type. A vapor flux from an e-beam evaporator impinges on the substrate at 45° to normal while a beam of high energy ions (30-200 keV) from an implanter bombards the film growing on the substrate, also at an angle of 45°. The film growth rate is measured with a quartz crystal monitor, the current on the sample is measured by a Faraday cup, and feedback to evaporator and implanter power supplies are provided by computer control. Metallic elements and gaseous elements are both available from the Freeman source used in the implanter. Deposition can be performed under ultra-high vacuum conditions or if desired, a controlled partial pressure of a reactive species can be added. The advantages of this type system are the ability to investigate the effects of high energy particles, the ability to homogenize or mix thicker layers of film, the production of an ion beam with single charge state and well-defined energy, versatility in the species which can be implanted, and potentially a cleaner vacuum system that can operate at lower pressure than a system with a Kaufman source. The disadvantages are the low beam current (<300 μA), the inability to reach low energies (unless a deceleration lens system is used), the non-normal incidence of both the ion beam and vapor flux, and the high cost of the system.

Greene and his co-workers have used the ionized and accelerated flux from an effusion cell in an MBE system (34). The ion source is mounted on a UHV flange approximately normal to the substrate surface and the flux is accelerated by a bias on the substrate. Standard MBE effusion cells provide one or more additional vapor fluxes for the synthesis or controlled doping of ultra-high purity flims of electronic materials. The advantages of this system are a beam with energies low enough to prevent damage in sensitive materials (<50 eV) and compatible with UHV operation of an MBE system. Ota 35 (50A) has also used a low-energy ion gun in a MBE system for accelerated doping but his system used a Colutron ion source for gaseous species and included analysis and acceleration capability.

420

3. FUNDAMENTAL PRINCIPLES OF ION BEAM ASSISTED DEPOSITION
3.1. Film Nucleation and Growth

The atomistic processes which can occur on the surface of a crystal are illustrated in Fig. 4 (36,37). Impingement of atoms from the vapor occurs at a rate given by,

$$R = p \, (2\pi MkT)^{-1/2}, \tag{1}$$

where p is the pressure, M is the molecular weight; k is Boltzmann's constant and T is the temperature. The single atoms striking the surface diffuse over the surface until lost by some process such as re-evaporation, formation of a critical size nucleus, capture by an existing cluster or trapping at a special site. Each of these processes has a characteristic time, controlled by the energetics of the process. The surface of a real solid is rarely perfect and defects such as ledges, kinks, dislocations and point defects modify the binding energy of an adatom or small cluster to the surface and hence modify the energetics of the processes, particularly nucleation. Finally, rearrangement of small clusters can occur by interdiffusion with the substrate, surface diffusion to form a more stable shape and annealing of defects.

Several modes of thin film growth have been observed in practice as illustrated in Fig 5. The layer or Frank-van der Merwe mode involves the growth of the film a layer at a time with complete coverage of the surface. The island on Volmer-Weber type of growth results in the formation and growth of islands several layers thick before complete surface coverage occurs. The mixed or Stranski-Krastanov mode is an intermediate case in which a monolayer first forms followed by the growth of islands. The growth mode is controlled by the film-substrate interactions, producing island growth when the condensing atoms are bound more strongly to each other than the substrate and layer by layer growth when the interaction with the substrate is stronger and monotonically decreases as each layer is added (36).

The early stages of film nucleation and growth ultimately influence the microstructure of a thick film or coating. Movchan and Demchishin (38) showed that the microstructure developed in coatings was strongly influenced by the substrate temperature expressed as a fraction of the absolute melting point. Thornton (39) has reviewed the microstructure development concepts for film growth by physical vapor deposition and examined the influence of the structure on mechanical properties of the film. The relationship between microstructure and processing conditions can be expressed concisely in the structure zone diagram as shown in Fig. 6. The diagram shows schematically how the microstructure and surface topography change with the substrate temperature (expressed as a fraction of the melting temperature in $^{\circ}$K) and Argon pressure used during sputtering.

Zone 1 is characterized by columnar grains with pipes along the grain boundaries and poor mechanical strength. The presence of high Ar pressures increases the temperature at which Zone 1 structures are stable possibly due to trapping of the gas at the boundary. The Zone 1 structure results when adatom diffusion is too limited to overcome the effects of shadowing. Shadowing produces open boundaries, particularly under conditions of oblique flux, because the high points receive more flux than the low points. Surface roughness may develop because of substrate roughness, preferential nucleation on inhomogeneities or from the shape of the initial nuclei. Zone T or transition microstructures are also dominated by shadowing effects but have finer structures which appear fibrous and have better mechanical integrity. Zone 2 microstructures form at $T_m=0.3-0.5$ and are dominated by adatom surface diffusion processes. The structure has columnar grains with good mechanical integrity at the boundaries, smooth surface facets and there is little sensitivity to the inert gas pressure. Zone 3 structures which form at $T_m \geq 0.5$ are dominated by bulk diffusion processes.   Recovery and recrystallization processes typically occur in this temperature regime, driven by the minimization of strain energy and surface energy of the grains.   Columnar grains recrystallize to form equiaxed grain in Zone 3.

3.2. Energetic Ion/Solid Interactions.

The processes by which energy is transferred from an energetic ion to a solid are important fundamentals for understanding how particle bombardment interacts with the growing film to modify the microstructure and properties of the film.   The interaction of high energy ions with solids has been extensively reviewed and treated by previous papers in this volume. Thompson (40) and Davies (41) review the production of collision cascades in the linear and high energy density regimes, Anderson (42) reviews sputtering and preferential sputtering, Greene (8,9) also treats sputtering in thin film deposition, Ziegler (43) provides a review of the fundamentals of ion-solid interactions and a practical guide to the use of the computer Code TRIM for such calculations, Averback (44) reviews the effects of the cascade in ion beam mixing, and Rehn (45) reviews segregation and diffusion resulting from point defect fluxes.

The impact of an energetic ion with a solid produces a variety of effects as illustrated schematically in Fig. 7.   The incoming ion loses energy by inelastic collisions which produce electronic excitations and by elastic collisions which displace atoms from their lattice positions.   The displaced atoms produce additional collisions in a short period of time and the collection of events is called a collision cascade.   Cascades which occur near the surface sputter atoms from the surface and cause desorption of adsorbed species.   Vacancies produced in the cascade increase the diffusion rate and interaction between interstitials or vacancies and solute atoms can lead to segregation as the complexes diffuse to sinks. The composition of the surface layers can be changed by implanting atoms, preferential sputtering, ion mixing and segregation.   Secondary electrons and photons may be emitted from excited atoms in the surface and these particles may in turn interact with atoms or molecules adsorbed on the surface.   The energy of the bombarding particle is an important parameter, particularly in the low energy regime, because some processes such as dissociation, desorption, defect production and sputtering have a threshold below which the process does not occur.

4. EFFECTS OF ION BOMBARDMENT ON NUCLEATION AND EARLY STAGES OF GROWTH OF THIN FILMS

It has been known since the earliest observations of plasma assisted deposition and sputtering that bombardment of a surface with energetic particles influences the nucleation and early growth stages of the film (2,46).   Early experiments with ion beam assisted deposition supplemented these qualitative observations.   Marinov (47) showed dramatic evidence for the effect of 1-10 keV Ar^+ ion bombardment of a Ag film deposited on amorphous carbon at room temperature using the experimental arrangement shown in Fig 8d.   Ag atoms from an evaporation source were deposited on the amorphous C substrate and a baffle was used to mask part of the substrate from the ion beam.   Figures 8a through 8c show an increase in island size and a decrease in number density as a consequence of exposure to the ion bombardment.   Bombardment at higher values of surface coverage showed the formation of crystallites with large denuded zones around them.   Other experiments showed ion bombardment increased the transport of deposited atoms under a knife edge, produced a preferred orientation in the film and created nucleation sites on a cleaved rock salt crystal. Marinov attributed the changes in nucleation and growth behavior to enhanced adatom mobility.

Babaev et al. (48) found that IBAD deposition of Zn films on Cu_2O and Sb on NaCl decreased the critical pressure at which Zn deposited on Cu_2O, increased the number of nuclei, increased the extent of surface coverage by the condensate and decreased the size of islands. Pranevicius (49-50) and Netterfied and Martin (52) found that films deposited by IBAD coalesced at an earlier time and for thinner films than for vapor deposition.

Lane and Anderson (52) conducted detailed studies of nucleation and the initial growth of Au films on NaCl using ion beam sputtering. The average sputtered particle energy was quoted as 5 eV. The island density was measured as a function of time, substrate temperature, and rate of deposition using TEM. The results were compared with data from thermal evaporation studies and analyzed using a rate theory approach. A plot of log N_s, the maximum island density, vs 1/T in Fig. 9 shows two regimes. For $T \leq 205^{\circ}C$ the adatom density was controlled by capture at islands while for $T \geq 205^{\circ}C$ the adatom density was controlled by desorption from the surface.

Krikorian and Snead (54) conducted an extensive study of the nucleation and growth of Ge films produced by thermal evaporation and ion beam sputtering, deposited on amorphous C, graphite and CaF_2, and as a function of substrate temperature. The sputtered flux increased the nucleation rate and the number of nuclei. The substrate also has a major effect on the nucleation kinetics with amorphous carbon producing greater acceleration than graphite. This effect was attributed to changes in the adsorption and diffusion energies of the adatoms. At low substrate temperatures the Ge deposit was amorphous and all islands consisted of a single atom. The island density was affected only by the deposition rate and sticking coefficient. At higher temperatures in the normal nucleation and growth regime, the effect of kinetic energy from the sputtered particles appeared as an increase in adsorption energy, migration energy and effective activation energy for nucleation.

Hasan et al. (55) using a partially ionized beam from an effusion cell studied the nucleation and early growth of In films on Si(100) and on amorphous Si_3N_4 in a UHV system. In on Si(100) follows a Stranski-Krastonov growth mode with deposition of 3 monolayers of In followed by growth of polyhedral islands of In oriented along <011> directions. Bombardment of the surface with 300 eV In^+ disrupted the preferred orientation, increased the nucleation rate of equiaxed islands, decreased the thickness for island coalescence, decreased the number of islands and increase the size of the remaining islands. Secondary nucleation, the formation of new nuclei between existing islands, was suppressed. The authors attributed these effects to sputtering and dissociation of small nuclei and possibly to increased adatom diffusivity. Barnett, Winter and Greene (56) showed unambiguously that preferential absorption sites are produced by low-energy ion bombardment and measured the energy for Sb_4 deposited on Si(100) surfaces at $425^{\circ}C$. The normal adsorption energy on this surface was 2.33 eV but the bombardment with 2 keV Ar^+ produced an additional site with an adsorption energy of 2.6 eV.

Sartwell (57) studied the growth mode of Cu deposited on Si (100) under ion bombardment using Auger analysis to monitor the relative signal strengths of Cu and Si. Ion bombardment of the substrate was provided by a second ion gun using 500 or 1000 eV Ar^+. The rate of attenuation of the Auger signal from the Si substrate during deposition of Cu depends on the mode of film growth. Island growth or Volmer-Weber mode would leave large areas of substrate uncovered prior to island coalescence, while layer by layer or Frank-van der Merwe growth quickly covers the substrate. The results of this experiment are shown in Fig. 10 (a-c) with the Cu/Si signal strength plotted as a function of equivalent monolayers deposited. Reference lines on the figures show the calculated trend for Frank-van der Merwe (FM) growth (solid line) and for Volmer-Weber (VM) growth assuming various fractions of surface coverage. The number of atoms deposited on the surface were determined by proton-induced x-ray emission (PIXE). Cu films deposited without bombardment followed a VW growth mode with between 25 and 50% surface coverage (Fig 10(a). Bombardment of the surface with 500 eV Ar^+ ions caused a transition from VM to FM growth mode at between 3 and 5 monolayers (Fig. 10(b). The transition was sharpest for an arrival rate ratio of 0.8 but with a discernable trend at an R of 0.51. An increase of ion energy to 1000 eV produced the surprising result that the growth mode had apparently reverted to VW, and the deposition at a R value of 0.56 showed a lower fraction of surface coverage than at R=0.34. A change from island growth to layer growth can be explained by an increase in nucleation sites or resputtering from the existing islands but the reversion to island growth mode at 1000 eV is not understood.

Gilmore et al. (58) have evaluated possible contributions of a thermal spike to surface processes during ion beam assisted deposition of Au on NaCl. Calculations led to the conclusion that a thermal spike created by a 100 eV Ar^+ ion would have little influence on adatom diffusivity during IBAD. The temperature pulse was limited to a radius of less than 1 nm for a period of the order of 10^{-12}s which would activate only 1.2×10^{-4} atoms per event. At normal adatom densities this would produce a negligible effect on adatom diffusivity.

In summary, there is clear evidence that ion bombardment of a growing film influences the nucleation and early stages of growth. There is however no single pattern of behavior with some observations showing an increase in nuclei density and smaller size and some observations showing fewer nuclei and larger sizes. This suggests nucleation is a kinetic process with several factors controlling the rate equations. Ion bombardment of crystalline surfaces does produce preferred sites with a higher binding energy than normal sites. The presence of strong traps produced by radiation damage would be expected to increase the number of nuclei and decrease the effective migration rate. The creation of preferred sites on amorphous surfaces would appear less likely and it is interesting to note that large increases in island size such as observed by Marinov occurred on amorphous substrates. Island growth can result from an increase in diffusivity or a removal of small size clusters from the population by sputtering. Momentum transfer to adatoms from the beam or from sputtered atoms which would increase mobility seems possible and has been seen in molecular dynamics simulations under specialized conditions (59). More work is clearly required to elucidate the factors in ion beam assisted deposition which control nucleation and early growth of thin films.

5.EFFECT OF ION BEAM ASSISTED DEPOSITION ON MICROSTRUCTURE OF THIN FILMS
5.1.Densification of Films

One of the most important microstructural modifications produced by IBAD is the densification of films deposited under conditions that would nominally fall within Zone 1 of the Structure Diagram. Films deposited in this regime have high porosity and low mechanical strength. Many film properties are indirectly influenced by high porosity so there is great interest in any process which will produce dense films at a lower substrate temperature than required for PVD.

One of the first definitive experiments showing film modification by IBAD was conducted by Hirsch and Varga (60). They observed that Ge films deposited on a glass substrate at room temperature could be made to adhere to the substrate when deposited under IBAD conditions whereas the films would normally spontaneously spall at thicknesses over 2 μm. They conducted a series of experiments in which the critical current density to produce adhesion was determined at Ar^+ ion beam energies from 65 to 3000 eV. The critical current density was determined from the diameter of the adherent film patch and the known beam profile (near-Gaussian). The critical current density was found to have an $E^{-3/2}$ energy dependence. The authors analyzed the experimental results in terms of a thermal spike model in which stress relief in the film was produced by thermally activated atomic rearrangements. Their model yielded an $E^{-5/3}$ dependence and correlated the critical current density with the point at which each atom in the film was subjected to an activation event. Brighton and Hubler (61) have reanalyzed the data using the Monte Carlo computer code MARLOWE. MARLOWE is based on a binary collision model of the cascade and eliminates many of the assumptions and approximations of the thermal spike model used by Hirsch and Varga. Brighton and Hubler found an $E^{-3/2}$ energy dependence that fit the Hirsch and Varga experimental data very well as shown in Fig. 11. The critical current density was found to correspond to the point at which each atom in the film had been displaced once in a collision cascade.

Optical thin films are particularly affected by porosity and low density because the adsorption of water vapor changes the refractive index. Martin et al. (62) deposited ZrO_2 films on borosilicate crown glass using electron beam evaporation, an ion beam of either 600 eV Ar^+ or 1200 eV O_2^+ ions, and substrate temperatures of room temperature and 300°C. These temperatures correspond to 0.1 T_m, and 0.2 T_m, both in Zone 1 of the Structure Diagram. Films deposited without IBAD showed a substantial shift in transmittance when removed from the vacuum system and exposed to air. Films prepared with 1200 eV O_2^+ beams showed a lesser shift which was dependent on current density but no shift for current densities above 200 μA/cm^2. Permeation of water vapor in the films was studied by a nuclear reaction analysis for H content and it was found that films prepared by Ar^+ IBAD at 300°C had a much lower H content than those prepared by thermal evaporation. The density and refractive index of ZrO_2 films prepared by IBAD were studied as a function of processing conditions. IBAD with 600 eV Ar^+ produced a maximum film density at a current density of 50 μA/cm^2 while the refractive index reached a maximum value at 40 μA/cm^2 before decreasing. IBAD with 1200 eV O_2^+ produced a maximum film density at 120 μA/cm^2 (as shown in Fig. 12) and the refractive index retained its maximum value beyond the critical current density. A similar study of CeO_2 at 300°C (63) showed IBAD with 1200 eV O_2^+ increased the packing density from 0.55 to 1.0. Targove et al. (64) explored the correlation of densification (refractive index) of LaF_3 with an energy dependent parameter and a momentum dependent parameter. He found the best correlation with the total momentum transfer, $P_{tot} = J(2m\gamma E)^{1/2}$ where m is the ion mass and γ is the energy transfer factor for elastic collisions.

Kant (65) observed densification of TiN films prepared by reactive IBAD using thermal evaporation of Ti in a N_2 partial pressure of 1.3×10^{-3} Pa, bombardment with 40 keV Ti^+ ions and an arrival rate ratio of Ti ions/Ti atoms of 0.014. A profilometer trace across an area of the film deposited under IBAD conditions and on a portion masked from the ion beam (PVD) showed a large step. Energy dispersive x-ray analysis showed that both the IBAD and PVD portions of the film had the same number of Ti atoms/unit area within experimental error. The step height difference corresponded to a

30% higher density in the IBAD film.

Computer modelling studies by Müller have provided insight into the mechanisms responsible for the densification. A recent conference paper (66) provides a concise exposition of the most important results from papers over the past three years. Müller started with a two-dimensional film growth model of the type developed by Dirks and Leamy (67) to demonstrate the development of a columnar microstructure with voids under the low adatom mobility conditions which lead to shadowing in Zone 1 (68,69) Müller's most recent work (70,71) has used a 2-d molecular dynamics simulation with a Lennard-Jones potential for atom interactions and a zero substrate temperature to examine microstructure, average density and epitaxy. The initial study examined deposition of atoms with low energies in the vapor deposition and sputtering ranges. Densification increased with energy and exceeded a relative density of 0.9 for energies typical of sputtered particles. Particle energies up to 100 eV were investigated to examine effects produced by IBAD using film parameters for Ni. Fig. 13 shows that atomic rearrangements are produced in a collision sequence which results in both a collapse of voids in the structure and transport of atoms on the surface. Packing density was found to increase with arrival rate ratio for a given energy in agreement with experimental observations of a critical current density. These MD simulations show that forward recoils of film atoms to fill voids, or keep them open until filled by the vapor flux, and some contribution to the mobility of surface atoms are the primary mechanisms by which ion bombardment promotes densification of films deposited under conditions of low adatom mobility.

5.2. Grain Size and Grain Morphology

It has previously been noted that the grain size and grain morphology of PVD films is a strong function of substrate temperature, or more specifically the dominant nucleation and growth process in a particular temperature regime. Since IBAD has been shown to modify the nucleation and early growth stages of film deposition, it would be expected to influence grain size and morphology.

Bland et al. (2) and Thornton (39) noted that ion bombardment disrupted the columnar structure and open boundaries in Zone 1, or if heating of the substrate occurred, produced some grain growth. Hibbs et al. (72) conducted a detailed study of the effect of substrate temperature, substrate bias and substrate material on microstructure development in TiN films deposited by magnetron sputtering, the results of which are summarized in Table I. Films grown on high speed steel had a bimodal grain size distribution which was not found in films grown on stainless steel substrates. The bimodal structure was shown to be a result of epitaxial nucleation on carbides in the HSS. Films grown without bias showed substantial growth of the grain population between 550 and 650°C. Addition of substrate bias produced bombardment of the growing film with 300 eV ions at an arrival rate ratio of 0.6. The ion bombardment disrupted the bimodal grain growth in HSS, produced a dense structure with no voids at the boundaries and reduced the grain size to 50 nm for both HSS and SS substrates. The addition of substrate bias reduces the grain size by causing continual renucleation of the grains.

426

TABLE I - Morphology and Microstructure of TiN Films Deposited by Magentron
Sputtering at Various Substrate Temperatures (72)

Bias and Substrate	200°C (0.15Tm)	400°C (0.21Tm)	500°C (0.26Tm)	650°C (0.29Tm)
0 HSS	70,700 nm F+V, Zone T	70,400 nm F+V, Zone T	80,500 nm C+V, Zone II	150 nm F+V, Zone T
-300V HSS	-	-	50 nm F+D, Zone T	-
0 SS	-	-	110 nm F+D, Zone T	-
-300V SS	-	-	50 nm F+D, Zone T	-

HSS = High Speed Steel
SS = Stainless Steel

F = Fibrous Structure
C = Columnar Structure
V = Void
D = Dense Structure

A few studies have examined the effect of IBAD parameters on grain size.
Parmigiani et al. (73) in studies of grain size in Cu and Ag films
deposited by IBAD found the grain size decreased with ion energy up to 50
eV/atom. Zieman and Kay (74) in a study of biased sputtering of Pd films
found that a minimum grain size was produced at an average deposited energy
of 80 eV/Pd atom. Roy et al. (75) found the grain in size of Cu decreased
with increasing R at 600 eV but was independent of R at 62 eV as shown in
Fig. 14. Data on AlN (76) prepared by dual ion beam sputtering showed an
increase in grain size of AlN with increasing ion energy in the range 100-
500 eV with an attendant change in morphology from elongated fibrous grains
to equiaxed grains.

Films grown under conditions of biased sputtering or ion beam assisted
deposition generally show a decrease in grain size in both Zone 1 and Zone
T temperature regimes and a disruption of columnar grain structures. This
reflects the influence of ion bombardment on nucleation of grains.
Occasional observations of an increase in grain size indicate other factors
which might promote grain growth, such as a higher substrate temperature or
high strain energy, are present and may dominate in certain materials.

5.3.Development of Texture in Thin Films

A frequently reported observation from experiments on the deposition of
thin films by IBAD is the development of a preferred orientation   or
texture in the films. A number of studies on texture development under ion
bombardment have been conducted, eg.,   Ag (47,28,77,78), Au (78), Cu (78-
80), Ni (81,82), Si (83), Ni-Fe alloys (84), AlN (76) and TiN (65,85).

Dobrev (78) recognized at an early stage that the development of texture
was correlated with the channeling directions of ions in the crystal
lattice and that the density of energy deposition would be inversely
related to the depth of channeling. Thus, in a FCC crystal, the ease of
channeling is in the order <110>, <200>, <111>. Dobrev suggested that in a
polycrystalline film the crystallites with easy channeling directions
aligned to the beam would remain the coolest in the resulting thermal spike
and would thus serve as nuclei for recrystallization of the surrounding
matrix as illustrated in Fig. 15.

Van Wyk and Smith (79) independently came to the same interpretation
based on their studies of the development of a preferred orientation in Cu

films which were vacuum evaporated and then bombarded with 40 keV Cu^+ ions.

Harper et al. (80) observed the development of a strong (110) texture, with azimuthal orientation as well, in bcc Nb films deposited in a dual ion beam system when the secondary 200 eV Ar^+ ion beam was oriented at 20° from the glancing angle. This orientation allowed (110) planar channeling to produce the constrained azimuthal texture without disrupting the (110) fiber texture normal to the substrate. The degree of orientation increased with current density (or arrival rate ratio) but was independent of temperature. It was noted that 1 ion per film atom was required to produce a preferred orientation as compared to 0.03 ions/atom to relieve the stress in a vapor deposited film.

Bradley et al. (86) have developed a model to explain the development of preferred orientation due to low energy ion bombardment during film growth which is based on the difference in sputtering yields for different orientations rather than reorientation during recrystallization. Both effects are of course based on the same phenomena of the variation of energy density with channeling direction. This model starts with the assumption that one crystal axis is fixed normal to the plane of the specimen but azimuthal orientations were random. Crystallites with high sputtering orientations are removed more rapidly and newly deposited material grew epitaxially on the low sputtering yield orientations.

Typical observations on ion beam induced texture development can be summarized as follows: Thin films of PVD deposited materials normally deposit with the planes of highest atomic density parallel to the substrate so FCC films have a <111> texture, BCC films have a <110> texture and hexagonal close packed films have a <0002> texture (for ideal c/a ratios). The easiest channeling directions in each structure are as follows (87):

> FCC - <110>, <100>, <111>
> BCC - <111>, <100>, <110>
> HCP - <1120>, <0002>

Ion bombardment causes a shift in the preferred orientation to alignment of the easiest channeling direction along the ion beam axis. Thus, an ion beam at normal incidence on an FCC film will cause a shift in orientation from <111> to <110> texture. A beam incident at an angle will produce a different texture depending on the crystallography. Kant (65) for example, saw a shift in texture from <111> to <100> when a TiN film (FCC) produced by IBAD was bombarded at 45° to normal and Nagai et al. (84) strengthened the <111> texture by bombarding Ni-Fe films at a glancing angle. The texturing effect appears to be most sensitive to high energy beams because of the larger volume affected per ion and the deeper penetration. There is a threshold for the onset of texturing as noted in the modeling done by Bradley et al. (86) and in studies with particularly low arrival rate ratios (28,77,74) no change in orientation was observed. Substrate temperature was also found to be important and at higher temperatures ($500^\circ C$) the preferred orientation of Cu films was lost due to large scale recrystallization (78).

5.4. Epitaxial Growth

Ion bombardment has been observed to influence the epitaxial deposition of thin films in numerous examples as noted by Greene (9). Babaev (48) reported a decrease in epitaxial temperature from $230^\circ C$ to $150^\circ C$ for Sb deposited on NaCl, Pranevicious (49) found a decrease in epitaxial temperature from $650^\circ C$ to $250-300^\circ C$ for 50 eV Si^+ on Si, Narusawa et al. (88) reported a decrease in epitaxial temperature of 100 to $150^\circ C$ with a increase in ionized flux in an ion plating system, Itoh et al. (89) reported a decrease in epitaxial temperature for Si on (111) Si above a critical current density and Ueda (90) reported several additional

428

examples. Yagi et al. (91), Ota (92) and Aleksandrov, (93) all report
lowering of the epitaxial temperature by ion beam deposition of Si or Ge.
Yamada et al. (94,95) have successfully grown epitaxial films of Al on
(111) and (100) Si at room temperature using the ionized cluster beam
technique. Perhaps the most complete examination of the effect of process
parameters on film properties was a study by Yamada and Tori (96) using an
accelerated beam of ionized SiH_4 from an ECR microwave source to deposit Si
on Si (111). The epitaxial temprature was lowered from 550°C to 425°C with
300 eV ions but increased to 475°C for 400 and 500 eV ions. Mechanisms
suggested for the lowering of temperature are break-up of native oxide film
on the surface which makes nucleation on clean metal easier, addition of
damage which produces nucleation sites and the addition of energy to
convert a physisorbed atom to a chemisorbed atom. Müller (71) has
suggested that the degree of perfection in films contributes to epitaxy and
has shown by 2-d molecular dynamics calculations that the degree of epitaxy
is influenced by arrival rate ratio and energy, with high perfection films
formed by 100 eV ions at an R=0.16.

5.5. Defect Production and Annealing in Ion Beam Assisted Deposition
Experiments

Relatively little work has been published on the production of defects
and their annealing during ion beam assisted deposition. This is more
likely due to the relative newness of the field than the absence of the
effects known to occur in neutron and charged particle irradiations. In
addition most IBAD films have been grown in a limited temperature range
because of experimental constraints, i.e., room temperature if the
apparatus had no capability for heating the substrate.

One series of experiments that has provided direct evidence of defect
production and annealing applicable to ion beam assisted deposition is the
direct ion beam deposition of Si and Ge films conducted by Appleton's group
at Oak Ridge National Laboratory (20,97,98). A 40 eV Ge^+ deposition on
Si(100) at room temperature produced no discernable damage in the Si
substrate thus indicating a possible damage threshold.

Another paper of relevance to defect production and annealing on IBAD
films is the work of Huttman et al. (99) in which epitaxial TiN films were
grown on MgO substrates by reactive biased magnetron sputtering. At 0
bias, the damage varied from a high density of loops ($5X10^{12}/cm^2$) at the
lowest temperature where epitaxial growth occurred (550°C) to a density of
$1.5X10^{10}/cm^{-2}$ at 850°C. The defect density was found to be a function of
both bias voltage and substrate temperature. At low bias, and low
temperature defects were primarily introduced through imperfections in the
crystal growth because of low adatom mobility. As the bias increased,
additional defects were produced by the ion bombardment and diffusion was
enhanced so that the dislocation structure annealed out until the defect
production rate became so high the defects could not anneal out and
extended defects nucleated and grew.

6.ION BEAM ASSISTED COMPOUND SYNTHESIS AND DEPOSITION

Ion beam assisted deposition provides the ability to precisely control
the material fluxes as well as the processing conditions so it has found
use in the synthesis of compound of thin films for a variety of
applications. The problems of film deposition for controlled
stoichiometry, are shown schematically in Fig. 16. The composition of the
growing film is the net result of the deposition of atoms, implantation of
energetic particles (ions and neutrals), the impingement of atoms from the
gas phase and the loss of material from the surface by ion reflection,
sputtering and desorption. The sputtered flux is of course dependent on
the surface composition and so internal diffusion and segregation, perhaps

enhanced by the radiation, must be considered. The deposition is monitored by probes such as Faraday cups and deposition rate monitors which may require corrections to give the actual fluxes on the film surface. Compound synthesis by IBAD tends to fall into one of three types, allowing some simplification of the general case. If an inert ion bombards the surface in the absence of a reactive gas the primary effect on composition is through sputtering (8,62). If one of the constituents is added in the ion beam by implantation, the ratio of fluxes in the ion beam and vapor stream control stoichiometry, at least to a first order approximation (27,100,101). If a high energy ion beam or an inert gas ion beam is used to stimulate the surface at low values of arrival rate ratio in the presence of a reactive gas, then the composition is primarily controlled by reaction of the gas on the surface (29.102,103,104). This case is termed reactive ion beam assisted deposition by analogy with reactive evaporation and reactive sputtering.

7.APPLICATIONS OF ION BEAM ASSISTED DEPOSITION

The control of microstructure and film properties afforded by ion beam assisted deposition offer many possibilities for thin film applications. Applications which have been investigated include stress control, adhesion enhancement, modification of surface mechanical properties, improvement of optical properties and modification of electrical properties including the synthesis of superconducting thin films. The most advanced of these applications is the deposition of thin films for optical applications.

7.1. Stress Control in Thin Films

Thin films deposited by PVD, CVD, and electrodeposition processes all have a residual stress in the film (105) which may be manifested by cracking or spalling of the film or by bending of a thin substrate. The stress may arise from a mismatch in the coefficient of thermal expansion between the substrate and the film if deposited at high temperature and measured at a lower temperature or from a mismatch in the elastic modulus if deposited on a substrate under stress. Another contribution to the residual stress is the intrinsic stress in the film which may be caused by a variety of effects such as contamination by impurities, the presence of defects, removal of defects after deposition and the occurrence of solid state transformations. The intrinsic stress can be influenced by film deposition conditions and control of stress is often one of the properties of concern when optimizing film deposition parameters. The residual stress in a film may also influence the adhesion of the film to the substrate.

The ability of ion bombardment to relieve the stress in vapor deposited thin films of Ge (60) has already been discussed in the section on as densification. Hirsch and Varga used the critical current density to prevent film spallation as a sensitive index of microstructural changes and stress relaxation in the film. They attributed the stress relaxation to annealing of the film by thermal spikes.

A number of studies have measured the stress in the film as a function of process parameters using bending of the substrate or x-ray diffraction techniques to determine the stress. Hoffman and Gaerttner (106) deposited Cr films while simultaneously bombarding the film with 11.5 keV Ar^+ and Xe^+ ions. The stress was found to vary as a function of ion to atom arrival rate ratio and at an R value of about 0.003 the stress reversed from tensile to compressive. The critical dose to produce stress reversal was studied as a function of substrate temperature (70, 300°C) ion mass (Ar^+, Xe^+) and energy (3.4, 11.5 keV) and found to correlate best with a momentum transfer parameter. Cuomo et al. (107) studied the stress changes in Nb films deposited by IBAD over a range of process conditions, 100-800 eV Ar^+ ion energy, 30-400°C substrate temperature, and arrival rate ratios from

.003 to 0.1. The stress was tensile in films deposited at 400°C with no bombardment and decreased to 0 or even became compressive with increasing current density. Specimens deposited at temperatures of 30-100°C had a compressive stress which became tensile with increasing ion current. Films deposited at 150 and 200°C had tensile stresses with no ion bombardment, reached a maximum tensile stress at low current densities and then decreased in stress at higher current densities. Cuomo et al. attributed the stress relief to two factors; a removal of oxygen impurities which produced tensile stresses and an annealing or compaction of the structure. Critical arrival rate ratios to produce stress changes were in the range .03 to 0.06 and an energy/atom value of 3 eV was calculated (for R=.03).

The IBM San Jose group has used x-ray diffraction techniques to measure lattice expansion, ($\Delta d/d$) and stress in Ag (28), Pd(74) and Cu, Ag, Pd and Ni films (108) produced by biased sputtering (74) and dual ion beam deposition (28,108). An expansion of the film normal to the substrate peaked at an average energy per atom of 40 eV, decreased slightly and then remained constant. The expansion normal to the film resulted in a contraction in the plane of the film to produce a compressive stress of 450 MPa in Ag films (28). Yee et al. (109) studied stress relaxation in WSi_2 films deposited by IBAD and found the film stress to depend on substrate temperature (amorphous or crystalline film) and arrival rate ratio.

A few measurements of stress have been made on dielectric films of SiO_2 and TiO_2. Allen (110) found the stress of amorphous SiO_2 films was compressive and decreased with increasing O_2^+ beam current, while amorphous TiO_2 films had a tensile stress which increased with beam current. McNeil et al. (111) measured stress in SiO_2 and TiO_2 films. SiO_2 had a compressive stress which was reduced slightly by increasing arrival rate at 30 eV and 50°C, 700 eV and 50°C, and 700 eV and 250°C. TiO_2 films also had a compressive stress which was not effected by increasing beam currents at 30 eV and 100°C. McNally et al. (112) measured the stress change in Ta_2O_5 films deposited by IBAD and found compressive stresses which increased with O_2^+ current density for bombardment with 300 and 500 eV ions at 100°C. Jacobs et al. (113) studied the effect of IBAD on adhesion and intrinsic stress in MgF_2 films deposited at 30°C with bombardment by 50 and 300 eV Ar^+ ions at an ion/atom arrival rate ratio of 0.45. Stresses in the IBAD films were relatively low and attributed by the authors to the incorporation of impurities in the film (7-10% oxygen).

The most detailed study of stress relaxation produced by IBAD was conducted by Wojciechowski (114) using a laser triangulation method to monitor deformation of a thin Kapton disk in situ. Deposition parameters explored included energy - 170-1500 eV, R-.022-2.2, energy/atom-37-2000 and temperature 100-250°C. The stress changes observed in a permalloy (Fe-Ni) film bombarded with 300 eV Ar^+ ions are summarized in Fig. 17. The stress in films deposited without IBAD was tensile at thickness up to 3.5 nm, became compressive from 5.5 to 8 nm and then became tensile again for all thicknesses above 8 nm. IBAD increased the tendency toward compressive stress and films deposited with R>0.22 had a compressive stress at thicknesses up to 100 nm. Similar trends were observed with Ge and SiO_2 films. In general there were 3 regions of film growth which exhibited unique stress behavior, (a) an interface region approximately 5 nm thick which was dominated by the substrate, (b) an intermediate region from 5 nm to several 10's of nm where the stress was compressive and quite sensitive to IBAD conditions, and (c) a bulk film growth condition where little change occurred in the intrinsic stress. A rather surprising result of these experiments was the finding that the critical current density to produce a compressive stress in the films was not sensitive to ion energy (or to energy/atom).

Stress modification by ion beam assisted deposition is a complex phenomena which under various conditions has been shown to include relaxation or annealing of defect structures in films deposited under conditions of low adatom mobility, the addition of impurities to the film, the removal of impurities from the film by preferential sputtering, and the implantation of the beam ions. Films deposited at elevated temperatures must of course be corrected for the thermal stresses to get the true intrinsic stress. Cascade annealing effects have been shown to follow an $E^{-3/2}$ energy dependence and to occur at R values of .01 to .001 (60,61). Stress changes due to ion implantation are expected at substantially higher R values and the depth dependence of stress would depend on factors such as the energy and the size of the implanted atom. Impurity effects are likely to occur at intermediate R values.

7.2. Improvement of Adhesion

One of the major advantages claimed for plasma based techniques such as sputtering and ion plating is the improvement in adhesion of the coating (2-4). This improvement in adhesion would be expected to apply to IBAD coatings as well, but only a few studies have been conducted to date and none of those have included a comprehensive study of the effect of processing conditions on adhesion. However, extensive work has been done on the effects of post-deposition ion bombardment and ion mixing on promoting adhesion. An excellent review of the subject and a list of 33 references to experimental data is given by Baglin (115). The adhesion of a coating has two major factors involved, the residual stress in the coating and the strength of the bonding between the coating and the substrate. The effect of IBAD on the intrinsic stress in a thin film was discussed in the previous section and the effect on bonding will be considered in this section.

Enhanced adhesion by ion bombardment was reported by Collins et al. (116) for Al on soda glass as early as 1969 and a patent was granted to Gukelburger and Kleinfelder for the process in 1972 (117). However systematic studies of adhesion improvement have only recently been performed. One of the first issues to be addressed was the relative importance of nuclear stopping processes compared to electronic stopping processes. Baglin et al. (118) bombarded Cu films on Al_2O_3 with 250 keV Ne^+ and 200 keV He^+ ions and measured the bond strength with a peel test. The energy deposited into electronic processes was nearly equal for the two ions while the energy deposited in nuclear (elastic) collisions was an order of magnitude larger for the Ne ions. The results are shown in Fig. 18 for both the initial experiments and a post-bombardment anneal. The peel strength for the initial bombardment is seen to increase with dose up to 1×10^{16} ions/cm^2 and then saturate. The Ne^+ produced a 10x larger increase in strength than the He^+ bombardment in accord with the ratio of nuclear stopping powers. A post-irradiation anneal produced an even greater increase in strength but with a reversal of the effectiveness of the ions. The increase in bonding after the anneal was attributed to a rearrangement of atoms at the interface to reform broken bonds into a stronger bonding configuration. The formation of Ne bubbles at the interface may weaken the bond for that system. Ingram and Pronko (119) found a very strong correlation between nuclear energy loss and the threshold dose for adhesion of the scotch tape test for Cu films deposited on Mo. The threshold dose for adhesion correlated with the magnitude of nuclear energy loss for ions ranging between 5 MeV Au^+ and 2 MeV He^+ and was between 1.1 and 2.9 displacements per atom (dpa) for all ions except He, for which the films did not adhere.

Tombrello and coworkers (120,121) have bombarded a number of systems of metal films on dielectric, semiconductor and metal substrates with MeV ions and observed an increase in adhesion using the threshold dose to pass the scotch tape test as the failure criteria. A model was proposed in which the improved adhesion was attributed to formation of new interface bonds as a result of electronic excitation. The formation of bonds requires electron transfer across the interface and the transfer was described by the Richardson-Dushman equation,

$$i \alpha T^2 exp(-\phi/kT), \qquad (2)$$

where T is a local temperature determined by the energy deposition and ϕ is the work function. Subsequent experiments have raised questions about the general validity of the equation and the failure to include nuclear collision effects limits the utility of this model.

The importance of contamination effects was demonstrated by Baglin (122) for Cu films deposited on Al_2O_3 after several surface treatments and ion bombardment. The treatments consisted of a sputter cleaning of the surface and (a) exposure to O_2, (b) wash with water or (c) wash with ethanol. The results, indicated that ion bombardment was ineffective in improving adhesion of a hydrocarbon contaminated interface, produced no improvement on a well bonded surface produced by heat treating and produced intermediate levels of enhancement for other conditions.

A final insight into the role of ion beam enhanced adhesion is the work of Dallaperta and Cros (123) who studied the adherence of Au films on Si and native oxide SiO_2 using a UHV system and electron irradiation of the surface. They found excellent adhesion of Au on clean Si with no enhancement by electron bombardment. Au films did not adhere to an oxidized surface at all unless bombarded by electrons. Auger electron spectroscopy showed the role of the electron bombardment was to reduce the SiO_2 and permit the formation of Au-Si bonds.

Among the few examples of ion beam assisted deposition in which adhesion has been addressed, Franks et al. (124) appears to have been the first to apply the technique with improvements noted for Au deposited on Si and Ge and Pt deposited on Ge using a scotch tape test and a pin pull-off test to measure adhesion. Hirsch and Varga (60), discussed in the previous section, studied the effect of current density and ion energy on the spontaneous spallation of Ge films on glass. Improved adhesion was attributed to relaxation of the intrinsic stress in the film by cascade annealing. Nandra et al. (125) studied the adhesion of Au films deposited on Cu substrates under various IBAD conditions. Adhesion was highest for coatings which were sputter cleaned prior to deposition or sputter cleaned and deposited under IBAD conditions.

McNally et al. (126) reported improved adhesion of SiO_2 and MgF_2 films deposited on heavy metal fluoride glasses which are very soft and sensitive to heat treatment. A film deposited by IBAD with 300 eV O_2^+ ions at 150°C produced abrasion resistant and adherent films on sputter cleaned substrates. Jacobs et al. (113) studied adhesion of MgF_2 films deposited on fused silica by IBAD using a pull-off test. The best adhesion was obtained with 10 eV Ar^+ ions at an R value of 0.11, while the lowest value was obtained for 600 eV ions at an R value of 0.9. Ensinger et al. (127) deposited Si films on steel substrates and compared several post-deposition ion bombardment conditions with IBAD using 6 keV Ar^+ ions. Samples cleaned with a chemical etch had poor adhesion while all samples with ion bombardment and ion beam assisted deposition exceeded the bond strength of the epoxy.

Kant et al. (102) reported substantial improvements in the adhesion of TiN films prepared by reactive IBAD with high energy (30 keV N_2^+) ions. A graphic illustration of the effect is shown in Fig. 19 where the film on the left deposited under PVD conditions delaminated when a diamond indenter was moved across the surface while the film on the right deposited with IBAD showed excellent adhesion. The film prepared by IBAD remained adherent to the AISI 52100 even after deformation to depths 20 times the film thickness at a load of 50 N. Subsequent tests under cavitation erosion conditions (29) also demonstrated the adherent nature of these films. Sato et al. (32) have also reported improved adhesion of TiN films on an Al alloy when prepared by ion beam synthesis with a 40 keV N_2^+ beam.

In summary, it is clear that good adhesion requires a clean surface as a starting point and sputtering during ion beam assisted deposition promotes this cleanliness. Atomic displacements at the interface can promote break-up of thin contaminant layers and cause rearrangement of bonds at the interface more effectively than electronic excitation by high energy ions although these too have an effect. The nature of the interface interactions is highly specific to the film/substrate systems involved and is difficult to generalize.

7.3. Modification of Surface Mechanical Properties

Ion beam assisted deposition has the potential to modify surface mechanical properties such as hardness, friction, wear and fatigue through several of the process characteristics already discussed. Improved adhesion is a critical factor for the development of hard coatings for wear and fatigue applications since non-adherent coatings generate debris that accelerates the failure process. The low process temperature is also of critical importance in depositing coatings on substrates that may be softened or otherwise damaged by overheating. The synthesis of unique or metastable materials in thin film form such as cubic boron nitride or diamond is also another potential attraction of IBAD. Finally, the control of microstructural variables such as columnar grain morphologies, grain size and film orientation, provide another incentive for use of ion beam assisted deposition.

Kennemore and Gibson (128) conducted qualitative abrasion tests on MgF_2 films deposited on quartz with and without IBAD and found films prepared at room temperature by IBAD and at 300°C without IBAD were hard and abrasion resistant while those deposited at room temperature without IBAD were soft and scratched easily. McNally et al. (126) deposited MgF_2 films on soft heavy metal fluoride films at 100°C using IBAD and found excellent abrasion resistance in the film. Colligan et al. (129) reported improved wear for Si-N films deposited on steel substrates.

The system which has received the most study for surface mechanical properties is TiN. Kant et al. (29,102) produced highly adherent films of TiN by reactive IBAD using high energy ions. These films had a coefficient of friction of 0.2 in contact with a steel slider and were found to be ductile and surprisingly soft (21). A study of friction, wear and abrasion resistance of 0.2 μm thick TiN films deposited on a variety of substrates (M50, 52100 steel, B_4C, Al_2O_3, Ni) showed that the IBAD films were 25% softer than reactively sputtered TiN and had an abrasion resistance with 3 μm diamond polishing compound only 1/25 that of reactively sputtered TiN (130). IBAD films deposited on SiC were tested in pin-on-disk wear tests at contact stresses greater than 1 GPa (140 ksi) and were found to have good wear resistance because of the excellent adhesion of the films. Thin coatings of reactively sputtered TiN failed by decohesion while IBAD films of comparable thickness maintained their integrity. No significant differences in surface mechanical properties have been seen for R values from 0.001 to 0.2, ion energy values from 30 to 190 keV, or bombardment with N^+ or Ti^+ ions. The only significant differences noted between samples were an increase in the breadth of the interface with increasing R values and a decrease in oxygen content of the film from 30 at% at R=0.002 to 10 at% at R=0.2 (29). Additional experiments on the synthesis of TiN by reactive IBAD (103) revealed the critical importance of control of the nitrogen gas flux to the growing film. A low energy IBAD system with a Kaufman gun producing a 500 eV Ar^+ beam was used to activate surface reactions with N_2 at a partial pressure of $2x10^{-4}$ Pa. Both the N_2 molecule to Ti atom arrival rate and the Ar ion/Ti atom arrival rate ratios had to be controlled to produce golden colored, hard TiN films. Optimum conditions were a N_2/Ti ratio of 0.5 and an Ar^+/Ti ratio of 0.3-0.4. Films produced under these conditions had very low oxygen contents and a Knoop hardness of 25.5 GPa with a 0.05 N load.

Sato et al. (32) deposited TiN films up to 15 μm thick on an Al-11 Si substrate by ion beam synthesis using a 20 keV beam of unanalyzed N_2^+ and N^+ ions to introduce the nitrogen. Microhardness and pin-on-disk wear tests showed the films to be hard and wear resistant. These films had less oxygen contamination than those made by Kant but were a mixture of TiN and TiN_x (X>1).

The hardness of TiN films is known to be affected by composition and microstructure (72,131). The hardest films are formed in the 2-phase region with Ti_2N and TiN. Pure stoichiometric TiN has a Vickers hardness of 19.6-21.6 GPa while films containing Ti_2N approach 39.2 GPa (131). Reactively sputtered films deposited under conditions where voids were created at grain boundaries were observed to soften the films (72). High magnification TEM micrographs of the soft films produced by Kant showed a fine grained structure of equiaxed grains 7.5 nm in diameter and a through-focus series showed the grains to be outlined by a network of micropores (21). A similar network of pores has been seen in amorphous Ge films produced by biased sputtering (132). This fine-grained microstructure outlined by pores appears to be responsible for the unusually high ductility and low hardness of these films. Hard TiN films with no porosity at the grain boundaries have recently been produced by reactive IBAD by precise control of the N_2 partial pressure (103).

A new area of interest in surface mechanical properties developed in 1985 with the discovery that ion beam treatment of solid lubricant films improved their properties. Hirano and Miyake (133) found that ion beam mixing of a WS_2 film deposited on 440C by sputtering and subsequently ion beam mixed with 50 keV Ar^+ showed a coefficient of friction in vacuum as low as 0.01, compared to .04 for the sputtered film, and showed double the wear life in a reciprocating slider wear test. The improvement was attributed to the formation of WS_2 crystallites in an amorphous matrix. Kuwano and Nagai (134) subsequently used dual ion beam sputtering to deposit dual layer coatings of MoS_2 and hard coats of B, TiB_2, B_4C and BN on steel substrates. Low coefficients of friction were observed even in air and were maintained for 10^4-10^5 cycles. All coatings bombarded with 1.5 keV Ar^+ ions during growth were crystalline while those not bombarded were amorphous. However, bombardment with 2.0 keV ions appeared to introduce some damage in the films. Kobs et al. (135) used high energy (up to 400 keV) ion beam mixing of sputtered MoS_x (x = 1.6-1.9) films deposited on M2 tool steel to improve the wear life of solid lubricant films in a dry nitrogen atmosphere. A 3-fold increase in wear life was found when the energetic particle penetrated the substrate interface. Other structural changes in the films were a 40% increase in density over the sputtered film and a reorientation of crystallites of MoS_2 so their c-axis was in the plane of the film. Mikkelsen et al. (136) have shown that the decreased friction is due to the formation of an amorphous film.

7.4. Improvement of Corrosion and Oxidation Resistance

Advantages in the use of ion beam assisted deposition for corrosion and oxidation protection would be expected to derive from the denser microstructure, improved bonding at the substrate and the possibility of depositing films with unique properties such as amorphous films with no grain boundaries. Porosity in electrodeposited and vapor deposited films is a particularly troublesome failure mode so the enhanced surface coverage observed for IBAD films appears promising.

Nandra et al. (125) studied the microporosity of Au films deposited on Cu by an electrodeposition technique and found that coatings deposited by IBAD with 5-6 keV Ar^+ bombardment at R values of 0.038 or higher produced a significant reduction in porosity. Martin et al. (135) reported that Al and Ag films overcoated with thin dielectric films of SiO_2, Al_2O_3 and ZrO_2 deposited by IBAD gave significantly longer protection against attack by 0.2 M NaOH (Al) or 1.0 M HNO_3 (Ag) than non-bombarded films. SiO_2 and Al_2O_3 films lasted 8 times longer while ZrO_2 films in both solutions lasted 200 times longer. Ensinger el al. (127) studied the corrosion resistance of Si films deposited on steel using ion mixing and IBAD techniques. An untreated Si coating gave a value of 280-300 $\mu A/cm^2$ for the anodic current density in a buffered acetate solution while ion beam mixed samples reduced the current density to 5-12 $\mu A/cm^2$. An IBAD film reduced the critical current density to 3-5 $\mu A/cm^2$, the best value obtained. Guzman et al. (138) used a technique involving sequential deposition and ion beam mixing of multilayers to produce BN and Cr_2N films on Fe which exhibited a critical current density 23x lower than the unprotected Fe and 4x less than a deposited BN film in 1M NaCl solutions. Colligan et al. (129) studied electro-chemical behavior of Pt-Ti and Au on glass films. Improvements were associated with better adhesion of the films.

McCafferty et al. (139) have studied the electrochemical properties of Cr_2O_3 films deposited on AISI 52100 steel by reactive IBAD using evaporation of Cr and bombardment with 40 keV Cr^+ ions in a partial pressure of $1.3X10^{-3}$ Pa O_2. The resulting films were stoichiometric Cr_2O_3, 200 nm thick and highly adherent. Anodic polarization tests were performed in both 1N H_2SO_4 and a 3 ppm NaCl solution. The current density in the NaCl solution was reduced by 10-100X relative to the steel specimen and in the H_2SO_4 solution the current density was reduced from 4500 $\mu A/cm^2$ to 125 $\mu A/cm^2$. Failure of the films was at isolated pits that penetrated the coating. These defects should be eliminated by using clean room conditions in the preparation of the films or by thicker coatings.

7.5. Improvement of Optical Properties

The major influence of ion beam assisted deposition on the optical properties of thin films is through effects on the microstructure, composition and defect structure of the films. Macleod (140) has reviewed the effects of microstructure on the optical properties of thin films. The most important parameter is the packing density, which can be related to the typical columnar grain structure of a thin film. The refractive index and absorption coefficient are both sensitive functions of the chemical composition of dielectric compounds and are affected by the processing conditions. Defects and impurities in dielectric films tend to smear the band edge and increase absorption in the visible (15). The attenuation coefficient characterizing absorption in the film, $I/I_o = e^{-\alpha x}$, is related to the imaginary part of the refractive index, $n^* = n - ik$, by the relation,

$$\alpha = 2\omega k/c. \tag{3}$$

Scatter from the surface of the film is influenced by surface roughness which in turn may be affected by the processing conditions.

Excellent reviews of ion beam assisted techniques (including plasma processes) applied to optical materials have been given by Martin and Martin and Netterfield (14) and by Gibson (15). One of the principal motivations for using ion beam assisted deposition is to reduce the change in transmittance from vacuum to air for optical films. The change in transmittance has been correlated with absorption of water vapor in the porous microstructure (62). A nuclear reaction analysis for H in the film, showed IBAD reduced H_2O absorption by the film, as was previously discussed. The refractive index, n, increased with ion current density. The vacuum to air property change was eliminated above a critical current density and the film reached a maximum n. The decrease in n above the maximum was attributed to preferential sputtering of 0 atoms from the surface layers by the 1200 eV Ar^+ ions. A comparable experiment using 1200 eV O_2^+ ions did not show a dip after reaching the critical current density and had a maximum index of 2.19 compared to 2.14 for the Ar^+ case. Increasing the substrate temperature from room temperature to 300°C produced an n of 2.23 for both Ar^+ and O_2^+ bombardments. Similar results showing a critical current density to produce the optimum value of n, eliminate vacuum to air shifts, and increase film density have also been observed for CeO_2(63), LaF_3(141), SiO_2(111), TiO_2(120,142), Al_2O_3(112) and Ta_2O_5(112).

The energy dependence of the IBAD process has been investigated in some detail for Ta_2O_5 and Al_2O_3(112). Fig. 20a. shows the change in n vs current density for 3 energies of O_2^+ ion beam for Ta_2O_5 films. The 200 eV energy beam produced less than optimum refractive indices while the 300 eV beam produced a maximum at 40 $\mu A/cm^2$ and then decreased substantially. The decrease in n was accompanied by an increase in extinction coefficient, k, at the same current density for a given energy (Fig. 20b). The degradation of the Ta_2O_5 films above the critical current density was attributed to preferential sputtering of oxygen from the films. A similar energy dependence has been seen for SiO_2 and TiO_2 films (111). Targove et al. (64) and Hwangbo et al. (143) have shown that the refractive index, as a measure of packing density correlates well with the total momentum transfer to the film, $p_{Tot} = J (2m\gamma E)^{1/2}$, where J is the current density, m is the mass of the ion, γ is the fraction of ion energy transferred to target atoms and E is the energy of the ion. This correlation holds well for LaF_3 films (64) and for Ag films (143). Martin (144) has compared refractive indices obtained by IBAD and by evaporation techniques and as can be seen in Table II, higher indices are produced by the IBAD process for all materials examined.

TABLE II. Refractive indices at 550 nm of films produced by evaporation and IBAD on cold and heated substrates. Differences in vacuum and air measurements indicate porosity in evaporated films (144).

Film	Vacuum (ambient)	Evaporation Air (after venting)	300°C	IBAD Ambient (vacuum and air)	300°C
SiO_2	1.45	1.46	...	1.55	...
Al_2O_3	...	1.54	1.63	1.67	...
TiO_3	...	1.90	2.56	2.48	...
ZrO_2	1.84	1.93	1.90	2.18	...
CeO_2	1.80	1.95	2.37	2.40	2.49
Ta_2O_5	1.90	1.99	2.10	2.13	...
MgF_2	1.32	1.39	1.39	1.39	...
Na_3AlF_6	1.34	...	...	1.37	...
In_2O_3:Sn	...	...	1.92	2.13	1.99

Martin and Netterfield (145) have also compared the critical current density in terms of ion to atom arrival rate ratios to produce the optimum refractive index for dielectric films prepared by IBAD as shown in Table III. Most values are between 0.1 and 0.3 with the exception of CeO_2 where an O_2^+ beam was used to maintain stoichiometry.

TABLE III.   Optimum ion-to-Molecular Ratio R for IAD Optical Films (145)

Film	R(oxygen)
SiO_2	0.2
MgF_2	0.16
TiO_2	0.12
ZrO_2	0.33
CeO_2	1.90

Several authors have reported data on the effect of processing variables on absorption in the films and/or the extinction coefficient k of Eq. 3. The results on Ta_2O_5 (112) have already been mentioned in the previous discussion on n_0 where the extinction coefficient increased with current density and with ion energy.   Similar observations have been made for an increase in extinction coefficient with increase in current density for 300 eV O_2^+ for MgF_2 (146). MgF_2 appears to be particularly sensitive to damage and both Ar^+ (500, 700 eV) and O_2^+ (350, 700 eV) IBAD lead to higher extinction coefficients than for evaporation on a room temperature substrate or at 449°C (147). Baking of MgF_2 films appeared to drive off water vapor and leave the films highly absorbing (148,149).   The compensation of bonds broken by the preferential sputtering of F tended to decrease the absorption.   The increased absorption was attributed to formation of MgO.   Indium tin oxide was also found to have relatively high absorption in crystalline films prepared by IBAD at elevated temperature. Amorphous films had a higher UV band edge but lower extinction coefficient above the edge (150). TiO_2 is also sensitive to damage and deposition conditions. Kuster and Ebert (151) found an increase in absorption index on increasing the temperature from 50 to 325°C and Ebert (152) found higher absorption for films bombarded with neutralized O_2 than with O_2^+. Ebert also provided a survey of absorption in SiO, SiO_2, Al_2O_3, Be, La_2O_3, ZrO_2, In_2O_3, Ta_2O_5, Ti, TiO and TiO_2 and Allen (110) provided additional data on silica and titania films.

Optical scatter from a reflecting metal surface is a function of surface roughness and is influenced by film deposition conditions while optical scatter from dielectric films represents contributions from both surface irregularities and volume defects.   Al-Jumaily et al (153) examined the scatter from Cu and Mo films prepared by IBAD, evaporation and sputtering and SiO_2 and TiO_2 films prepared by evaporation and IBAD.   The IBAD treatment resulted in a smoother surface for both the metal films and SiO_2. Similar data were obtained for Ta_2O_5 (112).   Total integrated scatter measurements on CeO_2 (63) showed a smoother surface for 300 and 600 eV bombardment with O_2^+ ions but a substantially rougher surface for 1200 eV bombardment.

Several efforts have been reported to use the improved properties of the films in optical applications.   Ebert (152) reported a BeO-SiO_2 laser mirror which had higher reflectivity and higher damage threshold than conventional Al_2O_3-SiO_2 coatings. Ebert also reported the production of a 3-layer anti-reflection coating using IBAD to deposit SiO_2, TiO_2 and BeO and a 4-layer conducting anti-reflection coating using MgF_2, In_2O_3, SiO_2 and MgF_2. Targove et al (141) reported the deposition of multilayer stacks of LaF_3/MgF_3 using IBAD. Bink et al. (154) reported the deposition of

438

waveguides of ZrO_2, Al_2O_3, CeO_2 and Ta_2O_5 using 1200 eV O_2^+ IBAD. The best results were obtained with Al_2O_3 where attenuation losses of 2-5 dbcm^{-1} were obtained compared to 15-20 dbcm^{-1} for evaporated films. Donovan et al. (155) report the fabrication of a 14-cycle Rugate interference filter using e-beam evaporation of Si and ion beam doping with 1000 eV N_2^+. The N composition determines the refractive index and with computer control during the deposition a precisely tailored composition profile can be made. Rugate filters with an optical density of 2 in the peak region have been produced.

The ionized cluster beam technique of Takagi and Yamada has also been used to deposit films of optical materials (156). Cu and Al films with high reflectivity have been deposited at low process temperature and continuous films were obtained with thinner films than by conventional evaporation techniques. A reflectance of 96% was obtained for an Al film at a wave length of 450 nm. Grain size and (111) texture increased with an increase in acceleration voltage. Dielectric films of MgF_2 2µm thick were deposited on glass and plastic substrates. The refractive index went through a minimum as a function of accelerating voltage with the minimum corresponding to the bulk value for MgF_2 (n=1.38). Semiconductor compounds deposited in multilayer stacks had low absorption losses. Changes in the band gap were produced by varying the deposition conditions.

The major advantages of ion beam assisted deposition for optical properties are the ability to control the microstructure and stoichiometry of the films. The densification of films by IBAD leads to an increase in refractive index and a decrease in extinction coefficient. The major difficulty arises from films which sputter preferentially or with materials which are subject to contamination (i.e. MgF_2). Materials with complex phase diagrams with multiple compounds, such as the Ti-O system are also difficult to control and lead to increased absorption at high energies and high beam currents. Work in the field of optical materials is the most advanced of any applications of IBAD at the present time.

7.6. Modification of Electrical Properties

The electrical properties of IBAD films have been studied primarily to provide insight into the perfection of the film and to synthesize unique materials such as superconductors. Dudonis and Pranevicuis (24) studied the resistivity and electron paramagnetic resonance as a function of ion dose in SiO_x films and found the resistance to increase as the EPR spin concentration decreased. The resistivity also increased as an Al film was doped with oxygen. Erler et al (157) measured resistivity as a function of ion energy for deposition of Si_3N_4 films and found a minimum at 3 keV N_2^+. The resistivity of Si_3N_4 and AlN films prepared by IBAD was lower than comparable films prepared by Rf sputter deposition.

Parmigiani et al. (77) found the trends shown in Table IV for films deposited by dual ion beam sputtering. The increase in resistivity was attributed to a decrease in grain size.

TABLE IV. Resistivities of Ag Films Deposited by Dual Ion Beam Sputtering (77).

Energy/Atom (eV/Atom)	Resistivity $(10^{-6}\Omega m)$
0	2.65
20	4.0
42	7.4
100	8.9
190	11.2

Hwangbo et al. (143) also found an increase in resistivity for Ag films deposited by IBAD which showed a good correlation with momentum flux density. The increase in resistivity was attributed to a smaller grain size and an increase in the Ar concentration in the film. Zieman and Kay (74) in a study of Pd films deposited by biased sputtering found a maximum in the resistivity curves at an average energy/atom of 75 eV, the same value at which a minimum in grain size was observed.

The resistivity in the Al-N system was found by Hentzell et al. (76) to be a strong function of composition in the system. The temperature coefficient of resistivity changed from metallic to insulator at a N/Al ratio of 0.45 while specimens prepared by magnetron sputtering showed the transition at a N/Al ratio of 0.25. The resistivity of IBAD specimens increased with N composition until it became too high to measure at approximately 0.7. The resistivity of the magnetron sputtered sample increased much more rapidly becoming unmeasurable at a N/Al ratio of 0.3.

The resistivity of Indium tin oxide was measured at several temperatures by Martin et al. (150) and found to exhibit a rapid decrease at about $300^{\circ}C$. This was correlated with a change from amorphous to crystalline which occurred above $100^{\circ}C$. Dobrowalski et al. (158) also measured the resistivity of ITO films as a function of temperature and beam current and found a 2-fold decrease in resistivity between $100^{\circ}C$ and $250^{\circ}C$. A minimum in resistivity was found as a function of current density and the optimum current density decreased with increasing temperature.

Fujimota et al. (159) used IBAD to synthesize the B1 structure of MoN which does not exist in the equilibrium phase diagram. Several efforts to synthesize B1-MoN by sputtering produced materials with T_c between $10-13^{\circ}K$. Films were deposited by IBAD with N_2^+ beams of 5 and 20 keV, current densities from 60-110 $\mu A/cm^2$ and temperatures of 25, 400 and $500^{\circ}C$. The best film was deposited at room temperature with a 20 keV N_2^+ beam. It was stoichiometric and had a T_c of $5.8^{\circ}K$. Films produced under other conditions had T_c values ranging from $4.2-5.3^{\circ}K$ with no particular dependence on deposition parameters. The low transition temperatures (relative to sputtered samples) were attributed to imperfections in the crystal. NbN_xC_y superconducting films have also been fabricated by Lin and Probes (160) using a dual ion beam system in which Nb was sputtered onto a Si substrate and bombarded with N_2^+ and CH_4^+ ions from a 1:1 gas mixture of CH_4 and N_2. The highest T_c of $13.0^{\circ}K$ was obtained for a 100 eV beam at $T<60^{\circ}C$. Higher energy (1500 eV) and lower energy (20 eV) beams reduced the T_c. Composition of the film was $NbN_{0.6}C_{0.6}$.

Ion beam assisted deposition has also been used to deposit films of the Y-Ba-Cu-O superconductors. Fujita et al. (161) have produced epitaxial films of Y-Ba-Cu-O using a dual ion beam method in which a composite target containing Y-Ba-Cu was sputtered onto a heated $SrTiO_3$ substrate while bombarding the growing film with 50 eV O_2^+ ions. The best film obtained had a T_c of $68^{\circ}K$ with onset of superconductivity at $88^{\circ}K$ and was grown at $630^{\circ}C$ with a current density of 0.63 A/cm^2, and a deposition rate of 0.055 nm/sec. The specimens were cooled from the deposition temperature at a rate of $25^{\circ}C/min$ in a 10 Pa atmosphere of oxygen. Superconducting films were not obtained for ion energies greater than 80 eV and for substrate temperatures below $600^{\circ}C$ for (100) $SrTiO_3$ and below $560^{\circ}C$ for (110) $SrTiO_3$. Films grown by IBAD had somewhat lower T_c than films grown using an O_2 gas jet thus indicating that some residual damage was left in the films. However films could be grown at 10x lower presure when O was introduced by the ion beam. The film orientation was found to be a function of both substrate temperature, beam current and substrate orientation.

Another growing application is the use of an ionized and accelerated

evaporant flux to dope growing films of semiconductor materials
(18,34,92,162,163). The main advantages of the technique are the formation
of sharper dopant profiles than can be achieved by diffusion or
implantation and higher incorporation probabilities with an accelerated
flux.

Electrical properties of films grown by IBAD are influenced by
composition, amorphous/crystalline structure, grain size and defect density
in the film. Resistivity has been used in a number of investigations to
monitor changes in the above variables as a function of processing
conditions.

8.CONCLUSIONS

Ion beam assisted deposition has tremendous potential as a technique to
produce thin films with controlled properties such as fully dense films
grown at low substrate temperature, films with low stress, adherent films,
films with refined microstructure, films with controlled texture and films
with optimum wear, corrosion protection and optical properties.

Ion bombardment of a growing film can effect the film and its properties
in many ways. The current research need is to identify the dominant and
controlling mechanisms of interaction between the bombarding particles and
the growing film and relate them to the basic process parameters so that
films with the desired properties can be made in a predictable and
reproduceable manner. At the same time recognition of the benefits of IBAD
have led to much empirical research aimed toward optimizing the properties
of films for a wide variety of applications. The future appears bright for
both advances in the science of IBAD and the development of many new
applications.

9.ACKNOWLEDGEMENTS

Financial support for this work was provided by the office of Naval
Research. The author gratefully acknowledges many stimulating discussions
with his colleagues at NRL including Graham Hubler, Dick Kant, Bruce
Sartwell, Ed Donovan and Jim Sprague and thanks them for the use of data
prior to publication.

REFERENCES
1. D.M. Mattox "Film Deposition Using Accelerated Ions," Sandia Corp.
    Report SC-DR-281-63 (1963).
2. R.D. Bland, G.J. Kominiak and D.M. Mattox, J. Vac. Sci. Technol. 11,
    (1974) 671.
3. J.L. Vossen and W. Kern, Thin Film Processes, Academic Press, New
    York, NY (1978).
4. R.F. Bunshah, Deposition Technologies for Films and Coatings, Noyes
    Publications, Park Ridge, N.J. (1982).
5. B. Chapman, Glow Discharge Processes, John Wiley and Sons, New York,
    NY (1980).
6. J.M.E. Harper, in Ref. 3 (1978) p. 175-206.
7. C. Weismantel, J. Vac. Sci. Technol. 18(2) (1981) 179; C. Weismantel,
    Thin Solid Films 92 (1982) 55.
8. J.E. Greene and S.A. Barnett, J. Vac. Sci. Technol. 21(2) (1982) 285.
9. J.E. Greene, CRC Critical Reviews in Solid State and Materials
    Sciences 11 (1983) 47.
10. W.A. Grant and J.S. Colligan, Vacuum 32 (1982) 675.
11. T. Takagi, Thin Solid Films 92 (1982) 1.
12. I. Yamada, Proc. International Ion Engineering Congress (1983) 1177.
13. J.M.E. Harper, J.J. Cuomo, R.J. Gambino, and H.E. Kaufman, Ch. 4 in
    Ion Bombardment Modification of Surfaces: Fundamentals and
    Applications ed. by O. Auciello and R. Kelly, Elsevier Science Publ.,

Amsterdam, Neth. (1984) p. 127-162.
14. P.J. Martin, J. Mater. Sci. <u>21</u> (1986) 1; P.J. Martin and R.P. Netterfield in <u>Progress in Optics XXIII</u> ed. by E. Wolf, Elsevier Science Publ., Amsterdam, Neth. (1986) 114-182.
15. U.J. Gibson, "Ion-Beam Processing of Optical Thin Films" in <u>Physics of Thin Films</u>, Vol. 13 ed by M.H. Francombe and J.L. Vorsen, Academic Press, Inc, San Diego, CA (1987) p. 109-150.
16. H.A. Macleod, in SPIE Vol. <u>652</u>, Thin Film Technologies II (1986) 221.
17. D.G. Armour, P. Bailey and G. Sharples, Vacuum <u>36</u>, (1986) 769.
18. J.E. Greene, T. Motocka, J.-E. Sundgren, A. Rockett, S. Gorbatkin, D. Lubben and S.A. Barnett, J. Crystal Growth <u>79</u> (1986) 19; J.E. Greene, T. Motocka, J.-E Sundgren, D. Lubben, S. Gorbatkin and S.A. Barnett, Nucl. Insts. and Meth. in Phys. Res <u>B27</u> (1987) 226; J.E. Greene, Solid State Technol. Apr. 1987, 115; J.E. Greene, A. Rockett and J.-E Sundgren in Vol. <u>74</u>, <u>Beam-Solid Interactions and Transient Processes</u>, ed. by M.O. Thompson, S.T. Picraux and J.S. Williams, Materials Research Soc., Pittsburgh, PA (1987) p. 463.
19. J.J. Cuomo and S. Rossnagel, Nucl. Inst. and Meth. in Phys. Res. <u>B27</u> (1987) 963; S. M. Rossnagel and J.J. Cuomo, MRS Bulletin <u>XII, No. 2</u> Feb-Mar 1987, 40.
20. B.R. Appleton, R.A. Zuhr, T.S. Noggle, N. Herbots, S.J. Pennycook and G.D. Alton, MRS Bulletin <u>XII, No. 2,</u> Feb-Mar 1987, 52; B.R. Appleton, S.J. Pennycook, R.A. Zuhr, N. Herbots and T.S. Noggle, Nucl. Insts. and Meth. in Phys. Res. <u>B19/20</u> (1987) 975.
21. F.A. Smidt, Proceedings of DOE Workshop on Coatings for Advanced Heat Engines, DOE, in press (1988).
22. A. Matthews, J. Vac. Sci. Technol. <u>A3</u> (1985) 2354.
23. C. Weissmantel, Thin Solid Films <u>32</u> (1976) 11.
24. J. Dudonis and L. Pranevicius, Thin Solid Films <u>36</u> (1976) 117.
25. D. Van Vechten, G.K. Hubler, and E.P. Donovan, Vacuum <u>36</u> (1986) 841.
26. J.R. McNeil, A.C. Barron, S.R. Wilson and W.C. Hermann Jr., Appl. Opt. <u>23</u> (1984) 559.
27. J.M.E. Harper, J.J. Cuomo, and H.T.G. Hentzell, J. Appl. Phys. <u>58</u> (1985) 550; J.M.E. Harper, J.J. Cuomo, and H.T.G. Hentzell, Appl. Phys. Lett. <u>43</u> (1983) 547.
28. T.C. Huang, G. Lim, F. Parmigiani and E. Kay, J. Vac. Sci. Technol. <u>A3</u> (1985) 2161.
29. R.A. Kant and B.D. Sartwell, Materials Sci. and Eng. <u>90</u> (1987) 357.
30. M. Satou and F. Fujimoto, Japan J. Appl. Phys. <u>22</u> (1983) L171.
31. B. Margesin, F. Giacomozzi, L. Guzman, G. Lazzari, and V. Zanini, Nucl. Inst. Meth. in Phys. Res. <u>B21</u> (1987) 566.
32. T. Sato, K. Ohata, N. Asahi, Y. Ono, Y. Oka, I. Hashimoto and K. Arimatsu, Nucl. Insts. Meth. in Phys. Res. <u>B19/20</u> (1987) 644.
33. G.K. Wolf, K. Zucholl, M. Barth and W. Ensinger, Nucl. Inst. Meth. in Phys. Res. <u>B21</u> (1987) 570.
34. A. Rockett, S.A. Barnett and J.E. Greene, J. Vac. Sci. Technol. <u>B2</u> (1984) 306
35. Y. Ota, J. Vac. Sci. Technol. <u>A2</u> (1984) 393.
36. J.A. Venables, G.D.T. Spiller and M. Hanbucken, Rep. Prog. Phys. <u>47</u> (1984) 399.
37. B. Lewis and J.C. Anderson, <u>Nucleation and Growth of Thin Films</u>, Academic Press, New York, NY (1978).
38. B.A. Movchan and A.V. Demchishin, Phys. Met. Metallogr. <u>28</u> (1969) 83.
39. J.A. Thornton, Ann. Rev. Mater. Sci. <u>7</u> (1977) 239.
40. D.A. Thompson, Rad. Eff. <u>56</u> (1981) 105.
41. J.A. Davies in <u>Ion Implantation and Beam Processing</u> ed. by J.S.

442

Williams and J.M. Poate, Academic Press, New York, NY (1984) p. 81-97.

42. H.H. Anderson, ibid. (1984). p. 128-187.
43. J.F. Ziegler, in Ion Implantation Science and Technology, ed. by J.F. Ziegler, Academic Press, New York, NY (1984) p. 51-108.
44. R.S. Averback, Nucl. Insts. and Meth. in Phys. Res. B15 (1986) 675.
45. L.E. Rehn and H. Wiedersich, in Surface Alloying by Ion, Electron and Laser Beams, ed. by L.E. Rehn, S.T. Picraux and H. Wiedersich, American Society for Metals, Metals. Park, OH (1985) p. 137-174.
46. E. Kay, Nucl. Insts. and Meth. 182/183 (1981) 259.
47. M. Marinov, Thin Solid Films 46 (1977) 267.49.48.
48. V.O. Babaev, J.V. Bykov and M.B. Guseva, Thin Solid Films 38 (1976)1.
49. L. Pranevicius, Thin Solid Films 63 (1979) 77.
50. L. Pranevicius, Nucl. Insts. and Meth. 182/183 (1981) 251.
51. L. Pranevicius and S. Tomulevichus, Nucl. Insts. and Meth. 209/210 (1983) 179.
52. R.P. Netterfield and P.J. Martin, Appl. Surf. Sci. 25 (1986) 265.
53. G.E. Lane and J.C. Anderson, Thin Solid Films 26 (1975) 5.
54. E. Krikorian and R.J. Sneed, Astrophysics and Space Science 65 (1979) 129.
55. M.A. Hasan, S.A. Barnett, J.-E. Sundgren, and J.E. Greene, J. Vac. Sci. Technol. A5 (1987) 1883.
56. S.A. Barnett, H.F. Winters and J.E. Greene, Surf. Sci. 181 (1987) 596.
57. B.D. Sartwell, to be published, J. Vac. Sci. Technol. (1988).
58. C.M. Gilmore, A. Haeri and J.A. Sprague, to be published Thin Solid Films, (1988).
59. B.W. Dodson and P.A. Taylor in Vol. 74, Beam-Solid Interactions and Transient Processes ed. by M.O. Thompson, S.T. Picraux and J.S. Williams, Materials Research Society, Pittsburgh, PA (1987) p. 463.
60. E.H. Hirsch and I.K. Varga, Thin Solid Films 69 (1980) 99.
61. D.R. Brighton and G.K. Hubler, Nucl. Insts. and Meth. in Phys. Res. B28 (1987) 527.
62. P.J. Martin, R.P. Netterfield and W.G. Sainty, J. Appl. Phys. 55 (1984) 235.
63. R.P. Netterfield, W.G. Sainty, P.J. Martin, and S.H. Sie, Appl. Opt. 24 (1985) 2267.
64. J.D. Targove, L.J. Lingg and H.A. Macleod, Optical Interference Coatings, 1988, Technical Digest Series, Vol. 6 Optical Society of America, Washington, DC (1988) 268.
65. R.A. Kant, (1987) Unpublished research cited in ref. 21.
66. K.H. Müller, in Materials Modification and Growth Using Ion Beams ed. by U.J. Gibson, A.E. White and P.P. Pronko, Materials Research Society, Pittsburgh, PA (1987) 275.
67. A.G. Dirks and H.J.Leamy, Thin Solid Films 47 (1977) 219.
68. K.-H Müller, J. Appl. Phys. 58 (1985) 2573.
69. K.-H Müller, J. Appl. Phys. 59 (1986) 2803.
70. K.-H Müller, Surf. Sci. 184 (1987) L375.
71. K.-H Müller, Phys. Rev. B35 (1987) 7906.
72. M.K. Hibbs, B.O. Johannsson, J.E. Sundgren and U. Helmerson, Thin Solid Films 122 (1984) 115.
73. F. Parmigiani, E. Kay, T.C. Huang, and J.D. Swalen, Appl. Opt. 24 (1985) 3335.
74. P. Zieman and E. Kay, J. Vac. Sci. Technol. A1 (1983) 512.
75. R.A. Roy, J.J. Cuomo and D.S. Yee, J. Vac. Sci. Technol. A6 (1988) 1621.

76. H.T.G. Hentzell, J.M.E. Harper and J.J. Cuomo, J. Appl. Phys. 58 (1985) 556.
77. F. Parmigiani, E. Kay, T.C. Huang, J. Perrin, M. Jurich and J.D. Swalen, Phys. Rev. B33 (1986) 879.
78. D. Dobrev, Thin Solid Films 92 (1982) 41.
79. G.N. Van Wyk and H.J. Smith, Nucl. Inst. & Meth. 170 (1980) 433.
80. J.M.E. Harper, D.A. Smith, L.S. Yu, and J.J. Cuomo, in Beam Solid Interactions and Phase Transformations ed by H. Kurz, G.L. Olson and J.M. Poate, Vol. 51 Materials Research Society, Pittsburgh, PA (1986) 343; L.S. Yu, J.M.E. Harper, J.J. Cuomo and D.A. Smith, J. Vac. Sci. Technol. A4 (1986) 443.
81. P. Wang, D.A. Thompson, W.W. Smeltzer, Nucl. Inst. Meth. in Phys. Res. B7/8 (1985) 97.
82. P. Wang, D.A. Thompson, and W.W. Smeltzer, Nucl. Insts. Meth. in Phys. Res. B16 (1986) 288.
83. K. T-Y. Kung and R. Reif, J. Appl. Phys. 59 (1986) 2422.
84. Y. Nagai, A. Togo, and T. Toshima, J. Vac. Sci. Technol. A5 (1987) 61.
85. M. Satou, Y. Andoh, K. Ogata, Y. Suzuki, K. Matsuda, and F. Fujimota, Jap. Jn. Appl. Phys. 24 (1985) 656.
86. R.M. Bradley, J.M.E. Harper, and D.A. Smith, J. Appl. Phys. 60 (1986) 4160; R.M. Bradley, J.M.E. Harper, D.A. Smith, J.Vac. Sci. Technol. A5 (1987) 1792.
87. B.R. Appleton and G. Foti, "Channeling" in Ion Beam Handbook for Materials Analysis, ed. by J.W. Mayer and E. Rimini, Academic Press, New York, NY (1977) p. 67.
88. T. Narusawa, S. Shimizu, and S. Komiya, J. Vac. Sci. Technol. 16 (1979) 366.
89. T. Itoh, T. Nakamura, M. Muromachi and T. Sugiyama, Jap. J. Appl. Phys. 16 (1977) 553.
90. R. Ueda, Thin Solid Films 39 (1976) 25.
91. K. Yagi, S. Tamura, and T. Tokuyama, Jap. J. Appl. Phys. 16 (1977) 245.
92. Y. Ota, J. Appl. Phys. 51 (1980) 1102.
93. I.N. Aleksandrov, A.S. Lutovich, E.D. Belorusets, phys. stat. sol (a) 54 (1979) 463.
94. I. Yamada, H. Inokawa, and T. Takagi, J. Appl. Phys. 56 (1984) 2746.
95. H. Inokawa, I. Yamada, and T. Takagi, Jap. J. Appl. Phys. 24 (1985) L173.
96. H. Yamada and Y. Tori, Appl. Phys. Lett. 50 (1987) 386.
97. N. Herbots, B.R. Appleton, T.S. Noggle, R.A. Zuhr and S.J. Pennycook, Nucl. Insts. and Meth. in Phys. Res. B13 (1986) 250; N. Herbots, B.R. Appleton, S.J. Pennycook, T.S. Noggle and R.A. Zuhr in MRS Vol. 51, Beam-Solid Interactions and Phase Transformations ed. by H. Kunz, G.L. Olson, and J.M. Poate, Materials Research Society, Pittsburgh, PA, (1986) p. 369; B.R. Appleton, R.A. Zuhr, T.S. Noggle, N. Herbots and S.J. Pennycook in MRS Vol. 74, Beam-Solid Interactions and Transient Processes ed. by M.O. Thompson, S. Thomas Picraux and J.S. Williams, Materials Research Society, Pittsburgh, PA, (1987) p. 45.
98. R.A. Zuhr, G.D. Alton, B.R. Appleton, N. Herbots, T.S. Noggle and S.J. Penneycook, in Materials Modification and Growth Using Ion Beams ed. by U.J. Gibson, A.E. White, and P.P. Pronko, Materials Research Soc. Pittsburgh, PA (1987) p. 243.
99. L. Huttman, V. Helmersson, S.A. Barnett, J-E. Sundgren and J.E. Greene, J. Appl. Phys. 61 (1987) 552.
100. E.P. Donovan, D.R. Brighton, G.K. Hubler and D. Van Vechten, Nucl.

Inst. and Meth. in Phys. Res. B19/20 (1987) 983.

101. G.K. Hubler submitted to J. Vac. Soc. 1988.

102. R.A. Kant, B.D. Sartwell, I.L. Singer and R.G. Vardiman, Nucl. Insts. and Meth. in Phys. Res. B7/8 (1985) 915.

103. R.A. Kant, Naval Research Laboratory, personal communication Mar. 1988. To be published.

104. G.K. Hubler, D. Van Vechten, E.P. Donovan and R.A. Kant, in Ion Implantation and Plasma Assisted Processes for Industrial Applications, ed. K. Legg and R. Hochman, ASM Metals Park, OH in press 1988.

105. D.S. Campbell in Handbook of Thin Film Technology ed. by L.I. Maissel and R. Gland, McGraw-Hill, New York (1970) p. 12.1.

106. D.W. Hoffman and M.R. Gaerttner, J. Vac. Sci. Tecnhol 17 (1980) 425.

107. J.J. Cuomo, J.M.E. Harper, G.R. Guarnieri, D.S. Yee, L.J. Attanasio, J. Angilello, C.T. Wu and R.H. Hammond, J. Vac. Sci. Technol. 20 (1982) 349.

108. E. Kay, F. Parmigiani and W.Parrish, J. Vac. Sci. Technol. A5 (1987) 44.

109. D.S. Yee, J. Floro, D.J. Mikalsen, J.J. Cuomo, K.Y. Ahn, and D.A, Smith, J. Vac. Sci. Technol. A3 (1985) 2121.

110. T.H. Allen, Proc. SPIE 325 Optical Thin Films, ed R.I. Seddon (1982) 93.

111. J.R. McNeil, A.C. Barron, S.R. Wilson and W.C. Herrmann, Appl. Opt. 23 (1984) 552.

112. J.J. McNally, F.L. Williams and J.R. McNeil, Proc. SPIE Vol. 678 Optical Thin Films II: New Developments, ed. R.I. Seddon (1986) p. 151.

113. S.D. Jacobs, A.L. Hryoin, K.A. Cerqua, C.M. Kennemore III and U.J. Gibson, Thin Solid Films 144 (1986) 69.

114. P.H. Wojciechowski, Proceedings 30th Annual Conference of Society of Vacuum Coaters, Washington, DC (1987) 247; P.H. Wojciechowski, J. Vac. Sci. Technol. A6 (1988) 1924.

115. J.E.E. Baglin, Ch. 15 in Ion Beam Modification of Insulators ed. by P. Mazzoldi and G.W. Arnold, Elsevier, Amsterdam (1986), 585.

116. L.E. Collins, J.G. Perkins and P.T. Stroud, Thin Solid Films 4 (1969) 41.

117. T.F.Gukelberger and W.J. Kleinfelder, U.S. Patent No. 3,682,729, Aug. 8, 1972.

118. J.E.E. Baglin, G.J. Clark, J. Bottiger, Materials Research Soc. Vol. 25, Thin Films and Interfaces II ed. by J.E.E. Baglin, D.R. Campbell and W.K. Chu, North Holland, New York (1984) p. 179.

119. D.C. Ingram and P.P. Pronko, Nucl Insts. and Meth. in Phys. Res. B13 (1986) 462.

120. J.E. Griffith, Y. Qiu and T.A. Tombrello, Nucl. Inst. and Meth. 198 (1982) 607.

121. T.A. Tombrello, Materials Research Soc. Vol. 25, Thin Films and Interfaces II ed. by J.E.E. Baglin, D.R. Campbell and W.K. Chu, North Holland New York, (1984) p. 123.

122. J.E.E. Baglin, Materials Research Soc. Vol. 47, Thin Films: The Relationship of Structure to Properties, ed. by C.R. Aila, K.S. Sree Harsha, Materials Research Soc., Pittsburgh, PA (1985) 3.

123. H. Dallaporta and A Cros, Appl. Phys. Lett. 48 (1986) 1357.

124. J. Franks, P.R. Stuart, and R.B. Withers, Thin Solid Films 60 (1979) 231.

125. S.S. Nandra, F.G. Wilson, and C.D. Des Forges, Thin Solid Films 107 (1983) 335.

126. J.J. McNally, G.A. Al-Jamaily, J.R. McNeil and B. Bendow, Apl. Optics 25 (1986) 1973.
127. W. Ensinger, M. Barth, and G.K. Wolf. private communication 1987.
128. C.M. Kennemore III and U.J. Gibson, Appl. Opt. 23 (1984) 3608.
129. J.S. Colligan, A.E. Hill and H. Kheyrandish, Vacuum 34 (1984) 843.
130. R.N. Bolster, I.L. Singer, R.A. Kant, B.D. Sartwell and C.R. Gossett, to be published, Surface and Coatings Technology, Nov. 1988.
131. J.-E. Sundgren, Thin Solid Films 128 (1985) 21.
132. R. Messier, A.P. Giri and R.A. Roy, J. Vac. Sci. Technol. A2 (1984) 500.
133. M. Hirano and S. Miyake, Appl. Phys. Lett. 47 (1985) 683.
134. H. Kuwano and K. Nagai, J. Vac. Sci. Technol. A4 (1986) 2993.
135. K. Kobs, H. Dimigen, H. Hübsch, H.J. Tolle, R. Leutenecker, and H. Ryssel, Appl. Phys. Lett. 49 (1986) 496.
136. N. Mikkelsen, J. Chevallier, G. Sorensen and C.A. Straede, Appl. Phys. Lett 52 (1988) 1130.
137. P.J. Martin, R.P. Netterfield, W.G. Sainty and C.G. Pacey, J. Vac. Sci. and Technol. A2 (1984) 341.
138. L. Guzman, F. Giacomozzi, B. Margesin, L. Calliari, L. Fedrizzi, P.M. Ossi and M. Scotini, Materials Sci. and Engr. 90 (1987) 349.
139. E. McCafferty, G.K. Hubler, P.M. Natishan, P.G. Moore, R.A. Kant and B.D. Sartwell, Materials Sci. and Engr. 86 (1987) 1.
140. H.A. Macleod, J. Vac. Sci. Technol. A4 (1986) 418.
141. J.D. Targove, J.P. Lehan, L.J. Lingh, H.A. Macleod, J.A. Leavitt, and L.C. McIntryre Jr., Appl. Opt. 26 (1987) 3733.
142. F.L. Williams, J.J. McNally, G.A. Al-Jumaily, and J.R. McNeil, J. Vac. Sci. and Tech. A5 (1987) 2159.
143. C.K. Hwangbo, L.J. Lingg, J.P. Lehan, M.R. Jacobson, H.A. Macleod, J.L. Makhous, and S.Y. Kim, Optical Interference Coatings, 1988, Technical Digest Series, Vol. 6, Optical Society of America, Washington, DC (1988) 272.
144. P.J. Martin, J. Vac. Sci. Technol. A5 (1987) 2158.
145. P.J. Martin and R.P. Netterfield, Appl. Optics 24 (1985) 1732.
146. J.D. Targove, M.J. Messerly, J.P. Lehan, C.C. Weng, R.H. Potoff, H.A. Macleod, L.C. McIntyre Jr. and J.A. Leavitt, Proc. of SPIE 678, Optical Thin Films II: New Developments, ed. by R.I. Seddon (1986) 115.
147. P.J. Martin, W.G. Sainty, R.P. Netterfield, D.R. McKenzie, P.J.H. Cockayne, S.H. Sie, O.R. Wood, and H.G. Craighead, Appl. Opt. 26 (1987) 1235.
148. U.J. Gibson and C.M. Kennemore III, Thin Solid Films 124 (1985) 27.
149. U.J. Gibson, and C.M. Kennemore III, Proc. SPIE 678, Optical Thin Films II: New Developments, ed. R.I. Seddon (1986) 130.
150. P.J. Martin, R.P. Netterfield, D.R. McKenzie, Thin Solid Films 137 (1986) 207.
151. H. Küster and J. Ebert, Thin Solid Films 70 (1980) 43.
152. J. Ebert, Proc. SPIE 325 Optical Thin Films, ed R.I. Seddon (1982) 29.
153. G.A. Al.-Jumaily, J.J. McNally, J.R. McNeil and W.C. Herrmann, Jr., J. Vac. Sci. Technol. A3 (1985) 651.
154. L. N. Binh, R.P. Netterfield and P.J. Martin, Appls. of Surf. Sci. 22/23 (1985) 656.
155. E.P. Donovan, D. Van Vechten, A. Kahn, C.A. Carosella, and G.K. Hubler, Optical Interference Coatings, 1988 Technical Digest Series, Vol. 6, Optical Society of America, Washington, DC (1988) 122.
156. T. Takagi and I. Yamada, Appl. Opt. 24 (1985) 879.

157. H.J. Erler, G. Reisse, and C. Weissmantel, Thin Solid Films, <u>65</u> (1980) 233.

158. J.A. Dobrowalski, F.C. Ho, D. Menagh, R. Simpson and A. Waldorf, Appl. Opt. <u>26</u> (1987) 5204

159. F. Fujimoto, Y. Nakane, M. Satou, F. Komori, K. Ogata and Y. Andoh, Nucl. Insts. and Meth. in Phys. Res. <u>B19/20</u> (1987) 791.

160. L.-J. Lin and D.E. Prober, Appl. Phys. Lett. <u>49</u> (1986) 446.

161. J. Fujita, T. Yoshitake, A. Kamijo, T. Satoh, and H. Igarashi, submitted to J. Appl. Phys. 1988; J. Fujita, T. Yoshitake, A. Kamijo, T. Satoh, and H. Igarashi, submitted to Nucl. Insts. and Meth. 1988 (IBMM88).

162. J.L. Zilko, S.A. Barnett, A.H. Eltouky and J.E. Greene, J. Vac. Sci. Technol. <u>17(2)</u> (1980) 595.

163. J.E. Greene, S.A. Barnett, A. Rockett and G. Bajor, Appl. of Surf. Sci. <u>22/23</u>, (1985) 520.

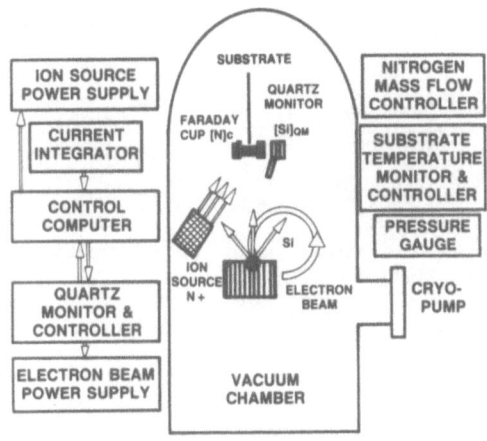

Fig. 1. Schematic drawing of an IBAD system using an electron beam evaporation source and a Kaufman ion gun operating under computer control (25).

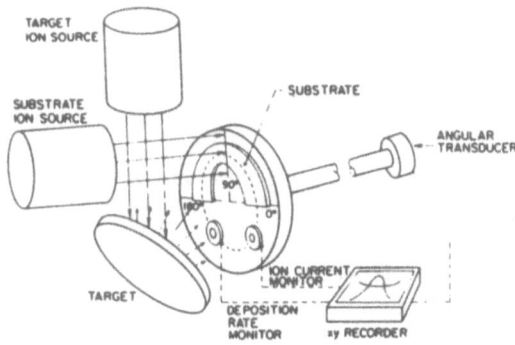

Fig. 2. Schematic drawing of a dual ion beam IBAD system using one ion gun to sputter atoms from a target and another ion gun to bombard the growing film on the substrate holder (27).

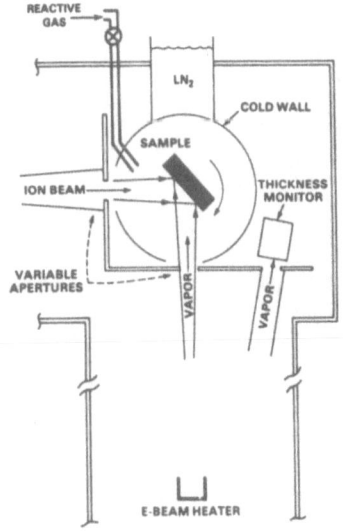

Fig. 3. Schematic drawing of an IBAD system using the 30-200 keV beam from an implanter to bombard the growing film at a 45° angle of incidence (29).

448

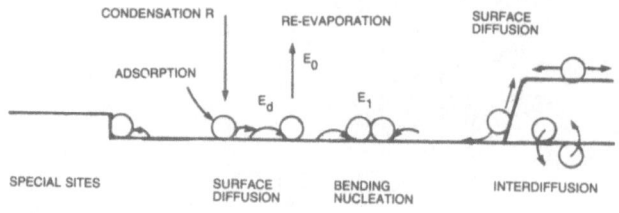

Fig. 4.   Illustration of atomic processes which occur on a surface during nucleation and growth of a thin film (36).

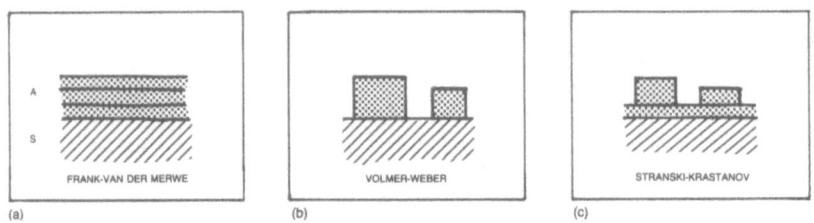

Fig. 5.   Illustration of the three common modes of film growth a) Frank Van der Merwe or layer by layer, b) Volmer-Weber or island growth and c) Stranski-Krastanov or mixed mode (36).

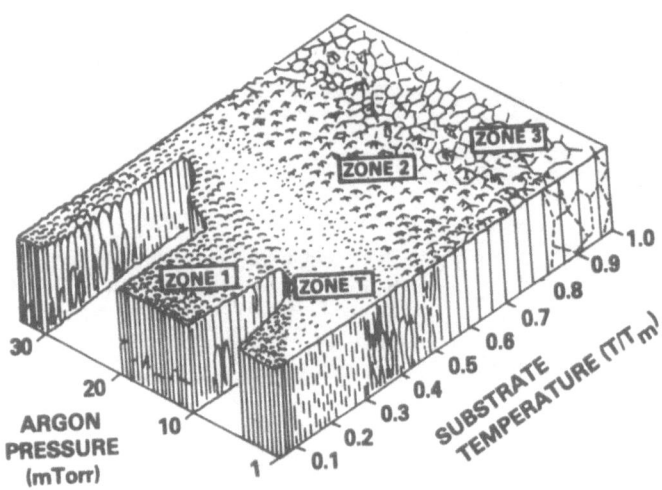

Fig. 6.   Thornton modification of the structure diagram for film growth illustrating the dependence of microstructure on substrate temperature and sputtering gas pressure (39).

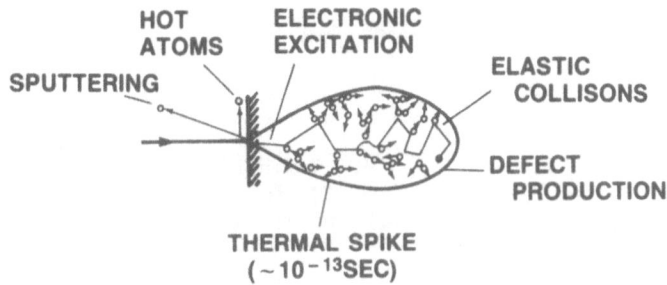

Fig. 7. Schematic of processes which occur during the impact of an energetic ion with a solid.

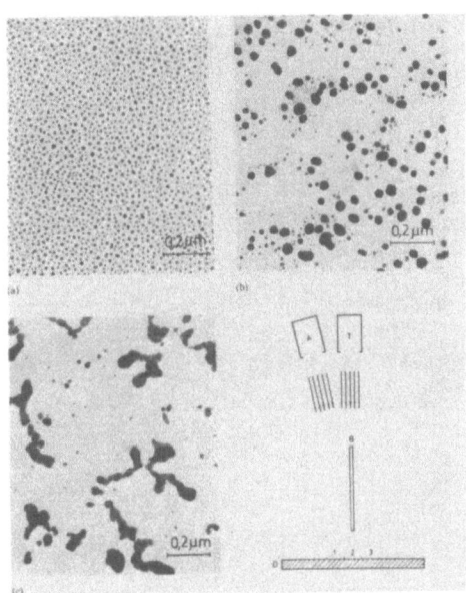

Fig. 8. The effect of bombardment with 10 keV Ar$^+$ ions on the nucleation and early growth of a Ag film deposited on amorphous C (47).

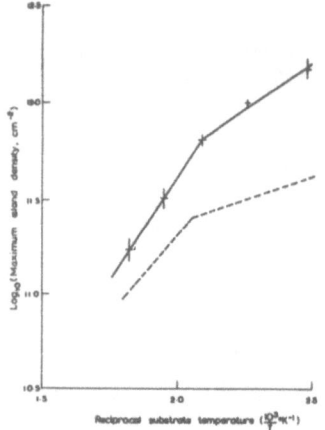

Fig. 9. Variation of island number density with (1/T) for sputtered Au (5eV) (upper curve) and thermally evaporated Au on NaCl (lower curve) (53). Nucleation is controlled by island capture at low temperatures and thermal desorption at high temperatures. Ion bombardment creates additional nucleation sites.

Fig. 10. An experiment studing the effect of ion bombardment on the mode of film growth for Cu deposited on Si(100), a) no bombardment, b) 500eV Ar$^+$ and c) 1000eV Ar$^+$ bombardment. The mode of growth changes from island to layer with 500eV ions but reverses again for 1000eV bombardment (57).

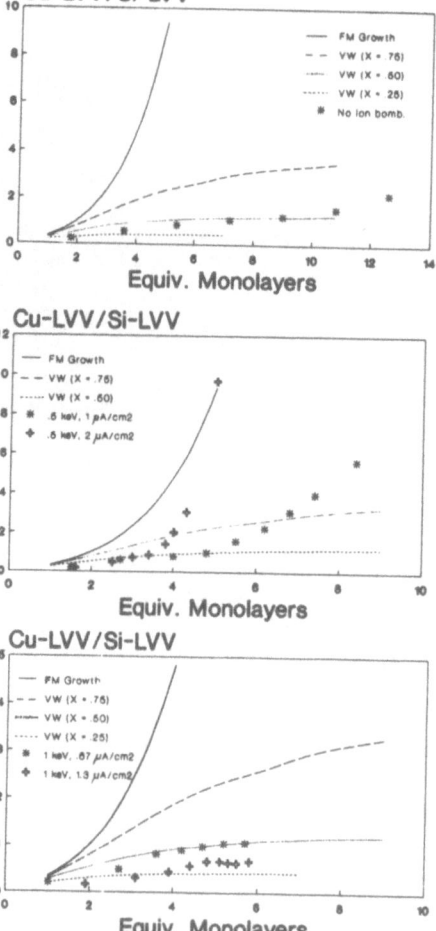

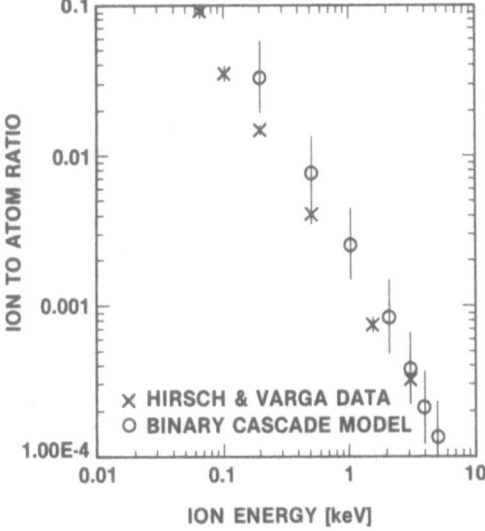

Fig. 11. A comparison of the ion to atom arrival rate ratio and energy required to produce adherent films of Ge in the Hirsch and Varga experiment (60) and calculated values of energy and R for which every atom is affected by a cascade (61).

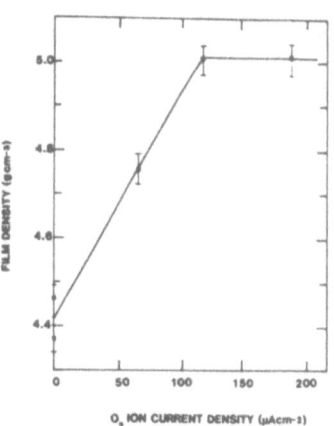

Fig. 12. Change in density with ion current of a ZrO_2 film deposited by IBAD at 300°C using a beam of 1200eV O_2^+ ions (62).

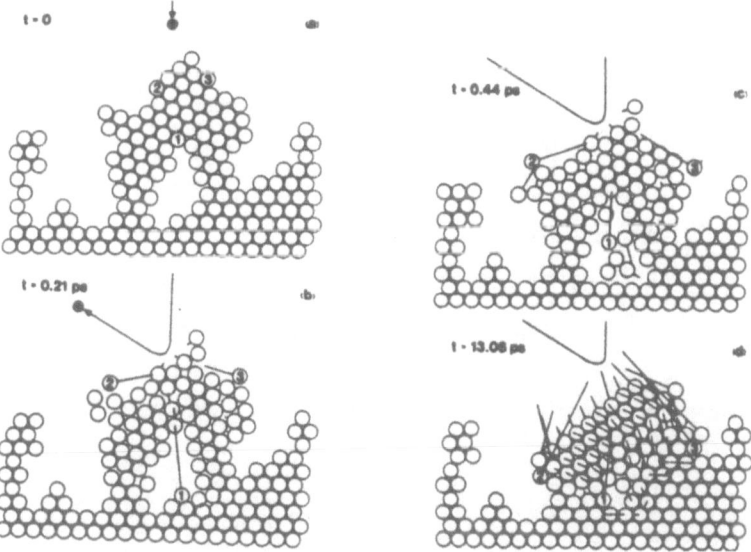

Fig. 13. Schematic diagrams of lattice rearrangements resulting from the impact of a 100eV Ar^+ ion on a Ni film with porous structure. Based on 2-dimensional molecular dynamics calculations by Müller (71). Primary effects are forward and lateral recoils which move lattice to more stable and higher density configuration.

452

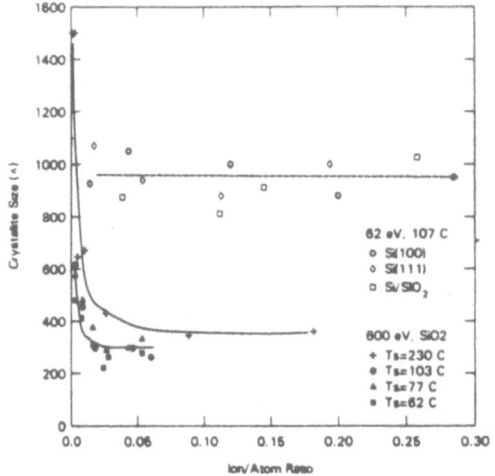

Fig. 14. The change in grain size during IBAD deposition of Cu films as a function of ion energy, deposition temperature and ion to atom arrival rate ratio (74).

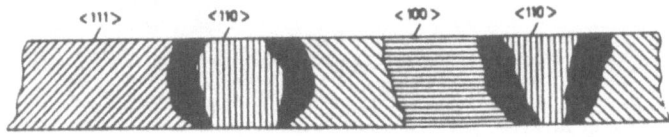

Fig. 15. A schematic illustration of the way in which <110> texture develops in FCC films during ion bombardment. <110> grains which provide deep channeling (and low energy density) serve as seeds for the growth of grains with higher energy density (and higher T) (77).

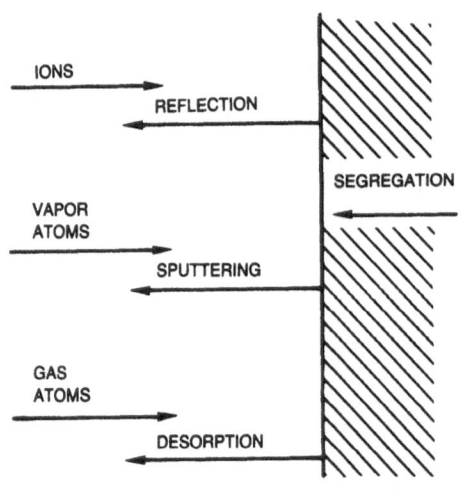

Fig. 16. Schematic of process which add and remove material on the surface during the IBAD deposition of compounds.

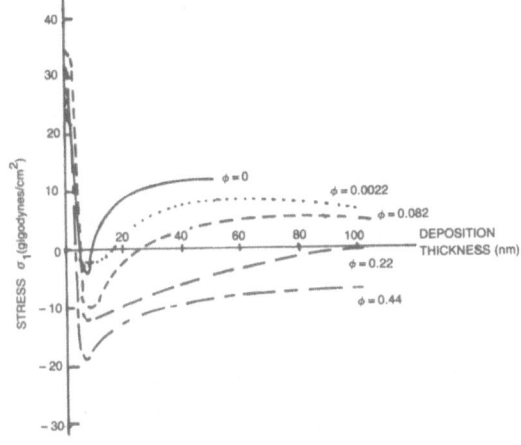

Fig. 17. Change in residual stress as a function of thickness for permalloy films deposited by IBAD with 300eV Ar$^+$ ions and ion to atom arrival rate ratios from 0 to 0.44 (113).

Fig. 18. The change in adhesion (peel strength) for Cu deposited on Al$_2$O$_3$ and then bombarded with 250 keV Ne$^+$ or 200 keV He$^+$ ions and after subsequent 1 hr. anneals at 450°C (117). The results show that energy transfer by nuclear collisions is more effective than electronic energy loss processes in promoting adhesion.

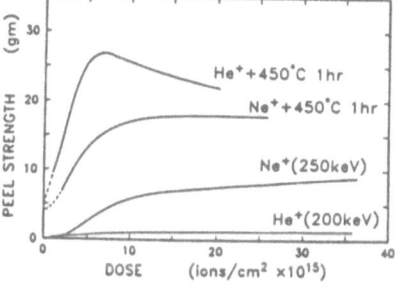

Fig. 19. Adhesion of TiN films to AISI 52100 substrate when prepared by PVD (left) and by 30 keV N$_2^+$ reactive IBAD (right) as indicated by the scratch adhesion test at a load of 50 N (101).

454

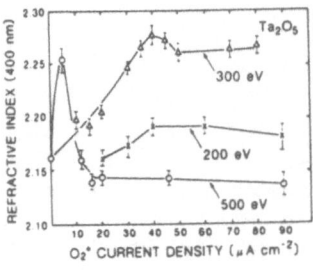
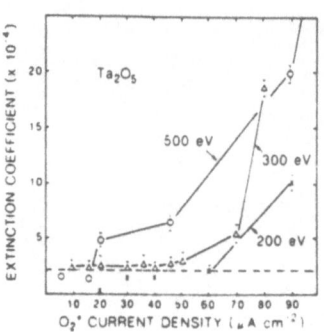

Fig. 20. Change in refractive index 20(a) and extinction coefficient 20(b) as a function of O_2^+ ion energy and current density for Ta_2O_5 films produced by IBAD (111).

ION-BEAM INDUCED DIAMOND-LIKE CARBON COATINGS

A. ANTTILA

Department of Physics, University of Helsinki,
SF-00170 Helsinki, Finland

1. INTRODUCTION
   Experiments during the last few years have clearly demonstrated that the
preparation of diamond and diamond-like films can be developed to a technically simple (and economically valuable) process. There are numerous
methods for the preparation of diamond and diamond-like films. Although all
have some clear advantages, they also have disadvantages and therefore at
this time no superior method is available.
   In order to distinguish this article from other existing ones, the emphasis is put on two methods used in our laboratory, i.e., on the use of mass-
separated carbon ion beams (1-6), and curved plasma beams. The methods are
presented in chronological succession. The first of these methods is
accurately controllable and therefore very pure foils can be produced. The
disadvantage of this method is the complicated equipment and the difficulty
of achieving a high production rate. The advantages of the plasma method
are the high production rates and rather simple equipment. The disadvantage
is that the energy distribution of the ions in the plasma stream varies and
the optimization of the plasma stream energy is difficult.
   As mentioned above, there are numerous methods for the preparation of
diamond and diamond-like films. For study of film properties, the number of
methods is even higher. Some typical analysis methods used in recent publications have been collected in Table 1. As in the case of the preparation
methods, the selection of the "right" method is difficult, e.g., there is no
generally accepted method even for distinguishing diamond from diamond-like
films, i.e., for the sp^2 and sp^3 bonding determination. (Graphite has sp^2
bonding and diamond has sp^3 bonding.) In our laboratory the main emphasis
has been on the preparation of hard wear-resistant coating materials and
therefore at least the tribologically important properties of the carbon
coatings have been analyzed. Another reason is that the price of the equipment needed for the measurement of the tribological properties is
reasonable.
   First, a short review on the properties of diamond and diamond-like films
is presented. Secondly, the preparation methods and the results from film
analyses are outlined and finally the applications and a short summary are
given. The abbreviations, D and DLC, will be used in this article for
diamond and diamond-like carbon, although our DLC is nearer to a diamond
film than a "typical" DLC film, which can have a high hydrogen
concentration.
   The publications found in a computer search with the key words "diamond"
and "diamond-like films" were used as the literature sources. The technical
data were taken mainly from the book, The Properties of Diamond (7), and the
recent review book, Diamond Films: Evaluating the Technology and
Opportunities (8), was used in evaluating the present status of this field.

C.J. McHargue et al. (eds.), Structure-Property Relationships in Surface-Modified Ceramics, 455–475.
© 1989 by Kluwer Academic Publishers.

TABLE 1. Some typical methods frequently used in the analysis of diamond and diamond-like films

---

Scanning electron microscopy (SEM)
Transmission electron microscopy (TEM)
Transmission electron diffraction (TED)
Reflection high energy electron diffraction (RHEED)
Electron spectroscopy for chemical analysis (ESCA)
Electron energy-loss spectroscopy (EELS)
Auger electron spectroscopy (AES)
X-ray photo-electron spectrometry (XPS)
Raman scattering spectrometry
Infra red (IR) spectrophotometry
Spectrophotometry
Optical polarizing microscopy (OPM)
Nuclear magnetic resonance (NMR) spectroscopy
Rutherford backscattering spectroscopy (RBS)
Vicker's hardness measurements

---

2. ON THE PROPERTIES OF DIAMOND AND DIAMOND-LIKE FILMS

The most common crystalline carbon allotrope is graphite. Its hexagonal crystalline structure is shown in Fig. 1. However, obviously the best known crystalline form of carbon is diamond (Fig. 1). In addition, carbon has two other allotropic forms, amorphous and "white carbon." According to the CRC Handbook of Chemistry and Physics (9), "little information is presently available about this (white carbon) allotrope."

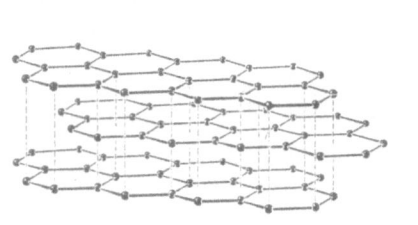 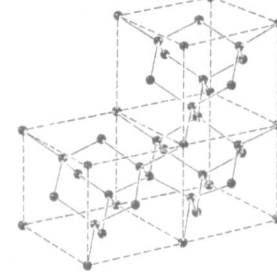

FIGURE 1. Structures of graphite and diamond.

A general belief that diamond is a very stable form of carbon is not exactly true (cf. "diamonds are forever"). In Fig. 2, the thermodynamic balance for crystalline carbon is presented. As can be seen, graphite has the lowest free energy and therefore it is the equilibrium form, which means that at room temperature in a certain time all diamond will transform into graphite. Accordingly, at elevated temperatures diamond and especially diamond-like carbon transforms into graphite. The DLC coatings transform to graphite at 400 to 600°C. It is evident that the transformation temperature

depends on the purity of the coatings. However, there is no general criterion for the boundary between a D and a DLC coating. The microstructure
varies depending on the deposition methods. Generally, the structure is
amorphous with possibly small (10-100 nm) microcrystals of diamond. The
carbon atoms are dominantly bonded by sp^3 bonds that gives the material many
physical and chemical properties very similar to natural diamond and
justified the name diamond-like carbon. At present, it seems that the limit
is set by the detection limit of the best methods available; if the sp^2
bonding is detected the coating is diamond-like.

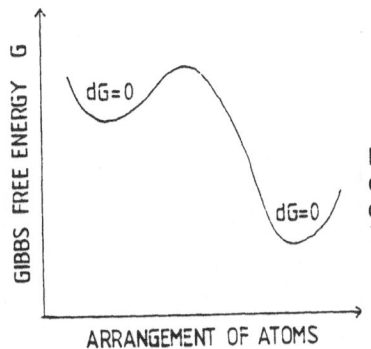

FIGURE 2. Schematic variation of Gibb's free
energy with the arrangement of atoms for
diamond (the upper minimum) and graphite (the
lowest free energy value).

Diamond has many exceptional properties; some are given in Table 2
(ref. 7). The competitive materials to diamond are other hard materials,
especially various ceramics. The comparison diagrams between some conventional ceramics, steel, hard metal, diamond, and DLC coatings are
illustrated in Fig. 3.

TABLE 2. Some general properties of diamond

Density	3.5 g/cm³
Hardness	75-90 GPa
Friction	0.1-0.05
Heat conductivity	10-20 W/cmK
Resistivity	10^{18} Ωm (in the dark)
Refractive index	2.4
Wear resistance	Excellent
Acid resistance	Excellent
Temperature resistance	900-1800 K (graphitization temperature)

3. MASS-SEPARATED CARBON ION-BEAM DEPOSITION
3.1. Equipment
All isotope separators with low ion energies available (<2 keV) can be
used without rearrangement for the production of the DLC coatings. In
Fig. 4 the scheme of the isotope separator in our laboratory is presented.
First, carbon is transformed to ion form in the ion source, after which the
ions are accelerated and focused. In order to remove impurities, the ion
beam is curved with the magnetic field. Finally, the ions are decelerated
before hitting the sample. The maximum ion current of our separator is

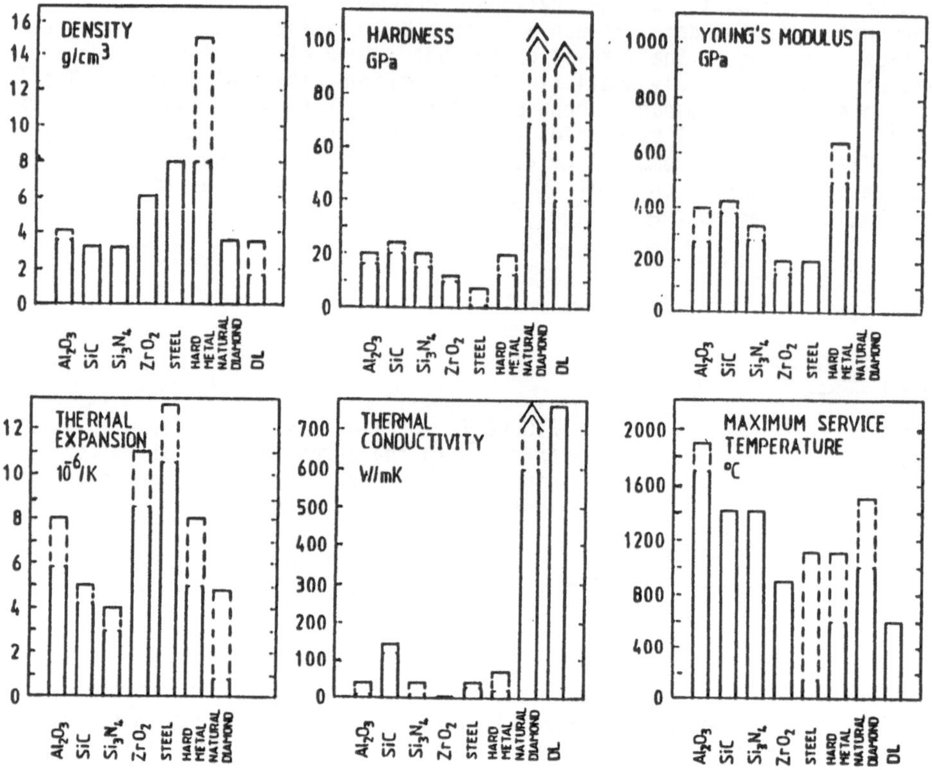

458

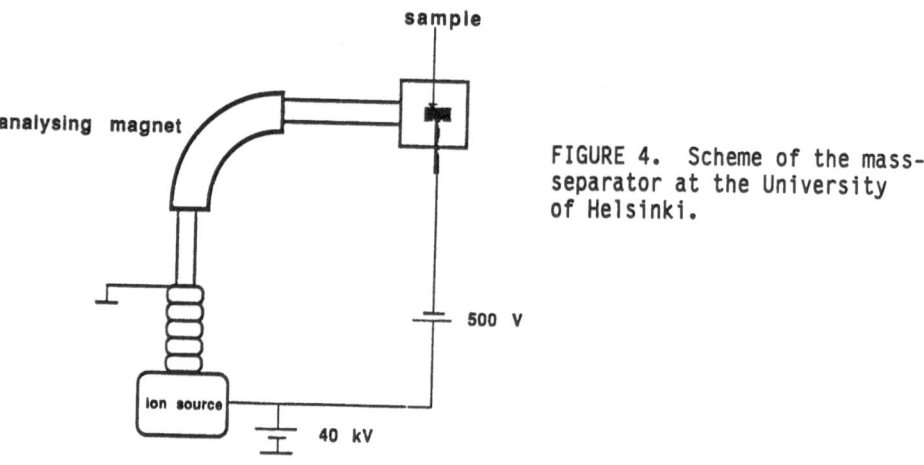

FIGURE 3. Selected physical properties of diamond compared to ceramics, steel, hard metal (10) and DL carbon (11-13).

FIGURE 4. Scheme of the mass-separator at the University of Helsinki.

some tens of microamperes, i.e., it is very low. The Kyoto group (14) uses a negative ion source and has achieved ion current of some hundreds of microamperes. However, even remarkably higher currents would be desirable for the production of DLC coatings. Although there are numerous carbon compounds, the generation of high intensity carbon beams has turned out to be a difficult problem. Carbon has a tendency to collect onto the surfaces into the ion source and in this way to preclude long runs.

3.2. Advantages of mass separation

The great advantage of the mass separator is its usefulness in the study of the DLC layers. The use of separated ion beams offers the best way of studying the deposition parameters under very controlled circumstances: ion energy, ion current density, angle of incidence, and amount of impurities can be controlled in a unique way. Due to the mass separation, the DLC films can be expected to be very pure. The only impurities are those which originate from residual gas in vacuo. Because the energy of the carbon ions is rather high, it can be expected that they sputter impurity atoms or molecules from the target. In Fig. 5 the backscattering spectrum taken from a self-supporting DLC foil is illustrated (3). As can be seen, there are no impurity elements heavier than carbon, except oxygen. The quantities of the $Z \geq 6$ impurities were ≤ 0.01 to 0.0001 at. % depending on the atomic number of an impurity. The quantity of hydrogen was lower than the detection sensitivity of forward recoiling spectroscopy (FRES), i.e., 0.5 at. % (Fig. 6).

The thickest DLC coatings prepared in our laboratory have been 15 µm, but there is evidently no upper limit to the growth of the film, which is proportional to the bombardment time. This is one advantage of ion beams over other methods, where there is a low maximum thickness. The areal current density value used, 1 mA/cm², indicates that rather high currents could be used without DLC film formation suffering. The maximum value of the areal current density for the production of high-quality films is not known, but it is higher than 1 mA/cm². This makes high production rates possible, in principle.

3.3. Experimental results

3.3.1. General. Although the use of the mass-separated ion beams has the advantages mentioned above, there are not many laboratories which use this method and report in the literature. Therefore this review is mainly based on the results of our group (1-6). The results have also been compared with those of the Kyoto group (14). In our laboratory, the properties that can be determined with existing analytical devices — such as hardness, friction, electrical conductivity, density, etc. — have been measured. The measurement of these properties is sufficient for practical uses but for the research purposes the determination of the ratio for the sp² and sp³ bondings would have been desirable.

3.3.2. Density. The density measurement of whatever material is a straightforward procedure if the material is available in sufficient quantity. However, in the DLC case, the quantity of material is very low, so that even typical low-density measurement methods failed — i.e., the flotation type; part of the DLC foil sank and part remained floating. A reason for this was that the foil adsorbed air differentially. Therefore, different nondestructive methods have been developed for the density measurements. The areal density of the layer was determined with two nuclear physical methods, i.e., with Rutherford backscattering spectrometry (RBS) and the nuclear resonance method (3). In Fig. 7, RBS spectra are shown. The areal density (g/cm²) can be obtained from the width of the carbon peak or from the shift of the edge of the substrate element in the spectra corresponding to the coated and uncoated samples. In Fig. 8, the nuclear resonance shift due to the carbon layer is shown. The advantage of the resonance shift method is its good resolving power in measuring thin layers. In addition, when the thickness of the DLC layer is determined with the profilometer (Fig. 9) the density in units of g/cm³ can be determined.

460

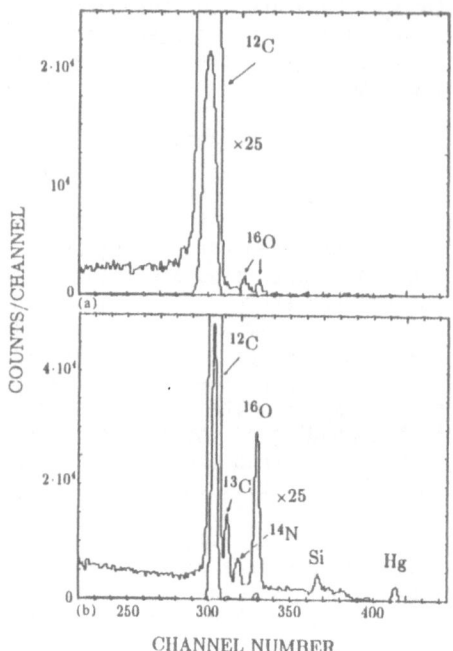

FIGURE 5. Backscattering spectra of 2.4 MeV protons from (a) self-supporting pure DLC ^{12}C and (b) evaporated carbon films. The impurities are mainly due to adsorption (4).

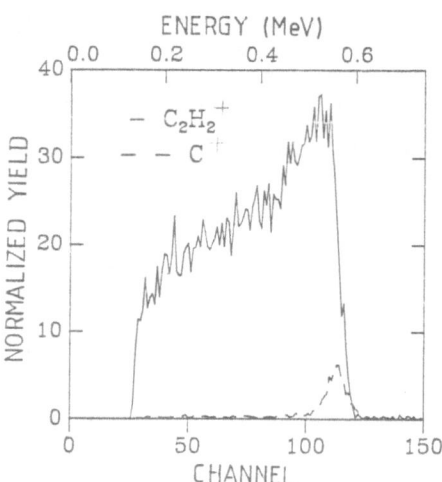

FIGURE 6. FRES spectra for the hydrogen profiles of the DLC films prepared with the 1.0 keV $C_2H_2^+$ and 0.5 keV C^+ ions.

461

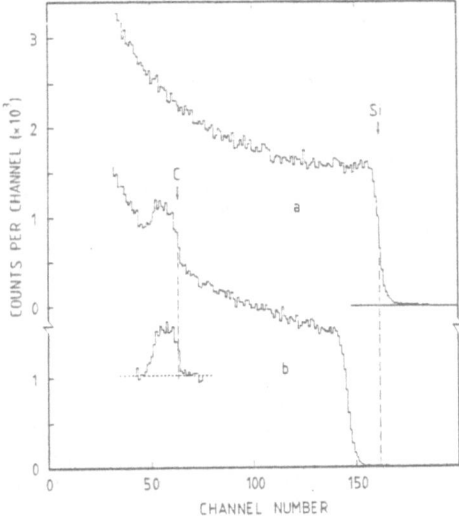

FIGURE 7. Backscattering spectra of the 1.0 MeV ⁴He⁺ ions from a pure silicon sample (curve a) and from a silicon sample onto which a 47 nm DLC layer had been deposited (curve b) (ref. 3).

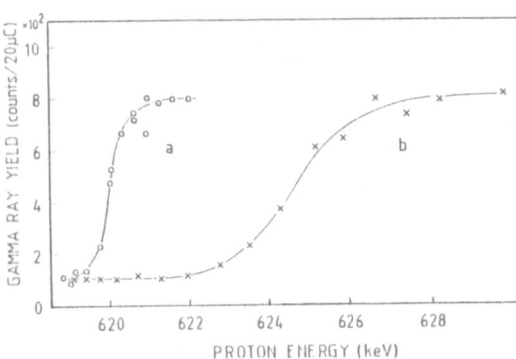

FIGURE 8. Shift in the E_p = 620 keV resonance of the $^{30}Si(p,\gamma)^{31}P$ reaction corresponding to the uncoated silicon sample (curve a) and the silicon sample coated with a DLC layer (curve b).

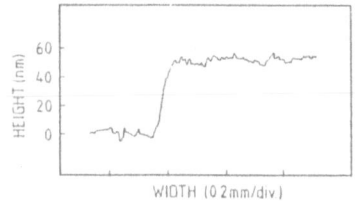

FIGURE 9. Profilometer plot from the edge of the carbon layer on the silicon sample.

462

Another method which was proposed in an earlier work (3) is based on the Doppler shift attenuation (DSA) method. In Fig. 10 the use of the DSA method is illustrated. As can be seen, the attenuated γ-ray peak profiles are quite different for the diamond and the graphite. By comparing the profiles deduced from the DLC layer, the density of the layer can be determined. However, the use of the method is not so simple because it requires rather complicated equipment and experience in nuclear physical measurements. Figure 11 shows the density measurements of the foil as a function of the ion energy. These results have been combined for references 6 and 14. As can be seen, the highest density occurs at an ion energy of about 0.5 keV.

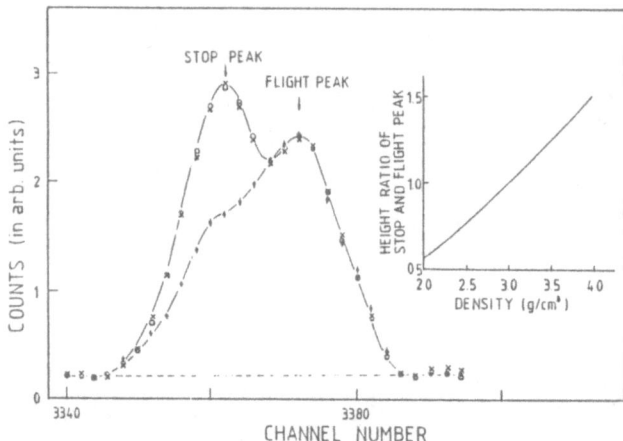

FIGURE 10. DSA measurement of the 2313 keV γ-rays from the $^{13}C(p,\gamma)^{14}N$ reaction using graphite. (•), diamond (x), and DLC (o) backings. The inset shows the height ratio of the stop and flight peak as a function of density.

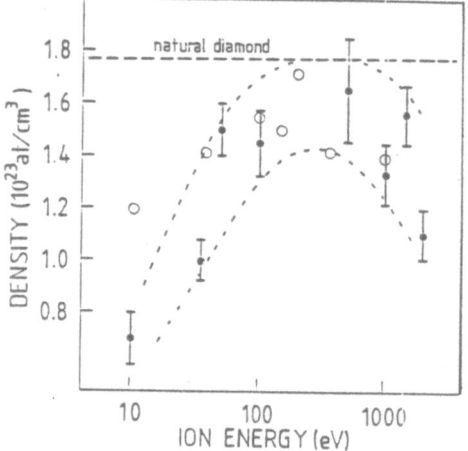

FIGURE 11. Atomic density of the DLC films as a function of ion energy. The black points are from ref. 6 and the open circles from ref. 14.

3.3.3. Hardness. The hardness value is good for the identification of diamond, because the hardness of diamond is remarkably higher than that of any other elemental form. However, there are difficulties in the hardness measurements and the comparison of the results. The hardness of natural diamond depends on the crystal face [the (111) face is the hardest for diamond]. The hardness also depends on the load of the indenter. With low loads (100 g, 0.98 N) Knoop values as high as 18,000 kg/mm² are obtained; with higher loads the values decrease (with 1 kg, 9.8 N) load; 7,000 kg/mm², 69 GPa) (ref. 7). One complicating factor is that the indenter tip is also made of diamond. When using high loads, the diamond tip has a tendency to break because of the brittleness of diamond. As a rule, the prepared D and DLC layers are thin, ranging from some nanometers to some micrometers. Therefore, the use of low loads is obligatory. For this reason, special devices called micro- and nanoindenters have been developed (e.g. ref. 15). In Fig. 12 the hardness results measured with the load dependent hardness testers are illustrated. In Fig. 13 the dependence of the hardness on the ion energy is shown. As can be seen the maximum hardness is reached at approximately 500 eV.

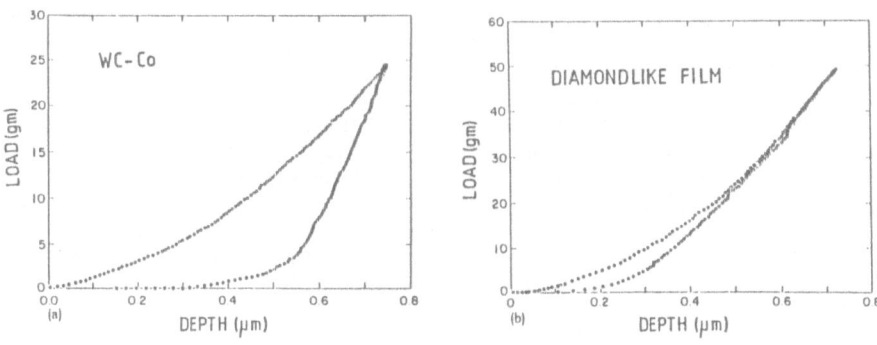

FIGURE 12. Loads versus indentation depth for the WC-Co substrate (a) and DLC film (b). The upper curves depict increasing load and the lower ones decreasing load (4).

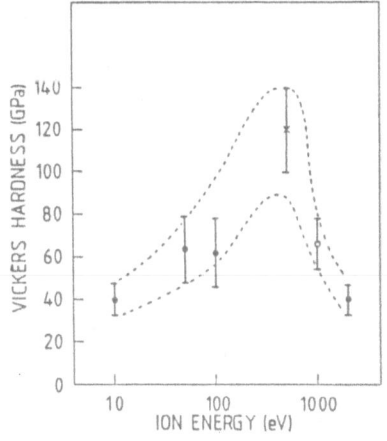

FIGURE 13. Vickers hardness of the DLC films as a function of ion energy. The used loads were 20 g (0.196 N) (•), 50 g (0.49 N) (o), and 200 g (1.96 N) (x) (ref. 6).

3.3.4. <u>Friction</u>. Because of the extreme hardness of the diamond, the friction coefficients are generally low, about 0.1. The friction of a steel tip against the DLC foil was measured with a pin-on-disc apparatus. The results are shown in Fig. 14. In Fig. 15, SEM pictures from the tips of the pins are shown. As can be seen very adhesive wear occurs in the case of the steel pin on WC-Co. The friction between steel and the DLC surface is very low. Due to the low friction, good wear resistance, and moderately good stability at elevated temperatures the DLC coatings are very promising candidates for bearings; especially for the cases where the use of lubrication is not possible.

3.3.5. <u>Wear resistance</u>. There are several ways to measure the wear rates of the materials. The wear rates and results depend on many different parameters such as the quality of the wearing material, whether the surface is dry, in liquid or slurry, the wear speed, etc. In Fig. 16 a typical wear test result is shown (4). The test was performed with a reciprocal motion machine.

The results shown in Figs. 17 and 18 were obtained with the arrangement where the sample was worn using a Cu wheel rotating on the sample (6). The sample was kept in a vessel filled with a slurry of castor oil and silicon carbide powder. The wheel was pressed with a constant force and it rotated at a constant speed. The electrical resistance between the substrate and the Cu wheel was monitored, and wearing was terminated when a hole was produced in the DLC film. As can be seen in Fig. 17, the minimum wear rate curve as a function of ion energy is at approximately 0.5 keV. Comparison of the wear rates between the DLC and three ceramics is shown in Fig. 18. The wear rate in the DLC foil is evidently drastically lower than of the other ceramics. The wear resistance of the DLC foil is also much better than that of the general wear resistant tool materials, e.g. WC-Co hard metal and TiN coating (see Fig. 16).

3.3.6. <u>Adhesion</u>. In the preparation of coatings, a difficult feature is the adhesion to the substrate. Especially in the case of the D and DLC coatings this can be expected to be a rather serious problem in many cases because the thermal expansion coefficient of diamond is very low as compared to other conventional materials. In the hardness measurements it turned out that the coatings deposited with an ion energy lower than 50 eV failed because of the poor adhesion (6). The coatings cracked off entirely in the adhesion test. It seems that rather high ion energies are required, at least at the beginning of the deposition process. Ions with an energy high enough can modify the coating-substrate interface in many ways: produce ion beam mixing at the interface, break adsorbed impurity layers remaining on the substrate, enhance chemical bonding of carbon atoms to the substrate and even enable an adherent interface layer to grow. The great advantage in the use of ion beam deposition is that high enough ion energies can be used to overcome adhesion problems. On the other hand, it is certain that serious adhesion problems are confronted in processes where sufficient ion energy is not available.

3.3.7. <u>Thermal stability</u>. One obvious drawback of the DLC coatings is their low thermal stability compared to that of diamond. In our laboratory, the stability was studied by heating the DLC coated WC-Co hard metal samples in a vacuum of 100 μPa for 1 h in the temperature range of 100 to 800°C in steps of 100°C. After each heating the hardness was measured and the results are shown in Fig. 19. Up to 600°C the hardness was constant, but after that the coating became softer. At 800°C the coating peeled off. Diamond transforms into graphite at 1500°C in vacuo but in the air a black coating can form on the surface of diamond at about 600°C (7). This temperature also corresponds to the softening temperature of the DLC coating.

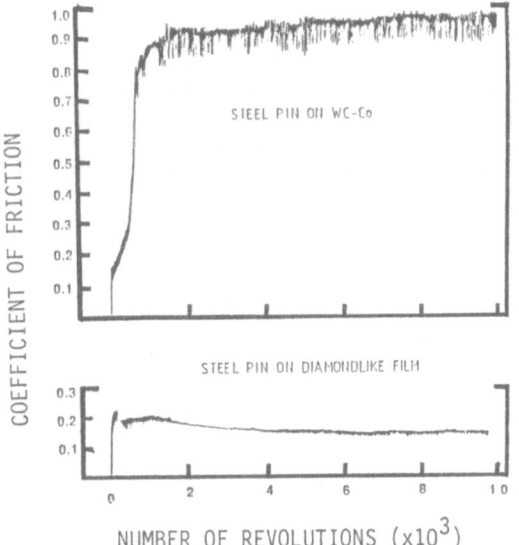

FIGURE 14. Friction of steel pins on WC-Co and DLC surfaces as a function of number of revolutions in the pin-on-disc apparatus.

NUMBER OF REVOLUTIONS (x10^3)

Pins

Steel pin on WC-Co   Steel pin on diamond-like film

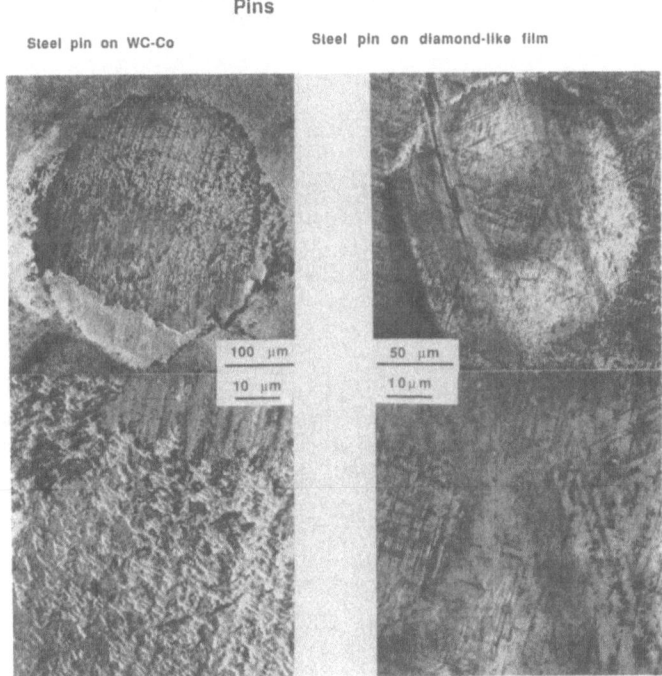

FIGURE 15. SEM photographs from the tips of the pins used in the experiments of Fig. 14.

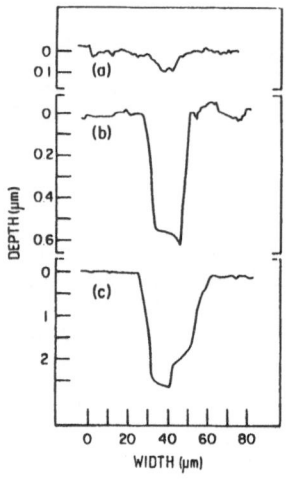

FIGURE 16. Profilometer diagrams from the
wear measurements performed with a 20-g load
after 28,000 passes on the mass-separated
ion beam deposited diamond-like ^{12}C surface
(a), TiN coating (b), and WC-Co hard metal
(c) (ref. 1).

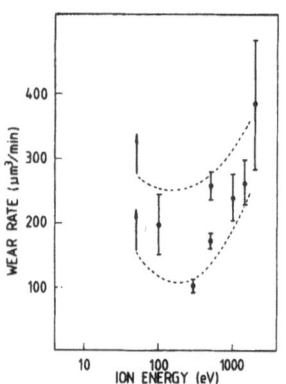

FIGURE 17. Abrasive wear rate of the DLC
films as a function of ion energy. The
deposits at ion energies $E_i \leq 50$ eV cracked
off immediately (6).

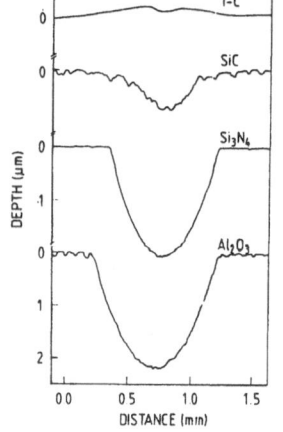

FIGURE 18. Wear track profiles in the DLC layer
deposited at $E_i = 1$ keV and in three different
bulk ceramic samples (6).

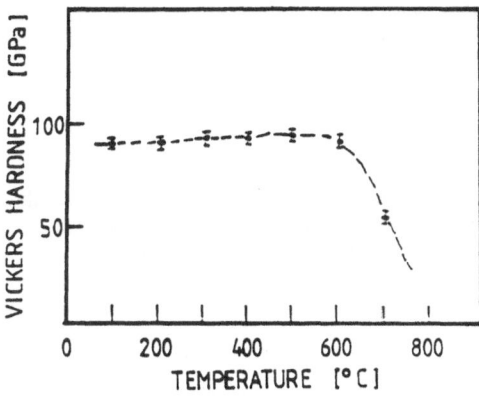

FIGURE 19. Hardness of the DLC foil prepared with the 0.5 keV ions as a function of temperature. The samples were annealed for 1 h in vacuo at 100 μPa (ref. 6).

3.3.8. Electrical resistivity. One rather easily measurable quantity for DLC coatings is the resistivity. As a rough test, any ohm-meter can be used to test whether the prepared carbon layer is close to graphite or diamond, because the resistivity of natural diamond at room temperature and in the dark is very high, about 10^{18} Ωm (ref. 7) [that of graphite is 0.14 Ωm (ref. 9)]. In Fig. 20 the electrical resistivity as a function of deposition energy is shown. This figure contains results from two separate works and as can be seen the results are very consistent. The highest resistivity is, as in the case of hardness, again at about 0.5 keV.

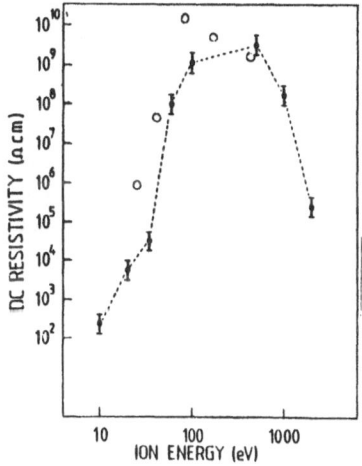

FIGURE 20. DC resistivity of the DLC films as a function of ion energy. The black points are from ref. 6 and the open circles from ref. 14.

3.3.9. Other properties. A typical feature of the DLC coating is its good acid resistance, e.g. in the experiments where the coating was kept in strong HF acid for 24 h at 20°C no solution of the DLC foil was observed. On the basis of x-ray diffraction, transmission electron diffraction and alpha-particle channeling experiments, the DLC-foils prepared are amorphous. This result also agrees with that of ref. 14.

In our laboratory no optical measurements were performed, but it is evident that they are close to those of ref. 14 because the other measured properties are very consistent. The maximum value reported for the optical gap was 1.5 eV (ref. 14) and was obtained with an ion energy of 100 eV, which corresponds well to the other properties of the foils as shown in Figs. 11, 13, 17, and 20.

4. PLASMA-ARC-DISCHARGE DEVICES
4.1. General

Although with the mass-separated ion-beam system, DLC films can be produced very controllably, the production rate is too low for practical applications. Therefore, more efficient ion sources were sought. There were numerous possibilities, but at first sight, there was no clearly superior system. Some failed systems are presented as examples. The main principle is naturally that carbon should be transformed to the ion form very effectively. Gaseous materials would be very suitable for this purpose because in many cases they can be ionized rather easily. An additional advantage is that gas can be fed continuously to the ion source. In addition, no droplets formed as occurs in the case of solid carbon. Carbon has numerous gaseous compounds. However, such compounds which would be easy for mass-separation purposes, i.e. containing mainly carbon atoms with somewhat heavier elements but no lighter elements, are difficult to find. Because such gas was not found acetylene was selected, although it contains 50 at. % hydrogen. Acetylene is exothermic and decomposes easily, e.g. in a heated tube. The drawback was that carbon contaminated the tube in a short time. It may be possible that by using high tube temperatures, over 3000°C, no contamination will occur. It is evident that the carbon beam produced could be ionized with the electron beam. Another possibility was to use electron guns. Unfortunately, it turned out that the maximum ionization cross-section of electrons to carbon lies at about 300 eV (ref. 16). The cross-section decreases rapidly with acceleration voltage. On the other hand, high voltages are needed to focus the electron beam to a small spot, i.e. some kV. Although this idea failed, it is not quite hopeless because carbon evaporations with a 4-kV and 3-kW electron gun, DLC deposits were found in some parts of the apparatus.

The final selection was the arc-discharge system. Because in this system the anode and cathode are made of solid carbon, i.e. of graphite, no impurities should be present. Many Russian groups (17-24) have used this method for the preparation of the DLC films and achieved very promising results, e.g. films which were harder and more wear resistant than natural diamond have been prepared. The ion source based on this principle is that presented by Brown, et al. (25). The maximum current obtained from this ion source is as high as 1 A. However, maximum operation time has not been reported.

One final factor in favor of the selection of the arc-discharge system for the DLC preparation was the observation by Dorodnov et al. (20) who reported ion energies up to 10 keV. This would mean that there are no difficulties with the adhesion of the coating to the sample because the ion source alone could act as the production source. However, it turned out later that carbon macroparticles are a serious difficulty in the sophisticated production of high quality DLC films. If the presence of macroparticles is accepted, the construction of the device which produces the DLC films with reasonable quality is a rather straightforward task. The elimination of macroparticles, however, is difficult if high energy ions are also required.

The final refinement of the arc-discharge technique is still in progress in our laboratory. In the following sections the main principles, difficulties, and results are presented.

4.2. Device construction

4.2.1. Theory.
There is no generally accepted theory of the behavior of arc-discharge plasma. This complicates the planning of an arc-discharge system. Dorodnov (18) has widely treated this field in his articles. According to him the acceleration of the plasma occurs with the Ampere force. Because in the plasma acceleration case the Ampere force is proportional to the square of the current; this means, a priori, that by increasing the current, higher plasma energies could be achieved.

4.2.2. Devices reported in the literature.
One of the most simple arc-discharge systems is that presented by Maslov et al. (19). It is based on pulsed arc-discharge techniques. Its main parts are: cathode, anode, igniter, and straight focusing solenoid. In Fig. 21 the scheme of the corresponding device, developed in our laboratory, is shown. The only essential difference is that the solenoid is curved. The main principle is as follows: in the igniter current circuit the capacitor of some tens of pF is charged with a voltage of some tens of kV. Between anode and cathode there is a much lower voltage, some hundreds or thousands of volts, but the capacitance of the capacitor bank is at least a thousand times higher. The advantages of the arc-discharge system are good reliability and a high production rate, but it also focuses well the carbon plasma streams. However, both solenoid types have disadvantages. If the straight solenoid is used, high energy (i.e. adhesive) plasma streams can be produced, but a part of the microparticles generated by the arc-discharges hit the sample and reduces the quality of the DLC coating. Microparticles are not a problem with the curved solenoid but curving of the energetic plasma is difficult.

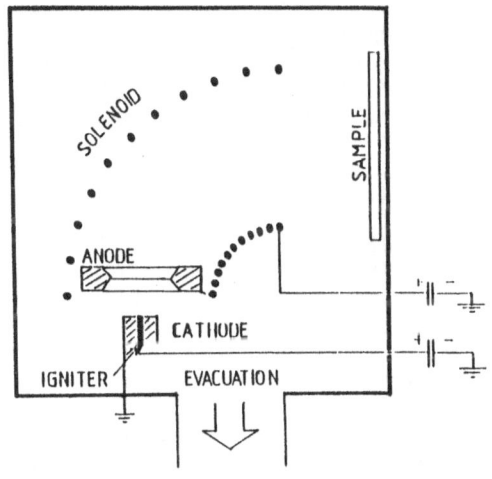

FIGURE 21. Scheme of the curved arc-discharge system.

Dorodnov et al. (18,20) have presented many different systems for the generation and acceleration of the plasma streams with arc-discharge techniques. In Fig. 22 one type is presented. Another type is described by the Kharkov group (17,21-24). It is based on a stationary curving magnetic field and a stationary current. Its main principle is presented in Fig. 23. Its advantage is that microparticles and neutrals can be rejected. The disadvantage is that the average energy of the ions in plasma is low, about 20 eV, which seemed to be near to the optimum because by increasing the current the ion energy decreased. Recently, a quite similar system has been constructed by an Australian group (26). They have confirmed the extreme hardness of the DLC coating achieved by the Kharkov group (17,21-24). The disadvantage of this system is its low ion energy, which in many cases results in low adhesion.

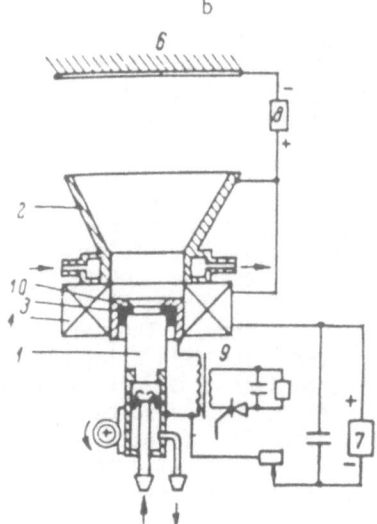

FIGURE 22. Scheme of the device for the DLC preparation using the pulsed current and the stationary current magnet presented by Dorodnov (18): (1) cathode, (2) anode, (3) insulator, (4) electromagnetic coil, (5) magnet system for stabilizing the spot, (6) condensation surface, (7) power supply for the main discharge, (8) biaspotential source, (9) ignition supply circuit, and (10) ignition electrode.

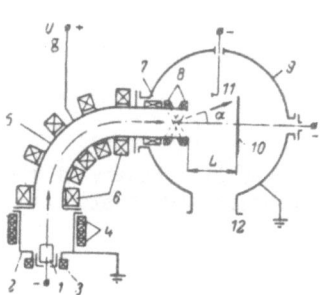

FIGURE 23. Scheme of the device for the DLC preparation using the stationary current and stationary curved magnet presented by Aksenov et al. (21): (1) cathode, (2) anode, (3) stabilization coil, (4) focus coil system, (5) plasma guide, (6) plasma-guide magnet system coil, (7) focus output coil, (8) scanning system coil, (9) vacuum chamber, (10) collector, (11) probe, and (12) vacuum pump connection.

4.2.3. Pulsed plasma stream and curved solenoid. In our laboratory in searching for a high efficiency carbon ion source, the first modification developed is shown in Fig. 21. It is a modified combination from those presented by Aksenov et al. (21) and Maslov et al. (19). For the curving of the plasma stream the pulsed anode-cathode current was led through the curved solenoid. A system was built where the curving magnetic field was stationary, as by Aksenov et al. (21) (Fig. 23), but where plasma was generated with pulsed arc-discharges.

The main feature in both these systems is that in a rather extensive region of the parameter combinations, DLC coatings can be produced. However, if curving of high energy plasma is required very strict parameter values should be used. No direct plasma ion energy measurements were performed, but a rough estimation of the average ion energy was obtained indirectly. The plasma spot on the sample was about 10 cm in diameter. The magnetic field compresses the plasma stream, but depending on the strength of the magnetic field it succeeds more or less, or in other words the high energy ions can escape and the ions with lowest energy hit the center of the plasma stream spot. As can be seen in Fig. 24, the plasma stream expands strongly immediately outside the solenoid. The divergence of the ion energies in the plasma can also be seen in the "failed" experiment as follows: in the outer sections of the spot the deposited carbon is bright, adhesive, with high electrical resistance and thus it is in the DLC form; in the center the carbon is black, nonadhesive and electrically conducting. Between these two regions there is white carbon, which is soft and loosens easily. This same energy effect can also be observed with a Scotch tape test in a "normal" experiment where the deposited carbon is generally bright and has high electrical resistance, but the layer from the center section can be lifted off with the tape, which means that its adhesion is poor and also the energy density of ions is evidently low.

FIGURE 24. Picture from the plasma explosion in the arc-discharge system presented in Fig. 21. Notice the expansion of the plasma stream outside the solenoid.

In the system shown in Fig. 21, the magnetic field is obtained with the cathode-anode current, which confirms that no delay or phase shift between it and the plasma stream exists. However, to couple strong magnetic fields with pulsed currents is complicated. The magnetic field, H, of the solenoid is approximately proportional to NI/ℓ, where N is the number of the coils of the solenoid, I is the current, and ℓ is the length of the solenoid. On the other hand the opposing electric voltage, U, , counteracts this magnetic field, H. U is proportional to $-LdI/dt$, where L is the inductance of the solenoid, and t is time. The inductance of the coil is proportional to N^2r^2/ℓ, where r is the radius of the coils. This means that in order to reach strong magnetic fields with the pulsed electric-arc discharges, N and r of the solenoid must be small and ℓ must be short. With the short (about

1 cm) solenoid system containing about ten coils of 5 cm diameter, the adhesion of the deposited layer was good in all tested materials. In the curving system six coils were used, which also yielded a rather good adhesion, but the sample distance was in a narrow region. This is due to the fact that although the magnetic field of the solenoid compresses plasma into a rather narrow stream, outside the solenoid the plasma stream is expanding very rapidly. This occurs because the magnetic field of the solenoid also decreases very rapidly outside the solenoid. The sample distance of about 5 cm from the solenoid was used in our geometry. If the shortest distances were used, electrical breakdown occurred between the sample and anode. In the case of longer distances the plasma stream starts separating, i.e. in the outer sections of the spot the ion energies measured with adhesion tests are higher than in the middle of the spot.

According to Dorodnov (18) the minimum current in the pulsed-arc-discharge sources is 2 kA, which agreed with the present results. The current and the shape of arc-discharge pulses were determined by measuring the voltage changes over the precision resistor with the oscilloscope. Surprisingly, high current values, even up to 15 kA, were possible to use without any radical changes in the deposited film. Many different capacitors were tested, and although all were aimed for the use of rapid pulses, in some cases the preparation of the DLC foils failed. Although the shapes of the current pulses were rather equal with different capacitors it is evident that the rise time of the current pulse should be rather high which means that the internal inductance of the capacitors should be low.

The ignition of the anode-cathode plasma pulse was performed with the igniter switch which was constructed by mechanically moving the point of the sparking-gap in the air. The sparking-gap, which is in the same circuit as the ignition gap, also causes the spark in vacuo, then igniting an anode-cathode plasma-pulse. The voltage was 20-30 kV and the capacitance of the capacitor was 20-50 nF. Although not studied systematically, the parameters of the ignition system had no radical significance of plasma formation. A totally electronic ignition system is possible, but expensive and could be destroyed by strong pulses. In Fig. 21 the solenoid coils encounter the anode but rather good results could be achieved without any coils. The disadvantage is that the spread is wide and the macroparticle content of the plasma stream is higher, e.g. according to Maslov et al. (19) and Martin et al. (27). However, without a solenoid, the current might have been higher and thus the macroparticle yield could be expected to be higher also. According to Ampere's law (18) the energy of the ions should be increased when the current increases, but it is evident that the strong pulses (the high current and long duration time of the pulse) have mainly an exploding effect and only a slight energy increasing effect. Thus, the main part of the material changes is due to microparticle bombardment rather than to the plasma. As a general remark the system should be mechanically rigid, electrically well isolated and effective water cooling is essential.

4.2.4. Pulsed plasma stream and curved stationary magnet. In the case of the curved solenoid the magnetic field is pulsed and difficult to measure. Therefore the system with a stationary magnetic field resembling that of Fig. 23 was constructed. It turned out that although the measurement of the strength of the magnetic field was easy, to get the magnetic field strong enough was not. With the maximum arrangement (15 A and 1500 coils) the available magnetic field strength was 1500 G. It is possible that if stationary fields with the desired strength would be available all adhesion problems could be avoided. However, the stationary magnet is an energy-wasting system and probably is not suitable for large-volume production purposes. The use of permanent magnets fails because the strong pulses demagnetize the magnetic field, at least in the case of conventional ceramic permanent magnets.

4.2.5. Properties of the films. Although the development of the plasma devices is still in progress, some properties of the films have been measured. Because the main emphasis has been to produce hard and wear resistant DLC coatings, the wear resistance was compared with that of the most conventionally used hard coating, i.e. TiN. The wear of the TiN layer by hardened steel pins was 10 times faster than that of the hard carbon layers. The friction of the hard carbon layer (0.1) was significantly lower than that of TiN (0.5). The DLC films had some other properties of diamond such as high hardness (about 40 GPa) and high electrical resistivity (10^8 Ωcm).

5. APPLICATIONS

The preparation of DLC films by some methods has reached the commercial stage. In the advertisement of a company manufacturing DLC films, extremely hard films are promised to be prepared on the surfaces of metal, ceramic, and glass components. The process has been used to treat cutting tools, plastic injection molds, metal stamping tools, and wear components such as gears and bearings.

These first applications relating to the tribological field are typical in the first stage. The DLC films are also used as different optical shield coatings. However, the electronics industry will be the largest commercial user. Although diamond is an excellent dielectric material, when doped, it is also a good semiconductor. In addition, the heat conductivity of diamond is four times better than copper and the radiation resistivity of diamond semiconductors is good. Consequently, the construction of densely packed high power, high temperature and radiation resistant electronic devices is in principle possible.

6. SUMMARY

Both the method presented in this article are competitive in their own fields. They are still in a stage of development as to the preparation of diamond or diamond-like films. The main disadvantage of the mass-separation method is the lack of a suitable ion source. However, although the mass-separation method is slow for production purposes the mass-separated ion beams are excellent for the study of diamond and diamond-like foils. The properties of the foils can be investigated over a wide range of deposition energies and currents. In addition, all parameters are accurately known and the deposition process can be easily repeated. Accordingly, difficulties such as poor adhesion and high temperatures of the sample frequently occurring in the coating processes can be easily overcome.

The advantages of the plasma-arc-discharge device are rather indisputable: it has simple construction, although it is difficult to find the optimum geometry. It has a high production rate, even with very slow pulse rates (1 pulse/s): the deposition rate corresponds to an ion current of about 1 A, which is a thousand times higher than that of the conventional "effective" ion sources. The process is clean because the plasma is made of pure solid graphite. The device withstands long-period runs and its service is simple compared to conventional ion sources. It is possible to produce ion energies high enough for good adhesion. The foils can be produced at low temperatures.

One striking disagreement between these two methods is the ion energies. The optimum ion energy in using the mass-separated system is 100-500 eV, but recently Martin et al. (26) reported that they have prepared, with the arrangement presented by Aksenov et al. (21), excellent DLC films with much

474

lower ion energies, 20 eV. The DLC foils prepared with the mass-separation method using the ion energies lower than 50 eV were very brittle and detached from the sample. This difference may be due to the charge equilibrium of the carbon plasma. In addition, quite recently Kasi et al. (28) have confirmed with Auger spectroscopy the formation of the "diamond-like" structure for mass-separated carbon ions in the energy range of 10 to 300 eV.

REFERENCES

1. Koskinen J, J-P Hirvonen, and A Anttila: Appl. Phys. Lett. 47 (1985) 47, 941.
2. Anttila A, J Koskinen, J Räisänen, and J-P Hirvonen: Nucl. Instr. Meth. B9 (1985) 352.
3. Anttila A, J Koskinen, M Bister, and J-P Hirvonen: Thin Solid Films 136 (1986) 129.
4. Anttila A, J Koskinen, R Lappalainen, J-P Hirvonen, D Stone, and C Paszkiet: Appl. Phys. Lett. 50 (1987) 132.
5. Hirvonen J-P, J Koskinen, A Anttila, D Stone, and C Paszkiet: Mater. Sci. Engr. 90 (1987) 343.
6. Koskinen J: J. Appl. Phys. 63 (1988) 2094.
7. Field JE (ed): The Properties of Diamond. London: Academic Press, 1971.
8. Diamond Films: Evaluation of the Technology and Opportunities. Englewood/Fort Lee, NJ: Technical Insights, Inc., 1987.
9. Weast RC and MY Astle (eds): Handbook of Chemistry and Physics. 61st edition. Boca Raton, FL: CRC Press, 1981.
10. Manufacturers' statements.
11. Weismantel C: Thin Films from Free Atoms and Particles. KJ Klabunde (ed). London: Academic Press, Inc., 1985, p. 153.
12. Jansen F, M Machoukin, S Kaplan, and S Hark: J. Vac. Sci. Technol. A3 (1985) 605.
13. Tsuji T, K Ogawa, J Ishikawa, and T Takagi: Proc. IPAT 87. Brighton, UK: 1987, p. 134.
14. Ishikawa J, Y Takeiri, K Ogawa, and T Takagi: J. Appl. Phys. 61 (1987) 2509.
15. Hannula S-P, D Stone, and C-Y Li: Proc. Materials Research Society. New York: North-Holland, 1985, Vol. 40, p. 217.
16. von Ardenne M: Tabellen der Elektronenphysik, Ionenphysik und Ubermikroskopie. Berlin: VEP, Deutscher Verlag der Wissenschafter, 1956, p. 489.
17. Strelnitskii VE, VG Padalka, and SI Vakula: Sov. Tech. Phys. 23 (1978) 222.
18. Dorodnov AM: Sov. Phys. Tech. Phys. 23 (1978) 1058.
19. Maslov AJ, GK Dmitriev, and YD Chistyakov: Instrum. Exp. Tech. 28 (1985) 662.
20. Dorodnov AM, SA Muboyadzhyan, YA Pomelov, and YA Strukov: J. Appl. Mech. Tech. Phys. 22 (1981) 28.
21. Aksenov II, SI Vakula, VE Strelnitskii, and VM Khoroshikh: Sov. Phys. Tech. Phys. 25 (1980) 1164.
22. Aksenov II, VA Belows, VG Padolka, and VM Khoroshikh: Sov. J. Plasma Phys. 4 (1978) 425.
23. Strelnitskii VE, II Aksenov, SI Vakula, VG Padalka, and VA Belows: Sov. Tech. Phys. Lett. 4 (1978) 546.
24. Brown I, J Galvin, and R MacGill: Appl. Phys. Lett. 47 (1985) 358.
25. Strelnitskii, VE VG Padalka, and SI Vakula: Sov. Phys. Tech. Phys. 23 (1978) 222.

26. Martin PJ, SW Filipczuk, PP Netterfield, JS Field, DF Whitnall, and
    DR McKenzie: J. Mat. Sci. Lett. 7 (1988) 410.
27. Martin PJ, DR McKenzie, RP Netterfield, P Swift, SW Filipczuk,
    KH Muller, CG Pacey, and B James: Thin Solid Filmsd 153 (1987) 91.
28. Kasi S, H Kang, and JW Rabalais: Phys. Rev. Lett. 59 (1987) 75.

ION BEAM ANALYSIS AND MODIFICATION OF THIN-FILM, HIGH-TEMPERATURE SUPERCONDUCTORS

Michael Nastasi

Materials Science and Technology Division
Los Alamos National Laboratory
Los Alamos, New Mexico 87545 USA

1. INTRODUCTION

The application of ion beams to ceramic processing, as the rest of these proceedings has shown, is a very powerful technique. In particular, ion beams have been shown to be very effective in producing novel material properties through their rather violent interactions with the component atoms of the ceramics. However, in all these cases ions are used to process and produce changes in ceramics, changes that must be monitored by other techniques. This paper shows that, in addition to modifying materials, ion beams can be used in a more gentle but very powerful way to explore what happens to a ceramic thin film as a result of processing. The following discussions are concerned exclusively with a new and exciting class of ceramic: the high-temperature superconductor (HTS). We will discuss the application of various ion beam backscattering techniques as well as examine the use of ion implantation in the processing of these materials.

2. BACKSCATTERING SPECTROSCOPY

This section reviews some of the basic physical concepts involved in backscattering spectroscopy. The reader should note that this is only a brief review and that a more complete treatment may be found in Refs. 1 through 3. The concepts discussed are the kinematic factor K, scattering cross section σ, and stopping cross section $[\epsilon]$. For a discussion on straggling, the reader is referred to Chapter 1 in Ref. 1 and Chapter 2 in Ref. 2.

2.1. Kinematic factor K

The perception of mass in a backscattering spectrum results from the energy transferred to a target atom by the projectile atom. The kinematic factor K, which is a measure of this, is the ratio of the projectile energy after an elastic collision to that before the collision. This factor can be derived by applying the principles of conservation of energy and momentum to an elastic collision between two masses M_1 and M_2. Applying these principles, we have for M_1 scattering from M_2 (4)

$$K \equiv \frac{E_1}{E_0} = \left[\frac{(M_2^2 - M_1^2 \sin^2 \theta)^{1/2} + M_1 \cos \theta}{M_2 + M_1} \right]^2 . \tag{1}$$

From Eq. (1) we see that the kinematic factor depends only on the projectile and target masses and the scattering angle θ. The scattering angle is defined in Fig. 1.

In many applications the target will contain more than one type of atom. The ability to resolve backscattering from atoms with similar masses requires that the mass difference ΔM produce a large difference in the energy of the backscattered projectile, ΔE. In the vicinity of $\theta = 180°$ and for $M_2 \gg M_1$, which is most often the case, it can be shown that (4)

C.J. McHargue et al. (eds.), Structure-Property Relationships in Surface-Modified Ceramics, 477–501.
© *1989 by Kluwer Academic Publishers.*

478

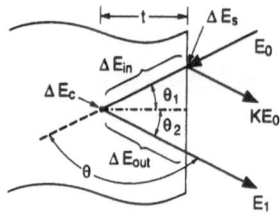

ENERGY LOSS

SCATTERING AT DEPTH t: **SCATTERING SURFACE:**

$$\Delta E_{in} = \frac{t}{\cos\theta_1} \left.\frac{dE}{dx}\right|_{in}$$ $$\Delta E_s = E_0 - KE_0$$

$$E_t = E_0 - \Delta E_{in}$$

$$\Delta E_c = E_t - KE_t$$

$$\Delta E_{out} = \frac{t}{\cos\theta_2} \left.\frac{dE}{dx}\right|_{out}$$

FIGURE 1. The energy loss components for a projectile scattered from a depth t in a single-element target. See Eqs. (3) through (6).

$$\Delta E = E_0(4 - \delta^2)\left(\frac{M_1}{M_2^2}\right)\Delta M \quad , \tag{2}$$

where $\delta = \pi - \theta$ in units of radians of arc. When the scattering geometry and ΔM are fixed, Eq. (2) indicates that good mass resolution (large ΔE) can be obtained by either increasing the primary energy E_0 or increasing the mass of the projectile, M_1. However, applying the latter situation requires some caution because a backscattering signal will not be produced from target atoms in which M_1 is greater than M_2.

2.2. Stopping cross section $[\epsilon]$

A highly energetic projectile traveling through a solid will lose energy primarily through electronic stopping. This energy loss arises from the viscous-like drag that the projectile experiences as it undergoes collisions with the electrons of the target atoms. The amount of energy lost in traversing the target is proportional to the distance traveled and gives the perception of depth in a backscattering spectrum.

Figure 1 presents the energy loss components for a projectile scattered from a depth t. For a projectile starting with an energy E_0, the initial energy loss will occur on the inward path, ΔE_{in}. At a depth t the projectiles energy will have been reduced to E_t. If at this point the projectile suffers a collision with a target atom, it will lose energy ΔE_c. The final energy loss occurs on the outward path, ΔE_{out}. The ultimate projectile energy reaching the detector will be $E_1 = E_0 - (\Delta E_{in} + \Delta E_c + \Delta E_{out}) = KE_t - \Delta E_{out}$. Alternatively, if the projectile has a backscattering collision at the surface without penetrating the solid, the only energy loss will be due to kinematics, and the detected projectile energy will be KE_0.

The total energy difference ΔE between projectiles scattered at the surface and at some depth t is related to the scattering depth through the relationship (4)

$$\Delta E = KE_0 - E_1 = K\Delta E_{in} - \Delta E_{out}$$
$$= t[S] = tN[\epsilon] \tag{3}$$

where $[S]$ is the energy loss factor. $[\epsilon]$ is the stopping power factor. and N is the the atomic density. The energy loss factor can be defined from Fig. 1 and Eq. (3) as (4)

$$[S] \equiv \left(\frac{K}{\cos\theta_1} \frac{dE}{dx}\bigg|_{\text{in}} + \frac{1}{\cos\theta_2} \frac{dE}{dx}\bigg|_{\text{out}} \right) \quad . \tag{4}$$

The stopping cross section factor is defined as (4)

$$[\epsilon] \equiv \left(\frac{K}{\cos\theta_1}\epsilon_{\text{in}} + \frac{1}{\cos\theta_2}\epsilon_{\text{out}} \right) \quad . \tag{5}$$

where ϵ is the stopping cross section. defined as

$$\epsilon \equiv \frac{1}{N} \frac{dE}{dx} \quad . \tag{6}$$

When the target is composed of more than one element. the total energy loss is assumed to be equal to the sum of the losses to the constituent elements. weighted by their abundances in the compound. This postulate is known as Bragg's rule (5). Thus. the stopping cross section for a mixture or molecule with the composition $A_m B_n$ is given by

$$\epsilon(A_m B_n) = m\epsilon(A) + n\epsilon(B) \quad . \tag{7}$$

where $\epsilon(A)$ and $\epsilon(B)$ are the stopping cross sections for elements A and B.

A schematic representation of the backscattering process from an idealized free-standing compound film is presented in Fig. 2. Energy losses in this situation are similar to those depicted in Fig. 1 with the added complexity of kinematic losses from two elements. For the A component, the energy loss components that arise from a projectile traversing a thickness t and then being backscattered out are

$$\Delta E_{\text{in}} = \frac{t}{\cos\theta_1} \frac{dE}{dx}\bigg|_{\text{in}}^{A_m B_n} \quad . \tag{8}$$

$$E_t = E_0 - \Delta E_{\text{in}} \quad . \tag{9}$$

$$\Delta E_c(A) = E_t - K_A E_t \quad . \tag{10}$$

and

$$\Delta E_{\text{out}}(A) = \frac{t}{\cos\theta_2} \frac{dE}{dx}\bigg|_{\text{out}.A}^{A_m B_n} \quad . \tag{11}$$

Similar expressions also exist for the B component of our compound target.

The relationship between energy width. $K_A E_0 - E_1(A)$. and the film thickness for the two-component target is

$$\Delta E_A = K_A E_0 - E_1(A) = t N^{AB} [\epsilon(A_m B_n)]_A \quad . \tag{12}$$

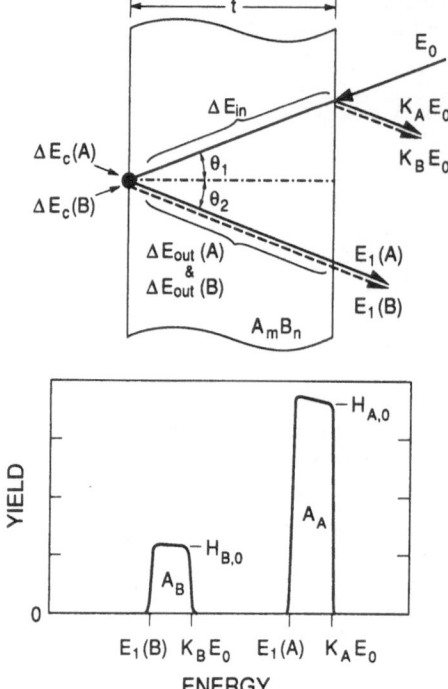

FIGURE 2. A schematic representation of the backscattering process from an idealized free-standing compound film with composition $A_m B_n$ and thickness t.

where N^{AB} is the number of $A_m B_n$ molecules per unit volume. The quantity $[\epsilon(A_m B_n)]_A$ is the stopping cross section factor for a projectile scattered from element A while traversing the medium $A_m B_n$ and has the form

$$[\epsilon(A_m B_n)]_A = \left[\frac{K_A}{\cos \theta_1} \epsilon(A_m B_n)_{in} + \frac{1}{\cos \theta_2} \epsilon(A_m B_n)_{out,A} \right] \quad . \tag{13}$$

An expression similar to Eq. (12) can also be written for the energy width of the B portion of the spectrum using the cross section factor for scattering from element B, which has the same form as Eq. (13). The energy width of a compound can also be expressed in terms of a compound energy loss factor $[S(A_m B_n)]_A$, which can be constructed from Eqs. (8) through (12).

2.3. Scattering cross section σ

In the above discussion we assumed that a collision would occur between the projectile and a target atom and that it would result in a backscattering event. The likelihood of such an occurrence leads to the concept of the scattering cross section and our ability to perform quantitative composition analysis.

For the experimental conditions in which a uniform beam of projectiles impinges at normal incidence on a uniform target of thickness t, the total number A of backscattered and detected particles is given by

$$A = \sigma N t Q \Omega \quad , \tag{14}$$

where σ is the scattering cross section, N is the atomic density (which makes Nt the number of target atoms per unit area). Q is the number of incident particles, and Ω is the detector solid angle.

Equation (14) shows that when Ω, Q, and σ are known, the number of atoms per unit area, Nt, can be calculated. Both Ω and Q can be experimentally measured, and if the repulsion between the projectile and target atom is coulombic, the scattering is Rutherford and σ is easily calculated (4). At higher energies the probability of the projectile interacting with the nucleus of the target atom increases, resulting in an interaction that is not strictly coulombic. In such situations large deviations from Rutherford scattering can occur, and Eq. (14) requires experimentally determined values of σ.

In the hypothetical situation of the free-standing alloy film with composition $A_m B_n$, depicted in Fig. 2, the total number of counts from element A is the area A_A. Assuming that only small changes in the projectile energy occur on the inward and outward paths (surface energy approximation), we can describe the area by (4)

$$A_A = \Omega \, Q \, \sigma_A(E_0) \, m N^{AB} \, \frac{t}{\cos \theta_1} \quad . \tag{15}$$

A similar expression can also be written for A_B using the atomic density of B atoms, $n N^{AB}$, and the surface energy scattering cross section for B atoms, $\sigma_B(E_0)$. The ratio of atomic densities is then given by

$$\frac{m}{n} = \frac{A_A/\sigma_A(E_0)}{A_B/\sigma_B(E_0)} \quad . \tag{16}$$

In some situations it may not be possible to resolve the full peak of a particular element in a backscattering spectrum. In this case Eq. (16), which uses a ratio of peak areas, cannot be used for composition analysis. However, composition analysis may be possible by examining the ratio of the surface counts of the backscattering yield. In an expression analogous to Eq. (15), the backscattering yield at the surface for element A in Fig. 2 is given by

$$H_{A,0} = \Omega \, Q \, \sigma_A(E_0) m N^{AB} \, \tau_{A,0} \cos \theta_1 \quad . \tag{17}$$

where $\tau_{A,0}$ is the thickness of a slab of the target at the surface and is defined by the energy width ξ of a channel in the detecting system. Projectiles scattered from within $\tau_{A,0}$ will have a depth scale at the surface given by

$$\xi = \tau_{A,0} \, N^{AB} \, [\epsilon(A_m B_n)]_A \quad . \tag{18}$$

where N^{AB} and $[\epsilon(A_m B_n)]_A$ have the same meaning as in Eq. (12). A similar set of equations can also be written for $H_{B,0}$ using $\tau_{B,0}$, $\sigma_B(E_0)$, and $n N^{AB}$. Combining these equations, we find that the ratio of atomic densities can be written as

$$\frac{m}{n} = \frac{H_{A,0} \, \sigma_B(E_0) \, [\epsilon(A_m B_n)]_A}{H_{B,0} \, \sigma_A(E_0) \, [\epsilon(A_m B_n)]_B} \quad . \tag{19}$$

3. BACKSCATTERING ANALYSIS IN A THIN-FILM HTS

One of the necessary conditions to fabricating a good HTS is stoichiometry. This is clearly evident when the pseudoternary Y_2O_3-BaO-CuO phase diagram is examined. The oxide phase diagram presented in Fig. 3 was determined at $950°C$ (6) and shows the compound $Y_1 Ba_2 Cu_3 O_x$ to be stoichiometric in its metal components. This indicates that small deviations from 1 :

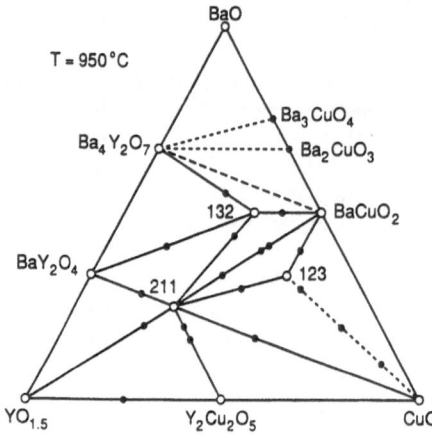

FIGURE 3. The pseudoternary Y_2O_3-BaO-CuO phase diagram determined at 950°C shows the compound $Y_1Ba_2Cu_3O_x$ to be stiochiometric in its metal composition. From Ref. 6.

2 : 3 in the Y-Ba-Cu stoichiometry will result in the formation of $Y_1Ba_2Cu_3O_x$ and two additional nonsuperconducting phases. If the stoichiometry is known, the fraction of the 1 : 2 : 3 superconducting phase expected under equilibrium-processing conditions can be calculated from the phase diagram by applying the lever rule (7).

In addition to the metal stoichiometry, oxygen content has been shown to be a critical factor in the formation of the orthorhombic superconducting phase (8) and affects the temperature at which the superconducting transition occurs (9,10).

From the above discussion it is clear that knowledge of the composition of an HTS sample can be a great assistance in the processing and fabrication of this material as well as provide fundamental information for the proper interpretation of results obtained from various physical and transport property experiments.

3.1. Rutherford backscattering spectroscopy

We have already established in Sec. 2 that backscattering spectroscopy is capable of supplying depth-sensitive composition information in the near-surface region, making it an attractive candidate for examining HTS thin films. However, traditional Rutherford backscattering spectroscopy (RBS) is much more sensitive to elements with a high atomic number. This results directly from the Rutherford scattering cross section σ_R, which is proportional to $(Z_1 Z_2 / E)^2$, where Z_1 is the atomic number of the projectile, Z_2 is the atomic number of the target atom, and E is the energy of the projectile. Examining Eq. (15) in light of this, we see that the Rutherford backscattered signal will be much greater for a high-Z element such as barium, yttrium, or copper relative to a low-Z element such oxygen. Oxygen RBS analysis is further complicated by the experimental observation that helium projectiles on oxygen start to show deviations from Rutherford scattering near 2.2 MeV (11), Fig. 4, thus limiting the maximum energy at which RBS analysis can be performed.

The major negative side effects associated with performing backscattering experiments in the low-energy regime, where the oxygen cross section is still Rutherford, are the limitations imposed on film thickness and composition analysis. As indicated by Eqs. (3) and (12), the backscattering peak width ΔE is related to the film thickness t. From Eq. (2), we also observe that mass resolution decreases with decreasing projectile energy. Together these points suggest that decreasing projectile energy and increasing film thickness will ultimately lead to signal overlap between the high-Z components in our superconducting film and making depth-dependent composition analysis difficult. This point is demonstrated in Fig. 5 with an RBS spectrum that was simulated with the program RUMP (11). The simulation was carried out

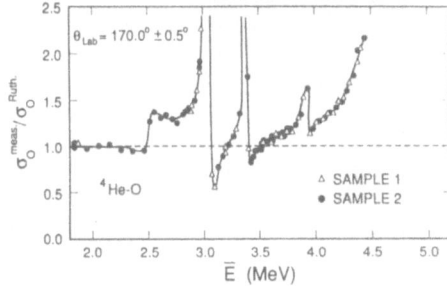

FIGURE 4. Measured backscattering cross section for helium on ^{16}O in the energy interval 1.8 to 4.8 MeV. The oxygen cross section starts to deviate from Rutherford scattering near 2.2 MeV. From Ref. 11.

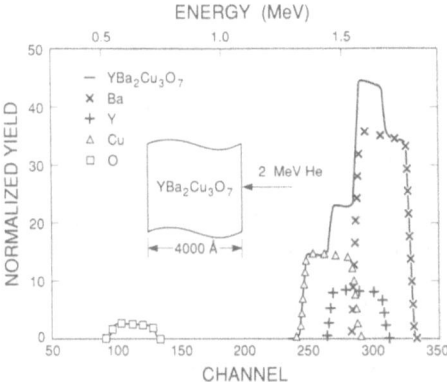

FIGURE 5. RBS spectra simulated with the program RUMP (11) for 2-MeV helium projectiles on a 4000-Å-thick free-standing $YBa_2Cu_3O_7$ film. Also shown are the component RBS signals from barium, yttrium, copper, and oxygen.

assuming 2-MeV helium projectiles and a 4000-Å-thick, free-standing $YBa_2Cu_3O_7$ film. Also shown are the component RBS signals from barium, yttrium, copper, and oxygen, which make up the total spectrum. Although these data ignore the additional spectrum complexities that will be present when a substrate is included in the simulation, they clearly demonstrate how complicated and convoluted the RBS data can be under the given conditions.

Two solutions to the problem of overlapping barium, yttrium, and copper peaks are presented in the RUMP backscattering simulations of Fig. 6. A possible but not very practical solution

FIGURE 6. RUMP backscattering simulations of two possible solutions to the problem of barium, yttrium, and copper peak overlap. The low-energy spectrum, which was simulates assuming 2-MeV helium, suggests a maximum $YBa_2Cu_3O_7$ thickness of 1200 Å at this energy. The high-energy portion shows that complete peak separation in a 4000-Å-thick film is possible at 4 MeV.

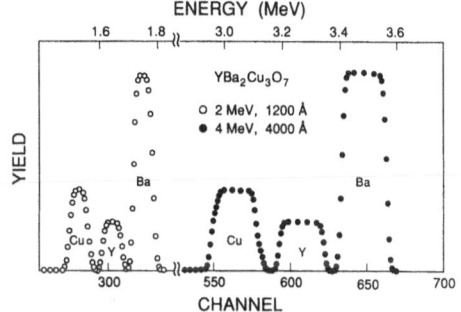

sible at 4 MeV.

CHANNEL

would be to work only at energies where oxygen is Rutherford and limit the thickness of the film to avoid peak overlap. The low-energy spectrum, which was simulated assuming 2-MeV helium, suggests a maximum $YBa_2Cu_3O_7$ thickness of 1200 Å in this case. An alternative approach would be to take advantage of the improved mass resolution [Eq. (2)] and the lower stopping cross section that occur at higher helium energies. The high-energy portion of Fig. 6 shows that complete peak separation in a 4000-Å-thick film is possible at 4 MeV. These last data suggest that a complete composition analysis may be possible by acquiring backscattering spectra from the same film at two different energies: a high-energy spectrum to determine the barium, yttrium, and copper composition profiles and a lower energy ($\leq$2.2-MeV) spectrum to determine the oxygen depth profile.

Although the two-step depth-profiling solution presented above is possible in theory, the low yield of backscattered oxygen at energies $\leq$2.2 MeV and the incorporation of backscattering signals from the substrate make accurate oxygen analysis difficult to carry out in practice. In Fig. 7, a RUMP-simulated RBS spectrum is presented for 2-MeV helium on a 4000-Å-thick $YBa_2Cu_3O_7$ film on a $SrTiO_3$ substrate. Also shown are the individual backscattering signals from barium, yttrium, copper, and oxygen in the surface film and signals from strontium, titanium, and oxygen in the substrate. Examining Fig. 7, we see that the small oxygen signal from the superconductor sits on top of a large background made up of the backscattering signals from the substrate components strontium and titanium. Note that the simulated spectrum is highly idealized and does not contain any of the background noise that is present in reality. The combination of the low oxygen yield (counting statistics) and our ability to properly subtract the background from the oxygen signal greatly limits the accuracy of oxygen composition analysis.

3.2. Non-Rutherford backscattering

3.2.1. Elastic resonance of 3.045-MeV helium with ^{16}O. Solutions to the limitation imposed on performing oxygen composition analysis in the Rutherford scattering regime have been offered by several research groups (13–16). One suggested solution is to take advantage of the enhanced, non-Rutherford, elastic resonant ^{16}O cross section that occurs for incident 3.045-MeV helium. This resonance cross section was first described by Cameron (17) and has recently been applied to oxide superconductors by Blanpain et al. in Ref. 16. The enhancement in the measured cross section relative to the Rutherford cross section at 3.045 MeV can be observed in Fig. 4 (11) and has been estimated to produce an oxygen yield 30 times greater than expected for Rutherford scattering (18). Backscattering data presented in Fig. 8 clearly show the effect of this enhancement on the oxygen backscattering yield from a 3050-Å-thick Ta_2O_3 film with 3.097-MeV helium relative to 2.030-MeV helium (16). However, because the resonant scattering

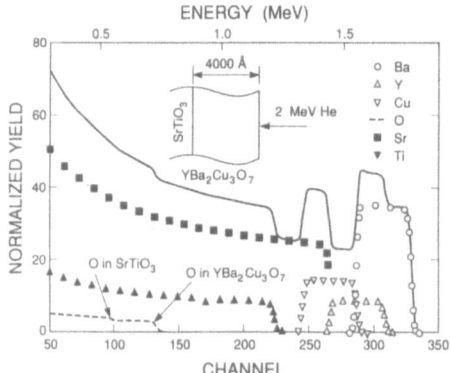

FIGURE 7. A RUMP-simulated RBS spectrum for 2-MeV helium on a 4000-Å-thick $YBa_2Cu_3O_7$ film on a $SrTiO_3$ substrate. Also shown are the individual backscattering signals from barium, yttrium, copper, and oxygen in the surface film and signals from strontium, titanium, and oxygen in the substrate.

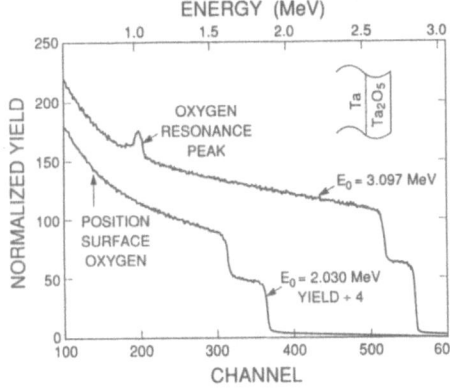

FIGURE 8. Backscattering spectrum from a 3050-Å-thick Ta_2O_5 film on tantalum. The data contrast the effect of an enhanced oxygen cross section, which exists for 3.097-MeV helium, compared with the Rutherford oxygen cross section for 2.030-MeV helium. From Ref. 16.

from ^{16}O occurs only over a narrow energy window of 11 keV (16), concentration depth profiles require that several backscattering spectra be taken at slightly increasing energies to probe the oxygen content at greater depths in the sample. An example of such a depth probe is given in Fig. 9 for a zirconia sample (from Ref. 16).

Figure 10 presents the results obtained by Blanpain et al. (16) from a 3.086 MeV helium backscattering experiment on a Y-Ba-Cu-O/zirconia film/substrate combination. Oxygen concentration and depth resolution under these operating conditions are quite good. Concentration accuracy is proposed to be less than 5%. The oxygen depth resolution, if we ignore straggling effects, will depend primarily on the resonance energy width, 11 keV. Assuming that the helium an energy loss is $dE/dx = 36$ eV/A in $YBa_2Cu_3O_7$ at 3.1 MeV suggests a depth resolution of approximately 300 Å at normal incidence. However, accurate oxygen depth profiling requires the use of a calibration sample, which is run concurrently with the unknown (16). Also, as the high-energy portion of Fig. 10 shows, additional spectra at higher helium energies will be necessary to achieve complete separation of the yttrium, barium, and copper peaks.

3.2.2. Elastic resonance of 2.5-MeV protons with ^{16}O. Although the use of the non-Rutherford elastic resonant ^{16}O cross section that occurs for incident 3.045-MeV helium has clear advantages, it does require the acquisition of multiple spectra at different energies to obtain the oxygen composition as a function of depth. Rauhala et al. (13) have recently described

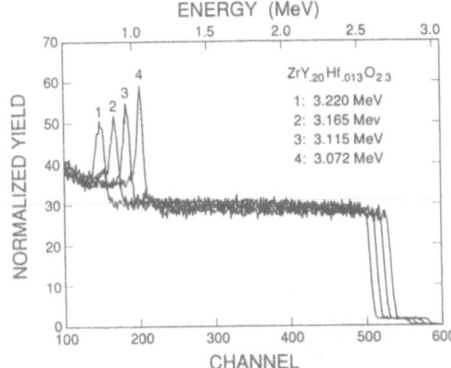

FIGURE 9. Multiple helium backscattering spectrum for helium projectiles on zirconia at increasing energy. From Ref. 16.

486

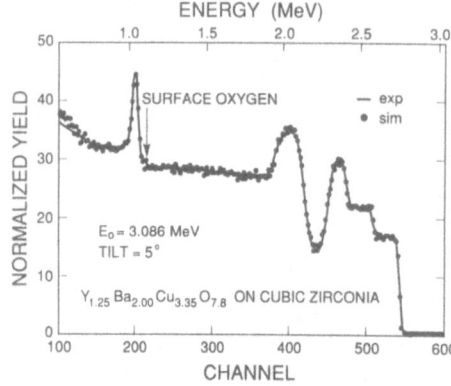

FIGURE 10. Backscattering data obtained from a 3.086-MeV helium projectile on a Y-Ba-Cu-O/zirconia film/substrate combination. From Ref. 16.

an alternative backscattering technique that also provides composition analysis in $YBa_2Cu_3O_7$ and does not require multiple spectra to determine the oxygen composition profile. The approach taken by these researchers was to combine the elastic non-Rutherford scattering of protons and the Rutherford scattering of helium to examine the composition of oxygen and the higher Z elements, respectively. This technique uses an enhanced oxygen scattering cross section that is relatively flat over a wide range of proton energies. The oxygen cross section occurs for proton projectiles of energy ≤ 2.5 MeV and is enhanced by a factor of 5 relative to the Rutherford cross section (19).

Barbour et al. (14) have also taken advantage of this cross section, and backscattering data from their work on bulk $YBa_2Cu_3O_7$ is presented in Fig. 11. The inset in Fig. 11 shows the behavior of the $^{16}O + p$ cross section as a function of energy. Examining both the cross section and energy loss factor for HTS material suggests that an oxygen analysis range of 4 μm and a depth resolution of 2000 Å is possible for 2.5-MeV incident protons (14). Although this technique is relatively simple to apply to oxygen analysis, the loss of depth and mass resolution that is observed in the high-energy portion of the spectrum clearly shows that accurate composition analysis is not possible for barium, yttrium, and copper. As proposed by Rauhala

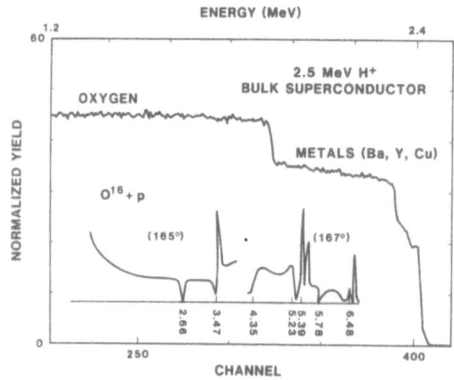

FIGURE 11. Backscattering data from 2.5-MeV protons on bulk $YBa_2Cu_3O_7$. The inset shows the behavior of $^{16}O + p$ cross section as a function of energy. From Ref. 14.

et al., additional experiments must also be performed with helium projectiles to obtain the composition information for the higher Z elements.

3.2.3. High-energy backscattering and the elastic resonance of 8.8-MeV helium with ^{16}O.

Up to this point we have primarily concerned ourselves with backscattering experiments at relatively low energies where complete composition analysis of a $YBa_2Cu_3O_7$ film required collecting multiple data, changes in projectile, or incident energy. As noted previously (Fig. 6), going to higher energies offers the advantage of producing greater mass separation between the high-Z components. However, it is well know that the likelihood for scattering to deviate from a Rutherford nature increases with increasing projectile energy. Deviations are expected when the projectile velocity is large enough to allow it to penetrate deep into the the orbitals of the atomic electrons and interact with the nucleus of the target atom. Mayer and Feldman (20) have shown that the onset energy for deviation from Rutherford scattering is roughly proportional to $Z_1Z_2/A^{1/3}$. where A is the mass number of the target atom and Z_1 and Z_2 are atomic numbers of the projectile and target atom, respectively. Thus, if any sense is to be made of the information obtained from higher energy backscattering experiments, we must first know the behavior of the scattering cross sections for all elements involved.

In the best of all possible situations, one would ideally like to obtain from a single spectrum both good mass separation and the depth-dependent composition information. Barbour et el. recently showed that this could be achieved by performing backscattering experiments with 8.7-MeV helium particles (14). The utility to oxygen analysis of working at this energy can be realized by examining the cross-section data of Hunt et al. (21), Fig. 12. The data in Fig. 12 show the high-energy behavior of the ^{16}O scattering cross section as a function of incident helium (alpha-particle) energy for a variety of scattering geometries. These data indicate that as the geometry becomes progressively more backscattering (increasing $\theta_{C.M.}$). a very broad plateau appears in the cross section centered near 8.6 MeV.

Figure 13 shows a typical backscattering spectrum, taken from the work of Martin et al. (15), for 8.8-MeV helium incident on a 7700-Å-thick film of $YBa_2Cu_3O_{7-x}$ sitting on a $SrTiO_3$ substrate. Indicated in the spectrum are the surface energies for barium, yttrium, copper, and oxygen. The unlabeled step edges at 2.80, 5.96, and 7.00 MeV correspond to subsurface oxygen. titanium, and strontium. respectively, in the substrate. Clearly evident in these data is the good mass separation between barium, yttrium, and copper as well as the large oxygen yield from the superconducting film. However, as noted above the absolute accuracy of determining the film composition from such a spectrum depends on how well the scattering cross sections are known.

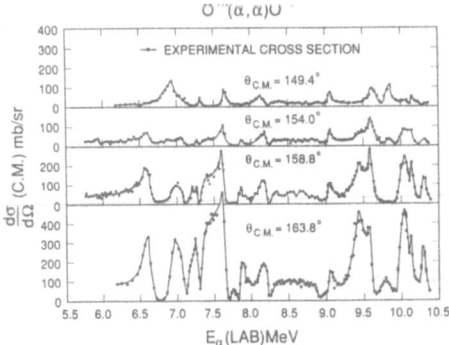

FIGURE 12. High-energy ^{16}O scattering cross section as a function of incident helium (alpha-particle) energy. The data indicate that as the geometry becomes progressively more backscattering. a very broad plateau appears in the cross section. centered near 8.6 MeV. From Ref. 21.

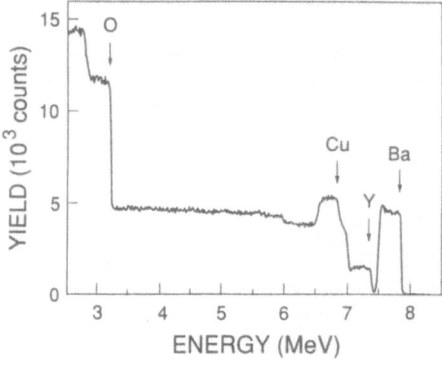

FIGURE 13. A typical backscattering spectrum for 8.8-MeV helium incident on a 7700-Å-thick film of $YBa_2Cu_3O_{7-x}$ sitting on a $SrTiO_3$ substrate. From Ref. 15.

To determine how the scattering cross sections behave as a function of helium energy, Martin and coworkers used a thin-film standard technique to measure the ratio cross section for several HTS elements. The production of the standards consisted of electron beam coevaporating a Y-Ba-Cu film onto SiO_2, which was capped with a thin layer of titanium, and evaporating barium in the presence of O_2 onto a titanium-coated graphite substrate, which was then capped with another layer of titanium. The substrates were chosen in each case because of their low kinematic factors. The titanium layers were employed to minimize environmental contamination and aid adhesion to the graphite. For each standard the film thickness was chosen to ensure elemental peak separation in a 2-MeV RBS spectrum. The RBS data for these two standards at 2 MeV are presented in Fig. 14 (22). Performing peak integration on these data and applying Eq. (16), we find that the the composition of yttrium and copper relative to barium and the composition of oxygen and barium can be determined to within a few per cent. Once the ratio m/n is determined, values for σ_A/σ_B at various energies can be measures by integrating the elemental peaks at the energy in question and again applying Eq. (16).

Figure 15 presents the results of this procedure for helium scattering from barium and oxygen in the energy range 8.2 to 9.1 MeV with a scattering angle of 166° (from Ref. 15). The dashed line in the figure represents the ratio of the cross sections assuming Rutherford scattering. The increased sensitivity to ^{16}O at 8.8 MeV is about 25 times greater than that for Rutherford scattering. The statistical uncertainty in measuring these data was less than 2%.

In Fig. 16 measured cross section ratios for Cu/Ba and Y/Ba are presented over the energy range 2 to 9 MeV (15). Again the dashed line represents the Rutherford cross section ratios for a scattering angle of 166°. Assuming the scattering cross section for barium to remain Rutherford up to 9 MeV (large Z) (14,15), we see that these data indicate that the copper and yttrium cross sections start to deviate from a Rutherford nature near 6.5 and 8.0 MeV, respectively.

4. APPLICATIONS OF HIGH-ENERGY BACKSCATTERING TO THIN-FILM PROCESSING AND ION IMPLANTATION OF HTS

To successfully analyze the thickness and composition of an HTS thin film with 8.8-MeV helium backscattering requires the cross section information presented in Figs. 15 and 16 as well as stopping data. Zigler et al. (24) have developed a reliable, consistent parameterization of stopping data for essentially all elements. Figure 17 presents stopping data calculated by the program TRIM (24) for the case of helium projectiles on a Y-Ba-Cu-0 target in the energy range 10 keV to 10 MeV. These data were calculated assuming a target atomic composition of 7% yttrium, 15% barium, 23% copper, 54% oxygen, and a mass density of 6.54 g/cm^3.

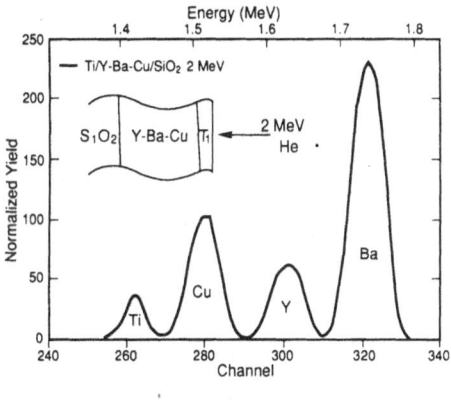

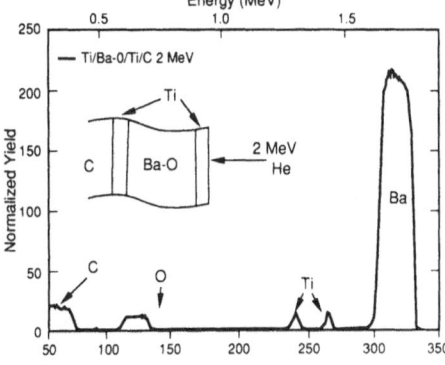

FIGURE 14. RBS data at 2 MeV from two films that were used to measure the relative cross sections of copper, yttrium, and oxygen relative to barium as a function of helium energy.

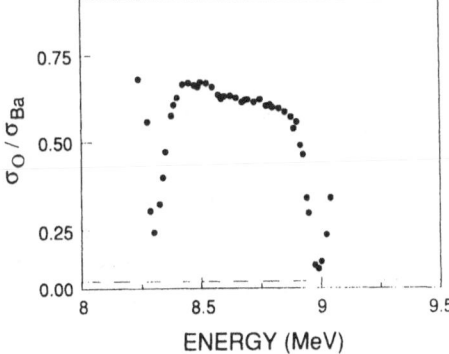

FIGURE 15. Ratio of scattering cross section for helium ions scattering from ^{16}O and barium as a function of energy. The dashed line shows the ratio of the Rutherford scattering cross sections. From Ref. 15.

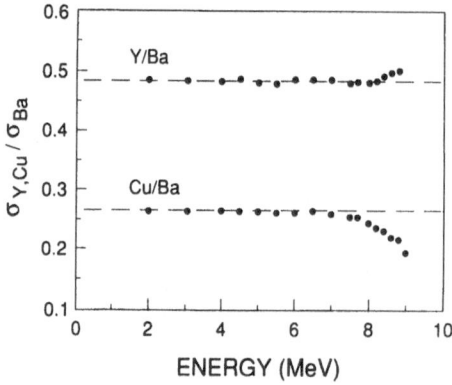

FIGURE 16. Ratio of scattering cross sections for helium ions from yttrium and copper relative to barium as a function of helium energy. The dashed line represents the ratio of the Rutherford scattering cross sections. From Ref. 15.

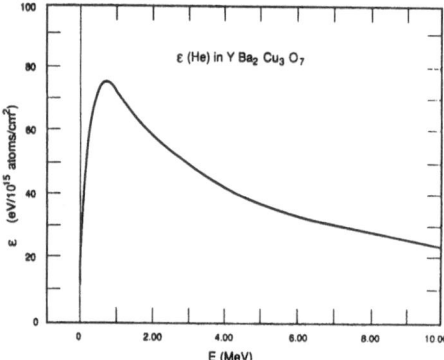

FIGURE 17. TRIM (Ref. 24) calculated stopping data for helium projectiles on a Y-Ba-Cu-0 target. These data were calculated assuming a target atomic composition of 7% yttrium, 15% barium, 23% copper, 54% oxygen, and a mass density of 6.54 g/cm^3.

4.1. Thin-Film Processing

Figure 18 presents 8.8-MeV helium backscattering data from Y-Ba-Cu films prepared by electron beam coevaporation in a vacuum of approximately 5×10^{-8} torr (22). Films were simultaneously deposited onto both SrTiO$_3$ and graphite substrates. An yttrium capping layer, approximately 150 Å thick, was deposited before the samples were removed from vacuum. This procedure was done in an attempt to minimize atmospheric contamination as the films were taken out of vacuum and placed in the O$_2$ annealing furnace.

The spectra in Fig. 18 are from films in their as-deposited state and allow us to examine the influence of the SrTiO$_3$ and graphite substrate signals on the barium, yttrium, copper, and oxygen peaks. These data indicate that the barium, yttrium, copper, and oxygen distributions for both substrates are identical, including the small oxygen peak that has formed in the thin yttrium capping layer. The step at the low-energy end of the yttrium peak, producing the background for the the copper peak, is backscattering from strontium in the SrTiO$_3$ substrate. The large step in the vicinity of channel 275 corresponds to backscattering from oxygen in the SrTiO$_3$ substrate. The graphite substrate offers a clear advantage in peak separation and allows rapid composition analysis of the as-deposited film by simply measuring the number of counts in the barium, yttrium, copper, and oxygen peaks.

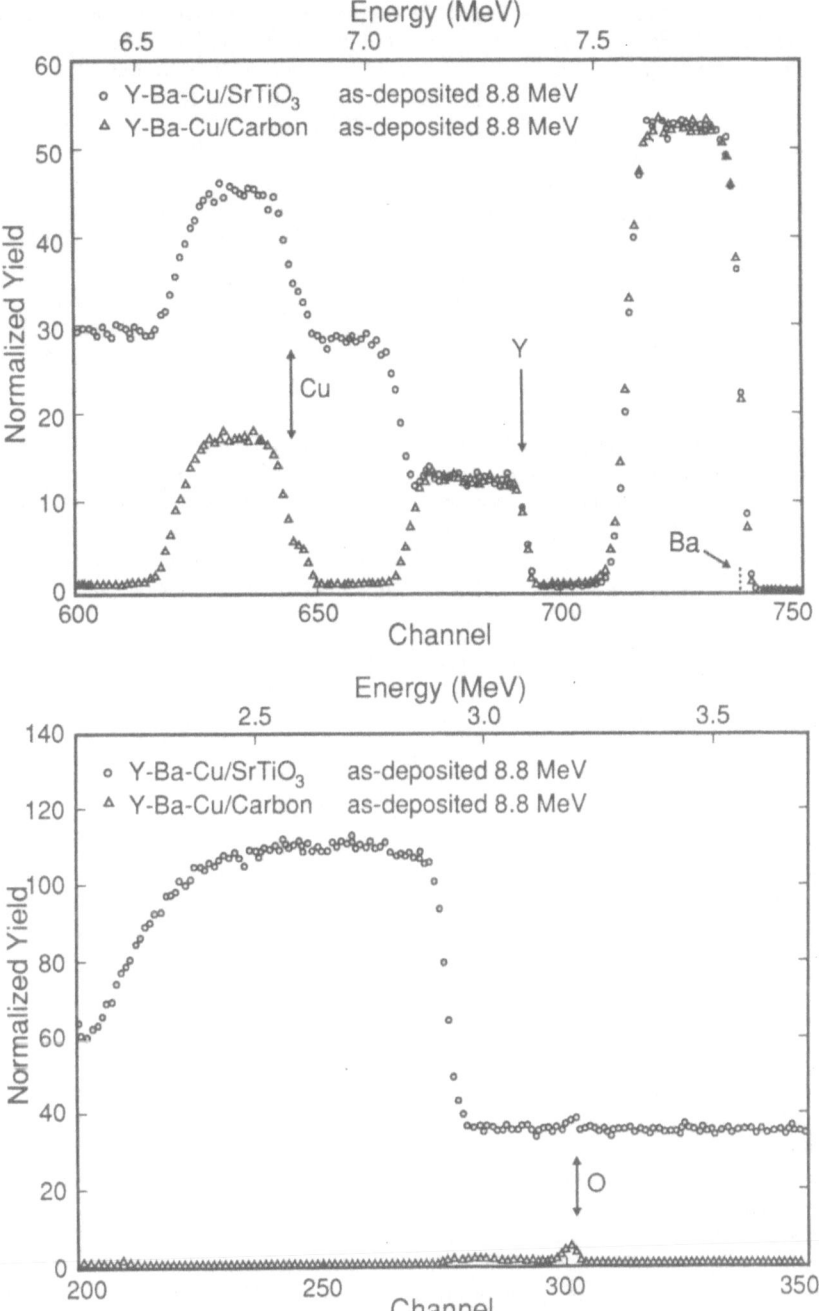

FIGURE 18. Backscattering data from 8.8-MeV helium on Y-Ba-Cu films prepared by vacuum electron beam coevaporation onto $SrTiO_3$ and graphite substrate.

To produce a superconductor out of the coevaporated Y-Ba-Cu films, an O_2 annealing sequence of 650°C for 1 h, 750°C for 1 h, 850°C 1 h, then 650°C for 1 h was followed by a slow cooling to room temperature. Only films deposited on $SrTiO_3$ were annealed. The graphite substrate samples, which are very effective for composition analysis of the as-deposited film, are not very practical for the actual formation of the superconductor. Graphite is readily converted into CO_2 under the annealing sequence described above. Figure 19 presents high-energy backscattering spectra from films on $SrTiO_3$ before and after O_2 annealing.

A comparison of the two spectra in Fig. 19 clearly shows the effects of the O_2 annealing procedure. Examining the data, we see that the heights of the barium, yttrium, and copper peaks have all undergone a reduction, whereas the height of the oxygen peak has grown larger than that of barium. In addition, the barium, yttrium, copper, and oxygen film peaks have all become wider, and the strontium and oxygen substrates signals have undergone a corresponding shift to lower energy positions. Qualitatively, these results are consistent with the production of a thicker film that has a lower barium, yttrium, and copper atomic density and a higher oxygen atomic density after annealing, relative to the as-deposited film.

To determine the composition of the annealed film and the as-deposited film on $SrTiO_3$, a peak height analysis, similar to that suggested by Eq. (19), will be require. However, instead of using the surface height analysis we will use an average height analysis, with the cross sections evaluated at an energy corresponding to the average energy of the incoming projectile, $\overline{E}_{in} = (E_0 + E_t)/2$ [see Fig. 2 and Eqs. (8) through (11)]. Correspondingly, an average stopping cross section factor will also be used, which for element A has the form (4)

$$[\overline{\epsilon}(A_m B_n)]_A = \left[\frac{K_A}{\cos\theta_1} \epsilon(\overline{E}_{in}) + \frac{1}{\cos\theta_2} \epsilon(\overline{E}_{out}) \right] \quad , \tag{20}$$

where the average energy of the outgoing particle, after a collision with element A is expressed as

$$\overline{E}_{out} = \frac{K_A E_t + E_1(A)}{2} \quad . \tag{21}$$

The average stopping cross section will also be used in Eq. (12) to determine the film thickness.

Though not explicitly stated, values of ϵ in Eq. (20) are for the compound target and should be calculated according to Eq. (7). For the superconducting compound $YBa_2Cu_3O_7$, the factor ϵ on an atomic fraction basis will be equal to

$$\epsilon(YBa_2Cu_3O_7) = \frac{1}{13}\epsilon(Y) + \frac{2}{13}\epsilon(Ba) + \frac{3}{13}\epsilon(Cu) + \frac{7}{13}\epsilon(O) \quad , \tag{22}$$

where the component stopping cross sections have been taken at the energies of interest, $\overline{E}_{in}$ and $\overline{E}_{out}$. The result of this calculation over a wide range of energies is presented in Fig. 17.

Iterative calculations of the film thickness, composition, E_{in}, E_{out}, and the average stopping cross section factor indicate that the annealed film is 8980 Å thick and has a composition of $Y_{1.11}Ba_{2.00}Cu_{2.91}O_{7.04}$, normalized to Ba = 2. Similar analysis on the unannealed film, using data from both the graphite and $SrTiO_3$ spectra (Fig. 18), gives us a film thickness of 7400 Å and a stoichiometry of $Y_{0.97}Ba_{2.00}Cu_{2.91}O_{0.21}$, normalized to Ba = 2. The above calculations assumed atomic densities of 4.17 and 7.69×10^{22} atoms/cm² for the unannealed and annealed films, respectively. The stopping cross section factors [Eq. (20)] calculated for the unannealed material are 82.8, 82.4, and 82.0×10^{-15} eV·cm² for the barium, yttrium, and copper peaks, respectively, and in the annealed material 53.2, 52.9, 52.8, and 59.9×10^{-15} eV · cm² for barium, yttrium, copper, and oxygen, respectively. This indicates that for a scattering

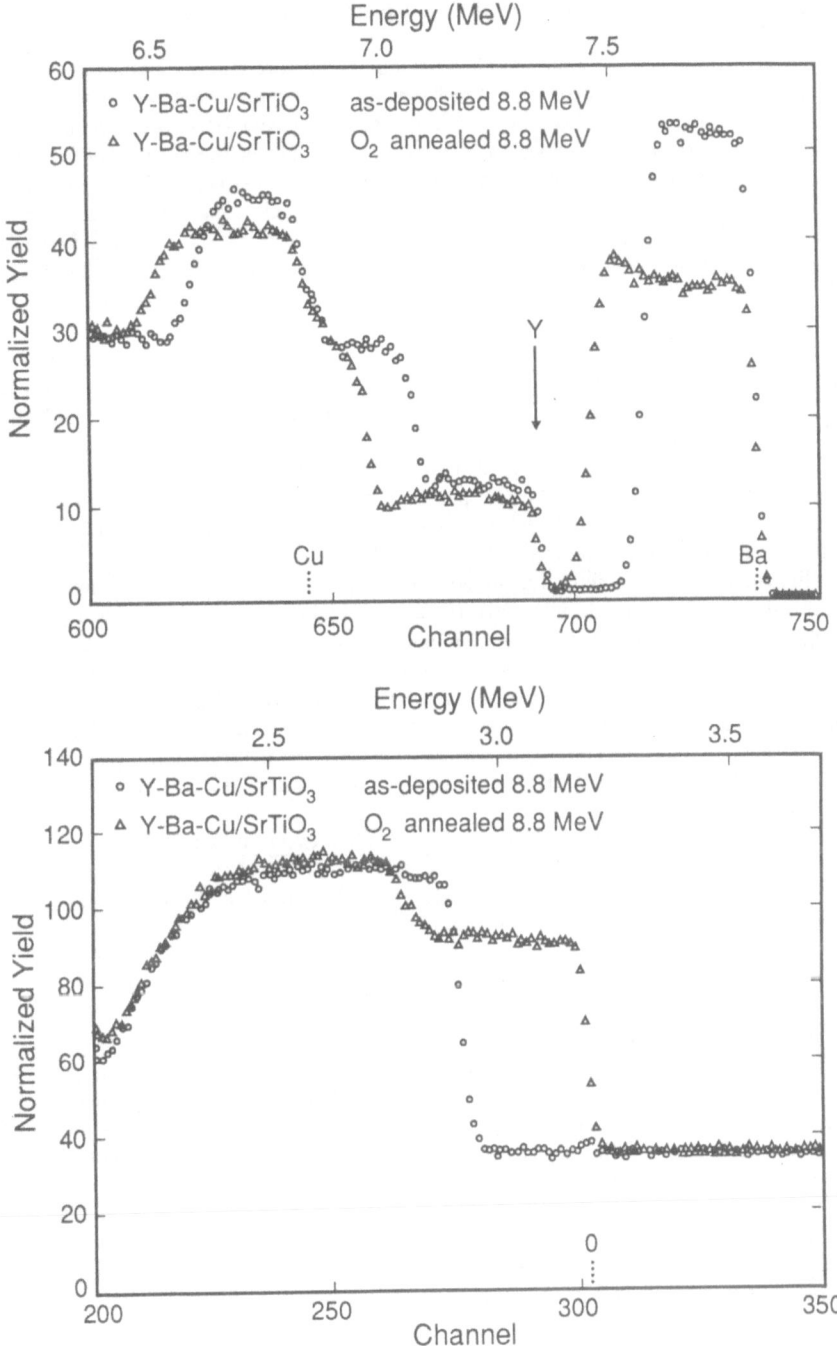

FIGURE 19. Backscattering data from 8.8-MeV helium on Y-Ba-Cu films on SrTiO₃ substrates before and after annealing.

geometry of $\theta_1 = 0°$ and $\theta_2 = 15°$ and a detector resolution of 15 keV, our depth resolution for 8.8-MeV helium is approximately 440 Å in the unannealed material and 370 Å in the annealed material.

A careful comparison of the metal stoichiometries before and after annealing indicates that the number of barium and copper atoms has remained constant, whereas the number of yttrium atoms has increased by 14%. This increase in the calculated yttrium stoichiometry results directly from a greater than expected increase in the height of the yttrium backscattered peak. Because the barium and copper data suggest that mass was conserved during the annealing procedure, the increased yttrium yield indicates that backscattering from near-surface strontium has occurred. This may be the result of strontium diffusion into the surface film or the formation of holes in the Y-Ba-Cu layer during annealing. If we assuming that the increase is the result of strontium contamination, the estimated strontium concentration in the film would be approximately 1.1 at.%.

To clearly show the advantage of working with 8.8-MeV helium, Fig. 20 contrasts the same high-energy data presented in Fig. 19 to RBS spectra taken on the same as-deposited and O_2-annealed samples. The RBS spectra were take at the same scattering geometry employed for the 8.8-MeV data. The 2-MeV helium RBS spectra also include data from oxygen backscattering.

The transport quality of the annealed film, as determined by four-point probe measurements, is presented in Fig. 21 (22). The film possesses a superconducting transition that is 8 K wide, reaching zero resistance at 79 K. The normal-state resistance shows mild metallic behavior but does not drop very quickly with decreasing temperature as would be expected from a high-quality superconducting film. The moderate transport quality of this film is consistent with the backscattering data, which has shown the metal stoichiometry to be less than perfect and has suggested strontium contamination. As indicated by the phase diagram in Fig. 3, deviations from a 1 : 2 : 3 atomic ratio for Y : Ba : Cu results in the formation of other, nonsuperconducting, phases. As in other ceramics, second phases in the $YBa_2Cu_3O_7$ superconductor will most likely accumulate at grain boundaries, producing Josephson junctions that tend to spread out the superconducting transition (24, 25). The addition of strontium has similar detrimental effects on HTS materials; studies have clearly demonstrated that T_c decreases with increasing strontium content (26,27).

4.2. Formation of $YBa_2Cu_3O_7$ superconducting films by ion implantation

One possible method of producing superconducting material that will be compatible with the fine structures required in electronic devices is ion implantation. Early work by by Koch et al. (28) demonstrated that the radiation damage resulting from ion implantation into $YBa_2Cu_3O_x$ thin films renders the new class of oxide superconductors normal at moderate doses (29,30). They used implantation and masking techniques to destroy the superconductivity in specific portions of a superconducting $YBa_2Cu_3O_x$ film to define the structure of a superconducting quantum interference device.

An alternative technique, which has proved to be quite effective in producing devices in semiconducting materials (31), is to use ion implantation to introduce additional elements into a base material and alter the electrical properties of the implanted region. Nastasi and coworkers (32) recently examined the effectiveness of implanting yttrium ions into a nonsuperconducting Ba-Cu-F base material to produce the superconducting $YBa_2Cu_3O_7$ phase.

The experimental procedure followed by these researchers was to form a base material for yttrium implantation by the electron beam coevaporation of BaF_2 and copper. A 4000-Å-thick alloy layer with an approximate composition of $Ba_2Cu_3F_4$ was deposited in a vacuum of $\sim 5 \times 10^{-8}$ torr onto room-temperature substrates of jeweler's-quality $SrTiO_3$ and single-crystal Al_2O_3. A 200-Å-thick layer of BaF_2 was used as a capping layer to minimize contamination and material loss from sputtering during ion implantation.

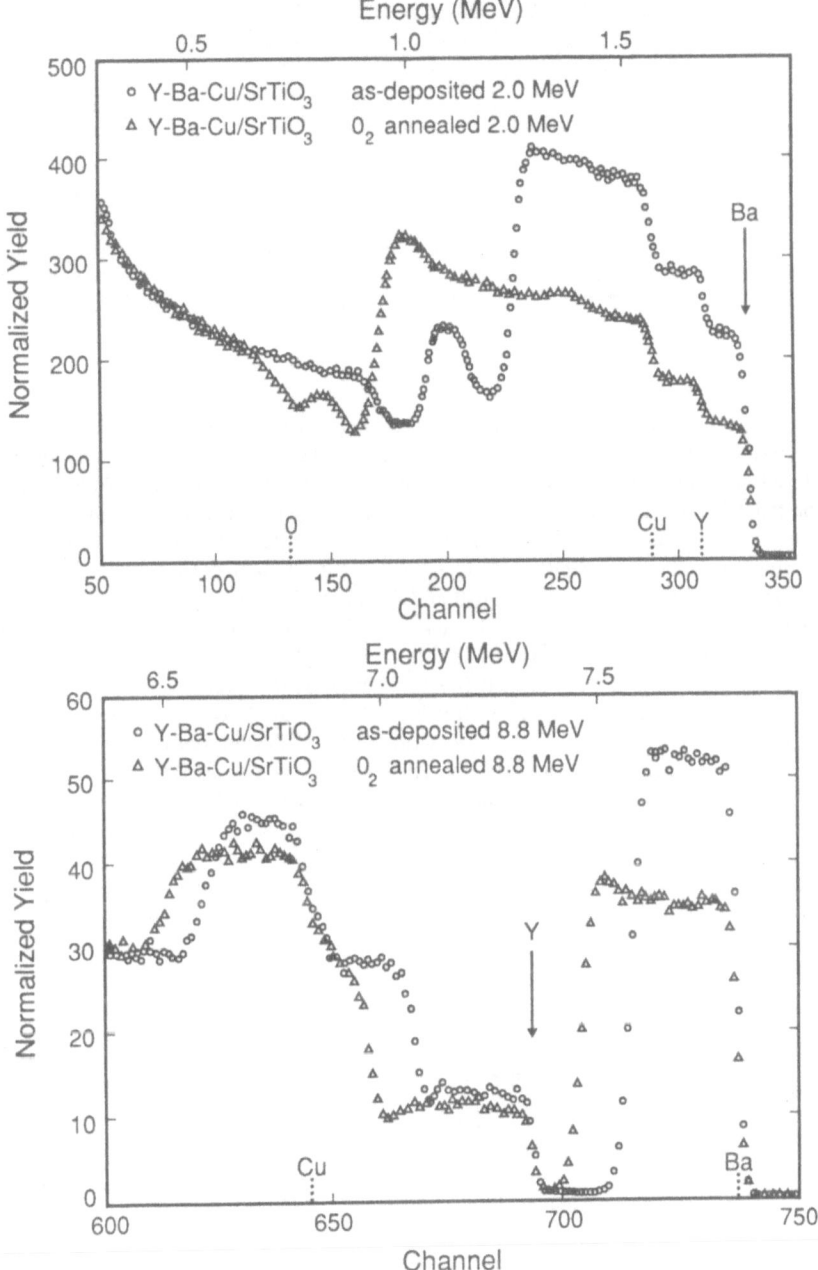

FIGURE 20. RBS spectrum of 2-MeV helium (top) and backscattering spectrum of 8.8-MeV helium taken from the same as-deposited and O_2-annealed samples. Both sets of data were taken at the same scattering geometry. The 2-MeV helium RBS spectrum also includes data from oxygen backscattering.

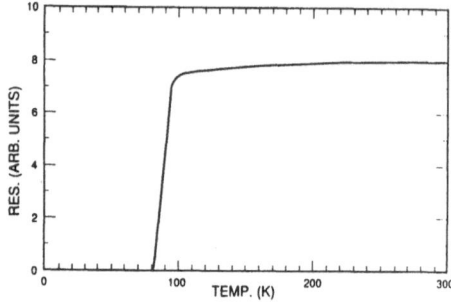

FIGURE 21. Four-point probe data following an O_2 anneal. The film possesses a superconducting transition that is 8 K wide, reaching zero resistance at 79 K.

During the yttrium ion implantation process, the coevaporated BaF_2-Cu films were held at 80 K in a vacuum of 5×10^{-7} torr. Yttrium implantations at two different energies were employed in an attempt to produce a more uniform yttrium distribution. The films were first implanted with 400-keV Y^{++} to a dose of 6×10^{16} atoms/cm^2, followed by 100-keV Y^+ to a dose of 2×10^{16} atoms/cm^2. TRIM Monte Carlo simulations (23), which did not consider sputtering, estimated the range and straggling of the implanted yttrium ion to be 1947 and 776 Å, respectively, at 400 keV, and 646 and 237 Å at 100 keV.

Figure 22 shows 8.8-MeV helium backscattering data from as-deposited and yttrium-implanted films on Al_2O_3. Analysis of the as-deposited spectra, using the cross-section data presented in Fig. 16 and Eq. (16), gives a total integrated barium and copper composition of Ba_2Cu_3. Backscattering studies at 2 MeV have determined that BaF_2 is evaporated stoichiometrically (22), indicating that the composition of the coevaporated film is $Ba_2Cu_3F_4$.

A comparison of the two spectra in Fig. 22 shows that ion implantation decreases the width of both the copper and barium peaks as well as decreasing the surface density of barium. An analysis using the total counts in the barium and copper peaks yields a postimplantation composition of $Ba_{1.8}Cu_3$, indicating that barium was preferentially sputtered during the implantation of yttrium. A further analysis using Eq. (19) and the surface heights of the barium, yttrium, and copper peaks gives a surface composition of $Y_{0.9}Ba_{1.6}Cu_3$ in the as-implanted

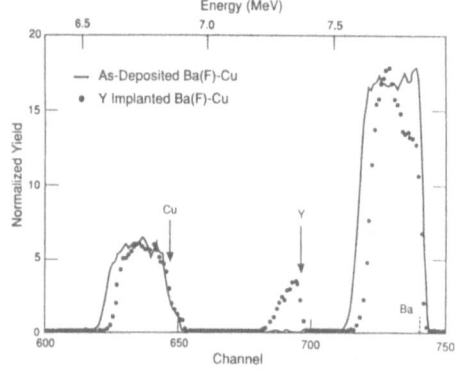

FIGURE 22. Backscattering data from 8.8-MeV helium on as-deposited and yttrium-implanted films on Al_2O_3. These spectra were taken with $\theta_1 = 60°$.

film. The implanted yttrium distribution was peaked at the film surface with a FWHM of approximately 1600 Å and a low-energy tail extending to a depth of approximately 2900 Å.

Both the as-deposited and yttrium-implanted samples were annealed using a wet-oxygen procedure similar to that proposed by Mankiewich et al. (33,34). The samples were annealed for 1 h at 750°C in dry oxygen, 1 h at 850°C in wet oxygen, 1 h at 650°C in dry oxygen, 1 h at 550°C in dry oxygen, and cooled slowly to 50°C over approximately 3 h in dry oxygen.

Figure 23 presents 8.8-MeV helium backscattering data from the as-deposited, yttrium-implanted, and yttrium-implanted plus oxygen-annealed samples on $SrTiO_3$ substrates. Evident in the low-energy portion of these data is the evolution of the fluorine and oxygen content as a result of processing. Both fluorine and oxygen are observed in the spectra from the as-deposited and yttrium-implanted films. Oxygen annealing was followed by a large increase in the oxygen signal and nearly complete attenuation of the fluorine signal.

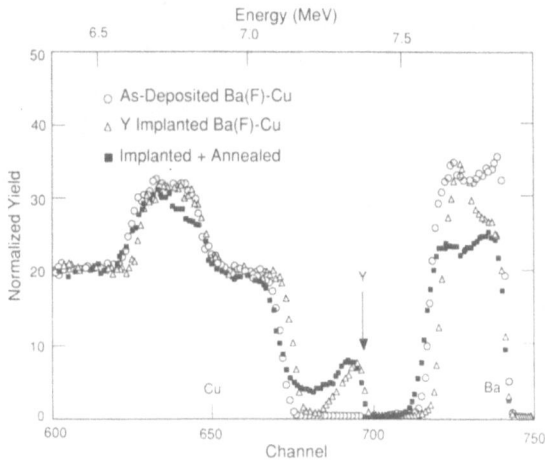

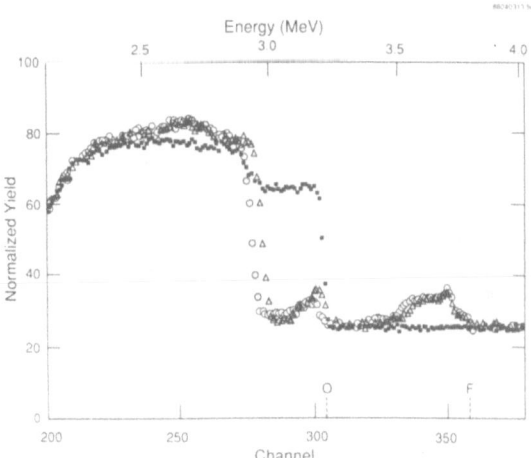

FIGURE 23. Backscattering data from 8.8-MeV helium on the as-deposited, yttrium-implanted, and yttrium-implanted plus oxygen-annealed samples on $SrTiO_3$ substrates. These spectra were taken with $\theta_1 = 60°$.

Examining the high-energy portion of Fig. 23, we see that annealing has increased the width of the barium, yttrium, and copper peaks. In addition, annealing has left the surface height of both the barium and yttrium peaks very close to their as-implanted heights, whereas the copper surface height has been reduced. However, a close examination of the yttrium peak region reveals that the total yttrium yield has increased after annealing. These data indicate the surface or near-surface scattering of strontium. Examination by scanning electron microscopy of this sample did detect some surface porosity, which leads us to conclude that strontium substrate backscattering contributes to the enhanced yttrium yield. Additional experiments are necessary to determine if strontium has also become incorporated into the surface layer. A surface height analysis of the barium, copper, and oxygen peaks after oxygen annealing and the yttrium as-implanted peak gives a surface composition of $Y_1Ba_{1.7}Cu_{2.2}O_{6.4}$.

Electron diffraction data, presented in Fig. 24, indicate that a minimum of two phases were formed during coevaporation; only BaF_2 could be positively identified. Additional diffuse rings from the carbon support film were also observed. Diffraction data from the implanted region, close to the surface (region 1 in Fig. 24b), show that low-temperature implantation of yttrium results in nearly complete amorphization of the surface. The diffraction pattern from a deeper region (region 2) is identical to that presented in Fig. 3a. The diffraction pattern from the yttrium-implanted and annealed film (Fig. 24c) indicates that substantial grain growth occurred during the oxygen annealing process. The indexed diffraction pattern suggests that at least two phases are present, $YBa_2Cu_3O_7$ and $BaCuO_2$.

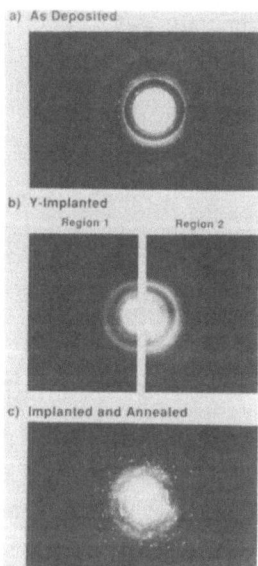

a) As Deposited

b) Y-Implanted

Region 1   Region 2

c) Implanted and Annealed

FIGURE 24. Electron diffraction patterns from the as-deposited (a), yttrium-implanted (b), and yttrium-implanted plus oxygen-annealed (c) films.

Data for resistance versus temperature are presented in Fig. 25 for both the as-deposited and yttrium-implanted $Ba_2Cu_3F_4$ films after oxygen annealing. As anticipated, oxygen annealing the as-deposited film results in the formation of an insulating material with a room-temperature resistance of 1.0×10^5 ohms, which increases to 6.4×10^5 ohms by 245 K. The yttrium-implanted film shows a moderately sharp superconducting transition at approximately 85 K, which is followed by broader transition starting at approximately 76 K. At higher temperatures the resistive quality of the film is semiconductor in nature.

The resistance data show for the first time that ion implantation alloying is capable of producing materials that possess a relatively high superconducting transition temperature. The semiconducting normal-state behavior and low-temperature tail in the superconducting transition indicate that multiple phases, both superconducting and nonsuperconducting, have formed in the yttrium-implanted region. This interpretation is consistent with the backscattering data of Fig. 23, which showed the annealed film to be off the ideal 1 : 2 : 3 stoichiometry and contaminated with strontium, and the electron diffraction data of Fig. 24c, which showed the presence of both superconducting and nonsuperconducting phases.

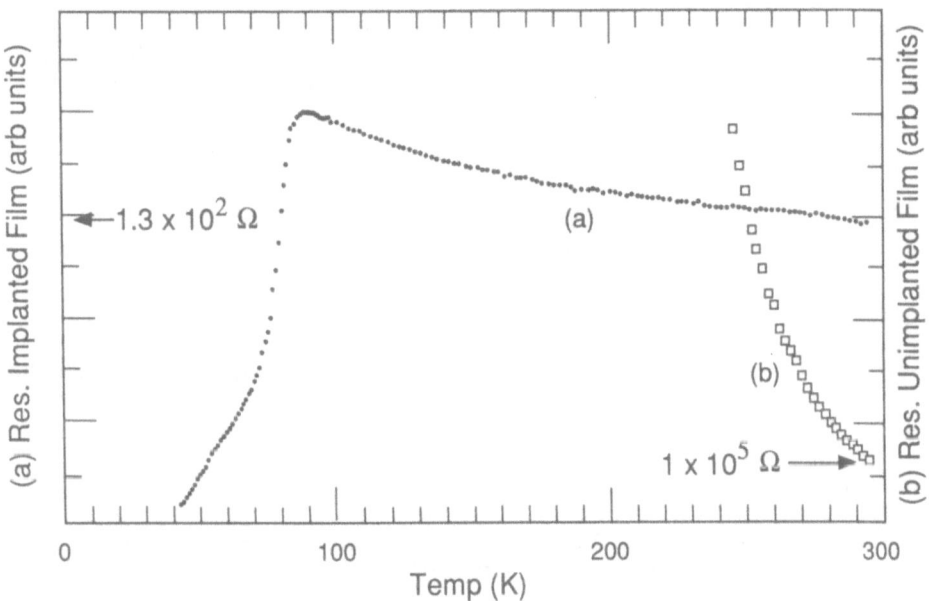

FIGURE 25. Data of resistance versus temperature for both the as-deposited and yttrium-implanted $Ba_2Cu_3F_4$ film after oxygen annealing.

5. SUMMARY

The preceding text has demonstrated that ion beams are quite capable of supplying a variety of depth-sensitive information from thin-film, high-temperature superconductors as well being useful in the production of these new materials. A variety of backscattering approaches have been discussed including classical Rutherford backscattering with helium ions and non-Rutherford backscattering techniques at both low and high energies, employing proton and helium projectiles. The utility of using 8.8-MeV helium backscattering was clearly demonstrated in the fabrication of thin-film superconductors by both conventional evaporation techniques and nonconventional ion implantation techniques. Finally, the application of ion implantation alloying to the production of high-temperature superconducting material was discussed.

6. ACKNOWLEDGMENTS

Discussions with my colleagues at the Los Alamos Ion Beam Materials Laboratory, Joe Martin, Joe Tesmer, Carl Maggiore, and Mark Hollander, as well as discussions with Charles Barbour of Sandia National Laboratories, Albuquerque, E. Rauhala of the University of Helsinki, and Jim Mayer of Cornell University are gratefully acknowledged. This work was performed under the auspices of the U.S. Department of Energy.

REFERENCES

1. Mayer J. W. and E. Rimini, *Ion Beam Handbook for Material Analysis* (Academic Press, New York, 1977).
2. Chu W-K., J. W. Mayer, and M-A. Nicolet, *Backscattering Spectrometry* (Academic Press, New York, 1978).
3. Feldman L. C. and J. W. Mayer, *Fundamentals of Surface and Thin Film Analysis* (North-Holland, New York, 1986).
4. Ref. 2, Chap. 4.
5. Bragg W. H. and R. Kleeman, *Phil. Mag.* **10** (1950) S318.
6. Frase K. G., L. G. Liniger, and D. R. Clarke, *J. Am. Ceram. Soc.* **70** (1987) C204.
7. Lupis C. H. P., *Chemical Thermodynamics of Materials* (North-Holland, New York, 1983), Chap. 10.
8. Jorgensen J. D., M. A. Beno, D. G. Hinks, L. Soderholm, K. J. Volin, R. L. Hitterman, J. D. Grace, I. K. Schuller, C. U. Segre, K. Zhang, and M. S. Kleefisch, *Phys. Rev.* **B36** (1987) 3608.
9. Cava R. J., B. Batlogg, C. H. Chen, E. A. Rietman, S. M. Zahurak, and D. Werder, *Phys. Rev.* **B36** (1987) 5719.
10. Shi D. and D. W. Capone II, *Appl. Phys. Lett.* **53** (1988) 159.
11. Leavitt J. A., P. Stoss, D. B. Cooper, J. L. Seerveld, L. C. McIntyre Jr., R. E. Davis, S. Gutierrez, and T. M. Reith, *Nucl. Inst. and Meth.* **B15** (1986) 296.
12. Doolittle L. R., *Nucl. Inst. and Meth.* **B9** (1985) 344.
13. Rauhala E., J. Keinonen, and R. Jarvinen, *Appl. Phys. Lett.* **52** (1988) 1520.
14. Barbour J. C., B. L. Doyle, and S. M. Myers, *Phys. Rev. B* **B38** (1988).
15. Martin J. A., M. Nastasi, J. R. Tesmer, and C. J. Maggiore, *Appl. Phys. Lett.* **52** (1988) 2177.
16. Blanpain B., P. Revez, L. R. Doolittle, K. H. Purser, and J. W. Mayer, *Nucl. Inst. and Meth. B*, in press.
17. Cameron J. R., *Phys. Rev.* **90** (1953) 839.

18. Mezey G., J. Gyulai, T. Nagy, E. Kotai, and A. Manuba, in *Ion Beam Surface Layer Analysis*, eds. O. Meyer, G. Linker, and Kappeler (Plenum Press, New York, 1976) p. 303.

19. Luomajarvi M., E. Rauhala, and M. Hautala, *Nucl. Inst. and Meth.* **B9** (1985) 255.

20. Ref. 3, Chap. 2.

21. Hunt W. E., M. K. Mehta, and R. H. Davis, *Phys. Rev.* **160** (1967) 782.

22. Nastasi M., J. A. Martin, J. R. Tesmer, M. G. Hollander, and C. J. Maggiore, unpublished data.

23. Zeigler J. F., J. P. Biersack, U. Littmark, *The Stopping and Range of Ions in Solids* (Pergamon Press, New York, 1985).

24. Naito M., R. H. Hammond, B. Oh, M. R. Hahn, J. W. P. Hsu, P. Rosenthal, A. F. Marshall, M. R. Beasley, T. H. Geballe, and A. Kapitulnik, *J. Mater. Res.* **2** (1987) 713.

25. White A. E., K. T. Short, D. C. Jacobson, J. M. Poate, R. C. Dynes, P. M. Mankiewich, W. J. Skocpol, R. E. Howard, M. Anzlowar, K. W. Baldwin, A. F. J. Levi, J. R. Kwo, T. Hsieh, and M. Hong, in *High Temperature Superconductors*, eds., M. B. Brodsky, R. C. Dynes, K. Kitazawa, and H. L. Tuller (Materials Research Society, Pittsburgh, 1988) p. 531.

26. Veal B. W., W. K. Kwok, A. Umezawa, G. W. Crabtree, J. D. Jorgensen, J. W. Downey, L. J. Nowicki, A. W. Mitchell, A. P. Paulokas, and C. H. Sowers, *Appl. Phys. Lett.* **51** (1987) 279.

27. Ono A., T. Tanaka, H. Nozaki, and Y. Ishizawa, *Jpn. J. Appl. Phys.* **26** (1987) L1687.

28. Koch R. H., C. P. Umbach, G. J. Clark, P. Chaudhari, and R.B. Laibowitz, *Appl. Phys. Lett.* **51**, 200 (1987).

29. Clark G. J., A. D. Marwick, R. Koch, and R. B. Laibowitz, *Appl. Phys. Lett.* **51**, 139 (1987).

30. Clark G. J., F. K. LeGoues, A. D. Marwick, R. B. Laibowitz, and R. Koch, *Appl. Phys. Lett.* **55**, 1462 (1987).

31. Mayer J. W., L. Eriksson, and J. A. Davis, *Ion Implantation in Semiconductors* (Academic Press, New York, 1970).

32. Nastasi M., J. R. Tesmer, M. G. Hollander, J. F.Smith, C. J. Maggiore, *Appl. Phys. Lett.* **52** (1988) 1729.

33. Mankiewich P. M., J. H. Scofield, W. J. Skocpol, R. E. Howard, A. H. Dayem, and E. Good, *Appl. Phys. Lett.* **51**, 1753, (1987).

34. Mankiewich P. M., R. E. Howard, W. J. Skocpol, A. H. Dayem, A. Ourmazd, M. G. Young, and E. Good, in *High Temperature Superconductors*, eds., M. B. Brodsky, R. C. Dynes, K. Kitazawa, and H. L. Tuller (Materials Research Society, Pittsburgh, 1988) p. 119.

SIMULATION OF PLASMA INDUCED MATERIAL DAMAGE WITH H-ION BEAMS

J. LINKE, H. BOLT, H. HOVEN, K. KOIZLIK, H. NICKEL, AND E. WALLURA

Nuclear Research Center Jülich, Institute for Reactormaterials, KFA-EURATOM
Association, P.O. Box 1913, 5170 Jülich, Federal Republic Germany

1. INTRODUCTION

Plasma wall interaction is a designation for several sub-processes
affecting the surface of the plasma-facing components in a thermonuclear
fusion reactor. Here besides electromagnetic radiation, arcing, and neutron
irradiation, two processes may play a life-limiting role for the so-called
first wall; namely, thermal fatigue due to the pulsed operation of a Tokamak
and thermal shock loading by plasma disruptions or run-away electrons during
unstable plasma states.

In order to determine the load limits for these two loading types for
various first-wall materials, systematic series of material tests have been
carried out in Tokamak machines and in laboratory experiments. To simulate
the thermal load to the first wall primarily energetic electrons or hydrogen
ions are used in different laboratories worldwide. In these tests the critical thermal loads for surface modifications such as erosion, grain growth,
crack formation, and melting or sublimation are determined. Besides the
quantification of the resulting damage, it is of significant importance to
improve the thermomechanical performance of the materials.

2. MATERIAL PROBLEMS IN THE FIRST WALL OF FUSION DEVICES

The toroidal confinement of hydrogen plasmas - the so-called Tokamak concept - has proved to be at present the most promising approach in connection
with worldwide efforts to realize thermonuclear fusion from a physical and
technical point of view. The term "plasma-wall interaction" summarizes all
the processes taking place between the hot plasma or its boundary layer and
the first wall during the pulsed operation of such devices (1). They
include particle bombardment, redeposition of eroded material, plasma
disruptions, so-called run-away electrons and arcing, i.e., processes which
stress, modify, damage, or even destroy the first wall, but which also
influence the plasma and thus interfere with plasma ignition. Figure 1
shows some of these damaging mechanisms on graphitic plasma-facing components.

2.1. Requirements

The last-mentioned point is of significance for the selection of materials
insofar as atoms of the wall material entering the plasma lead to considerable plasma cooling. In view of the Z^2-dependence of the radiation losses
(Z = atomic number), materials with low Z-numbers are therefore regarded as
candidates for the first wall (2). In addition to this fundamental requirement, the wall materials have to fulfill special requirements which are
listed in Table 1. Beside a high melting or sublimation temperature,
the material must exhibit a good thermal shock resistance, low erosion rates
due to physical or chemical sputtering, and favorable outgassing conditions.
In future machines which will be operated with D-T-plasmas beyond the

503

C.J. McHargue et al. (eds.), Structure-Property Relationships in Surface-Modified Ceramics, 503–511.
© 1989 by Kluwer Academic Publishers.

 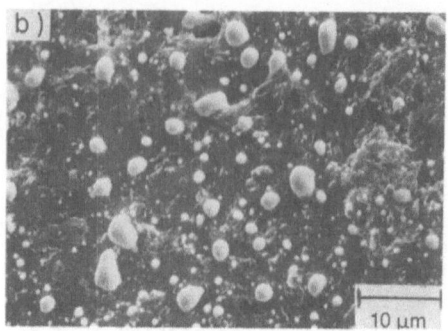

FIGURE 1. Typical material damage in a Tokamak reactor (graphite limiters in TEXTOR). (a) crack formation and erosion due to run-away electrons, (b) redeposition of metal on the limiter surface.

TABLE 1. Requirements for first-wall materials in thermonuclear fusion devices

- low atomic number (Z)
- high melting or sublimation temperature
- high thermal conductivity ⎫
- low thermal expansion coefficient ⎬ high thermal
- low Young's modulus ⎭ shock
- high strength ⎭ resistance
- low erosion rates (due to physical and chemical sputtering in hydrogen plasmas)
- good hydrogen recycling behaviour
- low outgassing rates
- favourable neutron irradiation behaviour (14 MeV neutrons)

break even point a good tritium retention and resistance against 14 MeV neutron irradiation is required.

Most of the existing confinement experiments use graphite or other carbon materials for their plasma facing components. On the basis of experience with graphite in fission reactors (3) the application of this material in fusion reactors appears very doubtful. The energetic neutrons of the fission spectrum (max. 2 MeV) already impair all relevant properties of the graphites; for example the thermal conductivity is considerably reduced by neutron-induced radiation damage. Additional effects on the strength of graphite and radiation-induced dimensional changes are also important.

2.2. Candidate materials

Taking into consideration the above mentioned requirements for first-wall materials, a large spectrum of materials has been proposed as candidates;

among these are metals, graphites, and ceramics (see Table 2). Generally
the monolithic materials are considered; in special applications composite
materials (e.g., fiber reinforced materials) or coatings have been tested
successfully. Stainless steel or Ni-base alloys are the "classical"
materials for plasma-facing components; today these materials have been
largely replaced by non-metallic, low-Z components. Here, graphites and
carbon fiber composites play the most important role. Due to the above mentioned problems in neutron-generating machines, radiation resistant C-
materials are required. In this field, beryllium or silicon carbide could
be a promising alternative, at least if the relatively poor thermal shock
resistance of the ceramic materials (4) can be improved by adequate methods
(e.g., alloying with BeO or by fiber reinforcement) and if the material temperature during operation remains sufficiently low (not exceeding about
1000°C for Be and below 1500°C for SiC).

TABLE 2. First-wall candidate materials

Metals:
  – stainless steel (e.g. 1.4311)
  – Ni-base alloys (e.g. INCONEL 600)
  – refractory metals (e.g. W, TZM)
  – beryllium

Carbon materials:
  – fine grain graphites (e.g. EK 98)
  – pyrolytic graphite
  – carbon fiber composite materials (e.g. 2D, 4D)

Ceramics:
  – monolithic silicon carbides (e.g with Al, AlN or
    BeO additives)
  – monolithic low-Z nitrides (e.g. BN, Si_3N_4, AlN)
  – coat-mix material on the basis of SiC
  – SiC-fiber reinforced SiC

Coatings:
  – PVD – Physical Vapor Deposition (e.g. amor-
    phous hydrocarbon or boron carbon layers)
  – CVD – Chemical Vapor Deposition (e.g. SiC,
    pyrocarbon)
  – LPPS – Low Presure Plasma Spray

3. HIGH HEAT FLUX TESTS

Aside from a purely theoretical analysis, e.g., by time dependent
temperature/stress analysis, there exist two different ways of investigating
plasma wall interaction experimentally (5): the first and most realistic
way is to expose material samples to a near-fusion plasma in one of the
existing large Tokamaks, the second way is to test materials in conventional
laboratory experiments. Concerning the high heat flux behavior of different
candidate first-wall materials the following methods have been applied.

3.1. Tests in TEXTOR

Extensive, systematic material tests in operating Tokamaks are not
possible at present, since the plasma machines are designed primarily for
other purposes, e.g., investigations of plasma physics, plasma diagnostics
etc. Nevertheless, Tokamak material tests are essential for the development
of improved first-wall materials, e.g., for studies of the material damage
caused by the real, complex system of plasma wall interaction, for the

adjustment of experimental parameters in simulation experiments, and for final decisive material tests.

A plasma machine which is well equipped for material tests is the Tokamak TEXTOR (6) in the Nuclear Research Center (KFA) Jülich. Here, up to 20 different material samples can be exposed to the plasma; this assembly, which is shown in Fig. 2, has a sandwiched structure with 10-mm thick slices of the test specimens (and reference samples). Today, a new sandwich limiter holder is available that can be inserted into the plasma by an air lock system without breaking the vacuum inside the torus. A variety of different diagnostics to determine the plasma parameters and the resulting surface temperatures are available during the plasma exposure. After removal from TEXTOR and disassembly the test specimens are investigated by suitable analytical methods. Figure 1(b) shows typical surface modifications observed in the sandwich limiter experiments.

3.2. Electron beam tests

To simulate the plasma wall interaction in laboratory experiments, electron beam devices (e.g., e^--beam welding machines, see Fig. 3) are used today. These machines in general are operated with beam currents in the range of some 100 mA and acceleration voltages of 30 to 150 kV; pulse durations from the ms-range up to quasi-stationary conditions are possible. These parameters make electron beam devices highly qualified to simulate plasma disruptions on the one hand, and to investigate the material failure due to thermal fatigue. In the former case the thermal shock behavior under extreme heat loads in the ms-range is of interest; these experiments are performed with focussed electron beams (beam diameter in the mm-range). For thermal cycling tests however, only moderate power densities on larger target areas are required; therefore, defocussed beams (loaded area ~1 cm²) or additionally scanned beams (loaded area up to ~100 cm²) are applied.

The heat load distribution in electron beam high heat flux tests is rather inhomogeneous due to the relatively small beam radius. Figure 4(a) shows a typical profile which was measured for a 150-kV beam with 100 mA beam current (7). The contour lines show an unexpected deviation from the axial symmetry; secondary maxima occur in a few millimeters distance from the central peak. The power density value in the beam center can even exceed 1 MW cm⁻², although the nominal mean beam power density (i.e., total beam power per area) is only in the range of some 10 kW cm⁻². This beam inhomogeneity will cause additional stresses in the loaded surface area; this is especially true for scanned electron beams which will result in subsequent thermal shocks whenever the beam strikes the same surface area.

3.3. Ion beam tests

To overcome this electron beam test deficiency — namely, the inhomogeneous heat distribution and the relatively small loaded areas — powerful ion beam facilities play an important role as a test bed for first-wall components. These devices - either specially designed for this purpose or modified neutral beam injector (NBI) test stands - with total beam powers of some MW and beam diameters in the range of some 10 cm are useful for simulation of plasma-surface interactions.

The ion beam material test facility installed at KFA (see Fig. 5) is a former test stand which was used for the development of the neutral beam injection heating of TEXTOR (8). With a total power of 4 MW, a beam cross-section of 40 x 45 cm and the capability to simulate short pulses (see specifications of the test facility in Table 3), it is a powerful tool for fusion material development. Similar to the sandwich limiter lock at TEXTOR, the test facility is provided with a sample manipulator which allows the changing of the test specimens without breaking the vacuum in the

FIGURE 2. Sandwich limiter for in-pile material tests in TEXTOR

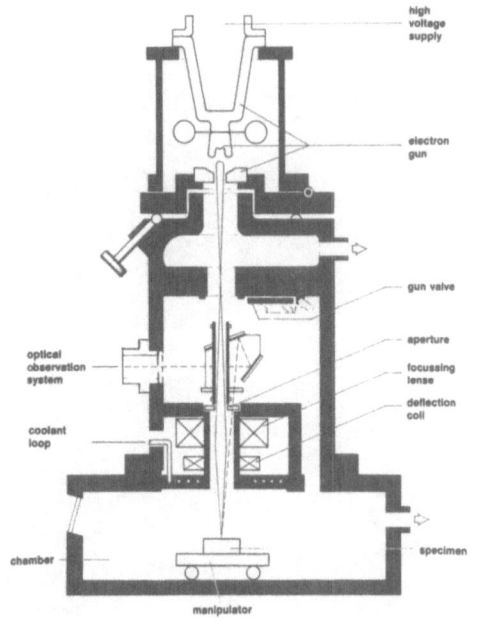

FIGURE 3. Electron beam welding machine for high heat flux tests.

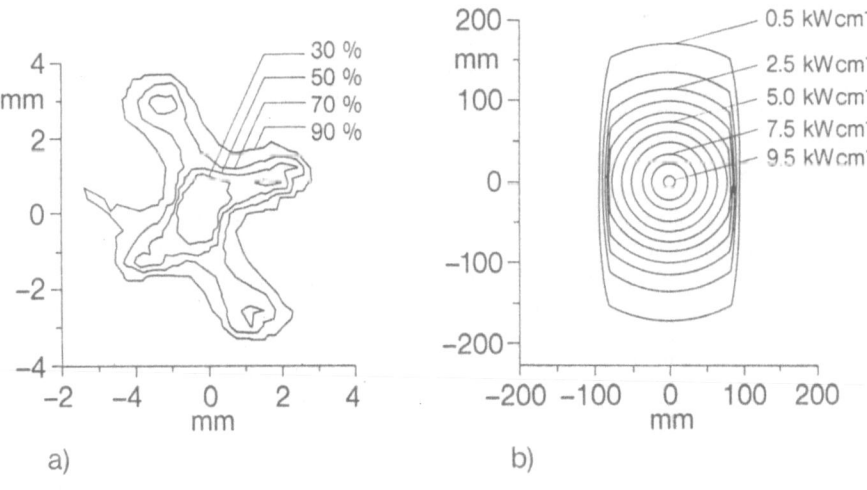

FIGURE 4. Beam profiles in high heat flux particle beam test facilities:
(a) KFA electron beam device (measured profile)
(b) KFA ion beam facility (calculated profile)

FIGURE 5. Ion beam high heat flux
materials test facility at KFA.

TABLE 3. Specifications of the KFA
ion beam materials test facility

total beam power:	4.0 MW (I_B = 78 A, E = 52 keV)
beam area:	height 40 cm, width 45 cm
sample holder:	15 cm x 10 cm sample size, actively or passively cooled
particle type:	H^+, H^0
particle energy:	maximum energy E_{max} = 20...52 keV energy distribution: 63 % E_{max} 27 % 1/2 E_{max} 10 % 1/3 E_{max}
power density:	maximum value F_{max} = 8...15 kWcm^{-2}
pulse length:	t_{min} = 10 ms t_{max} = 10 s
pulse time:	t_r = 2 ms
repetition rate:	1...5 min (depending on power density and pulse length)

vessel. In a first step only passively cooled samples (maximum sample size
= 10 x 15 cm) will be used; in a second phase tests on actively cooled
samples or on plasma facing components in real size are planned.
     Figure 4(b) shows a calculated beam profile for the KFA ion beam test
facility. In comparison to the e$^-$-beam tests, relatively homogeneous heat
load distributions can be realized here; within a sample size of 10 x 15 cm
the inhomogeneities (i.e., the deviations in the local heat load) are less
than 50%. Stresses that arise from peak heat load distributions in electron
beam tests do not play any significant role in ion beam tests.
     Typical material damage which was observed in ion beam high-heat-flux
tests (9) is shown in Fig. 6. The left micrograph shows the resolidified
surface of a stainless steel sample that was exposed to a pulse of
1.6 kW cm^{-2} for ~1 s. The surface shown in Fig. 6(b) belongs to an ultrafine grain graphite (P/A = 9.7 kW cm^{-2}; t = 290 ms) which developed a crack
pattern with undamaged islands of ~1 mm in diameter. Ceramographic sections
from this sample indicate that the crack depth reaches 2 mm in spite of the
moderate thermal shock load.
     To quantify the material damage due to the applied high heat fluxes, fast
diagnostics are necessary. The KFA ion beam test stand is equipped with two
pyrometers (see Fig. 7) which cover the temperature range from 300 to 3000°C
(temperatures above 800°C are measured by a two-color pyrometer with a time
resolution of 5 ms). To determine the bulk temperatures of the test
samples, the individual tiles are provided with Pt-PtRh-thermocouples. A
pressure gauge is available for the quantification of the pressure rise in
the vessel due to outgassing, evaporation, or sublimation. All data are

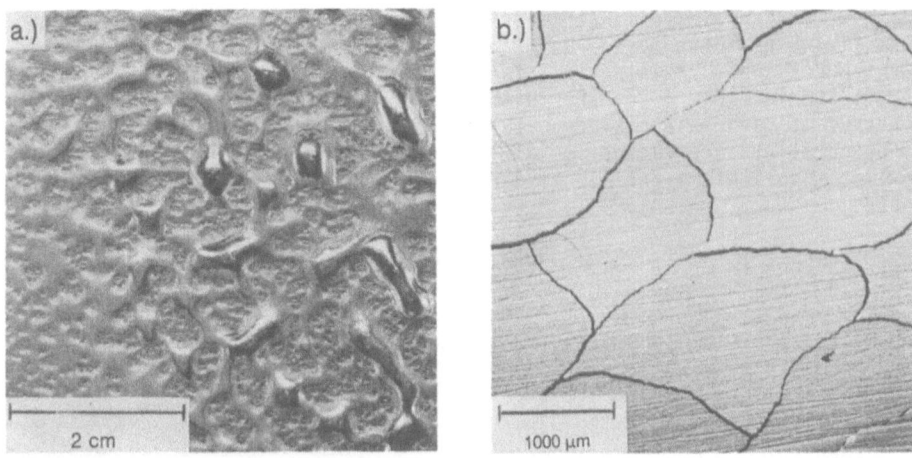

FIGURE 6. Typical damage de to high heat flux tests in the ion beam test stand: (a) structure of the resolidified surface of a 1.4311 stainless steel sample (P/A = 1.6 kw cm⁻², t = 951 ms); (b) cracks in the surface of an ultrafine grain graphite (P/A = 9.7 kW cm⁻², t = 290 ms).

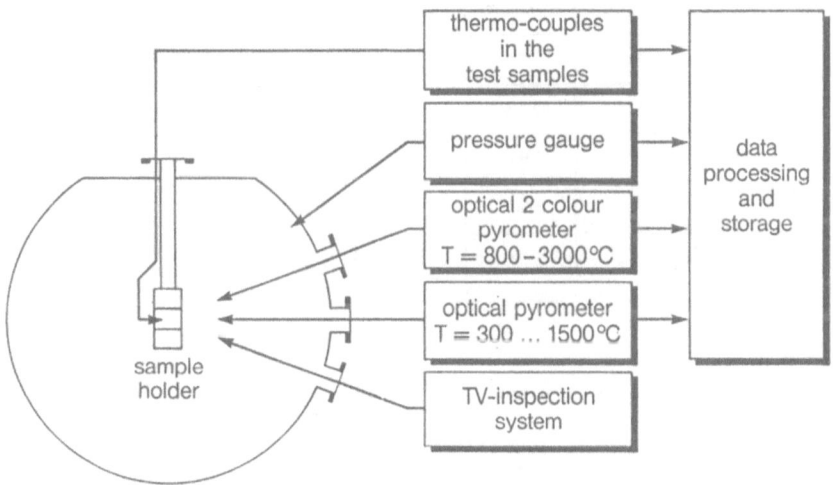

FIGURE 7. Diagnostics for the ion beam materials test facility.

processed and stored in a personal computer. The surface of the specimens can be observed by a CCD-TV-camera; the information is stored on a VTR. Thus, surface modifications such as cracking or melting as well as qualitative temperature distributions can be recorded.

510

4. SUMMARY

The performance of different first wall candidate materials under high heat fluxes has been tested both in electron beam and in ion beam experiments. For these tests several facilities are available (10); some of these are listed in Table 4 together with their operating parameters. The spectrum of materials tested so far ranges from metallic alloys (e.g., Fe-or Ni-based alloys, refractory metals) to graphites and ceramic materials; besides monolithic materials also coatings have been included in these tests.

TABLE 4. Specifications of some high heat flux test facilities

Location	IPP Nagoya Japan	JAERI Japan	KFA FRG	SNLA USA	KFA FRG	SNLA USA
Beam	H^+ H°	H^+ H°	H^+ H°	H^+ D^+	e^-	e^-
Particle Energy	120 keV	100 keV	50 keV	40 keV	150 keV	30 keV
Power density	150 MW/m²	200 MW/m²	150 MW/m²	40 MW/m²	500 MW/m²	1000 MW/m²
Pulse length	1 s	10 s	10 s	30 s	cont.	cont.
Loaded area	600 cm²	600 cm²	150 cm²		25 cm²	100 cm²
Total power	9 MW	8 MW	4 MW	800 kW	30 kW	30 kW
Thermal cycling	No	Yes	Yes	Yes	Yes	Yes
Status	operational	operational	under preparation	checkout	operational	operational

Typical damage effects observed in high heat flux tests are: (a) erosion due to melting, evaporation, sublimation or particle emission, (b) modifications in the grain morphology (such as twin formation or grain growth), and (c) formation of cracks and chipping effects.

Since first-wall materials with high atomic numbers (Z) are generally undesired from the plasma-physical viewpoint, non-metallic materials are the prime candidates for the inner wall of the next generation plasma machines. Graphites and especially ceramics are very sensitive to thermal shocks (e.g., by disruptions or run-away electrons); to meet the needs for a better thermal shock resistance of these materials different types of fiber-reinforced grades have been successfully tested.

5. REFERENCES

1. Engelmann F et al.: J. Nucl. Mater. 145-147 (1987) 154.
2. Mattas RF, DL Smith, and MA Abdou: J. Nucl. Mater. 122 & 123 (1984) 66.
3. Kelly BT: Physics of Graphite. London: Applied Science Publishers, 1981.
4. Gotoh Y et al.: J. Nucl. Mater. 133 & 134 (1985) 257.
5. Bolt H et al.: KFA-report JUL-2086, 1986.
6. Wolf GH: J. Nucl. Mater. 122 & 123 (1984) 1124.
7. Koizlik K et al.: Proceedings 15th Symposium on Fusion Technology, Utrecht, Netherlands, 1988 (to be published).
8. Pfister U: Proceedings 15th Symposium on Fusion Technology, Avignon, France, 1986, 361.
9. Bolt H: Thesis, RWTH Aachen, 1988.
10. Miyahara A, KL Wilson: Proceedings Japan-US Workshop on Plasma Material Interaction/High Heat Flux Data Needs, Nagoya, Japan, 1987.

NATO-ADVANCED STUDY INSTITUTE
Structure-Property Relationships in Ion-Beam Surface-Modified
Ceramics–Theory and Applications
Il Ciocco, Castelvecchio Pascoli, Italy
August 28–September 9, 1988

Eric Abonneau
Department de Physique des Materiaux
University Claude Bernard-Lyon I
69622 Villeurbanne, Cedex, France
Phone: 78898124, Ext. 4056

Terry Lynn Alford
Materials Science & Engineering Dept.
Cornell University
Ithaca, NY 14853 USA
Phone: 607-255-7179
607-255-8963 Home

Ahmad A.S. Al-Ghamdi
PG/PH M.A.P.S.
University of Sussex
Falmer Brighton BN1 9QH
England
Phone: 273-674404

William R. Allen
Metals and Ceramics Division
Oak Ridge National Laboratory
P.O. Box 2008
Building 5500, MS-6376
Oak Ridge, TN 37831-6376 USA
Phone: 615-574-7242

Asko Anttila
Accelerator Laboratory
University of Helsinki
SF-00550 Helsinki, Finland
Phone: 20-79455/20

Hacer Aygun
The Middle East Technical University
Metallurgical Engineering Department
Ankara, Turkey
Phone: 2237100/2528

Maria C. Batista
Estrada de Benfica, 523 F°-E
1500 Lisboa, Portugal
Phone: 01-7143105

Giancarlo Battaglin
Dipartimento di Fisica
Via Marzolo N.8
University of Padova
35131 Padova, Italy
Phone: 049-844-111

Ernst W. Behrens
Armstrong World Industries, Inc.
Basic Research
P.O. Box 3511
Lancaster, PA 17601 USA
Phone: 717-396-5817

Torsten Bremer
Universitat Osnabrück
Fachbereich Physik
Postfch 4469
D-4500 Osnabrück, FRG
Phone: 0541-608-2669

Chris H. Buchal
KFA-ISI
D-5170 Jülich, FRG
Phone: 02461-61-4431

C. R. A. Catlow
Department of Chemistry
University of Keele
Keele, Staffs. ST5 5 BG
England
Phone: 0782-021111

Lynann Clapham
Physics Department
Queen's University
Kingston, Ontario, Canada K7L 3N6
Phone: 613-545-2672

J-Marc Costantini
CEB-III/PTN
BP 12
91680 Bruyeres Le Chatel
France
Phone: 64-90-50-46

Patricia Cullen
Massachusetts Institute of Technology
Room 13-5141
Cambridge, MA 02139 USA
Phone: 617-253-6828

Gianantonio Della Mea
Department of Engineering
Mesiano
University of Trento
Trento, Italy
Phone:

Maria Erola
Accelerator Laboratory
University of Helsinki
SF-00550 Helsinki, Finland
Phone: 90-79455/20

Svend S. Eskildsen
Jutland Technological Institute
Teknologiparken
DK-8000 Aarhus C
Denmark
Phone: 06-142400, Ext. 5051

Hermann Ferber
c/o Daimler-Benz AG
Postfach 60 02 02
7000 Stuttgart 60, FRG
Phone: 0711-17-0

Gilbert Fuchs
Departement de Physique des Materiaux
Universite de Lyon-Lyon I
43 Boulevard du 11 Novembre
69622 Villeurbanne, Cedex, France
Phone:

Tatsumi Hioki
Toyota Central R & D Labs., Inc.
Nagakute, Aichi-gun,
Aichi, 480-11, Japan
Phone: 05616-2-6111
Telex: 04496023 TRDC J

Wolfgang O. Hofer
IPP
Kernforschungsanlage Jülich GmbH
Postfach 1913
D-5170 Jülich, FRG
Phone: 02461 61 6368

Debra L. Joslin
Metals & Ceramics Division
Oak Ridge National Laboratory
P.O. Box 2008, Bldg. 4500S
Oak Ridge, TN 37831-6118 USA
Phone: 615-574-4344

Roger Kelly
International Business Machines
Thomas J. Watson Research Center
P.O. Box 218
Yorktown Heights, NY 10598 USA
Phone: 914-945-3000

Wilto Kesternich
Institut für Festkorperforschung
Kernforschungsanlage Jülich
5170 Jülich, FRG
Phone: 49-2461-616089

Ram Kossowsky
Penn State University
Applied Research Laboratory
Box 30
State College, PA 16804 USA
Phone: 814-863-4481

J. Linke
Kernforschungsanlage Jülich GmbH
Institut für Reaktorwerkstoffe
Postfach 1913
5170 Jülich, FRG
Phone: 02461-613230

Bradley J. Luff
MAPS PG
University of Sussex
Falmer Brighton BN1 9QH
United Kingdom
Phone: 0279-501531

R. Dal Maschio
Universita Di Trento
Dipartimento di Ingegneria
Mesiano Di Povo
38050 Trento, Italy

Paolo Mazzoldi
University of Padova
Dipartimento di Fisica dell'Universita
Via Marzolo 8
35131 Padova, Italy
Phone: 049-844-111

Carl J. McHargue
Metals and Ceramics Division
Oak Ridge National Laboratory
P.O. Box 2008, Bldg. 4500S
Oak Ridge, TN 37831-6118 USA
Phone: 615-574-4344

David G. Mello
Tribology Programs
Office of Conservation and
  Renewable Resources
U.S. Department of Energy
CE-12, Room 5E066
Forrestal Building
1000 Independence Avenue
Washington, DC 20585 USA
Phone: 202-586-9345

Antonio Miotello
Dipartimento di Fisica
Universita di Trento
38050 Povo, Italy
Phone: 0461-881637

Nathalie Moncoffre
Institut de Physique Nucleaire
Universite Claude Bernard
43, Bd du 11 Novembre 1918
69622 Villeurbanne, Cedex, France
Phone: 78-89-81-24

Fred A. Nichols
Argonne National Laboratory
9700 S. Cass Avenue, Bldg. 212
Argonne, IL 60439 USA
Phone: 312-972-8292

Michael Nastasi
Center for Materials Science
Los Alamos National Laboratory
Mail Stop K765
Los Alamos, NM 87545 USA
Phone: 505-667-9243

Karur Padmanahan
Department of Physics & Astronomy
Wayne State University
Detroit, MI 48202 USA
Phone: 313-577-3005

Don M. Parkin
MS-K765
Los Alamos National Laboratory
Los Alamos, NM 87545 USA
Phone: 505-667-9243

Robert Parrish
P.O. Box 6730, Stat. B
Vanderbilt University
Nashville, TN 37235 USA
Phone: 615-322-3189

Janet E. Pawel
Metals & Ceramics Division
Oak Ridge National Laboratory
P.O. Box 2008, Bldg. 4500S
Oak Ridge, TN 37831-6118 USA
Phone: 615-574-4344

Alain Perez
Departement de Physique des Materiaux
Universite Claude Bernard-Lyon I
43, Bd du 11 Novembre 1918
69622 Villeurbanne, Cedex, France
Phone: 78-89-81-24, Ext. 4056

Stathis D. Peteves
CEC, JRC Petten
P.O. Box 2, 1755 ZG Petten
The Netherlands
Phone:

Nicolaos Platakis
Materials Science & Engineering Lab.
Chemical Engineering Department
University of Thessaloniki
Thessaloniki, Greece
Phone: 031-991547

J. T. A. Pollock
CSIRO, Applied Physics
P.O. Box 218
New South Wales, Australia
Phone: 02-467-6729

Renato Potenza
Physics Department
University of Catania
Corso Italia 52
195122 Catania, Italy
Phone: 95-322061

516

Lou Pyatt
Secretary
Metals and Ceramics Division
Oak Ridge National Laboratory
P.O. Box 2008, Bldg. 4500S
Oak Ridge, TN 37831-6118 USA
Phone: 615-574-4344

Stella Ramos
Instituto De Fisics DA USRGS
Av. Bento Goncalves 9500
91500 P. AlegRG/RS, Brasil
Phone:

Helen Rendell
GEOG Laboratory
University of Sussex
Falmer, Brighton BN1 9QN
United Kingdom
Phone: 0273-606755, Ext. 2213
       or Ext. 2897

Marc Riehm
Department of Engineering Physics
McMaster University
1280 Main Street, W
Hamilton, Ontario, Canada
Phone: 416-525-9140, Ext. 4683

J. Michael Rigsbee
Department of Materials Science
 and Engineering
University of Illinois
1304 W. Green Street
Urbana, IL 61801 USA
Phone: 217-333-6584

Lawrence Romana
BAT-203
Departement de Physique
 des Materiaux
University Claude Bernard-Lyon I
43 Bd du 11 Novembre 1918
69622 Villeurbanne, Cedex, France
Phone: 78-89-81-24, Ext. 4056

F. W. Saris
Director
F.O.M. Institute for Atomic
 and Molecular Physics
Kruislaan 407-1098SJ
Amsterdam, The Netherlands
Phone: 020-946711

Barbara Sawicka
Chalk River Nuclear Labs.
AECL, Chalk River K0J 1J0
Ontario, Canada
Phone: 613-584-3311, Ext. 3454

Jorge Sawicki
Chalk River Nuclear Labs.
AECL, Chalk River K0J 1J0
Ontario, Canada
Phone: 613-584-3311, Ext. 3153
           or 3123

Jorgen Schou
Physics Department
RISO National Laboratory
DK-4000 Roskilde
Denmark
Phone: 45-2371212, Ext. 4755

Philip Scott
Department of Materials Science
 and Engineering
University of Illinois
1304 W. Green Street
Urbana, IL 61801 USA
Phone: 217-333-8583

Wilbur C. Simmons
European Research Office
223 Old Marylebone Road
London, NW1, United Kingdom
Phone: 44-1-409-4423

Irwin Singer
Naval Research Laboratory
Code 6170
Washington, DC 20375 USA
Phone: 202-767-2327

Fred A. Smidt
Naval Research Laboratory
Code 4670
Washington, DC 20375 USA
Phone: 202-767-2327

Dinesh Sood
Microelectronics & Materials
 Technology Center
Royal Melbourne Institute of Technology
GPO Box 2476V
Melbourne, Vic. 3001, Australia
Phone: 03-660-2620

Paul Thevenard
Departement de Physique
  des Materiaux
Universite Claude Bernard-Lyon I
43 Bd du 11 Novembre 1918
69622 Villeurbanne, Cedex, France
Phone: 78-89-81-24, Ext. 3237

Peter Townsend
MAPS PG
University of Sussex
Falmer Brighton BN1 9QH
United Kingdom
Phone: 0279-501531

Demetre Tsatis
University of Patras
Patras, Greece
Phone: 061-993-133

Kazim Tur
Department of Metallurgical Engineering
Middle East Technical University
06531 Ankara, Turkey
Phone: 90-4-223-7100, Ext. 2523

Hans Ullmaier
KFA-IFF, Postfach 1913
D-5170 Jülich, FRG
Phone: 02461-613160

B. P. van Hassel
University of Twente
Faculty of Chemical Technology
P.O. Box 217
7500 AE Enschede, The Netherlands
Phone:

J. L. Whitton
Physics Department
Queen's University
Kingston, Ontario, Canada   K7L 3N6
Phone: 613-545-2672

Gerhard K. Wolf
Physikalische Chemie
Universitat Heidelberg
Im Neuenheimer Feld 500
6900 Heidelberg, FRG
Phone:

Lin Zhang
MAPS PG
University of Sussex
Falmer, Brighton BN1 9QH
United Kingdom
Phone: 0273-606755-2912

Steve Zinkle
Metals & Ceramics Division
Oak Ridge National Laboratory
P.O. Box 2008, Bldg. 5500
Oak Ridge, TN 37831-6376 USA
Phone: 615-576-7220

SUBJECT INDEX

AUTHOR INDEX